Optical Physics and Engineering

Optical Physics and Engineering

Edited by
Vladimir Latinovic

C WILLFORD PRESS

www.willfordpress.com

Published by Willford Press,
118-35 Queens Blvd., Suite 400,
Forest Hills, NY 11375, USA

ISBN: 978-1-68285-366-5

Cataloging-in-Publication Data

Optical physics and engineering / edited by Vladimir Latinovic.
    p. cm.
Includes bibliographical references and index.
ISBN 978-1-68285-366-5
1. Physical optics. 2. Optics. 3. Optical engineering. 4. Optical materials. 5. Photons. I. Latinovic, Vladimir.
QC395.2 .O68 2017
535.2--dc23

For information on all Willford Press publications
visit our website at www.willfordpress.com

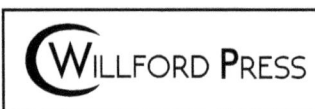

WILLFORD PRESS

Printed in the United States of America.

Contents

Preface..IX

Chapter 1   **Quantum interference in heterogeneous superconducting-photonic circuits on a silicon chip**..1
C. Schuck, X. Guo, L. Fan, X. Ma, M. Poot & H.X. Tang

Chapter 2   **Large elasto-optic effect and reversible electrochromism in multiferroic BiFeO$_3$**......8
D. Sando, Yurong Yang, E. Bousquet, C. Carrétéro, V. Garcia, S. Fusil, D. Dolfi,
A. Barthélémy, Ph. Ghosez, L. Bellaiche & M. Bibes

Chapter 3   **Graphene–ferroelectric metadevices for nonvolatile memory and reconfigurable logic-gate operations**...15
Woo Young Kim, Hyeon-Don Kim, Teun-Teun Kim, Hyun-Sung Park, Kanghee Lee,
Hyun Joo Choi, Seung Hoon Lee, Jaehyeon Son, Namkyoo Park & Bumki Min

Chapter 4   **Realization of mid-infrared graphene hyperbolic metamaterials**......................21
You-Chia Chang, Che-Hung Liu, Chang-Hua Liu, Siyuan Zhang, Seth R. Marder,
Evgenii E. Narimanov, Zhaohui Zhong & Theodore B. Norris

Chapter 5   **Frequency comb transferred by surface plasmon resonance**.........................28
Xiao Tao Geng, Byung Jae Chun, Ji Hoon Seo, Kwanyong Seo, Hana Yoon,
Dong-Eon Kim, Young-Jin Kim & Seungchul Kim

Chapter 6   **Amplification and generation of ultra-intense twisted laser pulses via stimulated Raman scattering**...35
J. Vieira, R.M.G.M. Trines, E.P. Alves, R.A. Fonseca, J.T. Mendonça, R. Bingham,
P. Norreys & L.O. Silva

Chapter 7   **Topological phase transitions and chiral inelastic transport induced by the squeezing of light**..42
Vittorio Peano, Martin Houde, Christian Brendel, Florian Marquardt
& Aashish A. Clerk

Chapter 8   **Super-crystals in composite ferroelectrics**...20
D. Pierangeli, M. Ferraro, F. Di Mei, G. Di Domenico, C.E.M. de Oliveira,
A.J. Agranat & E. DelRe

Chapter 9   **Guided post-acceleration of laser-driven ions by a miniature modular structure**......57
Satyabrata Kar, Hamad Ahmed, Rajendra Prasad, Mirela Cerchez,
Stephanie Brauckmann, Bastian Aurand, Giada Cantono, Prokopis Hadjisolomou,
Ciaran L.S. Lewis, Andrea Macchi, Gagik Nersisyan, Alexander P.L. Robinson,
Anna M. Schroer, Marco Swantusch, Matt Zepf, Oswald Willi & Marco Borghesi

Chapter 10   **Mapping multidimensional electronic structure and ultrafast dynamics with single-element detection and compressive sensing**..64
Austin P. Spencer, Boris Spokoyny, Supratim Ray, Fahad Sarvari & Elad Harel

Chapter 11   **Shaping the nonlinear near field**..**70**
Daniela Wolf, Thorsten Schumacher & Markus Lippitz

Chapter 12   **Optically controlled electroresistance and electrically controlled photovoltage in**
ferroelectric tunnel junctions..**75**
Wei Jin Hu, Zhihong Wang, Weili Yu & Tom Wu

Chapter 13   **Suppression law of quantum states in a 3D photonic fast Fourier transform chip**.............**84**
Andrea Crespi, Roberto Osellame, Roberta Ramponi, Marco Bentivegna,
Fulvio Flamini, Nicolò Spagnolo, Niko Viggianiello, Luca Innocenti, Paolo Mataloni
& Fabio Sciarrino

Chapter 14   **Wavelength-tunable sources of entangled photons interfaced with atomic vapours**...........**92**
Rinaldo Trotta, Javier Martín-Sánchez, Johannes S. Wildmann, Giovanni Piredda,
Marcus Reindl, Christian Schimpf, Eugenio Zallo, Sandra Stroj, Johannes Edlinger
& Armando Rastelli

Chapter 15   **Selectively enhanced photocurrent generation in twisted bilayer graphene with**
van Hove singularity...**99**
Jianbo Yin, HuanWang, Han Peng, Zhenjun Tan, Lei Liao, Li Lin, Xiao Sun,
Ai Leen Koh, Yulin Chen, Hailin Peng & Zhongfan Liu

Chapter 16   **Wavelength-tunable entangled photons from silicon-integrated III–V**
quantum dots..**106**
Yan Chen, Jiaxiang Zhang, Michael Zopf, Kyubong Jung, Yang Zhang, Robert Keil,
Fei Ding & Oliver G. Schmidt

Chapter 17   **Lasing in silicon–organic hybrid waveguides**...**113**
Dietmar Korn, Matthias Lauermann, Sebastian Koeber, Patrick Appel, Luca Alloatti,
Robert Palmer, Pieter Dumon, Wolfgang Freude, Juerg Leuthold and Christian Koos

Chapter 18   **Enhanced nonlinear interactions in quantum optomechanics via mechanical**
amplification..**122**
Marc-Antoine Lemonde, Nicolas Didier & Aashish A. Clerk

Chapter 19   **Frequency and bandwidth conversion of single photons in a room-temperature**
diamond quantum memory..**130**
Kent A.G. Fisher, Duncan G. England, Jean-Philippe W. MacLean, Philip J. Bustard,
Kevin J. Resch & Benjamin J. Sussman

Chapter 20   **Diffractive imaging of a rotational wavepacket in nitrogen molecules with**
femtosecond megaelectronvolt electron pulses..**136**
Jie Yang, Markus Guehr, Theodore Vecchione, Matthew S. Robinson, Renkai Li,
Nick Hartmann, Xiaozhe Shen, Ryan Coffee, Jeff Corbett, Alan Fry, Kelly Gaffney,
Tais Gorkhover, Carsten Hast, Keith Jobe, Igor Makasyuk, Alexander Reid,
Joseph Robinson, Sharon Vetter, Fenglin Wang, Stephen Weathersby,
Charles Yoneda, Martin Centurion & Xijie Wang

Chapter 21   **Size-dependent phase transition in methylammonium lead iodide perovskite**
microplate crystals...**145**
Dehui Li, Gongming Wang, Hung-Chieh Cheng, Chih-Yen Chen, Hao Wu,
Yuan Liu, Yu Huang & Xiangfeng Duan

Chapter 22   **Real-time high dynamic range laser scanning microscopy**..153
C. Vinegoni, C. Leon Swisher, P. Fumene Feruglio, R.J. Giedt, D.L. Rousso,
S. Stapleton & R. Weissleder

Chapter 23   **Mode engineering for realistic quantum-enhanced interferometry**..166
Michał Jachura, Radosław Chrapkiewicz, Rafał Demkowicz-Dobrzański,
Wojciech Wasilewski & Konrad Banaszek

Chapter 24   **Plasmonic piezoelectric nanomechanical resonator for spectrally selective**
infrared sensing...176
Yu Hui, Juan Sebastian Gomez-Diaz, Zhenyun Qian, Andrea Alù & Matteo Rinaldi

Chapter 25   **All-optical design for inherently energy-conserving reversible gates and circuits**.............185
Eyal Cohen, Shlomi Dolev & Michael Rosenblit

Chapter 26   **Magnetic-free non-reciprocity based on staggered commutation**..193
Negar Reiskarimian & Harish Krishnaswamy

Chapter 27   **Exciton localization in solution-processed organolead trihalide perovskites**........................203
Haiping He, Qianqian Yu, Hui Li, Jing Li, Junjie Si, Yizheng Jin, Nana Wang,
Jianpu Wang, Jingwen He, Xinke Wang, Yan Zhang & Zhizhen Ye

Chapter 28   **Statistical moments of quantum-walk dynamics reveal topological quantum**
transitions..210
Filippo Cardano, Maria Maffei, Francesco Massa, Bruno Piccirillo, Corrado de Lisio,
Giulio De Filippis, Vittorio Cataudella, Enrico Santamato & Lorenzo Marrucci

Chapter 29   **Estimation of a general time-dependent Hamiltonian for a single qubit**...............................218
L.E. de Clercq, R. Oswald, C. Flühmann, B. Keitch, D. Kienzler, H.-Y. Lo,
M. Marinelli, D. Nadlinger, V. Negnevitsky & J.P. Home

Permissions

List of Contributors

Index

Preface

The amalgamation of optical physics and engineering has advanced the designing of equipment like cameras, microscopes, telescopes, etc. Optical physics and engineering are well known and established branches of optics. While optical physics is focused on the discovery and applications of phenomena related to optics; optical engineering is the field of study that entirely focuses on application of optics. This book elucidates the concepts and innovative models around prospective developments with respect to optical physics. The readers would gain knowledge that would broaden their perspective about the inter-disciplinary aspects of optical physics and optical engineering. Researchers and students in this field will be assisted by this book.

Significant researches are present in this book. Intensive efforts have been employed by authors to make this book an outstanding discourse. This book contains the enlightening chapters which have been written on the basis of significant researches done by the experts.

Finally, I would also like to thank all the members involved in this book for being a team and meeting all the deadlines for the submission of their respective works. I would also like to thank my friends and family for being supportive in my efforts.

Editor

Quantum interference in heterogeneous superconducting-photonic circuits on a silicon chip

C. Schuck[1], X. Guo[1], L. Fan[1], X. Ma[1,2], M. Poot[1] & H.X. Tang[1]

Quantum information processing holds great promise for communicating and computing data efficiently. However, scaling current photonic implementation approaches to larger system size remains an outstanding challenge for realizing disruptive quantum technology. Two main ingredients of quantum information processors are quantum interference and single-photon detectors. Here we develop a hybrid superconducting-photonic circuit system to show how these elements can be combined in a scalable fashion on a silicon chip. We demonstrate the suitability of this approach for integrated quantum optics by interfering and detecting photon pairs directly on the chip with waveguide-coupled single-photon detectors. Using a directional coupler implemented with silicon nitride nanophotonic waveguides, we observe 97% interference visibility when measuring photon statistics with two monolithically integrated superconducting single-photon detectors. The photonic circuit and detector fabrication processes are compatible with standard semiconductor thin-film technology, making it possible to implement more complex and larger scale quantum photonic circuits on silicon chips.

[1] Department of Electrical Engineering, Yale University, New Haven, Connecticut 06511, USA. [2] Institute for Quantum Optics and Quantum Information, Austrian Academy of Science, A-1090 Vienna, Austria. Correspondence and requests for materials should be addressed to H.X.T. (email: hong.tang@yale.edu).

P roof-of-principle experiments have shown that quantum information processing has great potential for solving certain computational tasks, which are intractable with classical means[1]. Among the various approaches, integrated quantum photonics has emerged as a particularly interesting one for realizing optical quantum simulations[2-5], quantum information processing[6-9] and communication[10,11]. However, scaling current quantum technology to larger system sizes remains a significant challenge owing to the demanding requirements for high-fidelity signal processing at single-photon levels.

Advanced nanofabrication techniques have proven invaluable for ensuring scalability of electronic components used in classical information technology[12]. The corresponding complementary metal oxide semiconductor (CMOS) fabrication recipes have recently also been employed for realizing both nanophotonic waveguides[13], as well as superconducting single-photon detectors (SSPD)[14] on silicon chips. As most linear optics quantum logic schemes rely on non-classical interference and single-photon detection[15,16] it is crucial to realize both of these ingredients on a common scalable platform. Here we demonstrate such a quantum information processing platform by combining SSPDs with integrated silicon nitride photonic circuits to measure high-visibility quantum interference directly on-chip.

Highly efficient single-photon detection has previously been achieved with fibre-coupled SSPDs[17], which have found many exciting applications[18-20]. For integrated photonic technology, detection of photons inside a waveguide directly on-chip is required because it eliminates the chip-to-fibre interface, which is often a bottleneck in photonic device packaging. Optimal performance in this regard is achieved with nanowire SSPDs in travelling wave geometry[21,22]. This design is an excellent choice for integrated nanophotonic applications because large numbers of these compact detectors can be embedded directly in optical waveguide circuits[23]. On-chip detection efficiencies up to 90%[24,25] (system detection efficiencies up to 10% (refs 26,27)) at visible as well as telecom wavelengths have been demonstrated with such waveguide-coupled SSPDs. Furthermore, these detectors can operate at GHz rates, achieve <20 ps timing accuracy, sub-Hz dark count rate and extremely low-noise equivalent powers down to the 10^{-20} W Hz$^{-1/2}$ level[24,28].

Quantum interference can be observed when two indistinguishable photons impinge simultaneously on the inputs of a 50:50 beam splitter, that is, the probability of finding individual photons in separate output modes vanishes. This non-classical effect was first observed by Hong, Ou and Mandel (HOM)[29] and is the consequence of destructive interference of the probability amplitudes corresponding to both photons being transmitted/reflected at the beam splitter[30]. A waveguide implementation of an optical beam splitter is an optimal choice for realizing spatial mode matching[31-34], which is one of the limiting factors for achieving high-visibility quantum interference in free-space optics experiments.

The integration of photonic circuits and detectors on a silicon chip for demonstrating quantum interference has previously been attempted with surface plasmon polariton devices. Two-plasmon quantum interference with 93% visibility on a beam splitter has been demonstrated with off-chip detectors and for photons at visible wavelengths[35], which are not compatible with existing optical communication networks. Notably, the integration of plasmonic directional couplers with superconducting detectors on the same chip proved challenging and reduced the interference contrast below the classical limit[36].

Here we integrate low-noise niobium titanium nitride (NbTiN) nanowire SSPDs with dielectric silicon nitride (SiN) photonic circuits on a silicon chip. Using photons from spontaneous parametric down conversion (SPDC) we measure quantum interference with 97% visibility directly on-chip. Our circuit-detector approach is fully compatible with scalable, high-yield semiconductor microfabrication processes.

Results

Experimental set-up. The experimental set-up for measuring HOM interference with SSPDs directly on-chip is shown in Fig. 1. We produce energy–time correlated photon pairs via the process of type-II SPDC in a periodically poled potassium titanyl phosphate (ppKTP) crystal waveguide[37] and couple them into nanophotonic waveguides on a silicon chip. A continuous wave 775 nm pump laser is coupled directly from an optical single-mode fibre into the ppKTP waveguide and the generated 1,550-nm photon pairs are collected into a single-mode fibre. In type-II SPDC the generated photons of one pair have orthogonal polarization. Hence we use a fibre polarization beam splitter (PBS) to deterministically separate photons of each pair. To do so, we optimize the efficiency of the source by using a 1,550 nm telecom laser, which we send via both inputs/outputs of the fibre PBS into the ppKTP waveguide, that is in reverse, and monitor the second harmonic generation (SHG) of 775 nm light. First we maximize the SHG power by optimizing the spatial alignment between the ppKTP crystal waveguide and the input/output optical fibres. Similarly we optimize the phase matching of the nonlinear process by adjusting the crystal temperature for pump light of a given wavelength and waveguide geometry. We then minimize the SHG power using only the polarization controller between the ppKTP crystal and the fibre PBS (Fig. 1), thus ensuring orthogonal polarizations (H/V) at the PBS. Switching back to the 775 nm pump laser we first adjust the polarization controllers between 775 nm laser and ppKTP waveguide to maximize the photon pair generation efficiency. Then we adjust the polarization controllers behind the fibre PBS such that we achieve optimal coupling to the transverse electric mode of the on-chip waveguides. We use a 1,064-nm long-pass and a 1,550-nm band-pass filter, which efficiently suppress 775 nm, pump light in combination with on-chip grating couplers, effectively acting as additional band-pass filters. We introduce an optical delay line in one of the output ports of the fibre PBS that allows us to scan the relative arrival time between two photons of a pair at the silicon chip.

The chip with ~100 photonic integrated circuits and twice as many detectors (SSPDs) is mounted inside a closed-cycle cryostat, which provides continuous cooling to 1.7 K with <10 mK temperature variations[38]. We use a radiofrequency (rf) probe to make electrical contact to electrode pads on the chip, which connect to the SSPDs. The rf-probe lines are wired to a bias-T for supplying current from a low-noise source to the nanowires as well as reading out the voltage pulses upon photon detection by an SSPD[23]. Photons from the down conversion source are delivered to the on-chip photonic integrated circuits via an optical fibre array. Coupling loss from optical fibres to the on-chip waveguides is calibrated independently for each device input via monitor ports (Fig. 1) and is usually around 10 dB. The two device layouts shown in Figs 1 and 2a facilitate detector characterization, optical path-length measurements and fibre-to-device alignment. However, using 3 dB of the signal (idler) photons per input port for calibration purposes also leads to a quadratic decrease in the coincidence detection rates from correlated photon pairs. Future device designs could benefit from omitting these monitor ports. The chip is mounted on a stack of low-temperature compatible translation stages that allow us to position different devices under the fibre array and rf-probe-assembly for testing.

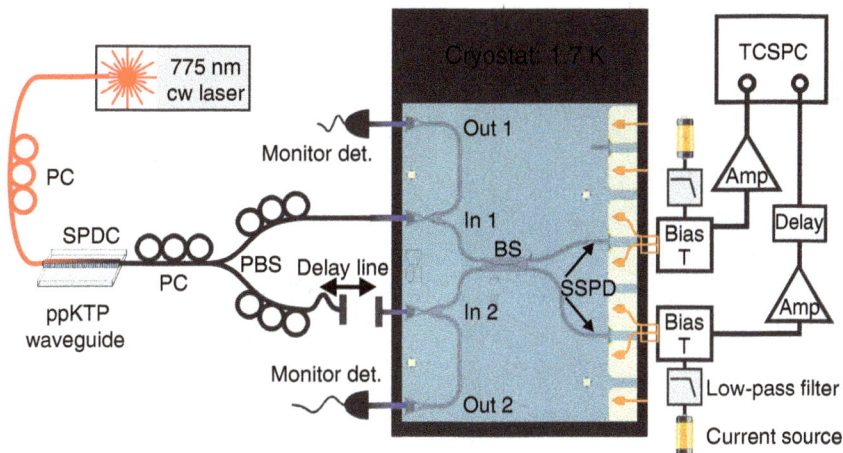

Figure 1 | Schematic of the on-chip quantum interference measurement set-up. A 775-nm continuous wave (cw) diode laser is used as a pump for generating orthogonally polarized 1,550 nm photon pairs via type-II spontaneous parametric down conversion in a 10.5-mm-long fibre-coupled periodically poled KTP waveguide. Input polarizations to the ppKTP source and a polarizing fibre beam splitter (PBS) are adjusted with fibre polarization controllers (PC) for optimal SPDC efficiency and deterministic splitting of photon pairs into separate PBS-output modes, respectively. Temporal delay between photons in separate output modes is set with a fibre-free-space-fibre optical delay line. Light is then guided into a closed-cycle cryostat where it is coupled from an optical fibre array into on-chip SiN photonic circuits via optical grating couplers at In 1 and In 2 (optical micrograph). The alignment of the chip to the fibre array with low-temperature nanopositioners (attocube) is aided by monitoring the optical transmission at the auxiliary ports Out 1 and Out 2. Photons interfere at a 33-μm-long directional coupler (beam splitter, BS) with 400 nm gap between waveguides in the coupling region. SSPDs on top of the beam splitter's output waveguides are supplied with a 10–15 μA bias from a low-noise current source and read out via a rf-probe connected to a bias-T outside the cryostat. After signal amplification (PPL 5828 & RF-Bay LNA-4050), photon statistics are recorded with a time-correlated single-photon counting unit (PicoHarp 300).

We amplify SSPD output signals before recording their arrival times with a time-correlated single-photon counting unit. In offline analysis we can thus identify detection events from correlated photon pairs by comparing arrival time lists for each of the two SSPDs[32].

Integrating photonic circuits and SSPDs. The device geometry for an on-chip HOM-interference measurement is shown in Fig. 2a. Fabrication starts from a commercial SiN on silicon dioxide (SiO$_2$) on silicon (Si) wafer onto which we sputter a thin film of NbTiN. Subsequently we define electrode pads, nanowire detectors and waveguides in standard electron-beam lithography followed by lift off and dry-etching chemistry, respectively (see Methods section). We use optical grating couplers to transmit light from the optical fibre array into waveguides of 1 μm width, designed for transverse electric single-mode propagation on-chip[24,25]. Photons are then guided to a beam splitter and detected by NbTiN nanowire SSPDs at the beam splitter's output.

The beam splitter is implemented as a directional coupler where two input waveguides are brought into close proximity over a coupling length L_c (Fig. 2b). With the waveguides acting as polarizers and spatial mode filters, the photons are indistinguishable when they arrive at the directional coupler. Finite element simulations of the coupling region show evanescent coupling between transverse electric modes of 1-μm-wide SiN waveguides. The resulting symmetric (Fig. 2c) and anti-symmetric (Fig. 2d) hybrid modes have a refractive index difference of Δn for a given gap. In simulations we find a coupling length of $L_c = \lambda/(4 \cdot \Delta n) = 28$ μm or a 400-nm gap to realize 50:50 splitting between the 330 nm high waveguides (Fig. 2e). In calibration devices (Mach–Zehnder interferometers and beam splitters, see Supplementary Fig. 1 and Supplementary Note 1) we observe that a slightly larger coupler length of $L_c = 33$ μm is required to achieve a 50:50 splitting ratio, which accounts for waveguide asymmetry, width and gap-offsets, as well as additional coupling in the input and output region of the

directional coupler, all of which are not taken into account in the finite element simulations. Owing to the high refractive index contrast of SiN on insulator (SiO$_2$), the footprint of our devices is approximately two orders of magnitude smaller than glass-based waveguide implementations and could in principle be even more compact if a smaller gap is chosen (see Supplementary Note 1 and Supplementary Fig. 2).

SSPDs were fabricated from sputter deposited 8.2 nm thin NbTiN films (see Methods section). We pattern 50 nm wide, 40 μm long U-shaped nanowires, which are connected via wider leads (Fig. 3a) to the electrode pads (Fig. 2a). We ensure that the SSPDs are precisely centred on top of the waveguides for optimal performance by aligning the nanowires to the same marks, which are subsequently used for patterning the SiN layer. We confirm that layer alignment between different lithography steps is better than 50 nm in scanning and transmission electron microscopy, as shown in Fig. 3. A calibration of the detector performance yielded an on-chip detection efficiency of 11.5% and a dark count rate of 0.7 and 2.2 Hz for a typical pair of SSPDs, when biasing close to the critical current (including black body radiation and stray light). We observe that our fabrication process features high yield of functional devices (see Supplementary Note 2 and Supplementary Fig. 3).

HOM-interference measurement. To measure quantum interference between photons produced in SPDC we first match the optical delays in the interferometer formed by the fibre PBS and the on-chip waveguide beam splitter (Fig. 1). We determine the zero-delay position using a 2.4-ps pulsed telecom wavelength laser (with corresponding 1.6 mm coherence length) and observing first-order interference between split pulses at one of the monitor ports with a fast O/E converter. Using this starting position we then send photon pairs from the SPDC source onto the chip and record detection events for both SSPDs at the outputs of the directional coupler.

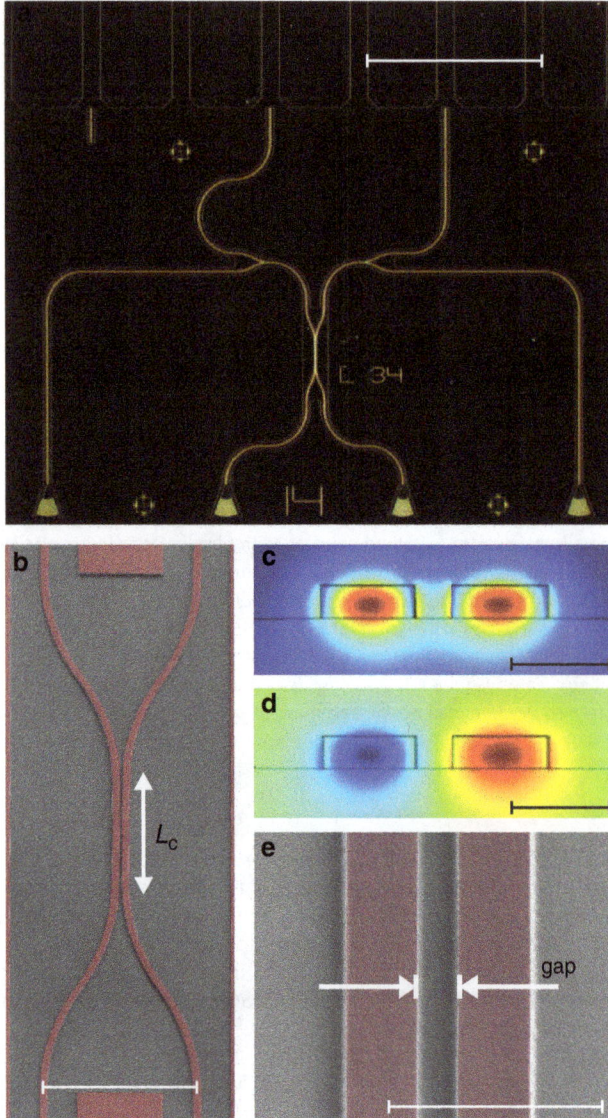

Figure 2 | Directional coupler device design. (a) Dark field micrograph of a HOM device with two input grating couplers (bottom centre), two output grating couplers for device alignment and calibration (bottom left/right), the directional coupler (DC) and waveguide-coupled SSPDs, which are contacted via Au-electrode pads (top) (scale bar, 250 μm); (b) SEM image of a directional coupler of nominal length L_c made from 1-μm-wide SiN waveguides (scale bar, 25 μm); (c) finite element simulation of the symmetric transverse electric field mode in the coupling region (max./min.: 1.8×10^{10} Vm^{-1}/0 Vm^{-1}; scale bar, 1 μm); (d) anti-symmetric transverse electric mode (max./min.: $\pm 1.8 \times 10^{10}$ Vm^{-1}; scale bar, 1 μm); (e) centre region of the directional coupler where evanescent coupling of the field modes occurs (scale bar, 3 μm).

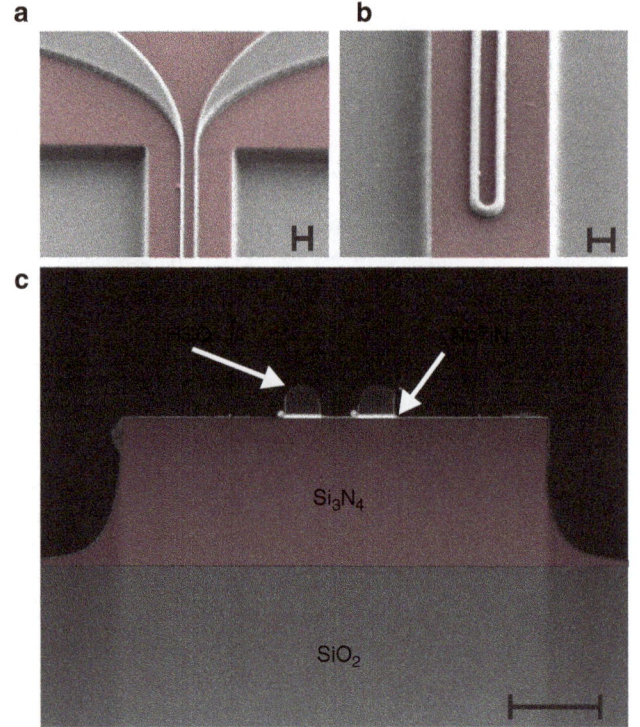

Figure 3 | Waveguide-coupled SSPD. (a) Scanning electron micrograph of the leads connecting a 50-nm narrow nanowire SSPD to Au-contact pads (see Fig. 2a); (b) U-shaped part of a 40-μm-long nanowire SSPD for optimal bias current distribution in the bending region where photons are incident from the SiN-waveguide underneath; (c) transmission electron micrograph of the SSPD-waveguide cross-section. The 8-nm-thin NbTiN nanowire (here 80 nm wide) is covered with electron-beam lithography resist (hydrogen silsesquioxane, HSQ) and centred on top of a 330-nm-high SiN waveguide on a buried oxide layer. All scale bars, 200 nm.

For delay positions larger than the coherence length, τ_c, of the down conversion photons we observe a coincidence rate of 4.2 ± 0.1 Hz for 256 ps binning of the photon arrival time data using a nominal 775 nm pump power of 10.5 mW (before coupling to the ppKTP waveguide). As we scan the delay line around the zero-delay position we observe how the coincidence rate drops almost to zero from its initial value for unmatched arrival times at the beam splitter and then recovers to its initial value of ∼4.2 Hz away from the zero-delay position, as shown in Fig. 4a. This is the expected behaviour for HOM interference of temporally correlated indistinguishable photons.

We approximate the spectral bandwidth of the down conversion photons incident on the beam splitter by a Gaussian function and fit the raw coincidence rate data with the corresponding function[39]:

$$C = C_n \left(1 - V e^{-d^2/(2\sigma^2)} \right) \qquad (1)$$

where C_n is the coincidence rate for photons with large arrival time delays $\Delta t \gg \tau_c$, V is the HOM-interference visibility, d is the delay position and σ the standard deviation. From the fit shown in Fig. 4a we extract the visibility $V = 96.9 \pm 5.3\%$ and a full width at half maximum of $w = 518 \pm 41$ μm, corresponding to a coherence time of 1.7 ± 0.1 ps from which we estimate a SPDC photon bandwidth of 2.1 ± 0.2 nm (ref. 40).

Note that the HOM-interference visibility was extracted directly from the raw data shown in Fig. 4a. Hence it contains contributions from accidental coincidence detection events, where each SSPD registered a count within the user-specified coincidence time window, although no photon pair was detected. Such accidental coincidence events occur statistically as photons belonging to different pairs and/or detector dark counts, are registered within a time interval shorter than the coincidence time window. Here we choose a coincidence time window of 256 ps, which is significantly longer than the jitter of our SSPDs (∼50 ps (ref. 28)) but short enough to avoid a significant background of accidental coincidences: at <10 kHz counting rate per SSPD (depending on pump power) we estimate an accidental coincidence contribution of 0.02 Hz, which is contained in the

a

b

c

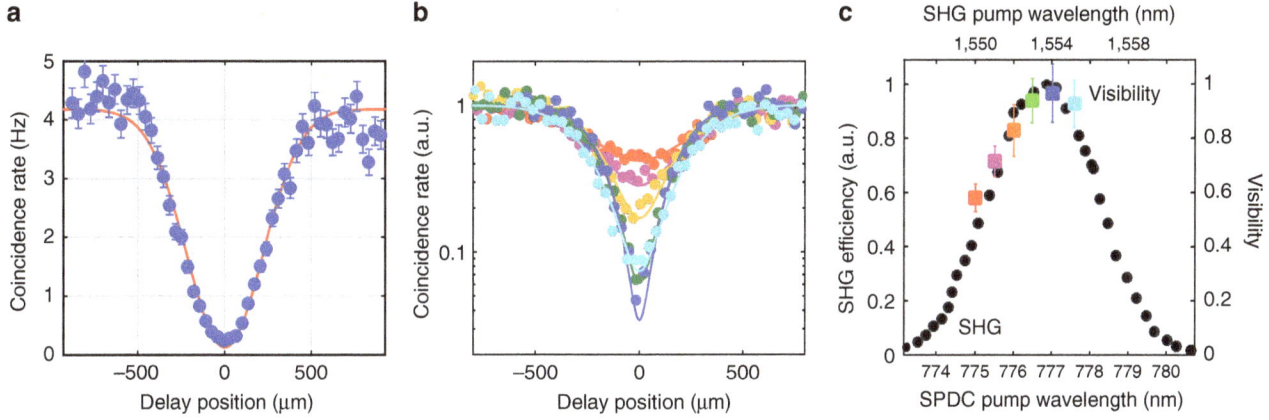

Figure 4 | On-chip HOM interference. (**a**) Coincidence detection rate (raw) between SSPDs coupled to the directional coupler's output waveguides as a function of photon arrival time delay at a nominal pump power of 10.5 mW. Error bars show the s.d. of the statistical ensemble of coincidence events. As we scan the position of the optical delay line (Fig. 1) the coincidence rate drops from 4.2 Hz close to zero when photons created in SPDC arrive simultaneously at the on-chip directional coupler. From a Gaussian fit to the data we extract a two-photon interference visibility of $V = 0.97 \pm 0.05$ and the coherence time of the SPDC photons as 1.7 ps; (**b**) measured coincidence rates as a function of relative delay between signal and idler photons for various SPDC pump laser wavelengths, $\lambda_p = 775 - 777.6$ nm. Error bars are similar to those shown in **a** but have been omitted for readability. From a fit to the data we find the visibilities $V_{red}(\lambda_p = 775$ nm$) = 58 \pm 5\%$, $V_{magenta}(\lambda_p = 775.5$ nm$) = 71 \pm 6\%$, $V_{orange}(\lambda_p = 776$ nm$) = 83 \pm 9\%$, $V_{green}(\lambda_p = 776.5$ nm$) = 94 \pm 8\%$, $V_{blue}(\lambda_p = 777.1$ nm$) = 97 \pm 10\%$, $V_{cyan}(\lambda_p = 777.6$ nm$) = 93 \pm 9\%$; (**c**) HOM-interference visibility as a function of SPDC pump wavelength as extracted from the fits in **b** where error bars denote 95% confidence bounds of the fit, and SHG efficiency as a function of pump wavelength at similar temperature of the ppKTP crystal.

rate shown in Fig. 4a. Our data clearly benefits from the low dark count rates of the two SSPDs used here (0.7 and 2.2 Hz, respectively), which only cause a negligible background contribution to the measured coincidence rate.

We repeat the two-photon interference measurement for a significantly lower pump power of 3.5 mW to avoid higher-order processes in the SPDC process but only observe a small improvement of HOM-interference visibility to 97.1% (see Supplementary Note 3 and Supplementary Fig. 4). We thus conclude that higher-order SPDC processes do not contribute appreciably to the observed coincidence rate at zero delay.

The fact that we observe a visibility slightly lower than 100% at $d = 0$ is thus mainly due to detection events from independent pairs that were created within the coincidence detection window, a slight imbalance in the splitting ratio of our on-chip directional coupler and the statistical photon counting noise. We anticipate that fine tuning of our fabrication recipes will improve the performance of our photonic circuits and detectors to allow for even higher interference visibilities, which comply with fault-tolerant quantum operations[41].

Characterization of photon indistinguishability. To achieve optimal visibility in HOM interference it is necessary that signal and idler photons are indistinguishable at the beam splitter, not only in arrival time but also in all other degrees of freedom. In nanophotonic implementations, as the one presented here, the spatial overlap between the input modes of the directional coupler is guaranteed by the high lithographic control over the waveguide dimensions on-chip. For the waveguide cross-section chosen here (330 nm × 1 μm) only a single transverse electric mode is supported such that other polarization and spatial modes are efficiently suppressed. We thus investigate the remaining spectral distinguishability between the down conversion photons in dependence of SPDC pump detuning, which can be varied off-chip.

The frequency correlation between signal and idler output modes in SPDC is described by a joint spectral amplitude, which is given by the product of a pump spectral amplitude and a phase-matching function. The latter can be approximated by a Gaussian

function with width σ_s around the degenerate (quasi) phase-matching frequency, ω_0, where $\Delta\mathbf{k} = 0$ with $\omega_s = \omega_i$. If the 775 nm pump frequency, ω_p, is detuned from ω_0, the signal and idler spectra around ω_s and ω_i, respectively, will shift according to the phase-matching conditions imposed by energy and momentum conservation: $\omega_p = \omega_s + \omega_i$ and $\Delta\mathbf{k} = \mathbf{k_p} - \mathbf{k_s} - \mathbf{k_i} - 2\pi/\Lambda$, with poling period Λ and wave vectors $k_m = n(\lambda_m, T) \cdot \frac{\omega_m}{c}$. Here T is the SPDC source temperature and the refractive indices $n(\lambda_m, T)$ for pump, signal and idler photons, $m = $ p,s,i, are given by the Sellmeier equations. A change in pump wavelength thus introduces spectral distinguishability between signal and idler photons, which reduces the visibility in a HOM-interference experiment.

We scan the delay between signal and idler photons around the zero-delay position for pump laser wavelengths of $\lambda = 775 - 777.6$ nm and observe the variation of interference visibility with pump wavelength, shown in Fig. 4b. For better comparison all data is normalized to the coincidence rate at $\Delta\tau \to \infty$ as determined from the respective fit. At $\lambda = 775$ nm pump wavelength the phase-matching conditions cause the signal and idler spectra to shift significantly such that the interference visibility drops to 58% (see red data in Fig. 4b). As the pump wavelength is increased to 777.1 nm the HOM-interference visibility gradually increases to 97% (see dark blue data in Fig. 4b) before it starts dropping again for $\lambda > 777.1$ nm (see cyan data in Fig. 4b). The variation of interference visibility follows roughly a Gaussian distribution (Fig. 4c) as expected from shifting the approximately Gaussian signal and idler spectra with respect to each other. High visibility is achieved over a relatively broad spectral range ($V > 90\%$ over $\Delta\lambda \approx 1.5$ nm around $\lambda = 777.1$ nm), which shows that signal and idler photon distinguishability is under accurate control in our experiment.

For comparison we show the SHG efficiency for pump wavelengths $\lambda = 1,546 - 1,562$ nm of the ppKTP crystal at similar temperature in Fig. 4c. Both SHG efficiency and HOM-interference visibility show similar behaviour as a function of the respective pump laser wavelength. This relation between SHG and SPDC phase matching is expected in a crystal of given material properties, waveguide length and cross-section[42].

Discussion

Recent experiments in quantum optics[3,4,5,43] manifest an ever more pressing need for a scalable solution to integrate photonic circuits and single-photon detectors. In particular, single-photon detection and high-visibility quantum interference have been identified as the two essential requirements for realizing scalable linear optic quantum computation[15]. Here we have shown how SSPDs embedded with nanophotonic circuits address these needs and achieve the requirements for scalable quantum technology on silicon chips. We observe high-visibility quantum interference of photons produced in SPDC with waveguide-coupled SSPDs and demonstrate that photon distinguishability is under accurate control in this architecture. All of the fabrication techniques used here can in principle be adapted to scalable technology developed for the CMOS industry, even at the front end of a CMOS line[44]. We anticipate that fine tuning of superconducting film and photonic circuit parameters and the implementation of photon-number resolving architectures[45,46] will further increase the functionality of our integrated quantum photonic system.

Recent progress in realizing sources of non-classical light directly on silicon chips ideally complements the integration of single-photon detectors and photonic circuits described here. Such integrated quantum light sources were realized employing spontaneous four wave mixing[31,47] and the excitation of waveguide-coupled quantum emitters[48,49] but could also be realized via SPDC in III-nitride waveguides[50,51]. Combining nanophotonic sources, circuits and single-photon detectors on a silicon chip will allow for generating, processing and detecting quantum information all on one scalable platform.

Methods

Device fabrication. We deposit NbTiN on commercial 330 nm stoichiometric Si_3N_4 on 3 μm thermally grown SiO_2 on Si wafers. The film thickness of the NbTiN layer is controlled via timed reactive ion sputtering from an NbTi alloy target in an Ar–N_2 atmosphere at room temperature. From atomic force microscopy, transmission electron microscopy and square resistance measurements we infer a film thickness of 8.2 nm and a deposition rate of 1.33 nm s^{-1}. Transmission electron micrographs (Fig. 3c) on a reference device show that our NbTiN films are slightly thicker than those used in previous device generations[25,26,28]. This could explain the somewhat lower detection efficiency and dark count rate as compared with those reported in ref. 28, which relied on NbTiN films sputtered elsewhere. We anticipate that fine tuning of our NbTiN sputter recipe and film thickness will yield SSPD performance on a par with previous demonstrations.

After NbTiN film deposition we define electrode pads and alignment marks (Fig. 2a) for subsequent layers in electron-beam lithography using double-layer polymethyl methacrylate positive-tone resist. After development in methyl isobutyl ketone and isopropyl alcohol we deposit an 8-nm Ti adhesion layer and 150 nm gold (Au) in electron-beam evaporation followed by lift off in acetone. In a second high-resolution (100 kV) electron-beam lithography step the detector nanowires are patterned in negative-tone hydrogen silsesquioxane resist. Each detector pair is aligned separately to the Au alignment marks in the write-field of the respective device. After development in tetramethylammonium hydroxide-based developer the pattern is transferred to the NbTiN layer in a timed reactive ion etching step employing tetrafluoromethane (CF4) chemistry. In a third and final electron-beam lithography step we expose the waveguide layer in positive-tone ZEP520A polymer resist. The patterns for each photonic circuit device are aligned to the same alignment marks used in the previous step for defining the respective nanowire detector pair. Following development in xylenes the waveguide patterns are transferred to the SiN film via carefully timed reactive-ion etching in fluoroform (CHF3). The resulting devices are shown in Figs 2 and 3.

References

1. Ladd, T. D. *et al.* Quantum computers. *Nature* **464**, 45–53 (2010).
2. Aspuru-Guzik, A. & Walther, P. Photonic quantum simulators. *Nat. Phys.* **8**, 285–291 (2012).
3. Spring, J. B. et al. Boson sampling on a photonic chip. *Science* **339**, 798–801 (2013).
4. Broome, M. A. et al. "Photonic boson sampling in a tunable circuit ". *Science* **339**, 794–798 (2013).
5. Spagnolo, N. et al. "Experimental validation of photonic boson sampling". *Nat. Photon* **8**, 615–620 (2014).
6. Politi, A., Cryan, M. J., Rarity, J. G., Yu, S. & O'Brien, J. L. "Silica-on-silicon waveguide quantum circuits". *Science* **320**, 646–649 (2008).

7. O'Brien, J. L., Furusawa, A. & Vuckovic, J. "Photonic quantum technology". *Nat. Photon* **3**, 687–695 (2009).
8. Shadbolt, P. J. et al. "Generating, manipulating and measuring entanglement and mixture with a reconfigurable photonic circuit". *Nat. Photon* **6**, 45–49 (2012).
9. Politi, A., Matthews, J. C. F. & O'Brien, J. L. "Shor's quantum factoring algorithm on a photonic chip". *Science* **325**, 1221 (2009).
10. Takesue, H. et al. "Quantum key distribution over a 40-dB channel loss using superconducting single-photon detectors". *Nat. Photon* **1**, 343–348 (2007).
11. Grassani, D. et al. Micrometer-scale integrated silicon source of time-energy entangled photons. *Optica* **2**, 88–94 (2015).
12. Moore, G. E. Cramming More Components onto integrated circuits. *Electronics* **38**, 114–117 (1965).
13. Bogaerts, W. et al. Nanophotonic waveguides in silicon-on-insulator fabricated with CMOS technology. *J. Lightwave Technol.* **23**, 401–412 (2005).
14. Gol'tsman, G. N. et al. Picosecond superconducting single-photon optical detector. *Appl. Phys. Lett.* **79**, 705–707 (2001).
15. Knill, E., Laflamme, R. & Milburn, G. J. A scheme form efficient quantum computation with linear optics. *Nature* **409**, 46–52 (2001).
16. Raussendorf, R. & Briegel, H. J. A one-way quantum computer. *Phys. Rev. Lett.* **86**, 5188–5191 (2001).
17. Marsili, F. et al. Detecting single infrared photons with 93% system efficiency. *Nat. Photon.* **7**, 210–214 (2013).
18. Hadfield, R. H. Single-photon detectors for optical quantum information applications. *Nat. Photon.* **3**, 696–705 (2009).
19. Saglamyurek, E. et al. Quantum storage of entangled telecom-wavelength photons in an erbium-doped optical fiber. *Nat. Photon.* **9**, 83–87 (2015).
20. Bussieres, F. et al. Quantum teleportation from a telecom-wavelength photon to a solid-stat quantum memory. *Nat. Photon.* **8**, 775–778 (2014).
21. Hu, X., Holzwarth, C. W., Masciarelli, D., Dauler, E. A. & Berggren, K. K. Efficiently coupling light to superconducting nanowire single-photon detectors. *IEEE Trans. Appl. Supercond.* **19**, 336–340 (2009).
22. Sprengers, J. P. et al. Waveguide superconducting single photon detectors for integrated quantum photonic circuits. *Appl. Phys. Lett.* **99**, 181110 (2011).
23. Schuck, C. et al. Matrix of integrated superconducting single-photon detectors with high timing resolution. *IEEE Trans. Appl. Supercond.* **23**, 2201007 (2013).
24. Pernice, W. H. P. et al. High-efficiency, ultrafast single-photon detectors integrated with nanophotonic circuits. *Nat. Commun.* **3**, 1325 (2012).
25. Schuck, C., Pernice, W. H. P. & Tang, H. X. NbTiN superconducting nanowire detectors for visible and telecom wavelengths single-photon counting on Si3N4 photonic circuits. *Appl. Phys. Lett.* **102**, 051101 (2013).
26. Schuck, C., Pernice, W. H. P., Ma, X. & Tang, H. X. Optical time domain reflectometry with low noise waveguide-coupled superconducting single-photon detectors. *Appl. Phys. Lett.* **102**, 191104 (2013).
27. Najafi, F. et al. On-chip detection of non-classical light by scalable integration of single-photon detectors. *Nat. Commun.* **6**, 5873 (2015).
28. Schuck, C., Pernice, W. H. P. & Tang, H. X. Waveguide integrated low noise NbTiN nanowire single-photon detectors with milli-Hz dark count rate. *Sci. Rep.* **3**, 1893 (2013).
29. Hong, C. K., Ou, Z. Y. & Mandel, L. Measurement of subpicosecond time intervals between two photons by interference. *Phys. Rev. Lett.* **59**, 2044–2046 (1987).
30. Zeilinger, A. General properties of lossless beam splitters in interferometry. *Am. J. Phys.* **49**, 882–883 (1981).
31. Silverstone, J. W. et al. On-chip quantum interference between silicon photon-pair sources. *Nat. Photon.* **8**, 104–108 (2014).
32. Xu, X. et al. Near-infrared Hong-Ou-Mandel interference on a silicon quantum photonic chip. *Opt. Express* **21**, 5014–5024 (2013).
33. Silverstone, J. W. et al. Qubit entanglement on a silicon photonic chip, Preprint at <http://arxiv.org/abs/ 1410.8332> (2014).
34. Gerrits, T. et al. Spectral correlation measurements at the Hong-Ou-Mandel interference dip. *Phys. Rev. A* **91**, 013830 (2015).
35. Fakonas, J. S., Lee, H., Kelaita, Y. A. & Atwater, H. A. Two-plasmon quantum interference. *Nat. Photon.* **8**, 317–320 (2014).
36. Heeres, R. W., Kouwenhoven, L. P. & Zwiller, V. Quantum interference in plasmonic circuits. *Nat. Nanotechnol.* **8**, 719–722 (2013).
37. Zhong, T., Wong, F. N. C., Roberts, T. D. & Battle, P. High performance photon-pair source based on fiber-coupled periodically poled KTiOPO₄ waveguide. *Opt. Express* **17**, 12019–12030 (2009).
38. Wang, C., Lichtenwalter, B., Friebel, A. & Tang, H. X. A closed-cycle 1K refrigeration cryostat. *Cryogenics* **64**, 5–9 (2014).
39. Bachor, H. A. & Ralph, T. C. *A Guide to Experiments in Quantum Optics* (Wiley-VCH, 2004).
40. Born, M. & Wolf, E. *Principles of optics* (Cambridge University Press, 1999).
41. Laing, A. et al. High-fidelity operation of quantum photonic circuits. *Appl. Phys. Lett.* **97**, 211109 (2010).

42. Helt, L. G., Liscidini, M. & Sipe, J. E. How does it scale? Comparing quantum and classical nonlinear processes in integrated devices. *J. Opt. Soc. Am. B* **29**, 2199–2212 (2012).

43. Crespi, A. *et al.* Anderson localization of entangled photons in an integrated quantum random walk. *Nat. Photon.* **7**, 322–328 (2013).

44. Tolpygo, S. K. *et al.* Fabrication process and properties of fully-planarized deep-submicron Nb/Al-AlO$_x$/Nb josephson-junctions for vlsi circuits. *IEEE Trans. Appl. Supercond.* **25**, 1101312 (2015).

45. Divochiy, A. *et al.* Superconducting nanowire photon-number resolving detector at telecommunication wavelengths. *Nat. Photon.* **2**, 302–306 (2008).

46. Sahin, D. *et al.* Waveguide photon-number-resolving detectors for quantum photonic integrated circuits. *Appl. Phys. Lett.* **103**, 111116 (2013).

47. Preble, S. F. *et al.* On-chip quantum interference from a single silicon ring resonator source. *Phys. Rev. Appl.* **4**, 021001 (2015).

48. Reithmaier, G. *et al.* On-chip time resolved detection of quantum dot emission using integrated superconducting single-photon detectors. *Sci. Rep.* **3**, 1901 (2013).

49. Mouradian, S. L. *et al.* Scalable integration of long-lived quantum memories into a photonic circuit. *Phys. Rev. X* **5**, 031009 (2015).

50. Xiong, C. *et al.* Integrated GaN photonic circuits on silicon (100) for second harmonic generation. *Opt. Express* **19**, 10462–10470 (2011).

51. Xiong, C. *et al.* Aluminum nitride as a new material for chip-scale optomechanics and nonlinear optics. *New J. Phys.* **14**, 095014 (2012).

Acknowledgements

C.S. acknowledges financial support from the Deutsche Forschungsgemeinschaft (SCHU 2871/2–1). X.M. is supported by a Marie Curie International Outgoing Fellowship within the 7th European Community Framework Programme. H.X.T. acknowledges support from a Packard Fellowship in Science and Engineering and a CAREER award from the National Science Foundation. We thank Dr Michael Rooks and Michael Power for their assistance in device fabrication.

Author contributions

H.X.T. and C.S. conceived the experiment, C.S. and X.G. performed the measurements and analysed the data, L.R.F. deposited the NbTiN films, C.S. wrote the manuscript, fabricated and characterized the devices, C.S., X.G., M.P., X.M. and H.X.T. discussed the results, H.X.T. supervised the work.

Additional information

Competing financial interests: The authors declare no competing financial interests.

2

Large elasto-optic effect and reversible electrochromism in multiferroic BiFeO$_3$

D. Sando[1,*,†], Yurong Yang[2,*], E. Bousquet[3], C. Carrétéro[1], V. Garcia[1], S. Fusil[1], D. Dolfi[4], A. Barthélémy[1], Ph. Ghosez[3], L. Bellaiche[2] & M. Bibes[1]

The control of optical fields is usually achieved through the electro-optic or acousto-optic effect in single-crystal ferroelectric or polar compounds such as LiNbO$_3$ or quartz. In recent years, tremendous progress has been made in ferroelectric oxide thin film technology—a field which is now a strong driving force in areas such as electronics, spintronics and photovoltaics. Here, we apply epitaxial strain engineering to tune the optical response of BiFeO$_3$ thin films, and find a very large variation of the optical index with strain, corresponding to an effective elasto-optic coefficient larger than that of quartz. We observe a concomitant strain-driven variation in light absorption—reminiscent of piezochromism—which we show can be manipulated by an electric field. This constitutes an electrochromic effect that is reversible, remanent and not driven by defects. These findings broaden the potential of multiferroics towards photonics and thin film acousto-optic devices, and suggest exciting device opportunities arising from the coupling of ferroic, piezoelectric and optical responses.

[1]Unité Mixte de Physique, CNRS, Thales, Univ. Paris-Sud, Université Paris-Saclay, 91767 Palaiseau, France. [2]Department of Physics and Institute for Nanoscience and Engineering, University of Arkansas, Fayetteville, Arkansas 72701, USA. [3]Theoretical Materials Physics, Université de Liège, B-5, B-4000 Sart-Tilman, Belgium. [4]Thales Research and Technology France, 1 Avenue Augustin Fresnel, 91767 Palaiseau, France. * These authors contributed equally to this work. † Present address: School of Materials Science and Engineering, University of New South Wales, Sydney 2052, Australia. Correspondence and requests for materials should be addressed to D.S. (email: daniel.sando@unsw.edu.au) or to M.B. (email: manuel.bibes@thalesgroup.com).

Bismuth ferrite ($BiFeO_3$—BFO) is multiferroic at room temperature with strong ferroelectric polarisation[1] and G-type antiferromagnetic ordering with a cycloidal modulation of the Fe spins[2]. Most research on this material has been driven by the prospect of electrically controlled spintronic devices[3]. More recently, however, BFO has revealed further remarkable multifunctional properties. Notable discoveries include conductive domain walls[4], a strain-driven morphotropic phase boundary[5] and a specific magnonic response that can be tuned by epitaxial strain[6] or electric field[7]. Moreover, with a bandgap (~ 2.7 eV) in the visible[8], large birefringence[9] (0.25–0.3), a strong photovoltaic effect[10] and sizeable linear electro-optic coefficients[11], BFO is garnering interest in photonics and plasmonics[12].

Most of these physical properties are intimately linked to structural parameters, and may thus be tuned in thin films by epitaxial strain. Strain engineering[13] is a powerful tool through which, for instance, ferroelectricity is strongly enhanced in $BaTiO_3$ (ref. 14), or induced in otherwise non-ferroelectric materials such as $SrTiO_3$ (ref. 15). In BFO, two structural instabilities are sensitive to epitaxial strain: the polar distortion—responsible for the ferroelectricity—and antiferrodistortive (FeO_6 octahedra) rotations. In strained BFO films, the competition between both instabilities and their coupling to ferroic order parameters yields rich phase diagrams, revealing new structural, ferroelectric and magnetic phases[6,16], as well as large variations in the ferroelectric Curie temperature[17] and the spin direction[6].

Here, we present a combined experimental and theoretical study demonstrating that strain induces a very large change in the refractive index of BFO, which corresponds to an effective elasto-optic coefficient larger than in any ferroelectric, and larger than that of quartz[18]. This effect is accompanied by a shift of the optical bandgap, reminiscent of pressure-induced changes in light-absorption[19], a phenomenon known as piezochromism in other materials systems[20]. The trends in the optical properties as a function of strain are well reproduced by our first-principles calculations, and we are able to clarify precisely why the optical bandgap of tetragonal-like BFO is larger than that of the rhombohedral-like phase. Finally, we show how an electric field can be used to toggle between two strain states with different light absorption, corresponding to an electrochromic effect that is intrinsic, reversible and non-volatile.

Results

Sample preparation and structural characterization.
Fully strained BFO thin films were grown using pulsed laser deposition on (001)-oriented substrates (in pseudocubic notation, which we use throughout this paper) spanning a broad range of lattice mismatch (from -7.0% to $+1.0\%$; Methods section). At low strain—compressive or tensile—the films crystallize in the so-called R-like phase of BFO, derived from the bulk rhombohedral (R3c) phase. At high compressive strain ($\leq 4\%$), the films grow in the T-like phase[16] with a large tetragonality ratio $c/a \approx 1.26$ (cf. Fig. 1b). Reciprocal space maps around the $(113)_{pc}$ or $(223)_{pc}$ reflections (Supplementary Fig. 1) reveal that all our BFO films possess a monoclinic structure (M_A or M_B for R-like, M_C for T-like[12,21]), and further scans (not shown) indicate the presence of two structural domain variants (see the sketch in Fig. 1a). The in-plane and out-of-plane pseudocubic lattice parameters are presented in Fig. 1b. The in-plane parameter shows a monotonic decrease with compressive strain, while the out-of-plane parameter concomitantly increases, albeit with a sharp jump at $\sim -3.5\%$ corresponding to the structural transition between the R-like and T-like phases.

First-principles calculations.
To explore the effect of strain on the optical properties of BFO thin films, we performed first-principles calculations, using the Heyd–Scuseria–Ernzerhof (HSE) hybrid functional (see Methods section for details). In the following, we denote the electronic bandgap as the energy difference between the valence band maximum (VBM) and the conduction band minimum (CBM), while the optical bandgap corresponds to the extrapolation of the linear region of the Tauc plot (cf. Fig. 2c); theoretically, the optical bandgap is computed from the complex dielectric function.

Figure 1c,d show the computed electronic density of states for the R-like phase of BFO (at 0, $+2$ and -3% strain) and for the T-like phase (-5 and -7%). The insets of Fig. 1c,d show that the electronic bandgap is lower for the T-like phase than for the R-like phase, particularly for -7% strain, consistent with previous studies[22,23]. In the R-like phase, both compressive and tensile strains yield an increase of the electronic bandgap, similar to the situation in $SrTiO_3$ (ref. 24).

The partial density of states of Fig. 1e-l shows that the VBM mostly consists of O 2p orbitals for any considered misfit strain, and the CBM mainly comprises Fe d_{xy}, d_{xz}, d_{yz} orbitals for both the R-like and T-like phases. In R-like BFO, strain-induced changes in the FeO_6 octahedra rotations and the polar modes conspire[24] to slightly lift the degeneracy of the Fe 3d orbitals, but the nature of the electronic states at the VBM and CBM is globally preserved. In contrast, the pyramidal coordination of the FeO_5 unit in highly elongated T-like BFO yields a large splitting of the 3d states with the d_{xy} orbital sitting 300 meV lower in energy (Fig. 1h). Figure 1l shows that this d_{xy} state is weakly hybridized with O states and the states near the CBM in the T-like phase have very little O 2p character. This suggests that optical transitions from the VBM to those states should be very weak, and that the main optical transitions in the T-like phase should occur from the VBM to d_{xz} and d_{yz} states that lie in energy ~ 300 meV above the CBM. In other words, the optical bandgap should be higher in T-like BFO than in R-like BFO despite the opposite trend in the electronic bandgap. This presumption is confirmed by the energy dependence of the extinction coefficient derived from our calculations (Fig. 2b): the absorption edge appears at least 200 meV higher in the T-like phase than in the R-like phase.

Optical characterization.
Figure 2a presents the experimental energy dependence of the extinction coefficient, extracted from spectroscopic ellipsometry measurements (Supplementary Fig. 2). These data confirm the theoretical prediction of a larger optical bandgap for the T-like phase, and are consistent with previous studies. The agreement between the experimental and calculated extinction coefficient curves (Fig. 2a,b) is very good, particularly for the onset of absorption. The corresponding experimental Tauc plots (cf. Methods section) for these three samples are shown in Fig. 2c, indicating that the optical bandgap for T-like BFO is 3.02 eV, while for R-like BFO, a compressive strain of 2.6% induces an increase in the bandgap from 2.76 to 2.80 eV. Figure 2d summarizes the strain dependence of the experimental and calculated optical bandgap. In the R-like phase, both compressive and tensile strains induce an increase of the optical bandgap, and the T-like polymorph exhibits an optical bandgap ~ 0.25 eV larger than the R-like phase, consistent with previous reports[25].

Our observation of a strain-induced change in optical bandgap and thus optical absorption is reminiscent of an effect called piezochromism[20], which corresponds to changes in light absorption driven by hydrostatic pressure. Piezochromic effects have been identified in several organic compounds, but for

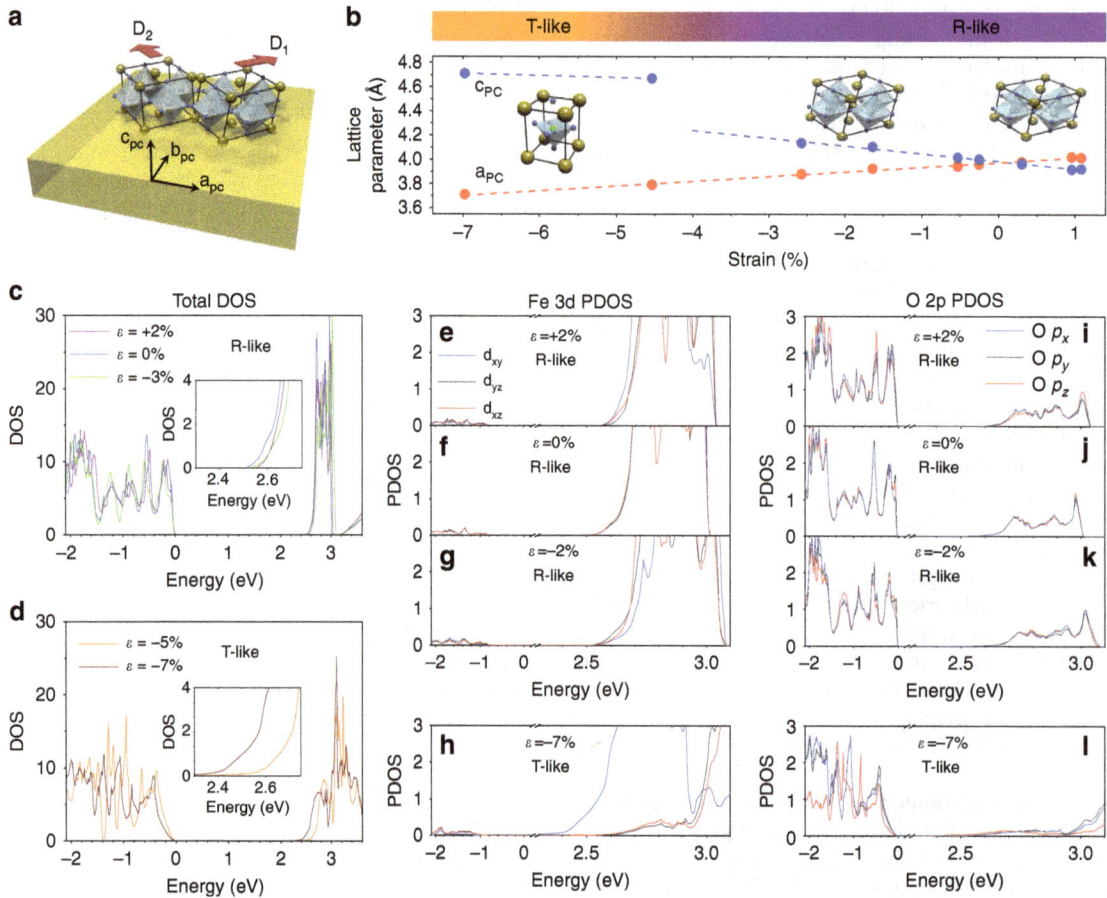

Figure 1 | Electronic structure of strained BiFeO₃ thin films. (**a**) Sketch of the two structural variants present in our monoclinic (M_A or M_B) R-like BiFeO₃ films. The red arrows indicate the direction of the monoclinic distortion for the two variants D_1 and D_2. (**b**) In-plane and out-of-plane lattice parameters of our strained BFO films. Total density of state (DOS) for R-like (**c**) and T-like phases (**d**). The insets show the DOS near the CBM. Partial density of states (PDOS) of iron 3d (**e–h**) and oxygen (**i–l**) states for R-like and T-like BFO. Note the break between 0.2 and 2.3 eV in the horizontal axes in (**e–l**). For all panels, only the spin-up channel states are shown; the spin-down channel states are the same as the spin up due to the antiferromagnetic order.

inorganics mainly in the CuMoO₄ family[26,27]. In this compound, the application of hydrostatic pressure triggers a first order transition between two polytypes having different optical absorption spectra due to changes in the oxygen cage surrounding the Cu ions[26]. Interestingly, in both CuMoO₄ and BiFeO₃ the absorption is stronger when the transition metal cation is in an octahedral oxygen environment, which suggests a possible trend, and strategies for engineering piezochromic effects in other perovskites.

In bulk BFO, the bandgap is known to decrease with pressure[19], particularly below 3.4 GPa and at the structural phase transition near 9.5 GPa. This corresponds to a piezochromic effect of amplitude 0.058 eV GPa⁻¹ at low pressure, and of 0.027 eV GPa⁻¹ on average between ambient pressure and 18 GPa (ref. 19). In our films, from the strain values and Young's modulus[28], we estimate the amplitude of the piezochromic effect in BiFeO₃ at ~0.12 eV GPa⁻¹. Importantly, working with thin films may be advantageous for several applications[29]. For instance, the thin film geometry allows the application of large electric fields to toggle between two optical polytypes, thereby producing potentially high-speed electrochromic effects.

We have explored this possibility in BFO thin films with coexisting R-like and T-like regions[5]. We applied an electric field to transform the mixed R − T BFO into nominally pure T-like BFO over $10 \times 10 \, \mu m^2$ regions (cf. Fig. 3a) and probed the local optical transmission in and out of this area using a conventional optical microscope. Figure 3b shows a transmission image with a

dielectric filter (bandwidth 10 nm) centred at 420 nm inserted between the white light source and the sample. Clearly, a $10 \times 10 \, \mu m^2$ square with a higher intensity than the background is visible in the image. Remarkably, this effect is reversible: applying a voltage with the opposite polarity restores a mixed R + T state (cf. Fig. 3c), which restores a stronger optical absorption, see Fig. 3d. This contrast is stable for several weeks.

We have acquired similar optical images using various dielectric filters, recorded the transmitted intensity in and out of the $10 \times 10 \, \mu m^2$ square and calculated the contrast difference as a function of wavelength. The contrast is maximal between 420 and 450 nm, see Fig. 3e. This dependence agrees very well with the expected contrast difference, calculated from the extinction coefficients of pure R-like BiFeO₃ and pure T-like BiFeO₃ films of Fig. 2a. This confirms that the contrast in the optical images is indeed due to the intrinsic modulation of the optical bandgap induced by the electrical poling, rather than by defect-mediated processes as in Ca-doped BFO (ref. 30) or WO₃ (ref. 20).

Finally, we focus on the influence of strain on the real part of the complex refractive index n. In Fig. 4a, we highlight representative results of the variation of n with wavelength for R-like BFO that is weakly strained (on SmScO₃, SSO), strongly compressively strained (on (La,Sr)(Al,Ta)O₃ (LSAT)), and T-like BFO (on LaAlO₃ (LAO)). Below the optical bandgap (that is, for wavelengths longer than ~460 nm), the refractive index systematically decreases with increasing strain. This is also visible in

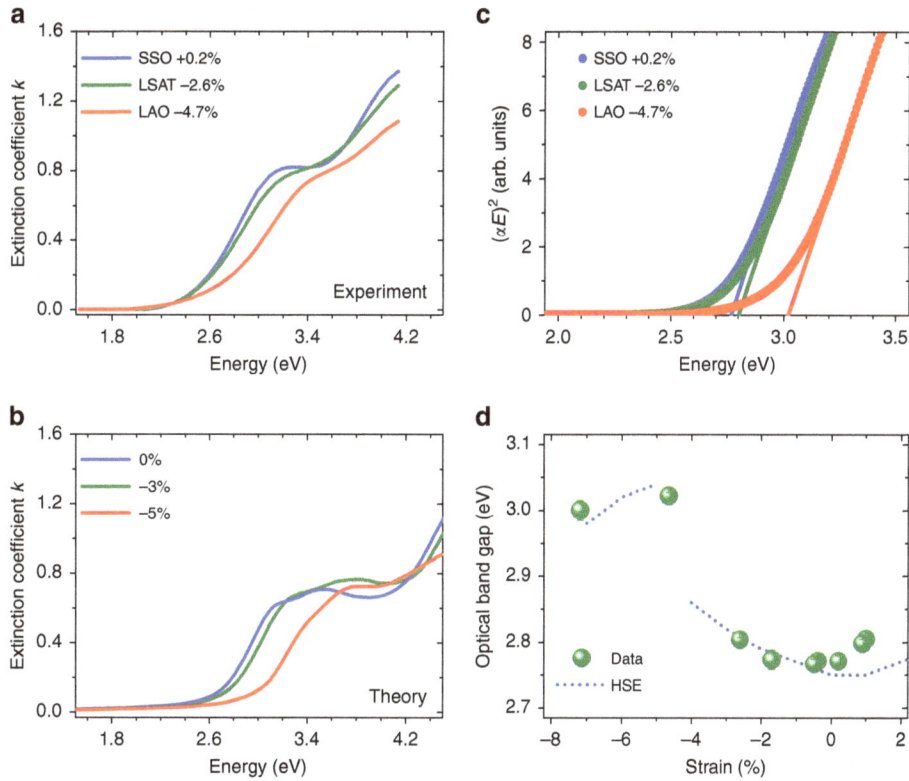

Figure 2 | Optical absorption properties of strained BiFeO$_3$ films. (**a**) Measured extinction coefficient for three representative strain levels. (**b**) Calculated extinction coefficient for strain levels comparable to those displayed in **a**. (**c**) Tauc plots generated from measurements for representative samples. (**d**) Summary of optical bandgap versus strain results, comparing theory and experiment. The error bars were determined by generating the dispersion laws using the upper and lower bounds of the Tauc–Lorentz oscillator parameters (from their uncertainties) and finding the resultant maximum variation in the bandgap.

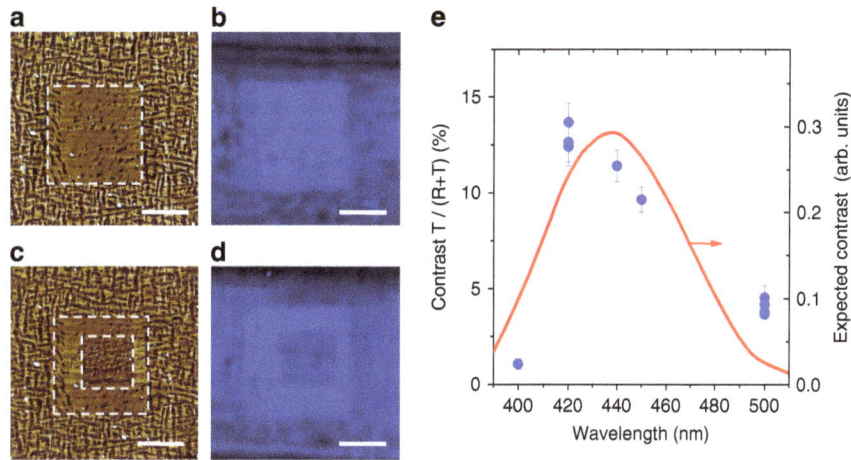

Figure 3 | Electrochromism in BiFeO$_3$ thin films. (**a**) Topography image after poling a $10 \times 10\,\mu m^2$ square, locally transforming R + T BFO into T-like BFO. (**b**) Transmission optical image acquired in the same region with a dielectric filter centred at 420 nm (bandwidth 10 nm). (**c**) Topography image of the same area after poling a $5 \times 5\,\mu m^2$ region with an opposite voltage, restoring the R + T structure. (**d**) Transmission optical image with a 420-nm filter. The horizontal dark features are due to twin boundaries in the LaAlO$_3$ substrate. All white scale bars are $5\,\mu m$. (**e**) Blue symbols: normalized difference in transmitted light in (T) and out (R + T) of the square in **a** with dielectric filters centred at different wavelengths. Red line: expected contrast calculated from the transmission of pure R-like and T-like films and the transmission function of the dielectric filters. The error bars in **e** are derived from the s.d. of the image pixel values in zones in and out of the T and (R + T) regions.

Fig. 4b, which displays the strain dependence of n at various wavelengths for all samples. The strain-induced change in refractive index measured at 633 nm is reproduced in Fig. 4c

and compared with first-principles calculations. The refractive index is higher in the R-like phase and globally decreases with strain, both compressive and tensile.

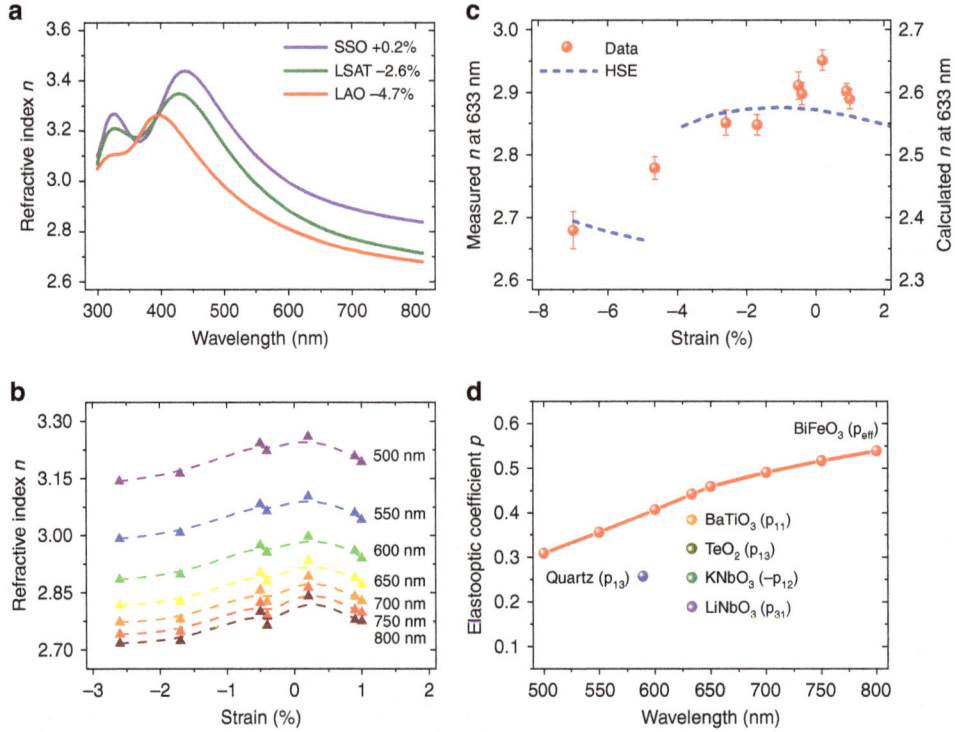

Figure 4 | Optical refractive index and elasto-optic coefficients in strained BiFeO₃ films. (**a**) Measured refractive index n as a function of wavelength for various strain levels. (**b**) Measured refractive index as a function of strain for various wavelengths, for the R-like phase only. The lines serve as guides to the eye. (**c**) Measured and calculated refractive index at 633 nm as a function of strain. The error bars were determined by generating the dispersion laws using the upper and lower bounds of the Tauc–Lorentz oscillator parameters (from their uncertainties) and finding the resultant maximum variation in the refractive index. (**d**) Effective elasto-optic coefficient of BFO as a function of wavelength. Representative reported largest elasto-optic coefficients of various other crystalline materials are plotted for comparison.

Discussion

The experimental results of Fig. 4 indicate that BFO exhibits a strong elasto-optic effect (change in refractive index on physical strain). Taking the slope of the change in $1/n^2$ with strain (Supplementary Fig. 3) for the weakly compressively strained BFO samples at various wavelengths larger than the BFO bandgap (at ~460 nm), we obtain an effective elasto-optic coefficient for BFO, as shown in Fig. 4d (Supplementary Note 1). In this figure we also plot the elasto-optic tensor element with the largest magnitude for various elasto-optic media[18,31]. The results suggest that BFO should be a robust elasto-optic medium and, more specifically, that BFO has an effective elasto-optic coefficient at least twice as large as LiNbO₃.

Combined with its relatively low Young's modulus[28] and sound speed[32], the large elasto-optic coefficient of BFO yields an acousto-optic figure-of-merit[18] $M = \frac{n^6 p^2}{E v_{ac}}$ as large as $M = 365 \times 10^{-15}\,s^3\,kg^{-1}$, a value much larger than for any other material for longitudinal acoustic waves (cf. in TeO₂, $M = 23 \times 10^{-15}\,s^3\,kg^{-1}$ and in LiNbO₃ $M = 1.8 \times 10^{-15}\,s^3\,kg^{-1}$, ref. 18). Importantly, as BFO may easily be grown[33] on isostructural perovskites with giant piezoelectric responses (such as PMN-PT or PZN-PT)[34], this huge figure-of-merit opens the way towards thin film acousto-optical components[35] with potential performance orders of magnitude greater than those currently based on single crystals. This would extend the potential of BFO to devices to deflect or modulate optical fields, and towards the emerging field of optomechanics, from back-action and laser cooling, to highly integrated sensors and frequency references[36].

More generally, our work has implications for the design of multifunctional devices exploiting the magnetic, ferroelectric or piezoelectric response of BiFeO₃ in conjunction with these unique optical properties. Importantly, the mechanisms that we identify to modulate bandgap and absorption are not specific to BiFeO₃ and can be transposed to many perovskites. Strain-induced elasto-optic and piezochromic effects even larger than in BiFeO₃, possibly by one order of magnitude or more, could be awaiting discovery in other oxide materials, particularly in Mott insulators[37] in which the bandgap falls between $3d$ states and is lower than in BiFeO₃. Giant, electric field-controllable optical absorption in the visible range could thus be exploited, opening the way towards devices harvesting both mechanical and solar energies.

Methods

BiFeO₃ thin film growth and structural characterization. Single phase R-like or T-like films of BFO were prepared by pulsed laser deposition (using the conditions of ref. 38) on the following single crystal substrates: YAlO₃, LAO, LSAT, SrTiO₃, DyScO₃, TbScO₃, SmScO₃, NdScO₃ and PrScO₃. The scandates and YAlO₃ were (110)-oriented (orthorhombic notation) while cubic SrTiO₃, LSAT and rhombohedral LAO were (001)-oriented. The nominal biaxial strain induced by these substrates ranges from -7.0% (compressive) to $+1.0\%$ (tensile). The thickness of the films was 50–70 nm as determined by X-Ray reflectometry, and confirmed by spectroscopic ellipsometry measurements. High-angle X-Ray diffraction 2θ–θ scans, collected with a Panalytical Empyrean diffractometer using CuK$_{\alpha 1}$ radiation, indicated that the films were epitaxial and grew in a single phase. Mixed R + T phase BFO (nominal thickness 100 nm) were grown using a KrF excimer laser at 540 °C and 0.36 mbar on (001)-oriented LAO substrates after the growth of a 10-nm-thick LaNiO₃ bottom electrode at 640 °C and 0.36 mbar.

Optical characterization. The films were characterized using spectroscopic ellipsometry with a UVISEL spectral-scanning near infrared spectroscopic phase-modulated ellipsometer from HORIBA Jobin-Yvon. The incidence angle was 70° and the wavelength range was 300–840 nm (0.62–4.13 eV). This range was imposed by the ellipsometer (maximum energy ~4.13 eV), while absorption peaks arising from colour centres in the scandate substrates limited the maximum wavelength to 840 nm. These boundaries do not adversely affect the present analysis since the optical bandgap of BFO is well within the explored spectral range. The raw ellipsometry data were fitted to a multilayer model consisting a semi-infinite substrate, BFO layer and roughness layer implemented by the Bruggeman approximation with a void and BFO mixture. The dispersion law of the BFO layer was described by three Tauc–Lorentz oscillators[8], where the central energies correspond to charge transfer transitions. An example of a typical fit (in this case for BFO on NdScO$_3$) is shown in Supplementary Fig. 2b. The complex dispersion law ($\tilde{n} = n + ik$) of BFO was determined for each sample, and for all fits the mean square error, χ^2, was <2 (as indicated in Supplementary Table 1). To extract the bandgap from the dispersion laws, Tauc plots of $(\alpha E)^2$ versus E were constructed, and the linear region was extrapolated to the E axis (Fig. 2c), yielding the gap value. For each sample, ellipsometric data were collected and the dispersion law and bandgap calculated a minimum of four times, and the results averaged. The error bars displayed in Figs 2d and 4c were determined by generating the dispersion laws using the upper and lower bounds of the Tauc–Lorentz oscillator parameters (from their uncertainties) and finding the resultant maximum variation in the bandgap and refractive index.

First-principles calculations. Calculations were performed within density-functional theory, as implemented in the Vienna ab initio simulation package[39,40]. An energy cutoff of 550 eV was used, and the set of projector-augmented wave potentials was employed to describe the electron–ion interaction. We considered the following valence electron configuration: $5d^{10}6s^26p^3$ for Bi, $3p^63d^64s^2$ for Fe and $2s^22p^4$ for O. Supercells containing 20 atoms were used, and G-type antiferromagnetism was adopted. Electronic relaxations converged within 10^{-6} eV and ionic relaxation was performed until the residual force was <1 meV Å^{-1}. We used the PBEsol + U functional[41] (selecting $U = 4$ eV for the Fe ions) to relax the structures, and used both this PBEsol + U functional and the HSE hybrid functional[42] to calculate physical properties such as electronic structure and the dielectric function. These two methods yielded very similar results (hence we only report results for HSE in Figs 1, 2 and 4), with the exception that the PBEsol + U functional underestimated the electronic bandgap by 0.4 eV, while HSE overestimated this bandgap by 0.8 eV. The imaginary part of the dielectric tensor was obtained via

$$\varepsilon''_{\alpha\beta}(\omega) = \frac{4\pi^2 e^2}{\Omega} \lim_{q \to 0} \frac{1}{q^2} \sum_{c,v,\mathbf{k}} 2\omega_k \delta(\epsilon_{ck} - \epsilon_{vk} - \omega) \times \langle u_{ck+e_\alpha q} \mid u_{vk} \rangle \langle u_{ck+e_\beta q} \mid u_{vk} \rangle^*,$$

(1)

where the indices c and v refer to conduction and valence band states, respectively, u_{ck} is the cell periodic part of the orbitals at the k-point $\mathbf{k}$, and $\mathbf{e}_\alpha$ is a unit vector along the α Cartesian direction[43]. Finally, the real part of the dielectric tensor $\varepsilon'_{\alpha\beta}$ was obtained through the Kramers–Kronig transformation $\varepsilon'_{\alpha\beta}(\omega) = 1 + \frac{2}{\pi} P \int_0^\infty \frac{\varepsilon''_{\alpha\beta}(\omega')\omega'}{\omega'^2 - \omega^2} d\omega'$, where P denotes the principal value. We then obtained the extinction coefficient k and refractive index n by $\tilde{\varepsilon} = \varepsilon' + i\varepsilon'' = (n + ik)^2$. Note that local field effects were neglected in our calculations. The optical bandgap determined from the calculated dielectric function was seen to overestimate the experiment by 0.8 eV; therefore, in all figures in this manuscript, the conduction band has been systematically shifted by 0.8 eV with respect to the VBM to reflect this scissors correction.

For the refractive index we find a systematic quantitative difference of ~0.3 between experiment and theory, which can be understood by the fact that first-principles calculations consider defect-free samples, neglect local field and temperature effects, and only incorporate the average between the different components of the dielectric function tensor (which is, additionally, a quantity rather difficult to simulate precisely by ab initio methods).

References

1. Lebeugle, D. et al. Room-temperature coexistence of large electric polarization and magnetic order in BiFeO$_3$ single crystals. *Phys. Rev. B* **76**, 024116 (2007).
2. Sosnowska, I., Peterlin-Neumaier, T. & Steichele, E. Spiral magnetic ordering in bismuth ferrite. *J. Phys. C Solid State Phys.* **15**, 4835–4846 (1982).
3. Heron, J. T. et al. Deterministic switching of ferromagnetism at room temperature using an electric field. *Nature* **516**, 370–373 (2014).
4. Seidel, J. et al. Conduction at domain walls in oxide multiferroics. *Nat. Mater.* **8**, 229–234 (2009).
5. Zeches, R. J. et al. A strain-driven morphotropic phase boundary in BiFeO$_3$. *Science* **326**, 977–980 (2009).
6. Sando, D. et al. Crafting the magnonic and spintronic response of BiFeO$_3$ films by epitaxial strain. *Nat. Mater.* **12**, 641–646 (2013).
7. Rovillain, P. et al. Electric-field control of spin waves at room temperature in multiferroic BiFeO$_3$. *Nat. Mater.* **9**, 975–979 (2010).
8. Allibe, J. et al. Optical properties of integrated multiferroic BiFeO$_3$ thin films for microwave applications. *Appl. Phys. Lett.* **96**, 182902 (2010).
9. Rivera, J.-P. & Schmid, H. On the birefringence of magnetoelectric BiFeO$_3$. *Ferroelectrics* **204**, 23–33 (1997).
10. Choi, T., Lee, S., Choi, Y. J., Kiryukhin, V. & Cheong, S.-W. Switchable ferroelectric diode and photovoltaic effect in BiFeO$_3$. *Science* **324**, 63–66 (2009).
11. Sando, D. et al. Linear electro-optic effect in multiferroic BiFeO$_3$ thin films. *Phys. Rev. B* **89**, 195106 (2014).
12. Sando, D., Barthélémy, A. & Bibes, M. BiFeO$_3$ epitaxial thin films and devices: past, present and future. *J. Phys. Condens. Matter* **26**, 473201 (2014).
13. Schlom, D. G. et al. Elastic strain engineering of ferroic oxides. *MRS Bull.* **39**, 118–130 (2014).
14. Choi, K. J. et al. Enhancement of ferroelectricity in strained BaTiO$_3$ thin films. *Science* **306**, 1005–1008 (2004).
15. Haeni, J. H. et al. Room-temperature ferroelectricity in strained SrTiO$_3$. *Nature* **430**, 758–761 (2004).
16. Béa, H. et al. Evidence for room-temperature multiferroicity in a compound with a giant axial ratio. *Phys. Rev. Lett.* **102**, 217603 (2009).
17. Infante, I. C. et al. Bridging multiferroic phase transitions by epitaxial strain in BiFeO$_3$. *Phys. Rev. Lett.* **105**, 057601 (2010).
18. Dieulesaint, E. & Royer, D. *Elastic Waves in Solids II* (Springer-Verlag Berlin Heidelberg, 2000).
19. Gómez-Salces, S. et al. Effect of pressure on the band gap and the local FeO$_6$ environment in BiFeO$_3$. *Phys. Rev. B* **85**, 144109 (2012).
20. Bamfield, P & Hutchings, M. G. *Chromic Phenomena: Technological Applications of Colour Chemistry*. 2nd edn (Royal Society of Chemistry, 2010).
21. Vanderbilt, D. & Cohen, M. H. Monoclinic and triclinic phases in higher-order Devonshire theory. *Phys. Rev. B* **63**, 094108 (2001).
22. Ju, S. & Cai, T. Ab initio study of ferroelectric and nonlinear optical performance in BiFeO$_3$ ultrathin films. *Appl. Phys. Lett.* **95**, 112506 (2009).
23. Dong, H., Liu, H. & Wang, S. Optical anisotropy and blue-shift phenomenon in tetragonal BiFeO$_3$. *J. Phys. D Appl. Phys.* **46**, 135102 (2013).
24. Berger, R. F., Fennie, C. J. & Neaton, J. B. Band gap and edge engineering via ferroic distortion and anisotropic strain: the case of SrTiO$_3$. *Phys. Rev. Lett.* **107**, 146804 (2011).
25. Chen, P. et al. Optical properties of quasi-tetragonal BiFeO$_3$ thin films. *Appl. Phys. Lett.* **96**, 131907 (2010).
26. Rodríguez, F., Hernández, D., Garcia-Jaca, J., Ehrenberg, H. & Weitzel, H. Optical study of the piezochromic transition in CuMoO$_4$ by pressure spectroscopy. *Phys. Rev. B* **61**, 16497–16501 (2000).
27. Gaudon, M. et al. Unprecedented 'one-finger-push'-induced phase transition with a drastic color change in an inorganic material. *Adv. Mater.* **19**, 3517–3519 (2007).
28. Redfern, S. A. T., Wang, C., Hong, J. W., Catalan, G. & Scott, J. F. Elastic and electrical anomalies at low-temperature phase transitions in BiFeO$_3$. *J. Phys. Condens. Matter* **20**, 452205 (2008).
29. Greenberg, C. B. Optically switchable thin films: a review. *Thin Solid Films* **251**, 81–93 (1994).
30. Seidel, J. et al. Prominent electrochromism through vacancy-order melting in a complex oxide. *Nat. Commun.* **3**, 799 (2012).
31. Weber, M. J. *Handbook of Optical Materials* (CRC Press, 2003).
32. Smirnova, E. P. et al. Acoustic properties of multiferroic BiFeO$_3$ over the temperature range 4.2–830 K. *Eur. Phys. J. B* **83**, 39–45 (2011).
33. Biegalski, M. D. et al. Strong strain dependence of ferroelectric coercivity in a BiFeO$_3$ film. *Appl. Phys. Lett.* **98**, 142902 (2011).
34. Park, S.-E. & Shrout, T. R. Ultrahigh strain and piezoelectric behavior in relaxor based ferroelectric single crystals. *J. Appl. Phys.* **82**, 1804–1811 (1997).
35. Lean, E. G. H., White, J. M. & Wilkinson, C. D. W. Thin-film acoustooptic devices. *Proc. IEEE* **64**, 779–788 (1976).
36. Aspelmeyer, M., Kippenberg, T. J. & Marquardt, F. Cavity optomechanics. *Rev. Mod. Phys.* **86**, 1391–1452 (2014).
37. Imada, M., Fujimori, A. & Tokura, Y. Metal-insulator transitions. *Rev. Mod. Phys.* **70**, 1039–1263 (1998).
38. Béa, H et al. Influence of parasitic phases on the properties of BiFeO$_3$ epitaxial thin films. *Appl. Phys. Lett.* **87**, 072508 (2005).
39. Kresse, G. & Furthmüller, J. Efficient iterative schemes for ab initio total-energy calculations using a plane-wave basis set. *Phys. Rev. B* **90**, 11169–11186 (1996).
40. Kresse, G. & Hafner, J. Ab initio molecular dynamics for liquid metals. *Phys. Rev. B* **47**, 558–561 (1993).
41. Perdew, J. et al. Restoring the density-gradient expansion for exchange in solids and surfaces. *Phys. Rev. Lett.* **100**, 136406 (2008).
42. Krukau, A. V., Vydrov, O. A., Izmaylov, A. F. & Scuseria, G. E. Influence of the exchange screening parameter on the performance of screened hybrid functionals. *J. Chem. Phys.* **125**, 224106 (2006).
43. Gajdoš, M., Hummer, K., Kresse, G., Furthmüller, J. & Bechstedt, F. Linear optical properties in the projector-augmented wave methodology. *Phys. Rev. B* **73**, 045112 (2006).

Acknowledgements

This work was supported by the French Research Agency (ANR) projects 'Méloïc', 'Nomilops' and 'Multidolls,' the European Research Council Advanced Grant 'FEMMES' (Contract No. 267579) and the European Research Council Consolidator Grant 'MINT' (Contract No. 615759). Y.Y. and L.B. thank the financial support of ONR Grant No N00014-12-1-1034 and DARPA Grant No. HR0011-15-2-0038 (Matrix program), and the Arkansas High Performance Computer Center for the use of its supercomputers. Ph.G. acknowledges a Research Professorship from the Francqui Foundation, financial support of the ARC project 'AIMED' and F.R.S.-FNRS PDR project ' HiT4FiT' as well as access to Céci-HPC facilities funded by F.R.S.-FNRS (Grant No 2.5020.1) and the Tier-1 supercomputer of the Fédération Wallonie- Bruxelles funded by the Walloon Region (Grant No 1117545). This work was also supported (E.B.) by F.R.S.-FNRS Belgium, and calculations were partly performed within the PRACE projects TheoMoMuLaM and TheDeNoMo. We thank J.-L. Reverchon for assistance with the ellipsometry measurements.

Author contributions

M.B. conceived and supervised the study with the help of A.B. and D.D. D.S. and C.C. grew the samples and characterized them with X-ray diffraction. D.S. performed ellipsometry measurements and analysed the data. V.G., S.F. and M.B. characterized the electrochromic response. Y.Y, E.B., Ph.G. and L.B. performed first-principles calculations. D.S. and M.B. wrote the manuscript with input from all authors.

Additional information

Competing financial interests: The authors declare no competing financial interests.

3

Graphene–ferroelectric metadevices for nonvolatile memory and reconfigurable logic-gate operations

Woo Young Kim[1,*], Hyeon-Don Kim[1,*], Teun-Teun Kim[1,2,*], Hyun-Sung Park[1], Kanghee Lee[1], Hyun Joo Choi[1], Seung Hoon Lee[1,†], Jaehyeon Son[1], Namkyoo Park[3] & Bumki Min[1]

Memory metamaterials are artificial media that sustain transformed electromagnetic properties without persistent external stimuli. Previous memory metamaterials were realized with phase-change materials, such as vanadium dioxide or chalcogenide glasses, which exhibit memory behaviour with respect to electrically/optically induced thermal stimuli. However, they require a thermally isolated environment for longer retention or strong optical pump for phase-change. Here we demonstrate electrically programmable nonvolatile memory metadevices realised by the hybridization of graphene, a ferroelectric and meta-atoms/ meta-molecules, and extend the concept further to establish reconfigurable logic-gate metadevices. For a memory metadevice having a single electrical input, amplitude, phase and even the polarization multi-states were clearly distinguishable with a retention time of over 10 years at room temperature. Furthermore, logic-gate functionalities were demonstrated with reconfigurable logic-gate metadevices having two electrical inputs, with each connected to separate ferroelectric layers that act as the multi-level controller for the doping level of the sandwiched graphene layer.

[1] Department of Mechanical Engineering, Korea Advanced Institute of Science and Technology(KAIST), Daejeon 305-701, Republic of Korea. [2] Metamaterial Research Centre, School of Physics and Astronomy, University of Birmingham, Birmingham B15 2TT, UK. [3] Photonic Systems Laboratory, School of EECS, Seoul National University, Seoul 151-744, Republic of Korea. * These authors contributed equally to this work. † Present address: Department of Applied Physics and Materials Science, California Institute of Technology, California 91125, USA. Correspondence and requests for materials should be addressed to B.M. (email: bmin@kaist.ac.kr).

Metamaterials are artificial media that exhibit unusual electromagnetic properties such as anomalous refraction[1-5], invisible cloak[6,7] and strong chirality[8,9]. Light–matter interaction can be dramatically intensified with the use of electromagnetic responses around the resonance of specifically designed meta-atoms (MAs). Most of the wave properties, such as amplitude[10], phase[11] and polarization states[12], and even the direction of light[13,] can be manipulated by the use of metamaterials or metasurfaces. Furthermore, variability in light–matter interaction is manifested through the external application of electrical[10,14], mechanical[15], optical[16,17] and thermal[18] stimuli when they are hybridized with natural active media. However, application of persistent stimuli is still required to sustain the transformed metamaterial properties unless there are memory functionalities. In this aspect, memory metamaterials are unique because long-lasting modification of the effective properties is possible even with the application of impulsive stimuli. This may lead to additional saving of energy resources that may otherwise have been used in sustaining the transformed properties. Previous memory metamaterials were implemented with natural memory media, such as phase-change materials[19-25], that show memory behaviour with respect to electrically or optically induced thermal stimuli. The memory metamaterials have shown the possibility for unique performances, such as dynamic resonance tuning, multi-level data storage and bi-directionality[19-25]. However, they require a thermally isolated environment to obtain a longer retention of transformed properties or a strong optical pump for phase-change.

Here, we demonstrate electrically programmable nonvolatile memory metadevices operating at room temperature that are made possible by the hybridization of graphene, a ferroelectric and MAs/meta-molecules. The doping level of graphene embedded in the metadevices exhibits hysteretic behaviour with respect to external gate voltage[26] that, in turn, leads to the hysteretic response in the effective properties of metadevices. Hence, their amplitude, phase and polarization states of light could be stored to multi-states with an appropriate MA structure. All stored states were stably maintained over 10^5 s even in the absence of additional electrical stimuli, from which a retention time over 10 years can be anticipated at room temperature. Beyond the single electrical input memory function, the operational principle has been proven to be scalable to a multi-input system; for example, two-input logic-gate operations such as AND, OR and XOR (complementary NOR, NAND and XNOR, respectively) were implemented. Furthermore, it is shown that the same logic-gate metadevice can be reconfigured to operate as a two-bit digital-to-analogue convertor by a slight change in the operating condition.

Results

Graphene–ferroelectric nonvolatile memory metadevice.

A schematic representation of the graphene–ferroelectric nonvolatile memory metadevice (GF-NMM) is depicted in Fig. 1a. An array of hexagonal MAs, single-layer graphene, a ferroelectric polymer (poly(vinylidene fluoride-co-trifluoroethylene) or P(VDF-TrFE)) and a terahertz transparent electrode (TTE) composed of periodical subwavelength-scale metallic strips are placed sequentially on a polyimide substrate (fabrication details are described in the Methods, Supplementary Fig. 1 and Supplementary Note 1). An array of the hexagonal metallic pattern exhibiting polarization-independent inductance-capacitance (LC) resonance was chosen as the MA structure to intensify light–matter interaction. A large-size single-layer graphene synthesized by chemical vapour deposition[27] on

a Cu foil is transferred onto the array of MAs on polyimide using a ferroelectric polymer as a mechanically supporting film[28]. Since the dipoles in the ferroelectric polymer, P(VDF-TrFE), consist of weakly electronegative hydrogen atoms and strongly electronegative fluorine atoms, the application of an external gate voltage (V_G) over the coercive voltage (V_C) aligns the dipoles in the ferroelectric. These aligned dipoles, as a result, induce polarization (P) at the surface of the ferroelectric and exert an electrostatic force consistently on the charge carriers in the graphene layer (Fig. 1b). The TTE was carefully designed to apply a uniform electric field to the ferroelectric and transmit broadband terahertz waves vertically incident on the TTE without much loss as in previous works[29]. Pulsed external gate voltage ($V_{G,pulse}(V)$) lasting for 1 s was applied between the TTE and the graphene/MAs in the measurement.

Terahertz time domain spectroscopy (THz-TDS) was carried out to characterize the fabricated GF-NMM as shown in Fig. 1c. On applying $V_{G,pulse}(+200 V)$ P, corresponding to positive remanent polarization ($+P_R$), depletes the same polar charges out of graphene; THz transmission spectra through the GF-NMM showed a resonance dip at 1.1 THz. With the subsequent application of $V_{G,pulse}(-200 V)$, P changes to negative remanent polarization ($-P_R$); the resonance frequency then shifted to 0.8 THz and the bandwidth was observed to slightly broaden.

Figure 1d shows the measured transmission amplitude (T_A) through the GF-NMM and the quantitatively estimated Fermi level (E_F) of graphene in the GF-NMM as a function of $V_{G,pulse}$. Experimentally observed hysteretic variation of the spectral features is attributed to the change in graphene doping level resulting from the reversal of ferroelectric P (Supplementary Figs 2,3 and Supplementary Discussion). It is worthwhile to note that gradual transmission change is observed near positive and negative V_C and multi-state memory operation can be achieved by utilizing this gradual change near V_C. To validate multi-state memory operation, a timing diagram of T_A was recorded at 0.5 THz for various $V_{G,pulse}$ values of -80, -105, -115 and $-130 V$ and plotted in Fig. 1e. From the highest T_A, data states were designated as 00, 01, 10 and 11. Before addressing each state, $V_{G,pulse}(+200 V)$ was applied to reset ferroelectric polarization to $+P_R$.

Reliable nonvolatile memory operation requires not only data separation but also long retention time. Figure 1f depicts the retention time for each stored T_A state; the GF-NMM has been observed to retain state information for all the states for over 10^5 s (the standard variation for T_A, σ_{state} is in the range between 5.62×10^{-3} and 7.58×10^{-3}). The extrapolated T_A for all states do not cross each other for over 8.5×10^8 s (~ 10 years), which confirms that the demonstrated GF-NMM has indisputable nonvolatility. An additional measurement on THz-TDS system stability revealed that the slight variation in the T_A was mostly a result of the fluctuations of the femtosecond laser output that was used for the THz-wave generation (the fluctuation of our femtosecond laser, σ_{laser}, was 4.43×10^{-3}; Supplementary Fig. 4 and Supplementary Table 1). As the MA structure used in the measurement is of a resonant type, the phase tunability and nonvolatility were also measured. In the case of phase response, $V_{G,pulse}$-dependent hysteretic behaviour and multi-state stable retention properties were found to be similar to the transmission amplitude response (Supplementary Fig. 5).

Doping concentration in graphene is more resistant to change with time when graphene–ferroelectric hybrid devices operate in the accumulation mode compared with the depletion mode, especially at the beginning of the retention measurement[30]. Because the graphene used in our experiments is inherently p-doped and the intrinsic doping concentration ($6.17 \times 10^{12} cm^{-2}$) is larger than the extrinsic doping

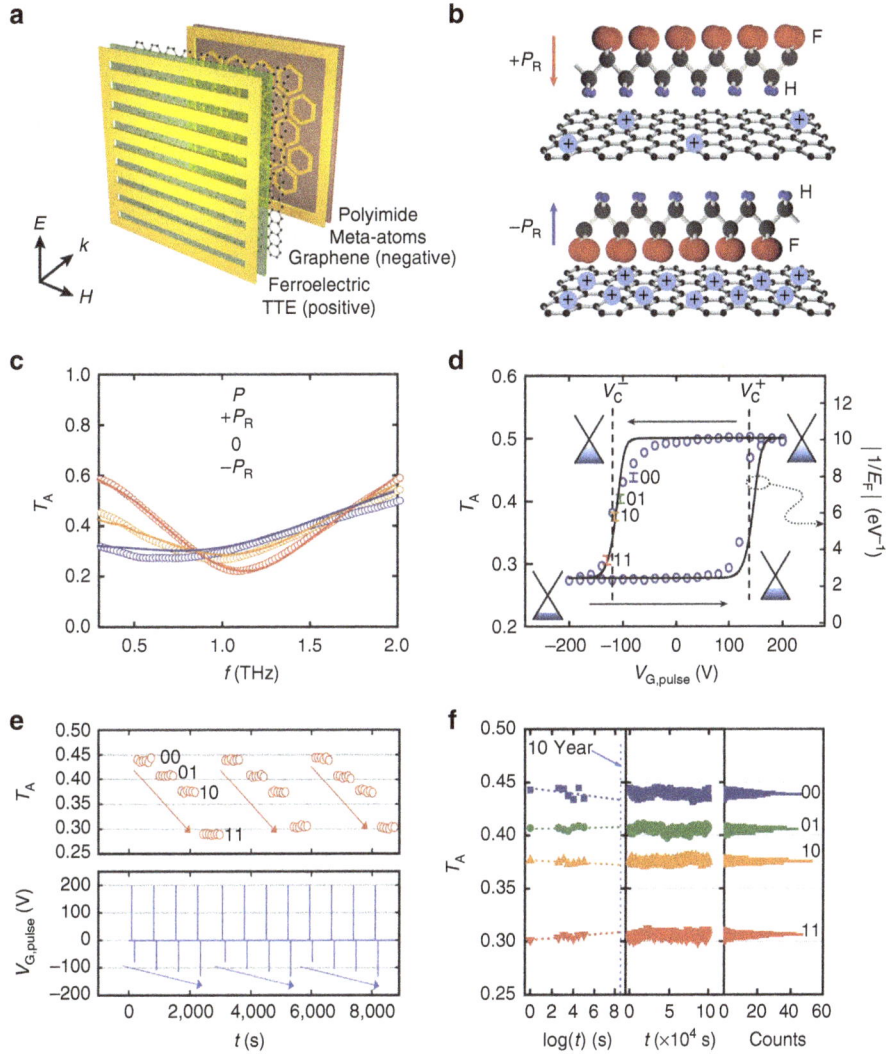

Figure 1 | Graphene–ferroelectric amplitude and phase memory metadevice. (**a**) Schematic representation of the graphene–ferroelectric memory metadevice composed of a THz transparent electrode (TTE) with periodical metallic lines (4 μm linewidth and 2 μm spaces between the lines), ferroelectric polymer layer (2.1 μm, represented by green), single-layer graphene, hexagonal MAs and polyimide (1 μm, represented by light red) as substrate. Polarization of the incident THz is perpendicular to the TTE lines. (**b**) Schematic representation of the principle of nonvolatile doping in inherently p-doped graphene by ferroelectric polarization (P). Positive external voltage induces $+P_R$ at the surface of a ferroelectric, resulting in hole depletion in graphene while negative external voltage changes P to $-P_R$, thereby resulting in the accumulation of holes in graphene. (**c**) Measured (open circles) and simulated (solid lines) THz transmission spectra for external pulsed gating voltage ($V_{G,pulse}$) lasting for 1s. The red, blue and yellow lines and circles represent the results of application of $V_{G,pulse}(+200\,V)$, $V_{G,pulse}(-200\,V)$ and $V_{G,pulse}(-120\,V)$, respectively. (**d**) Hysteresis in the measured transmission amplitude (T_A) and the calculated Fermi level of graphene for $V_{G,pulse}$ within a range of $+200$ and $-200\,V$ at a specific frequency of 0.5 THz. V_C^+ and V_C^- are the positive and negative coercive voltages, respectively. Arrow refers to the $V_{G,pulse}$ sweep direction. Logic states denoted as 00, 01, 10 and 11 correspond to the multi-level transmission amplitudes for the retention time measurement. Error bars indicate the variation of each measured logic state. (**e**) Timing diagram of the transmission amplitude (T_A) measured at 0.5 THz for various $V_{G,pulse}$ values of -80, -105, -115 and $-130\,V$. Counting from the highest transmission amplitude, data states were designated as 00, 01, 10 and 11. Before each $V_{G,pulse}$ was applied, $V_{G,pulse}(+200\,V)$ was applied to reset the ferroelectric polarization to $+P_R$. (**f**) Transmission amplitude (T_A) retention time measured at 0.5 THz for 1×10^5 s and a histogram for each state.

concentration modulated by the ferroelectric polarization ($6.08 \times 10^{12}\,cm^{-2}$), our device operates exclusively in the p-doped regime (Supplementary Fig. 6 and Supplementary Note 2). Moreover, all multi-states were addressed by applying a negative $V_{G,pulse}$ for further accumulation mode, so that the time-dependent drift in the E_F of graphene in the GF-NMM is negligible, and, therefore, stable operation is possible. Ferroelectric polarization switching depends on the magnitude and duration of $V_{G,pulse}$. For an electric field of $1\,MV\,cm^{-1}$ corresponding to $V_{G,pulse}(+200\,V)$, previous studies have reported a polarization switching time of ~1 ms (refs 31,32). Approximately 10 times faster switching would be

expected by a $V_{G,pulse}$ with a one and a half times greater amplitude[32].

The principle of amplitude and phase memory operation can also be applied to light polarization memory with a chiral metamaterial (chiral GF-NMM), in which the plane of linearly polarized incident light is rotated as the light travels through. Recently, it was shown that with the integration of active materials into metamaterials, polarization switching and modulation can be dynamically controlled[33]. However, optical activity such as circular dichroism and polarization rotation in most metamaterials can so far only be tuned by continuous external optical stimuli. With the incorporation of graphene and a ferroelectric into the chiral metamaterials, the polarization states of light passing through a chiral GF-NMM can be stored by $V_{G,pulse}$(V). Strongly coupled chiral meta-molecules (MM) (ref. 34) were employed in the fabrication of the chiral GF-NMM as shown in Fig. 2a. The rest of the chiral GF-NMM is identical with the amplitude and phase GF-NMM described above (Supplementary Fig. 7 and Supplementary Note 3).

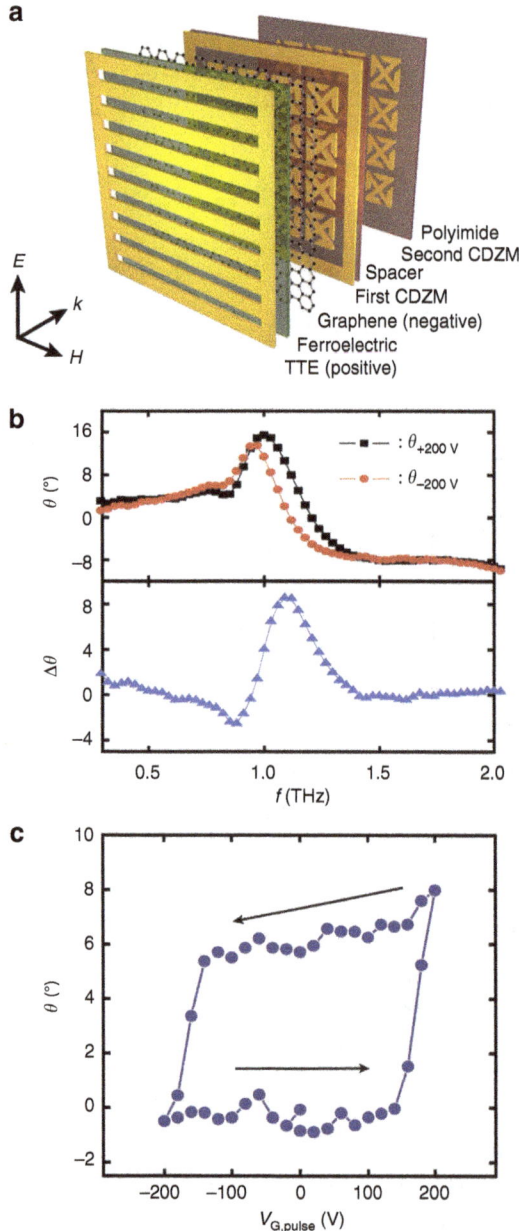

Figure 2 | Graphene-ferroelectric chiral memory metadevice.
(**a**) Schematic representation of the graphene-ferroelectric chiral memory metadevice composed of THz transparent electrode (TTE), ferroelectric polymer layer (2.1 μm, represented by green), single-layer graphene, CDZM with a spacer of 2 μm and a polyimide (1 μm, represented by light red) as the substrate. Polarization of the incident THz is perpendicular to the TTE lines. (**b**) Polarization rotation angle (θ) through chiral memory metadevice after the application of an external pulsed gating voltage ($V_{G,pulse}$) lasting for 1 s, and the difference ($\Delta\theta$) between θ_{+200V} and θ_{-200V}. (**c**) Polarization rotation angle (θ) in the $V_{G,pulse}$ within a range of $+200$ and -200 V at 1.1 THz. Arrows refer to the $V_{G,pulse}$ sweep direction. CDZM, conjugated double Z meta-molecules.

Figure 2b shows the azimuthal polarization rotation angle (θ) for two distinct $V_{G,pulse}$ values. Here, θ is extracted from the phase difference between the two circular polarizations (see Methods). On applying $V_{G,pulse}$($+200$ V), θ_{+200V} shows a maximum value of $15°$ at 1.0 THz, while θ_{-200V} exhibits a maximum value of $14°$ at 0.9 THz. It can be seen from Fig. 2b that $\Delta\theta$ ($=\theta_{+200V}$ to θ_{-200V}) attains the maximum value of $8°$ at 1.1 THz. To trace the hysteretic behaviour in the polarization states more clearly, θ measurement was carried out at 1.1 THz and is plotted in Fig. 2c. Because of ferroelectricity, θ changes gradually near positive and negative V_C. Multi-level polarization states can also be stored for over 10^5 s without much degradation as shown in the operation of three different polarization states (Supplementary Fig. 8).

Graphene-ferroelectric reconfigurable logic-gate metadevice. The underlying concept for the operation of GF-NMM can be extended to a multi-input system such as a reconfigurable logic-gates metadevice (graphene-ferroelectric reconfigurable logic-gate metadevice (GF-RLM)). For example, a two-input system can be implemented by encapsulating graphene within two controllable ferroelectric layers as shown in Fig. 3a (Supplementary Fig. 9 and Supplementary Note 4). Independent pulsed gate control ($V_{G,pulse}$) of each ferroelectric layer and the resulting combination of polarizations offered by the individual ferroelectric layers can lead to an increase in the degree of freedom in the manipulation of carrier concentration in graphene (N_G) when compared with the GF-NMM having a single ferroelectric layer. Corresponding to the combination of two electrical inputs (the top electrode (T) and the bottom electrode (B)), N_G as well as the THz transmission through the two-input system is expected to give unique logic outputs. Figure 3b shows the combination of polarization values in the ferroelectric layers that correspond to four kinds of logic inputs, (0, 0), (0, 1), (1, 0) and (1, 1). Input logic states 1 and 0 are prepared by applying $V_{G,pulse}$($+200$ V) and $V_{G,pulse}$(-200 V) to the corresponding ferroelectric layers, respectively. If graphene in GF-RLM is inherently p-doped, logic input (1, 1) depletes majority carriers (holes) and logic input (0, 0) accumulates holes in the graphene. In the two intermediate states, (0, 1) and (1, 0), the hole concentration in graphene will be set to the values that are between those corresponding to input states (0, 0) and (1, 1). The variation in N_G results in a change in the transmission spectrum as shown schematically in Fig. 3c, in which the resonance frequency is red-shifted as N_G increases[29]. With the appropriate frequency choice for data reading (f_{READ}), the logic output can be decoded by comparing with T_{REF1} for the AND (complementary NOR) operation or T_{REF2} for the OR (complementary NAND) operation. Furthermore, XOR (complementary XNOR) operation can also be realized by simply rearranging the electric connections Supplementary Figs 10,11). All the logic-gate operations are measured using THz-TDS and shown in Fig. 3d (Supplementary Fig. 12 and Supplementary Table 2).

In addition to the logic operations, the device can also be configured as a digital-to-analogue converter if the remanant polarization in each of the ferroelectric layers assumes a different value. If the two ferroelectric layers supply two different P_R values, four kinds of N_G levels corresponding to ($+P_{R1}$, $+P_{R2}$), ($+P_{R1}$, $-P_{R2}$), ($-P_{R1}$, $+P_{R2}$) and ($-P_{R1}$, $-P_{R2}$) are possible, which implies that $f_{R(0,1)}$ is different from $f_{R(1,0)}$ as shown in Fig. 3c. The effective method to control polarization switching was demonstrated in a prior work[35], in which ferroelectric switching was controlled by setting the current limitation (I^C, compliance current). By setting different compliance

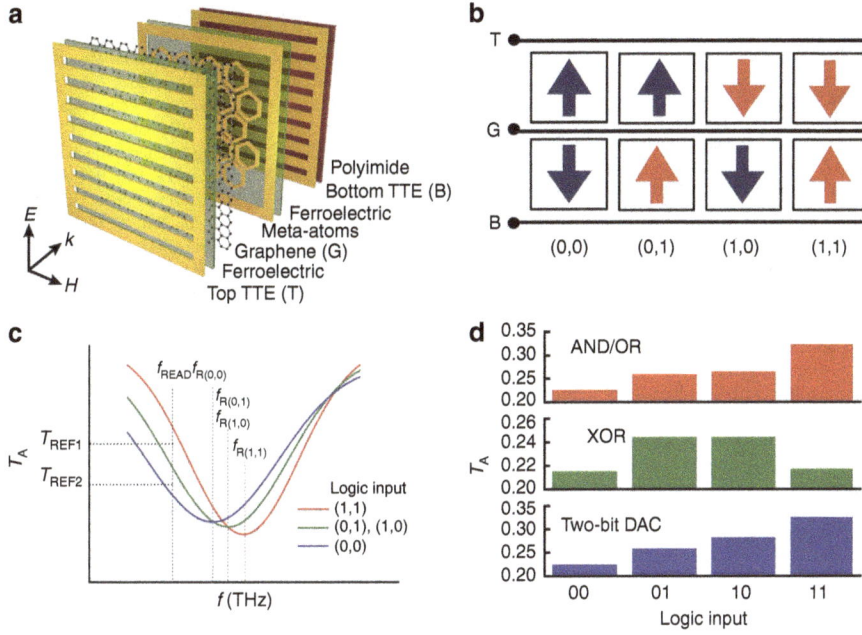

Figure 3 | Graphene–ferroelectric reconfigurable logic-gate metadevice. (a) Schematic representation of the graphene–ferroelectric reconfigurable logic-gate metadevice composed of a top THz transparent electrode (top; TTE, T), a ferroelectric polymer layer (2.1 µm, represented by green), a single-layer graphene, hexagonal MAs, a ferroelectric polymer layer (2.1 µm, represented by green), a bottom THz transparent electrode (bottom; TTE, B) and a polyimide layer (1 µm, represented by light red) as the substrate. Polarization of the incident THz is perpendicular to the two TTE lines. **(b)** Schematic representation of the four kinds of polarization alignments for input logic states. Red arrow implies an application of positive pulsed gating voltage for logic state 1 and blue arrow refers to the application of negative pulsed gating voltage for logic state 0. **(c)** Schematic representation of the transmission spectra for input logic states. For each input logic state, the relationship $f_{R(0,0)} < f_{R(0,1)} = f_{R(1,0)} < f_{R(1,1)}$ is satisfied in the graphene–ferroelectric reconfigurable logic-gate metadevice because of p-doped graphene. A frequency of f_{READ} was designated for data reading and two reference transmission amplitudes, T_{REF1} and T_{REF2} were defined to execute AND (complementary NOR) and OR (complementary NAND) gate operations. For AND gate operation, the reference transmission amplitude is set to T_{REF1}. For OR gate operation, the reference is T_{REF2}. **(d)** Experimental transmission amplitude (T_A) measured at 0.5 THz for the four types of logic inputs in the AND/OR gates, the XOR gate and the two-bit DAC.

current for bottom and top TTE, four distinct levels of transmission amplitude were measured as shown in Fig. 3d (Supplementary Fig. 12). The combination of two digital inputs resulting in four levels of optical analogue states validates the two-bit digital-to-analogue convertor operation. Although the transmission loss and optoelectric conversion should be considered in the compact integration of diverse functional metadevices, the platform and the principle of operation provided here might be extended to a certain class of multi-input systems in principle.

Discussion

In this work, electrically programmable nonvolatile memory and reconfigurable logic-gate metadevices were demonstrated with the hybridization of graphene, a ferroelectric and MAs/molecules. These functional metadevices are the first demonstration of superior nonvolatility and logic-gate operation at room temperature, which offer new pathways for emerging optoelectronic applications. Nonvolatile memory function liberates active metadevices from immovable power supply units, leading to saving of energy resources. Reconfigurability presents a user-oriented general-purpose metadevice that can be configured through electrical programming. More complex functionality can be made possible by the compact integration of diverse functional metadevices and/or by the design changes of the demonstrated metadevices. Ultimately, high-end metasystems to perform advanced functionalities may be developed by grafting the concept of GF-NMM and GF-RLM onto the diverse architectures of graphene-based nonvolatile memory, as recently demonstrated in the field of electronic devices[36].

Methods

Fabrication processes for the GF-NMM and GF-RLM. All metallic parts of the hexagonal MAs/molecules and the TTE were made of 100-nm thick Au with a 10 nm thick Cr layer for enhanced adhesive strength. Single-layer graphene was synthesized by chemical vapour deposition on a Cu foil (G/Cu). Poly(vinylidene fluoride-trifluoroethylene), P(VDF-TrFE), manufactured by MSI Sensors Inc. was chosen as the ferroelectric polymer.

For the nonvolatile memory metadevice, a polyimide (PI, PI-2610, HD MicroSystems) was spin-coated and cured on a Si wafer. An array of hexagonal MAs was deposited by a photolithography, thermal evaporation and lift-off process (MA/PI/Si). On a SiO₂/Si wafer, a ferroelectric polymer (FP) was spin-coated, annealed at 130 °C for 1 h, and cooled down to room temperature slowly. A TTE was patterned by a photolithography, thermal evaporation of Cr/Au and lift-off process (TTE/FP/SiO₂/Si). By etching the SiO₂ with an HF aqueous solution, the TTE/FP was transferred onto graphene (G) on Cu foil. By etching the Cu foil with the Cu etchant APS-100, the TTE/FP/G was transferred onto the MA/PI/Si and thermally treated for adhesion. Finally, Si was detached mechanically. For the polarization state memory metadevice, MA/PI on Si was replaced with MM/PI, which was fabricated by stacking conjugated double Z patterns with a polyimide spacer of 2 µm.

For the reconfigurable logic-gate metadevice, an MA layer was deposited on G/Cu. An FP was spin-coated on the MA/G/Cu. In addition, a TTE as the top electrode was deposited on the FP/MA/G/Cu. By etching the Cu foil with a Cu etchant of APS-100, a TTE/FP/MA/G hybrid film was prepared. On a Si wafer sacrificial substrate, PI was spin-coated and TTE as the bottom electrode was deposited on PI/Si (TTE/PI/Si). An FP was spin-coated onto the TTE/PI/Si (FP/TTE/PI/Si). Reconfigurable logic-gate metadevice was fabricated by transferring the TTE/FP/MA/G onto the FP/TTE/PI/Si. Finally, Si was detached mechanically.

All metadevices were mounted on a punched printed circuit board for THz-TDS measurements.

THz-TDS system. To generate the terahertz signal, we used a low-temperature grown GaAs THz emitter (Tera-SED, Gigaoptics) illuminated by a femtosecond Ti:sapphire laser pulse train of wavelength 800 nm and 80 MHz repetition rate, respectively. An electro-optic sampling method was used to detect the transmitted terahertz signals in the time domain by using a (110) oriented ZnTe crystal of 1 mm thickness. The THz-TDS system has a usable bandwidth of 0.3–2.5 THz and a signal to noise ratio (S/N) of over 10,000:1.

Measurement of the chiral GF-NMM. The chiral GF-NMM was characterized using conventional THz-TDS. The metadevice was positioned between two wire-grid THz polarizers, that are mounted on motorized rotational stages with parallel or crossed configurations, to measure the co-polarized ($T_{\parallel}$) and cross-polarized ($T_{\perp}$) transmission coefficients. The sample was carefully aligned to assure the TTE of the metadevice remains parallel to the front polarizer. From the measured transmission coefficients, the right and left circularly polarized transmission coefficients can be obtained as $T_+ = T_{\parallel} + iT_{\perp}$ and $T_- = T_{\parallel} - iT_{\perp}$. The azimuthal rotation angle can be calculated by the phase retardation between two circularly polarized waves as $\theta = \frac{1}{2}[\arg(T_-) - \arg(T_+)]$.

References

1. Huang, X. *et al.* Dirac cones induced by accidental degeneracy in photonic crystals and zero-refractive-index materials. *Nat. Mater.* **10**, 582–586 (2011).
2. Moitra, P. *et al.* Realization of an all-dielectric zero-index optical material. *Nat. Photon.* **7**, 791–795 (2013).
3. Smith, D. R. *et al.* Composite medium with simultaneously negative permeability and permittivity. *Phys. Rev. Lett.* **84**, 4184–4187 (2000).
4. Smith, D. R. *et al.* Metamaterials and negative refractive index. *Science* **305**, 788–792 (2004).
5. Choi, M. *et al.* A terahertz metamaterial with unnaturally high refractive index. *Nature* **470**, 369–373 (2011).
6. Schurig, D. *et al.* Metamaterial electromagnetic cloak at microwave frequencies. *Science* **314**, 977–980 (2006).
7. Pendry, J. B. *et al.* Controlling electromagnetic fields. *Science* **312**, 1780–1782 (2006).
8. Pendry, J. B. A chiral route to negative refraction. *Science* **306**, 1353–1355 (2004).
9. Zhang, S. *et al.* Negative refractive index in chiral metamaterials. *Phys. Rev. Lett.* **102**, 023901 (2009).
10. Chen, H.-T. *et al.* Active terahertz metamaterial devices. *Nature* **444**, 597–600 (2006).
11. Chen, H.-T. *et al.* A metamaterial solid-state terahertz phase modulator. *Nat. Photon.* **3**, 148–151 (2009).
12. Wang, B. *et al.* Chiral metamaterials: simulations and experiments. *J. Opt. A: Pure Appl. Opt.* **11**, 114003 (2009).
13. Yu, N. *et al.* Light propagation with phase discontinuities: generalized laws of reflection and refraction. *Science* **334**, 333–337 (2011).
14. Ju, L. *et al.* Graphene plasmonics for tunable terahertz metamaterials. *Nat. Nanotechnol.* **6**, 630–634 (2011).
15. Fu, Y. H. *et al.* A micromachined reconfigurable metamaterial via reconfiguration of assymmetric split-ring resonators. *Adv. Funct. Mater.* **21**, 3589–3594 (2011).
16. Padilla, W. J. *et al.* Dynamical electric and magnetic metamaterial response at terahertz frequencies. *Phys. Rev. Lett.* **96**, 107401 (2006).
17. Chen, H.-T. *et al.* Experimental demonstration of frequency-agile terahertz metamaterials. *Nat. Photon.* **2**, 295–298 (2008).
18. Tao, H. *et al.* Reconfigurable terahertz metamaterials. *Phys. Rev. Lett.* **103**, 147401 (2009).
19. Driscoll, T. *et al.* Memory metamaterials. *Science* **325**, 1518–1521 (2009).
20. Dicken, M. J. *et al.* Frequency tunable near-infrared metamatereials based on VO_2 phase transition. *Opt. Express* **17**, 18330–18339 (2009).
21. Goldflam, M. D. *et al.* Two-dimensional reconfigurable gradient index memory metasurface. *Appl. Phys. Lett.* **102**, 224103 (2013).
22. Samson, Z. L. *et al.* Metamaterial electro-optic switch of nanoscale thickness. *Appl. Phys. Lett.* **96**, 143105 (2010).
23. Gholipour, B. *et al.* An all-optical, non-volatile, bidirectional, phase-change meta-switch. *Adv. Mater.* **25**, 3050–3054 (2013).
24. Wang, Q. *et al.* 1.7 Gbit/in.² grey-scale continuous-phase-change femtosecond image storage. *Appl. Phys. Lett.* **104**, 121105 (2014).
25. Michel, A.-K. U. *et al.* Reversible optical switching of infrared antenna resonances with ultrathin phase-change layers using femtosecond laser pulses. *ACS Photonics* **1**, 833–839 (2014).
26. Zheng, L. *et al.* Gate-controlled nonvolatile graphene-ferroelectric memory. *Appl. Phys. Lett.* **94**, 163505 (2009).
27. Kim, K. S. *et al.* Large-scale pattern growth of graphene films for stretchable transparent electrodes. *Nature* **457**, 706–710 (2009).
28. Ni, G.-X. *et al.* Graphene-ferroelectric hybrid structure for flexible transparent electrodes. *ACS Nano* **6**, 3935–3942 (2012).
29. Lee, S. H. *et al.* Switching terahertz waves with gate-controlled active graphene metamaterials. *Nat. Mater.* **11**, 936–941 (2012).
30. Raghavan, S. *et al.* Long-term retention in organic ferroelectric-graphene memories. *Appl. Phys. Lett.* **100**, 023507 (2012).
31. Naber, R. C. G. *et al.* High-performance solution-processed polymer ferroelectric field-effect transistors. *Nat. Mater.* **4**, 243–248 (2005).
32. Furukawa, T. *et al.* Factors governing ferroelectirc switching characteristics of thin VDF/TrFE copolymer films. *IEEE Trans. Dielectr. Electr. Insul.* **13**, 1120–1131 (2006).
33. Zhou, J. *et al.* Terahertz chiral metamaterials with giant and dynamically tunable optical activity. *Phys. Rev. B* **86**, 035448 (2012).
34. Kim, T.-T. *et al.* Optical activity enhanced by strong inter-molecular coupling in planar chiral metamaterials. *Sci. Rep.* **4**, 5864 (2014).
35. Lee, D. *et al.* Multilevel data storage memory using deterministic polarization control. *Adv. Mater.* **24**, 402–406 (2012).
36. Wang, X. *et al.* Graphene based non-volatile memory devices. *Adv. Mater.* **26**, 5496–5503 (2014).

Acknowledgements

This work was supported by the BK21 Plus Program, and Nano · Material Technology Development Program (2015036205), the World Class Institute (WCI) Program (No. WCI 2011-001) and the Pioneer Research Center Program (2014M3C1A3052537) through the National Research Foundation of Korea (NRF) and the Center for Advanced Meta-Materials (CAMM-2014M3A6B3063709) funded by the Ministry of Science, ICT and Future Planning (MSIP) (Nos 2012R1A2A1A03670391 and 2015001948). T.-T.K was also supported by Marie-Curie IIF (ref. 626184).

Author contributions

W.Y.K., T-T.K., S.H.L. and B.M. conceived the original ideas for nonvolatile memory and logic-gate metadevices. W.Y.K., H-D.K., H-S.P. and H.J.C. fabricated the metadevices and characterized the graphene. H-D.K., T-T.K. and H-S.P. performed simulations. W.Y.K., T-T.K., H-S.P., K.L. and J.S. performed THz-TDS. W.Y.K., H-D.K., T-T.K., H-S.P, K.L., H.J.C., S.H.L., N.P. and B.M. analysed the data and discussed the results. W.Y.K., H-D.K., T-T.K., N.P. and B.M. wrote the paper and all authors provided feedback.

Additional information

4

Realization of mid-infrared graphene hyperbolic metamaterials

You-Chia Chang[1,2], Che-Hung Liu[1,3], Chang-Hua Liu[3], Siyuan Zhang[4], Seth R. Marder[4], Evgenii E. Narimanov[5], Zhaohui Zhong[1,3] & Theodore B. Norris[1,3]

While metal is the most common conducting constituent element in the fabrication of metamaterials, graphene provides another useful building block, that is, a truly two-dimensional conducting sheet whose conductivity can be controlled by doping. Here we report the experimental realization of a multilayer structure of alternating graphene and Al_2O_3 layers, a structure similar to the metal-dielectric multilayers commonly used in creating visible wavelength hyperbolic metamaterials. Chemical vapour deposited graphene rather than exfoliated or epitaxial graphene is used, because layer transfer methods are easily applied in fabrication. We employ a method of doping to increase the layer conductivity, and our analysis shows that the doped chemical vapour deposited graphene has good optical properties in the mid-infrared range. We therefore design the metamaterial for mid-infrared operation; our characterization with an infrared ellipsometer demonstrates that the metamaterial experiences an optical topological transition from elliptic to hyperbolic dispersion at a wavelength of 4.5 μm.

[1]Center for Photonics and Multiscale Nanomaterials, University of Michigan, 2200 Bonisteel Blvd., Ann Arbor, Michigan 48109, USA. [2]Department of Physics, University of Michigan, 450 Church St, Ann Arbor, Michigan 48109, USA. [3]Department of Electrical Engineering and Computer Science, University of Michigan, 1301 Beal Avenue, Ann Arbor, Michigan 48109, USA. [4]School of Chemistry and Biochemistry, Georgia Institute of Technology, 901 Atlantic Drive, Atlanta, Geogia 30332, USA. [5]School of Electrical and Computer Engineering and Birck Nanotechnology Center, Purdue University, 1205 West State Street, West Lafayette, Indiana 47907, USA. Correspondence and requests for materials should be addressed to T.N. (email: tnorris@umich.edu).

Hyperbolic metamaterials (HMMs) are artificially structured materials designed to attain an extremely anisotropic optical response, in which the permittivities associated with different polarization directions exhibit opposite signs[1-3]. Such anisotropic behaviour results in an isofrequency surface in the shape of a hyperboloid, which supports propagating high k-modes and exhibits an enhanced photonic density of states. Many interesting applications have been enabled by HMMs. For example, the spontaneous emission rate of quantum emitters can be modified if they are brought close to a HMM[4], and similarly, the scattering cross-section of small scatterers near a HMM is enhanced[5]. The near-field radiative heat transfer associated with HMMs becomes super-Planckian[6]. Also, the propagating high k-modes supported by HMM are exploited to achieve sub-diffraction-limited images using a hyperlens[7]. Some natural materials such as bismuth, graphite and hexagonal boron nitride exhibit hyperbolic dispersion in specific spectral ranges[8-10], while artificial HMMs are most commonly realized with two categories of structures such as metal-dielectric multilayers[4,7] and metallic nanorod arrays[11]. The former structure can be fabricated layer by layer using vapour deposition, and the latter is often obtained by electrochemical deposition of a metal on porous anodic aluminium oxide. In both cases, metal is the essential element to provide the conducting electrons that make the extreme anisotropicity possible. Metals can also be replaced by doped semiconductors for realizing HMMs in the infrared range[12].

In this paper, we explore the realization of a particular HMM, in which the role of the metal in providing a conducting layer is taken over by graphene[13-21]. Graphene is a two-dimensional (2D) semi-metal with a thickness of only one atom[22,23]. It has been shown that doped graphene is a good infrared plasmonic material in terms of material loss[24]. As a truly 2D material that only conducts in the plane, graphene by nature has the anisotropicity required for HMMs. As the thinnest material imaginable, graphene also makes an ideal building block for multilayer structures, as it enables the minimum possible period and therefore the highest possible cutoff for the high k-modes[14,25], which has been limited in metal and semiconductor-based HMMs by the non-negligible thickness of those materials. The conductivity of graphene, unlike that of metals, can be effectively modulated by electrical gating (see Supplementary Fig. 1 and Supplementary Note 1) or optical pumping[26,27]. This unique advantage has been demonstrated in other graphene-based metamaterials[28], and can potentially be exploited to realize a tunable HMM, in which the photonic density of states can be controlled electronically on demand. In addition, graphene shows much richer optoelectronic behaviour than metals, and the massless Dirac quasi-particles in graphene also give rise to very different carrier dynamics compared with other semiconductors. Various photodetection mechanisms, such as thermoelectric, bolometric, photovoltaic, photo-gating and photo-Dember effects, have been demonstrated with graphene[29-32]. Graphene multilayer structures can therefore serve as a unique platform in optoelectronics, incorporating the unusual photonic behaviour of HMMs into graphene detectors or other optoelectronic devices. For example, an ultrathin super-absorber enabled by HMM could be incorporated into graphene detectors to enhance the light absorption[18]. A brief summary of this report is as follows. The design criterions and material choices for realizing the graphene HMM are discussed. Chemical vapour deposited (CVD) graphene is identified as a good practical choice in the mid-infrared range when it is heavily doped. A chemical doping method is developed to obtain the desired high carrier density and ellipsometry is used to characterize the optical conductivity of monolayer graphene. The metamaterial with

multilayer structure is fabricated by repetitive graphene transfer and dielectric deposition. We characterize the effective permittivities of the fabricated metamaterial with ellipsometry to demonstrate the hyperbolic dispersion in the mid-infrared range.

Results

Design of graphene HMM. Figure 1 shows the structure of the graphene-based HMM, which consists of alternating dielectric and graphene layers. Similar graphene-dielectric multilayer structures have been proposed and analysed theoretically by different groups and shown to function as a HMM operating at terahertz (THz) and mid-infrared frequencies[13-21]. Various applications have also been discussed. For example, in our previous work we have calculated theoretically the Purcell factor of a graphene-based HMM with a finite number of layers[17], and we have simulated numerically the light coupling from free space into a graphene-based HMM slab with a metallic grating[18]. In spite of the large body of theoretical work on graphene-based HMM, no experimental demonstrations have yet been reported, the primary reason being the challenge in obtaining a sufficiently high level of doping in the graphene layers in the required multilayer structure.

The graphene-dielectric multilayer structure can be homogenized and viewed as a metamaterial using the effective medium approximation (EMA). The effective out-of-plane and in-plane permittivities of this metamaterial can be derived by taking the long-wavelength limit of the Bloch theory[13-16]:

$$\varepsilon_{\text{eff},\perp} = \varepsilon_{\text{d}},$$
$$\varepsilon_{\text{eff},\parallel} = \varepsilon_{\text{d}} + i \frac{\sigma Z_0}{2\pi} \left(\frac{\lambda}{d} \right). \tag{1}$$

Here ε_{d} is the permittivity of the dielectric layer, d is the dielectric thickness and σ is the optical conductivity of graphene. Z_0 is the vacuum impedance. Here graphene, as a 2D material, is treated as an infinitely thin layer described by its in-plane sheet conductivity. As indicated by equation 1, the graphene-dielectric multilayer system forms a uniaxial anisotropic metamaterial. $\varepsilon_{\text{eff},\perp}$ is the same as the constituent dielectric and is always positive. On the other hand, the real part of $\varepsilon_{\text{eff},\parallel}$ becomes negative if

$$\text{Im}\,\sigma > 2\pi(d/\lambda)(\varepsilon_{\text{d}}/Z_0). \tag{2}$$

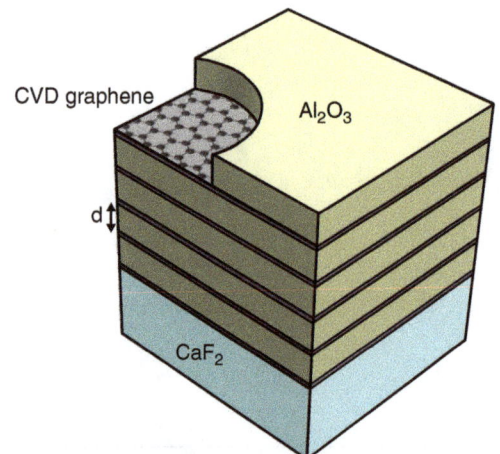

Figure 1 | The schematic representation of the graphene-dielectric multilayer structure that turns into a HMM at mid-infrared frequencies. It consists of five periods of alternating CVD graphene sheets and Al_2O_3 layers on a CaF_2 substrate. The thickness d of the Al_2O_3 layer is ~10 nm.

When this criterion is satisfied, the isofrequency surface becomes a hyperboloid and we obtain HMM. Such an isofrequency surface allows the existence of propagating high k-modes, which can be traced back to the coupled plasmon modes in the graphene-dielectric multilayer structure[17]. The criterion described by equation 2 determines the wavelength at which the optical topological transition between elliptical and hyperbolic dispersions occurs[4].

While most previous theoretical work has concentrated on using high-mobility graphene that may be obtained from mechanically exfoliated or epitaxially grown samples, we use CVD graphene because it is the most realistic choice for practical fabrication of a multilayer structure[33]. Growth of large-area CVD graphene is well established, and it can be transferred onto arbitrary surfaces using poly(methyl methacrylate) (PMMA) as the carrier material. In spite of its advantage in fabrication, CVD graphene often has a higher degree of disorder, which is typically manifested by a reduced mobility (usually on the order of thousands $cm^2 V^{-1} s^{-1}$). As a result of the lower crystal quality, the stronger carrier scattering in typical polycrystalline CVD graphene enhances the free-carrier absorption at THz frequencies, which can be understood from the theoretical optical conductivity of graphene[34–36]

$$\sigma(\omega) = \frac{\sigma_0}{2}\left(\tanh\frac{\hbar\omega+2E_F}{4k_BT}+\tanh\frac{\hbar\omega-2E_F}{4k_BT}\right)$$
$$-i\frac{\sigma_0}{2\pi}\log\left[\frac{(\hbar\omega+2E_F)^2}{(\hbar\omega-2E_F)^2+(2k_BT)^2}\right]+i\frac{4\sigma_0}{\pi}\frac{E_F}{\hbar\omega+i\hbar\gamma},$$

(3)

where σ_0 equals to $e^2/(4\hbar)$, E_F is the Fermi energy relative to the Dirac point and γ is the intraband scattering rate. In this expression, the first two terms correspond to interband transitions, while the third term is the Drude-like intraband conductivity. Figure 2 shows a plot of the theoretical optical conductivity given by equation 3 with parameters typical for doped polycrystalline CVD graphene. To realize a good HMM, we need graphene with a large positive imaginary conductivity to

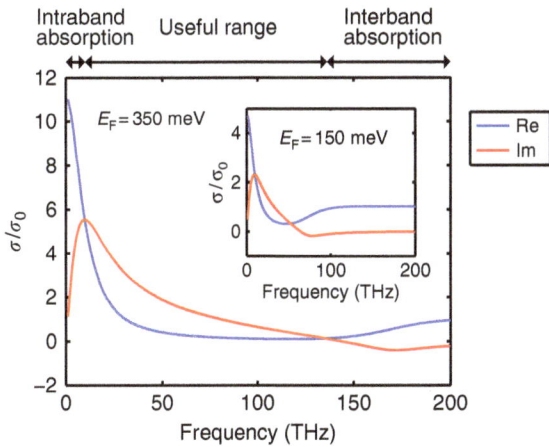

Figure 2 | The theoretical optical conductivity of graphene. It is plotted with $E_F = 350$ meV and $\hbar\gamma = 40$ meV. These numbers correspond to heavily doped CVD graphene. At the high-frequency end of the spectrum, graphene is lossy because of the interband absorption. At the low-frequency end, graphene is again lossy because of the intraband free-carrier absorption. There is a useful spectral range in between, where the imaginary part of the optical conductivity exceeds the real part. In this particular example, the useful wavelengths range from 2 to 30 μm in the mid-infrared range. The inset shows another example of lightly doped CVD graphene with $E_F = 150$ meV and $\hbar\gamma = 40$ meV. The useful wavelength range is smaller when the doping is lower.

interact with light, but with a small real conductivity to minimize the material loss. As indicated by Fig. 2, graphene is lossy at high frequencies when $\hbar\omega > 2E_F$ because of interband transitions. On the other hand, at low frequencies when $\hbar\omega \lesssim \hbar\gamma$, graphene also exhibits a large loss because of the intraband free-carrier absorption enabled by scattering. Because CVD graphene typically has a $\hbar\gamma$ of tens of meV, it is a lossy material at THz frequencies[37]. As shown by Fig. 2, however, there is a spectral range between the two lossy regions, such that the imaginary part of the conductivity exceeds the real part. As this spectral range lies in the mid-infrared part of the spectrum, CVD graphene-based HMM operates better in the mid-infrared than the THz region. Also, Fig. 2 indicates that doping can improve the properties of graphene for realizing a HMM. A large E_F can turn off the interband absorption by the Pauli blocking and increase the Imσ required for achieving negative $\varepsilon_{eff,\parallel}$. Furthermore, doping can also suppress the intraband scattering by screening charged impurities[37,38].

Characterization of the optical conductivity of graphene. Because graphene is the key building block of the metamaterial, it is important to have an accurate measurement on the optical conductivity of the actual CVD graphene layers used to fabricate the sample. Although the theoretical optical conductivity given by equation 3 provides a good guideline for designing the graphene HMM, real CVD graphene layers can have imperfections or extrinsic properties that are not taken into account by equation 3. We therefore need to characterize actual graphene samples and examine the scope of validity of equation 3.

In our previous work, we have developed a technique based on ellipsometry to measure the optical conductivity of truly 2D materials[39]. In this technique, the analysis used in conventional ellipsometry is modified to handle the infinitely thin 2D material whose properties are fully described by the 2D optical conductivity. To characterize actual CVD graphene samples with this technique, we have prepared two kinds of samples, unintentionally doped and the chemically doped CVD graphene, on CaF$_2$ substrates by the standard PMMA transfer method. Even without chemical treatment, unintentionally doped CVD graphene is p-type because of adsorbed gas molecules and residual ammonium persulfate from the transfer process[40,41]. The chemically doped CVD graphene is prepared by a solution process that leaves a sub-monolayer of Tris (4-bromo-phenyl)ammoniumyl hexachloroantimonate (also known as 'magic blue'), a somewhat air-stable p-type dopant, on the surface (see Methods section, Supplementary Fig. 2 and Supplementary Note 2)[42,43]. Figure 3a shows the optical conductivities of both samples measured with ellipsometry. The optical conductivities shown here are mathematically described by cubic splines without assuming an a priori theoretical expression like equation 3. Consistent with Fig. 2, in the mid-infrared range the chemically doped graphene has a larger imaginary conductivity, which is necessary for creating the extreme anisotropicity in the metamaterial.

Although the spline-fitted conductivity of actual CVD graphene sample shown in Fig. 3a is useful in many applications, a conductivity model based on a theoretical expression such as equation 3 provides more physical insight and requires fewer unknown parameters to perform the fit. The latter is important when we want to parameterize the homogenized metamaterial, which will be discussed in next section. In Fig. 3b, we examine how well equation 3 works for our chemically doped CVD graphene samples. In fitting the ellipsometer data, we express the optical conductivity $\sigma(\omega)$ by the model given by equation 3 with E_F and γ being the only two unknown fitting parameters. We also

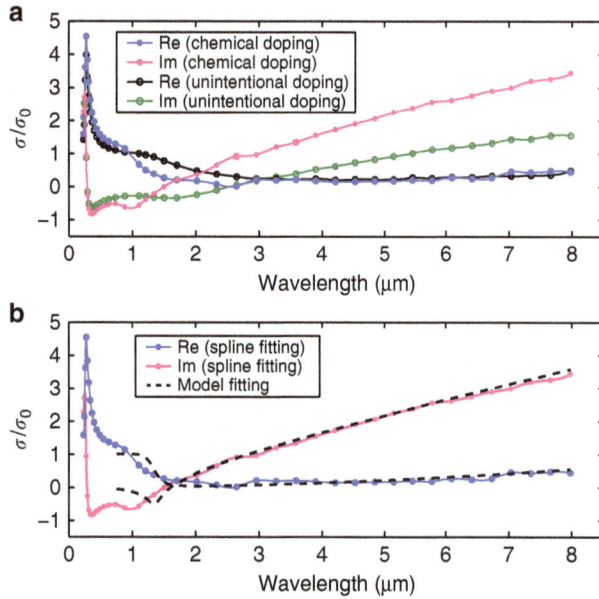

Figure 3 | The optical conductivity of CVD graphene measured by ellipsometry. (a) The real and imaginary part of the optical conductivity of the chemically doped CVD graphene (blue and magenta curves) and the unintentionally doped CVD graphene (black and green curves). These curves are mathematically expressed by cubic splines, and the markers denote the control points of the splines. The chemically doped CVD graphene has a larger imaginary conductivity in the mid-infrared range. (b) The real and imaginary part of the optical conductivity of the chemically doped CVD graphene. The blue and magenta curves are obtained by fitting with cubic splines, and the black dash lines are obtained by using the model given by equation 3. The model fitting is consistent with the spline fitting in the mid-infrared range. The extracted E_F and $\hbar\gamma$ from the model fitting are 460 and 23 meV, respectively, which corresponds to a mobility of $\sim 2,000\,\mathrm{cm^2\,V^{-1}\,s^{-1}}$.

show in the same figure the spline-fitted conductivity obtained from the same set of data. It is apparent that the resulting conductivity based on equation 3 overlaps very well with the spline-fitted conductivity throughout the mid-infrared range, assuring the validity of using equation 3 for the mid-infrared metamaterial. We extract from the fit that $E_F = 460$ meV and $\hbar\gamma = 23$ meV. A mobility of $\sim 2,000\,\mathrm{cm^2\,V^{-1}\,s^{-1}}$ can be calculated from these numbers using the relationship $\mu = e\pi\hbar\,V_F^2/(\hbar\gamma E_F)$, where μ is the mobility and V_F is the Fermi velocity.

In the mid-infrared range, the optical conductivity is mostly determined by intraband transitions, which are described by the Drude-like term in equation 3. Our result is consistent with ref. 37, which shows that the Drude model can successfully fit the measured absorption spectrum of CVD graphene over a broad range of infrared wavelengths. We do not apply equation 3 in the ultraviolet to visible wavelength range because the many-body correction has been shown to be important[44,45]. There is some discrepancy between the model and spline fits in the near-infrared ($\sim 1.5\,\mu$m, that is, near the wavelength corresponding to interband transitions close to the Fermi level). The origin of this discrepancy is not quantitatively understood, but may be related to spatial inhomogeneity in the Fermi energy or other disorder effects. Since the optical topological transition wavelength of our HMM is very far from this spectral region, and the fit is excellent over the entire mid-infrared range, the failure of the simple model in the near-infrared region does not affect the behaviour of the material in the mid-infrared, which is the region of concern in

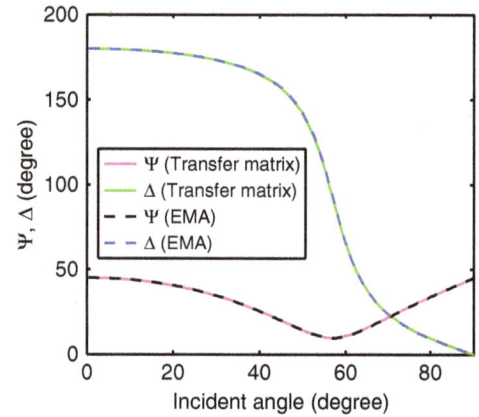

Figure 4 | Calculation of ellipsometric angles with exact transfer-matrix method and EMA. Ellipsometric angles Ψ and Δ are defined by $r_p/r_s = (\tan\Psi)e^{i\Delta}$, where r_p and r_s are the reflection coefficients for p and s light, respectively. They are the quantities an ellipsometer measures. The transfer-matrix method calculates the response of five periods of graphene-dielectric multilayer structure, while the EMA simulates a homogenized anisotropic layer with the permittivities given by equation 1. This calculation shows that the EMA is an accurate approximation for the structure. The wavelength used in this simulation is 6 μm. The material properties are $\varepsilon_d = 2.1$ and $\sigma = (0.43 + 1.98i)\,\sigma_0$. Thickness $d = 10$ nm. The substrate has a refractive index of 1.39.

this work. Equation 3 thus provides an excellent description for the mid-infrared conductivity. Other imperfections that are typically present in transferred CVD graphene samples, such as the existence of small multilayer graphene patches and holes (see Supplementary Fig. 3 and Supplementary Note 3), can also contribute to the deviations observed in Fig. 3b (ref. 46).

Measurement of the effective permittivity of graphene HMM. We have fabricated the multilayer structure shown in Fig. 1, which consists five periods of alternating CVD graphene and Al_2O_3. The CVD graphene is transferred by the PMMA method and doped with Tris (4-romophenyl) ammoniumyl hexachloroantimonate ('magic blue'). The Al_2O_3 dielectric layer is grown by atomic layer deposition (ALD). We choose Al_2O_3 as the dielectric material, because it has negligible loss at the mid-infrared wavelengths up to 8 μm. The dielectric thickness is chosen to be ~ 10 nm to create an optical topological transition in the mid-infrared range.

To characterize the metamaterial, we use infrared ellipsometry, which is appropriate to probe the effective permittivity of a metamaterial, since it measures the sample with free-space plane waves and the transverse wave vector ($k_0\sin\theta$) associated with the free-space plane waves is very small ($k_0\sin\theta d \ll 1$, where θ is the angle of incidence). We are therefore probing the low k-modes of the metamaterial, ensuring the validity of the long-wavelength approximation. Although the long-wavelength approximation is evidently satisfied for our metamaterial ($d/\lambda < 1/300$ in our case), we still need to confirm the validity of the EMA with a rigorous transfer-matrix calculation, since the EMA is derived for an infinite periodic system, while our metamaterial has only five periods. In Fig. 4 we show the transfer-matrix calculation of five periods of graphene-dielectric multilayer structure and the EMA calculation with the structure homogenized into an anisotropic layer, with the permittivities of the homogenized anisotropic layer given by equation 1. Here we calculate the ellipsometric angles Ψ and Δ, the quantities an ellipsometer acquires directly, at different incident angles. Ψ and Δ are defined by $r_p/r_s = (\tan\Psi)e^{i\Delta}$, where

r_p and r_s are the reflection coefficients for p and s light, respectively. Numbers used in the simulation are chosen according to measured material properties of the individual layers. As demonstrated by Fig. 4, the two methods give very close results, confirming that the five-period graphene-dielectric structure, in the low k-regime probed by ellipsometry, can be accurately treated as a metamaterial with the effective permittivities given by equation 1. In fact, in the low k-regime, even one period of the graphene-dielectric unit cell can be homogenized by the same EMA formula given by equation 1 and still reproduce the optical properties accurately (see Supplementary Figs 4 and 5 and Supplementary Note 4). However, the high k-regime is where the real interest of HMM lies, and as discussed in Supplementary Note 4, the high k optical properties depend on the number of unit cells in the metamaterial. The five-period structure in our experimental realization of graphene HMM is chosen to create desirable high k optical properties.

The results of infrared ellipsometry, ellipsometric angles Ψ and Δ for our HMM sample, are shown in Fig. 5a,b, from which we extract the effective permittivities by fitting the acquired data. A robust and physical fitting in ellipsometry requires correct prior knowledge about the sample parameters, which allows us to use a minimal number of unknowns. Since our simulation in Fig. 4 demonstrates that the EMA is an accurate description for the multilayer structure, we can apply equation 1 in fitting the data. More precisely, we fit the experimental data to a layer of an anisotropic material on a CaF$_2$ substrate with the permittivities of the anisotropic material given by equation 1. In equation 1, we know everything except the optical conductivity of graphene σ, as we have measured the thickness d independently after depositing each Al$_2$O$_3$ layer, and we have measured the refractive index of the ALD-grown Al$_2$O$_3$ in the relevant spectral range independently on a reference sample (see Methods section). Furthermore, as shown by Fig. 3b, considering the mid-infrared range with only the intraband response, the expression of equation 3 is a good description for the optical conductivity of the actual CVD graphene layers. Therefore, we can apply equation 3 and parameterize the optical conductivity with only E_F and γ. As a result of this independent knowledge of the sample, only two unknowns, E_F and γ, are sufficient to fit the experimental data of the multilayer metamaterial.

The fitted results of the ellipsometric angles Ψ and Δ are plotted as the blue dash lines in Fig. 5a,b. We restrict the wavelengths range of the fitting to 3.5–8 μm, where the lower bound is limited by the requirement of intraband-only response in the application of equation 3, and the upper bound is because of the limited transparent spectral range of Al$_2$O$_3$. As shown by Fig. 5, we are able to reproduce all six Ψ and Δ curves acquired at different incident angles with only two free parameters in the fitting. The extracted E_F is 365 meV, and the extracted $\hbar\gamma$ is 41 meV. The extracted E_F is lower than the value we typically obtain from chemically doped monolayer CVD graphene, because some dopants are lost in the ALD process because of the vacuum environment and the elevated temperature. The obtained

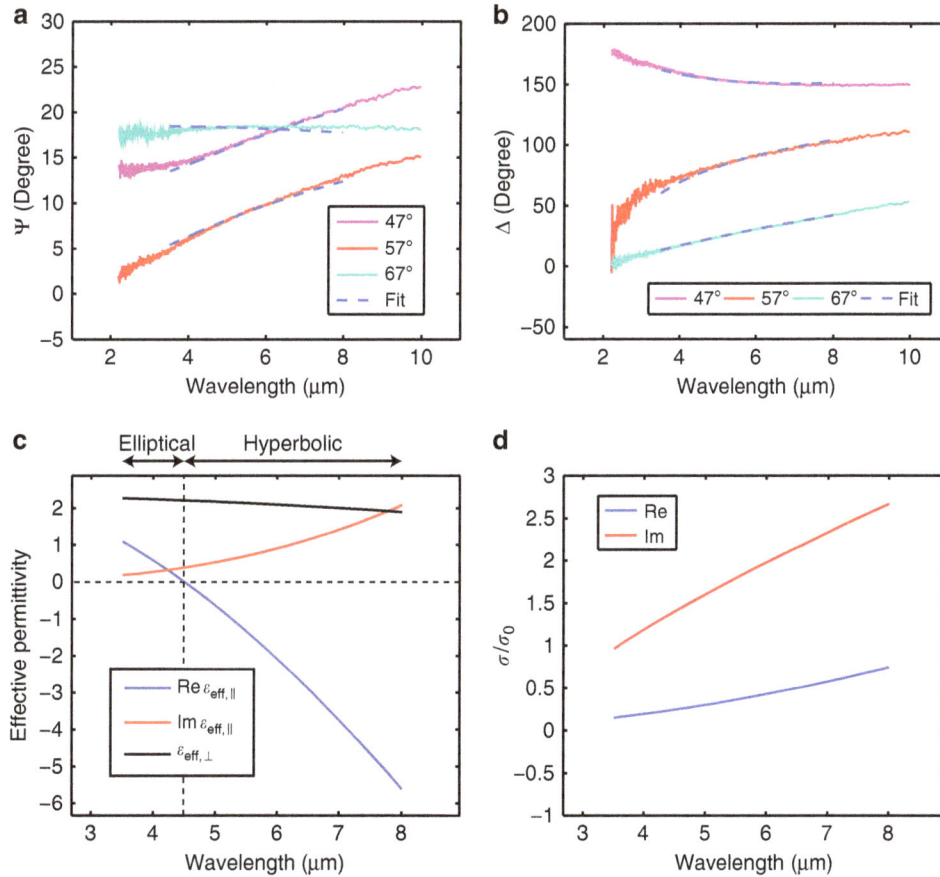

Figure 5 | Extraction of the effective permittivity of the graphene HMM. (a,b) The ellipsometric angles Ψ and Δ acquired from the graphene-dielectric multilayer structure. The measurement is performed at incident angles of 47°, 57° and 67°. The blue dash lines show the fitting by homogenizing the multilayer structure into a metamaterial with the effective permittivities given by equation 1. We extract from the fitting that $E_F = 365$ meV and $\hbar\gamma = 41$ meV. (c) The extracted effective permittivity of the metamaterial, which exhibits an optical topological transition from elliptical to hyperbolic dispersion at 4.5 μm. When the wavelength is at 6 μm, $\varepsilon_{eff,\parallel}$ equals $2.1 + 0.9i$ and $\varepsilon_{eff,\perp}$ equals 2.1. (d) The extracted optical conductivity of the constituent CVD graphene in the metamaterial.

scattering rate $\hbar\gamma$ is higher than the value of graphene on CaF_2 substrate shown in Fig. 3. This can be explained by the fact that the carrier scattering in graphene depends on the surrounding environment, from which we conclude that sandwiching graphene between Al_2O_3 increases the carrier scattering.

Figure 5c shows the effective permittivity of the graphene metamaterial given by the extracted values of E_F and γ. $\varepsilon_{eff,\perp}$ is always positive because it equals the permittivity of Al_2O_3. On the other hand, the real part of $\varepsilon_{eff,\|}$ changes from positive to negative at 4.5 μm indicating an optical topological transition from an elliptical metamaterial to a HMM. This graphene metamaterial is therefore a transverse epsilon-near-zero metamaterial at the wavelength of 4.5 μm (ref. 14). The imaginary part of $\varepsilon_{eff,\|}$ is several times smaller than the real part in most of the spectral range with hyperbolic dispersion, indicating that the loss of this HMM is reasonably low. In Fig. 5d, we plot the optical conductivity of the constituent graphene sheet of the metamaterial using the extracted E_F and γ.

Discussion

Our characterization by the infrared ellipsometry demonstrates that the graphene-dielectric multilayer structure indeed experiences an optical topological transition from an elliptical to a hyperbolic dispersion in the mid-infrared range, confirming the theoretical predictions in previous works[13-21]. Our metamaterial sample has an optical topological transition at a wavelength of 4.5 μm, and maintains good hyperbolic properties up to 8 μm. The upper bound of the wavelength range is limited by the absorption in Al_2O_3 and CVD graphene. While the absorption in the dielectric layer can be overcome by replacing Al_2O_3 with other infrared transparent materials such as ZnSe, the absorption in CVD graphene is limited by the quality of graphene. Recently, there have been reports of the growth of large-area CVD graphene with the quality of a single crystal[47], and new transfer process for CVD graphene without degrading the mobility[48]. With higher quality CVD graphene, the intraband absorption resulted from scattering could potentially be suppressed. The transition wavelength, as determined by equation 2, can be shifted by choosing the dielectric thickness or controlling the doping of graphene. The latter is especially useful if it can be done by the electrical gating. Shifting the transition wavelength farther into the infrared can be done by using lightly doped graphene or thicker dielectric. We have also realized a graphene HMM with the same structure except that the CVD graphene layers were not chemically doped (see Supplementary Fig. 6 and Supplementary Note 5), resulting in a transition wavelength red-shifted to 7.2 μm. On the other hand, blue shifting the transition wavelength is limited by the highest doping and the thinnest dielectric layers achievable in practice. While the structure reported in this work has only five periods, the procedure developed here can be repeated to scale up the graphene HMM. Some applications of HMMs do not require a large number of periods; for example, only a few periods is sufficient to produce a Purcell factor close to a semi-infinite structure, according to the theoretical calculations in ref. 17.

Methods

Sample fabrication. The graphene-dielectric multilayer structure with five periods is fabricated on a CaF_2 wedge. The CVD graphene is grown on copper foil (Graphenea Inc) and transferred to the substrate using the standard PMMA transfer technique[33,46]. The copper foil is etched using an ammonium persulfate solution. The size of the CVD graphene we transfer is $\sim$ 10 mm by 10 mm. After transferring each graphene layer, we dope the graphene by soaking the sample in a 0.25 mM solution of Tris (4-bromophenyl)ammoniumyl hexachloroantimonate 'magic blue' in dichloromethane for 10 min, and then rinse the sample with dichloromethane (see Supplementary Fig. 2 and Supplementary Note 2). The Al_2O_3 dielectric layer is deposited by the ALD at 150 °C using trimethylaluminium as the

Al precursor and H_2O as the oxygen precursor. The number of cycles used in the ALD process is calibrated to grow $\sim$ 10 nm of Al_2O_3 on graphene, with the thickness characterized by an ellipsometer (Woollam M-2000). The procedure is repeated to fabricate five periods of the graphene-Al_2O_3 unit cell. We have also confirmed that the chemical doping with Tris (4-bromophenyl) ammoniumyl hexachloroantimonate does not affect the Al_2O_3 layer and the substrate. We have found that although nitric acid can also p-dope graphene effectively[37,39], it is not a good dopant for making the multilayer structure because of damage to the thin Al_2O_3 layer. The substrate is wedged to avoid back side reflection in the ellipsometry measurement. We also characterize the graphene-dielectric multilayer structure with the Woollam M-2000 ellipsometer after depositing each Al_2O_3 layer and after transferring each graphene layer. With the acquired ellipsometry data, we extract an average Al_2O_3 thickness of 10.4 nm.

Ellipsometry characterization. The optical conductivity of monolayer graphene is measured by the ellipsometric analysis method described in ref. 39. Two ellipsometers designed for different spectral ranges, Woollam M-2000 and Woollam IR-VASE, are used for the wavelengths from 230 nm to 1.64 μm and the wavelengths above 2 μm, respectively. The data are acquired at three angles of incidence: 47°, 57° and 67°. The spot sizes of M-2000 and IR-VASE are 3 mm by 5.5 mm and 8 mm by 20 mm, respectively when the incident angle is 57°. We mask the samples for the IR-VASE measurement because the spot size is larger than the graphene area. To obtain the refractive index of Al_2O_3, we have prepared a sample with ALD-grown Al_2O_3 film on a CaF_2 wedge. We measure the sample with both ellipsometers, and fit the refractive index of Al_2O_3 with the Sellmeier equation. The measurement of the effective permittivities of the graphene-dielectric multilayer structure is performed by the IR-VASE ellipsometer with the same settings described above.

References

1. Smith, D. R. & Schurig, D. Electromagnetic wave propagation in media with indefinite permittivity and permeability tensors. *Phys. Rev. Lett.* **90**, 077405 (2003).
2. Poddubny, A., Iorsh, I., Belov, P. & Kivshar, Y. Hyperbolic metamaterials. *Nat. Photon.* **7**, 948–957 (2013).
3. Jacob, Z. *et al.* Engineering photonic density of states using metamaterials. *Appl. Phys. B* **100**, 215–218 (2010).
4. Krishnamoorthy, H. N., Jacob, Z., Narimanov, E., Kretzschmar, I. & Menon, V. M. Topological transitions in metamaterials. *Science* **336**, 205–209 (2012).
5. Guclu, C., Campione, S. & Capolino, F. Hyperbolic metamaterial as super absorber for scattered fields generated at its surface. *Phys. Rev. B* **86**, 205130 (2012).
6. Biehs, S. A., Tschikin, M., Messina, R. & Ben-Abdallah, P. Super-Planckian near-field thermal emission with phonon-polaritonic hyperbolic metamaterials. *Appl. Phys. Lett.* **102**, 131106 (2013).
7. Liu, Z., Lee, H., Xiong, Y., Sun, C. & Zhang, X. Far-field optical hyperlens magnifying sub-diffraction-limited objects. *Science* **315**, 1686–1686 (2007).
8. Narimanov, E. E. & Kildishev, A. V. Naturally hyperbolic. *Nat. Photon.* **9**, 214–216 (2015).
9. Dai, S. *et al.* Tunable phonon polaritons in atomically thin van der Waals crystals of boron nitride. *Science* **343**, 1125–1129 (2014).
10. Caldwell, J. D *et al.* Sub-diffractional volume-confined polaritons in the natural hyperbolic material hexagonal boron nitride. *Nat. Commun.* **5**, 5221 (2014).
11. Noginov, M. A. *et al.* Bulk photonic metamaterial with hyperbolic dispersion. *Appl. Phys. Lett.* **94**, 151105 (2009).
12. Hoffman, A. J. *et al.* Negative refraction in semiconductor metamaterials. *Nat. Mater.* **6**, 946–950 (2007).
13. Iorsh, I. V., Mukhin, I. S., Shadrivov, I. V., Belov, P. A. & Kivshar, Y. S. Hyperbolic metamaterials based on multilayer graphene structures. *Phys. Rev. B* **87**, 075416 (2013).
14. Othman, M. A. K., Guclu, C. & Capolino, F. Graphene–dielectric composite metamaterials: evolution from elliptic to hyperbolic wavevector dispersion and the transverse epsilon-near-zero condition. *J. Nanophoton.* **7**, 073089–073089 (2013).
15. Othman, M. A. K., Guclu, C. & Capolino, F. Graphene-based tunable hyperbolic metamaterials and enhanced near-field absorption. *Opt. Express* **21**, 7614–7632 (2013).
16. Wang, B., Zhang, X., García-Vidal, F. J., Yuan, X. & Teng, J. Strong coupling of surface plasmon polaritons in monolayer graphene sheet arrays. *Phys. Rev. Lett.* **109**, 073901 (2012).
17. DaSilva, A. M., Chang, Y. C., Norris, T. B. & MacDonald, A. H. Enhancement of photonic density of states in finite graphene multilayers. *Phys. Rev. B* **88**, 195411 (2013).
18. Chang, Y. C. *et al.* Mid-infrared hyperbolic metamaterial based on graphene-dielectric multilayers. *Proc. SPIE* **9544**, 954417 (2015).
19. Nefedov, I. S., Valaginnopoulos, C. A. & Melnikov, L. A. Perfect absorption in graphene multilayers. *J. Opt.* **15**, 114003 (2013).

20. Sreekanth, K. V., De Luca, A. & Strangi, G. Negative refraction in graphene-based hyperbolic metamaterials. *Appl. Phys. Lett.* **103**, 023107 (2013).

21. Andryieuski, A., Lavrinenko, A. V. & Chigrin, D. N. Graphene hyperlens for terahertz radiation. *Phys. Rev. B* **86**, 121108 (2012).

22. Novoselov, K. S. *et al.* Electric field effect in atomically thin carbon films. *Science* **306**, 666–669 (2004).

23. Novoselov, K. S. A. *et al.* Two-dimensional gas of massless Dirac fermions in graphene. *Nature* **438**, 197–200 (2005).

24. Jablan, M., Buljan, H. & Soljacic, M. Plasmonics in graphene at infrared frequencies. *Phys. Rev. B* **80**, 245435 (2009).

25. Kidwai, O., Zhukovsky, S. V. & Sipe, J. E. Effective-medium approach to planar multilayer hyperbolic metamaterials: strengths and limitations. *Phys. Rev. A* **85**, 053842 (2012).

26. Liu, M. *et al.* A graphene-based broadband optical modulator. *Nature* **474**, 64–67 (2011).

27. Bao, Q. *et al.* Atomic-layer graphene as a saturable absorber for ultrafast pulsed lasers. *Adv. Funct. Mater.* **19**, 3077–3083 (2009).

28. Lee, S. H., Choi, J., Kim, H. D., Choi, H. & Min, B. Ultrafast refractive index control of a terahertz graphene metamaterial. *Sci. Rep.* **3**, 2135 (2013).

29. Gabor, N. M. *et al.* Hot carrier–assisted intrinsic photoresponse in graphene. *Science* **334**, 648–652 (2011).

30. Yan, J. *et al.* Dual-gated bilayer graphene hot-electron bolometer. *Nat. Nanotechnol.* **7**, 472–478 (2012).

31. Liu, C. H., Chang, Y. C., Norris, T. B. & Zhong, Z. Graphene photodetectors with ultra-broadband and high responsivity at room temperature. *Nat. Nanotechnol.* **9**, 273–278 (2014).

32. Liu, C. H. *et al.* Ultrafast lateral photo-Dember effect in graphene induced by nonequilibrium hot carrier dynamics. *Nano Lett.* **15**, 4234–4239 (2015).

33. Li, X. *et al.* Large-area synthesis of high-quality and uniform graphene films on copper foils. *Science* **324**, 1312–1314 (2009).

34. Falkovsky, L. A. & Pershoguba, S. S. Optical far-infrared properties of a graphene monolayer and multilayer. *Phys. Rev. B* **76**, 153410 (2007).

35. Falkovsky, L. A. & Varlamov, A. A. Space-time dispersion of graphene conductivity. *Eur. Phys. J. B* **56**, 281–284 (2007).

36. Stauber, T., Peres, N. M. R. & Geim, A. K. Optical conductivity of graphene in the visible region of the spectrum. *Phys. Rev. B* **78**, 085432 (2008).

37. Yan, H. *et al.* Tunable infrared plasmonic devices using graphene/insulator stacks. *Nat. Nanotechnol.* **7**, 330–334 (2012).

38. Hwang, E. H., Adam, S. & Sarma, S. D. Carrier transport in two-dimensional graphene layers. *Phys. Rev. Lett.* **98**, 186806 (2007).

39. Chang, Y. C., Liu, C. H., Liu, C. H., Zhong, Z. & Norris, T. B. Extracting the complex optical conductivity of mono- and bilayer graphene by Ellipsometry. *Appl. Phys. Lett.* **104**, 261909 (2014).

40. Schedin, F. *et al.* Detection of individual gas molecules adsorbed on graphene. *Nat. Mater.* **6**, 652–655 (2007).

41. Bae, S. *et al.* Roll-to-roll production of 30-inch graphene films for transparent electrodes. *Nat. Nanotechnol.* **5**, 574–578 (2010).

42. Tarasov, A. *et al.* Controlled doping of large-area trilayer MoS$_2$ with molecular reductants and oxidants. *Adv. Mater.* **27**, 1175–1181 (2015).

43. Paniagua, S. A. *et al.* Production of heavily n-and p-doped CVD graphene with solution-processed redox-active metal–organic species. *Mater. Horiz.* **1**, 111–115 (2014).

44. Mak, K. F., Shan, J. & Heinz, T. F. Seeing many-body effects in single-and few-layer graphene: observation of two-dimensional saddle-point excitons. *Phys. Rev. Lett.* **106**, 046401 (2011).

45. Yang, L., Deslippe, J., Park, C. H., Cohen, M. L. & Louie, S. G. Excitonic effects on the optical response of graphene and bilayer graphene. *Phys. Rev. Lett* **103**, 186802 (2009).

46. Liang, X. *et al.* Toward clean and crackless transfer of graphene. *ACS Nano* **5**, 9144–9153 (2011).

47. Hao, Y. *et al.* The role of surface oxygen in the growth of large single-crystal graphene on copper. *Science* **342**, 720–723 (2013).

48. Banszerus, L. *et al.* Ultrahigh-mobility graphene devices from chemical vapor deposition on reusable copper. *Sci. Adv.* **1**, e1500222, (2015).

Acknowledgements

This work was supported by the National Science Foundation (NSF) Center for Photonic and Multiscale Nanomaterials (DMR 1120923). This work was performed in part at the Lurie Nanofabrication Facility, a member of the National Nanotechnology Infrastructure Network, which is supported in part by the National Science Foundation. This research was also funded by the National Science MRSEC Program, DMR-0820382. Z.Z. thanks the support from NSF CAREER Award (ECCS-1254468).

Author contributions

Y.-C.C. and T.B.N. conceived the experiments. Y.-C.C., Che.-H.L. and Cha.-H.L. fabricated the samples. S.Z and S.R.M identified suitable dopants and suggested procedures for doping studies. Y.-C.C. performed the measurements. All authors discussed the results. Y.-C.C. and T.B.N. co-wrote the manuscript and all authors provided comments.

Additional information

Competing financial interests: The authors declare no competing financial interests.

Frequency comb transferred by surface plasmon resonance

Xiao Tao Geng[1,2], Byung Jae Chun[3], Ji Hoon Seo[4], Kwanyong Seo[4], Hana Yoon[5], Dong-Eon Kim[1,2], Young-Jin Kim[3] & Seungchul Kim[1,2]

Frequency combs, millions of narrow-linewidth optical modes referenced to an atomic clock, have shown remarkable potential in time/frequency metrology, atomic/molecular spectroscopy and precision LIDARs. Applications have extended to coherent nonlinear Raman spectroscopy of molecules and quantum metrology for entangled atomic qubits. Frequency combs will create novel possibilities in nano-photonics and plasmonics; however, its interrelation with surface plasmons is unexplored despite the important role that plasmonics plays in nonlinear spectroscopy and quantum optics through the manipulation of light on a subwavelength scale. Here, we demonstrate that a frequency comb can be transformed to a plasmonic comb in plasmonic nanostructures and reverted to the original frequency comb without noticeable degradation of $<6.51 \times 10^{-19}$ in absolute position, 2.92×10^{-19} in stability and $1\,Hz$ in linewidth. The results indicate that the superior performance of a well-defined frequency comb can be applied to nanoplasmonic spectroscopy, quantum metrology and subwavelength photonic circuits.

[1] Max Planck Center for Attosecond Science, Max Planck POSTECH/KOREA Res. Initiative, Pohang, Gyeongbuk 376-73, South Korea. [2] Department of Physics, Center for Attosecond Science and Technology (CASTECH), POSTECH, Pohang, Gyeongbuk 376-73, South Korea. [3] School of Mechanical and Aerospace Engineering, Nanyang Technological University (NTU), 50 Nanyang Avenue, Singapore 639798, Singapore. [4] Department of Energy Engineering, Ulsan National Institute of Science and Technology (UNIST), Ulsan 689-798, South Korea. [5] Energy Storage Department, Korea Institute of Energy Research (KIER), Daejeon 305-343, South Korea. Correspondence and requests for materials should be addressed to Y.-J.K. (email: yj.kim@ntu.edu.sg) or to S.K. (email: inter99@postech.ac.kr).

The frequency comb of mode-locked femtosecond lasers has led to remarkable advances in high-resolution spectroscopy[1,2], broadband calibration of astronomical spectrographs[3,4], time/frequency transfer over long distances[5,6], absolute laser ranging[7–10] and inter-comparison of atomic clocks[11,12]. It provides millions of well-defined optical modes over a broad spectral bandwidth with high-level phase coherence referenced to an atomic clock. Recently, the potential of frequency comb has expanded to microscopic applications; high inter-mode coherence within a short pulse duration enabled manipulating atomic qubits[13], operating quantum logic gates and performing high-speed molecular detection by coherent Raman spectroscopy through harnessing inter-mode beat frequencies between two frequency combs at different repetition rates[14].

Coupling surface plasmons (SPs)[15,16], collective charge oscillations produced by the resonant interaction of light and free electrons on the interface of metallic and dielectric materials, to frequency comb creates numerous advantages. First, SP can allow for the frequency comb to access nanoscopic volumes that surpass the diffraction limit[17]. Second, the field enhancement by localized SP enables the highly sensitive detection of weak signals, even from a single molecule (for example, surface-enhanced Raman scattering)[18]. Third, next-generation photonic devices and circuits can be implemented within a small subwavelength volume by all-optical control of light properties (amplitude, phase and polarization state) in plasmonic nanostructures within ultrafast time scales[19–22]. However, the superior performance of the frequency comb, such as absolute frequency uncertainty, high-frequency stability and narrow linewidth, could deteriorate during the photon-plasmon conversion process. For exploring novel combination of frequency comb and SP resonance, it is prerequisite to verify that frequency comb maintains its performance under plasmonic resonance; however, there have been no studies to date.

In the following, we report that frequency comb successfully maintains core performances in photon-plasmon conversion by exploiting plasmonic extraordinary transmission through a subwavelength plasmonic hole array. This implies that the original frequency comb can be transformed into a form of plasmonic comb on metallic nanostructures and reverted to an original frequency comb without noticeable degradation in absolute frequency position, stability and linewidth. The superior performance of well-defined frequency combs can therefore be applied to various nanoplasmonic spectroscopy, coherent quantum metrology and subwavelength photonic circuits.

Results

Frequency comb transferred by SP resonance. Figure 1 shows the experimental apparatus to characterize the conservation of frequency comb for the conversion from photon to SP. The frequency comb is split into reference and measurement beams; one part of the beam transmits through an acousto-optic modulator (AOM) for a frequency shift of 40 MHz to construct a reference frequency comb and the other part of the beam passes through the plasmonic sample. The frequency comb structure in SP resonance was generated by the exploitation of a metallic nanohole array used for extraordinary optical transmission (EOT) that converted photon into SP. The small diameter of each hole prevents light passing through the sample based on classical optics. However, the SP-mediated tunnelling effect of nanohole array drastically enhances optical transmittance[23]. These intriguing optical phenomena have been studied widely for high-resolution chemical sensing, ultrafast optical modulation, wavelength-tunable optical filtering and subwavelength lithography[24,25]. The physical origin of EOT has been attributed to resonant SP

polaritons (SPPs)[26]. The appropriate geometrical and material parameters of nanohole array excite the SPP mode that allows the transmission of light that contains plasmonic information inside an EOT sample. The resonant nature of the SP changes the transmitted spectral distribution, depending on sample design, input polarization and incident angle. Plasmonic EOT can also induce wavelength-dependent changes in optical frequency and phase in addition to wavelength-dependent transmittance. The optical frequency of a single frequency comb mode transmitted through the plasmonic sample via SP resonance (f_{MEA}) can be expressed as

$$f_{\text{MEA}} = nf_{\text{r}} + f_{\text{ceo}} + \Delta f_{\text{sp}} \qquad (1)$$

where f_{r} is the pulse repetition frequency, f_{ceo} the carrier-envelope offset frequency, and Δf_{sp} the frequency and phase change generated by SP resonance. Meanwhile, the optical frequency of the single mode passing through the reference path (f_{REF}) can be expressed as

$$f_{\text{REF}} = nf_{\text{r}} + f_{\text{ceo}} + f_{\text{AOM}} \qquad (2)$$

where f_{AOM} denotes the intentional frequency shift by AOM. The detection of the heterodyne beat-frequency generated by the interference between the reference and measurement beams enables the measurement of optical frequency difference, ($f_{\text{REF}} - f_{\text{MEA}}$) at a radio-frequency (RF) regime using a fast avalanche photodiode. This resultant frequency difference can be simplified to $f_{\text{AOM}} - \Delta f_{\text{sp}}$, where f_{AOM} works as the high-frequency carrier to isolate Δf_{sp} from the relatively strong low frequency noise components.

Plasmonic extraordinary transmission. For transmitting frequency combs through the subwavelength holes by SP resonance, there are three important geometric parameters: hole diameter (d), hole pitch (l) and Au film thickness (t; Fig. 2a). For maximum optical transmission at a wavelength of 840 nm, three parameters were optimized by solving Maxwell's equations using finite-difference time-domain (FDTD) method. Figure 2b,c show the calculated plasmonic field distribution through the optimized sample. The electric field around the hole was significantly enhanced by SP in the periodic apertures, delivering the optical energy through the hole. Figure 2a shows the scanning electron microscope image of the fabricated nanohole array; all dimensions were matched with optimized design parameters within a geometric error of < 5%. Figure 2d shows that the transmitted optical spectrum coincided with the numerical FDTD results and validated the numerical analysis. Minor deviations between the two spectrums are expected by focusing geometry onto the plasmonic sample. The plasmonic resonance conditions are dissimilar in given transverse electric-transverse magnetic polarization if the angle of incidence is not surface normal. As a result, plasmonic sample shows different transmission spectra for transverse electric-transverse magnetic polarization at the incidence angle of 45° (Fig. 2e); therefore, optical transmission of our sample is dominated by the plasmonic EOT, not classical diffraction theory.

Frequency comb structure after plasmonic transmission. The transmitted frequency combs through the plasmonic sample results in an interference with the reference frequency comb to verify the frequency comb structure after the photon-plasmon mode conversion by the EOT (Fig. 3a). For comparative analysis, interference signals were obtained at three different wavelength regimes with optical band-pass filters, representing on-resonance (840 nm) and off-resonance (800 and 900 nm) positions.

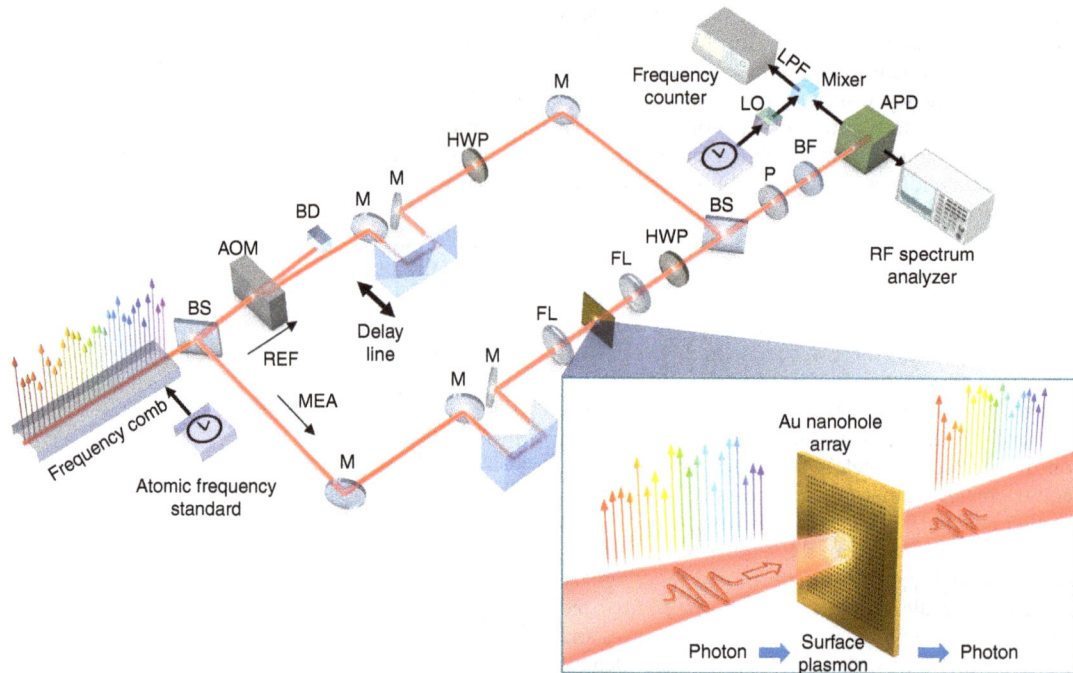

Figure 1 | Generation and characterization of plasmonic frequency comb. Part of the frequency comb experiences plasmonic mode conversion by passing through the plasmonic sample. The sample consists of a subwavelength nanohole array on an Au thin-film, enabling the conversion from photon to SP (from SP to photon). The other part of the frequency comb is used as a reference beam to compare with the frequency comb passed through the plasmonic sample. The frequency combs at two different paths are combined and monitored by APD. The characteristics of the frequency comb at measurement path are analysed by an RF spectrum analyser and a frequency counter. APD, avalanche photo-detector; BD, beam dumper; BF, band-pass filter; BS, beam splitter; FL, focusing lens; HWP, half-wave plate; LO, local oscillator; LPF, low-pass filter; M, mirror; MEA, measurement beam path; P, polarizer; REF, reference beam path.

The coherence of a large number of frequency comb modes can be deteriorated by temporal and spectral plasmonic dispersion, phase noise and frequency noise during the propagation through the plasmonic EOT sample. The frequency comb fundamentally suffers from phase and frequency noises when passing through the optical medium (for example, ambient air and optical fibre) exposed to environmental variations, such as vibration, temperature variation and humidity change. Therefore, it has been an important task to monitor and compensate the temporal and spectral dispersion, phase noise and frequency noise generated in the medium, as reported through long optical-fibre[6] and through ambient air[27]. SPs also suffer from the dispersion and phase change by the medium and environmental disturbances, which have not been investigated with the frequency comb for their quantitative or qualitative analysis. Propagating SPs through the EOT sample experience phase delay depending on their wavelengths and spatial locations before and after tunnelling through each subwavelength hole; this phase delay can be additionally induced by the plasmonic dynamic damping, imperfect sample geometry, surface roughness of the metal film or air refractive index change around the sample. Therefore, the total summation of the electromagnetic waves at the output side of each hole may contain temporal and spectral dispersion, phase distortion and frequency change.

Most noise sources of the frequency comb can be categorized into intra-cavity and extra-cavity sources; intra-cavity noise sources (including cavity length change, cavity loss fluctuations and pump noise) cause frequency noise whereas extra-cavity noise sources (induced by path-length fluctuation, shot noise from the limited power or noise generated during super-continuum generation) result in time-varying phase noise floor[5]. In this investigation, plasmonic mode conversion by the EOT was

considered as an extra-cavity noise source that provided wavelength-dependent power attenuation, phase shift and frequency noise, similar to the supercontinuum generation process. Noise contributions should be observed at $f_{AOM} - \Delta f_{SP}$ in the form of linewidth broadening, frequency shift, signal-to-noise (S/N) ratio reduction, increased phase noise or a higher Allan deviation if the plasmonic frequency comb suffers from phase or frequency noise during the plasmonic mode conversion.

Linewidth broadening and S/N ratio reduction in plasmonic mode conversion process was initially evaluated by measuring RF beat linewidth of $f_{AOM} - \Delta f_{SP}$ at three different wavelength regimes (Fig. 3b). With different resolution bandwidths (RBWs), there was no substantial degradation in the linewidth at 840 nm before and after the installation of the plasmonic sample in the beam path. The high-level S/N ratio of ~60 dB beat signal indicates that the plasmonic EOT provide no significant phase noise to the frequency comb.

Phase noise and frequency stability was measured for the quantitative analysis of frequency-dependent noise contributions. Figure 4a shows the phase noise spectrum obtained by monitoring one of high harmonics of the beat frequencies at ~1.2 GHz with and without the plasmonic sample; this confirms that there was no noticeable frequency noise inclusion. For high-precision frequency position measurement, the beat frequency between reference and measurement frequency comb was measured by a frequency counter for 3,000 s, resulting in 0.24 mHz frequency difference with a s.d. of 61 mHz (Fig. 4b). This corresponds to 6.51×10^{-19}, which proves that plasmonic mode conversion provides no substantial degradation in the frequency accuracy of the frequency comb. The stability of the beat signal was measured to be 4.08×10^{-18} without the plasmonic sample, 4.37×10^{-18} with the plasmonic sample at

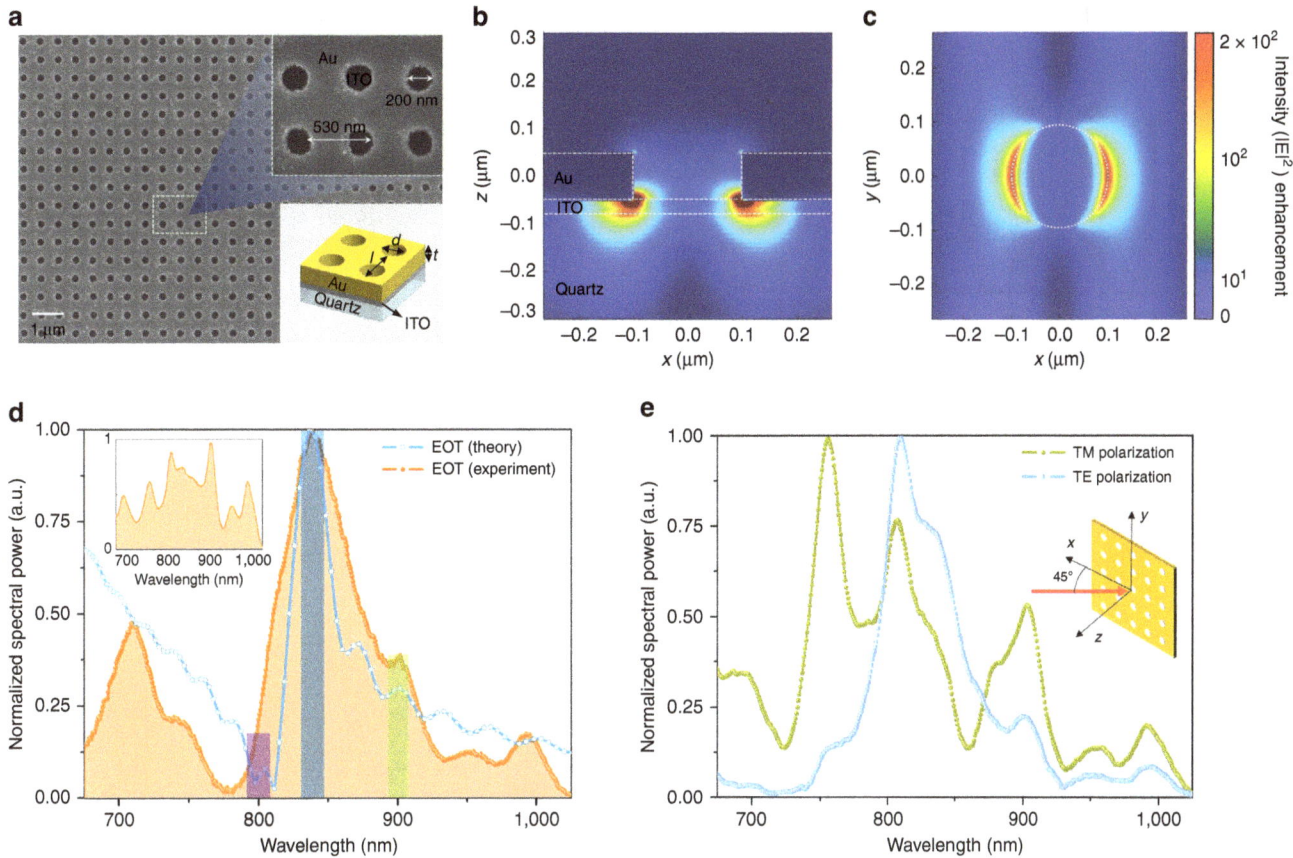

Figure 2 | Numerical simulation and characterization of fabricated plasmonic sample. (**a**) Scanning electron microscope image of the fabricated subwavelength nanohole array for plasmonic EOT. The fabricated nanohole has the diameter (*d*) of 200 nm, pitch (*l*) of 530 nm and thickness (*t*) of 100 nm on 25-nm-thick ITO-coated quartz substrate. (**b**) Calculated intensity distribution of an plasmonic sample taken from the side. (**c**) Calculated intensity distribution at the interface between Au and ITO layer. (**d**) Theoretical (blue line) and experimental (orange line) spectrum of transmitted frequency combs through the plasmonic sample. Purple, blue and green bars represent the selected spectral components (800, 840 and 900 nm) to characterize frequency comb, respectively. Inset (top left) shows the original spectrum of the frequency comb. (**e**) Polarization dependent transmission spectrum through the plasmonic sample at an incident angle of 45°.

resonance wavelength of 840 nm for an averaging time of 100 s, respectively (Fig. 4c). At the off-resonance wavelength, the stability of beat signal was 4.59×10^{-18}, signifying almost no difference between on- and off-plasmonic resonance stabilities. All the experiments pointed that plasmonic mode conversion causes no substantial degradation to the frequency comb in terms of linewidth, frequency position, S/N ratio and frequency stability.

Discussions

All hundreds of thousands optical modes in the frequency comb were firstly converted from photonic to plasmonic mode at the input side of the plasmonic EOT sample and then reverted to photonic mode at the other output side of the sample. It is known to be practically difficult to directly measure the optical frequency of the plasmonic mode so the characteristics of the plasmonic comb were measured here in the far field. Because the plasmonic and photonic modes are assumed to be mutually coherent, if there is any change in the frequency comb characteristics during the plasmonic propagation (in plasmonic mode) through the sample, it should be monitored at the output side in the far field (in photonic mode). Therefore, the beat-frequency detection using the transmitted photonic mode in the far-field regime enabled us to compare the qualities of the plasmonic comb with the original frequency comb, which cannot be implemented in the

near-field regime. As the result of the comparison, there were no noticeable degradation in linewidth, frequency shift, S/N ratio, phase noise and Allan deviation. This implies that SP, the collective electrons, can be regarded as information carrier as precise as the optical frequency comb.

The frequency comb passing through the plasmonic EOT sample experiences the different physical process with the light reflection at a metallic mirror. Although both of the SP resonance and the surface reflection are governed by free-electron oscillation in conduction band of metals, the SP resonance additionally requires the specific momentum matching between incident photon and SP, whose relationship is determined by the plasmonic dispersion relation. Therefore, it is natural to maintain the coherence during the light reflection at metal surface (governed by frequency conservation), which is not the case in plasmonic structures (governed by frequency and momentum matching). Once the incident photon (in photonic mode) is converted into SP, it will propagate through the metal as the form of SPPs (in plasmonic mode). This plasmonic propagation causes temporal and spectral dispersions, phase variations and frequency changes, which may degrade the inherently high coherence of the optical frequency comb.

Plasmonic EOT is governed by not only the hole geometry[28] but also hole pitch. Therefore, the incidence angle tuning of the input beam can provide the change in plasmonic coupling mode without dimensional changes, which can possibly cause some

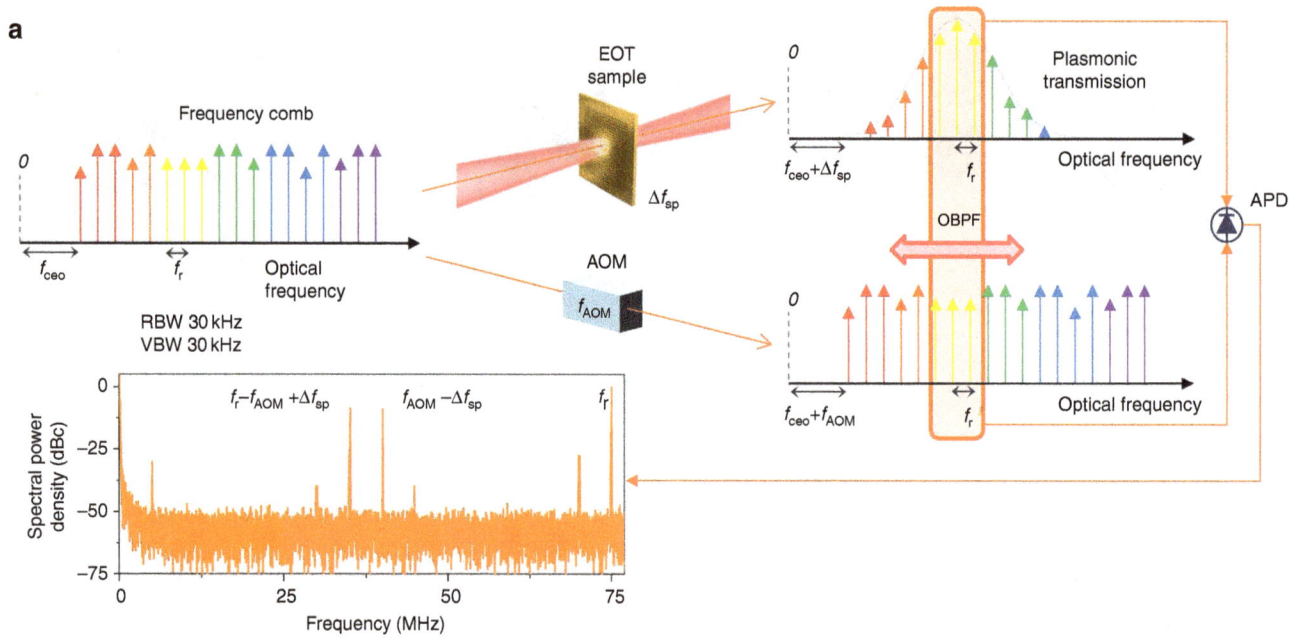

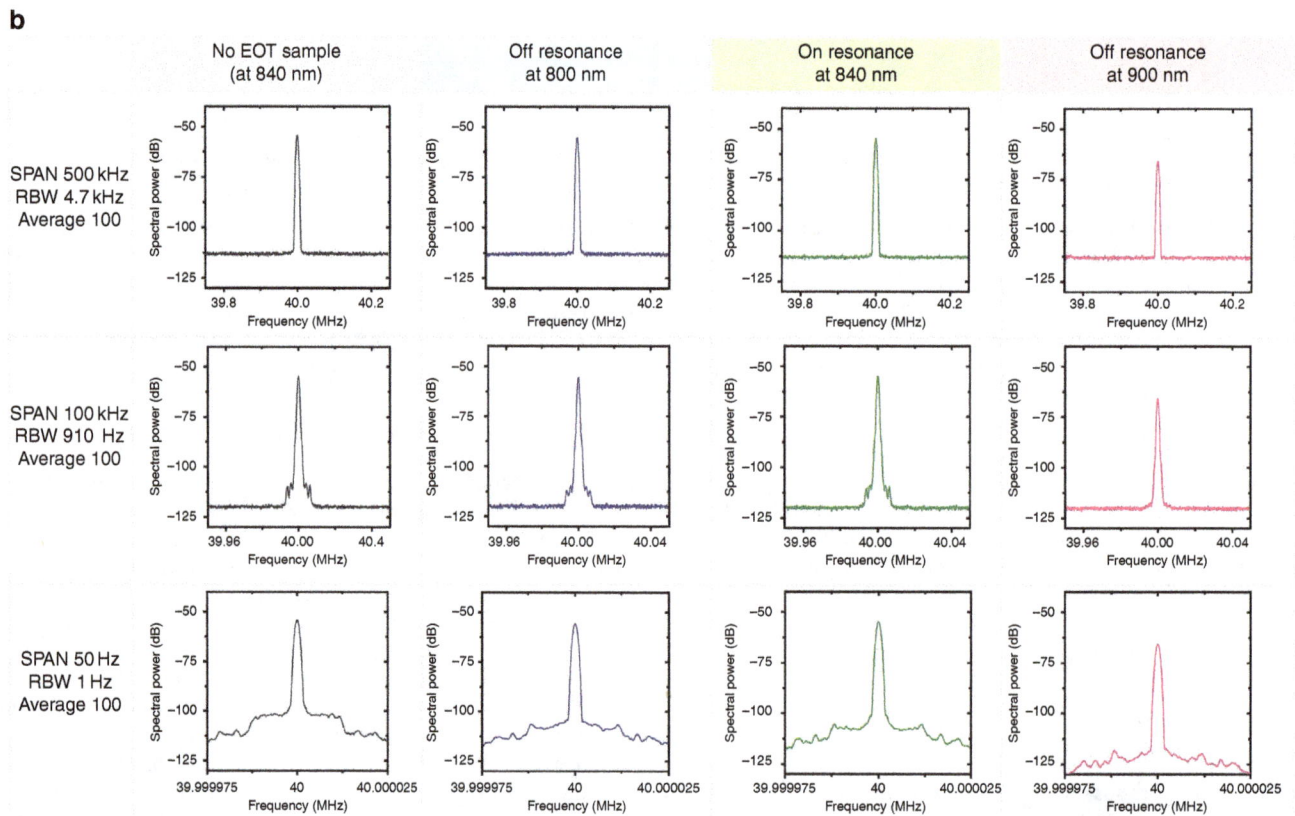

Figure 3 | Evaluation of the plasmonic frequency comb by EOT. (**a**) Generation of RF beats by the interference between frequency-shifted (40 MHz) reference combs and plasmonic EOT combs. The beat spectra of plasmonically transmitted frequency comb and the reference comb are measured at three wavelengths: one at a strong plasmonic resonance position (a 840-nm centre wavelength with a 10-nm bandwidth), two at off-resonance positions (a 800-nm centre wavelength with a 40-nm bandwidth and 900 nm with a 10-nm bandwidth) using three optical band-pass filters. These are compared with a beat spectrum at a 840-nm wavelength, acquired without the plasmonic sample. (**b**) Linewidth measurement of RF beats with different span, RBWs and VBWs. There was no noticeable linewidth degradation by the plasmonic transduction (<1 Hz, limited by RBW of the instrument). APD, avalanche photo-detector; OBPF, optical band-pass filter; VBWs, video bandwidths.

degradation in the frequency characteristics of the frequency comb by providing different plasmonic field distribution and enhancement. To test this, the beat spectrum was monitored while the sample was rotated by up to 45° (for transverse magnetic wave) as shown in Figure 2e. For the given condition, all frequency characteristics were maintained in the same level with

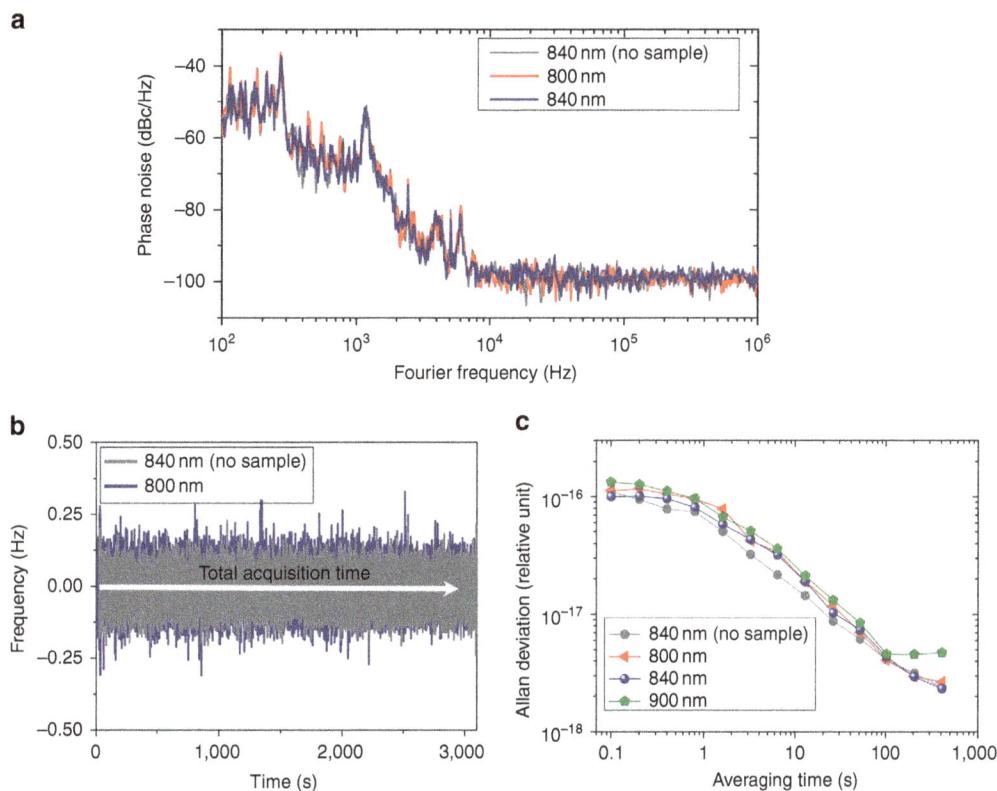

Figure 4 | Plasmonic frequency comb: phase noise and frequency stability. (**a**) Phase noise spectra at a 1.2-GHz RF carrier at on- and off-resonance wavelengths. (**b**) Time trace of the beat frequency with and without the plasmonic sample over 3,000 s. (**c**) Allan deviations of frequency stability with varying average time at the positions of on-resonance and off-resonance wavelengths.

normal incidence case, which shows that no performance degradation exist depending on plasmonic coupling or geometrical parameters of the sample.

The linewidth broadening by plasmonic EOT was evaluated to be <1 Hz, which is limited by RBW of the instrument in use (Fig. 3b). A single RF beat-frequency corresponds to the superposition of small RF beat contributions of $>10^4$ frequency comb modes, which proves that there is no significant wavelength-dependent frequency or phase noise during the plasmonic EOT. There was minor increase in spectral power in the pedestal peaks at 12, 17 and 21 Hz when the frequency comb passed through the plasmonic sample; this is expected to be caused by the vibrational and thermal noises at the plasmonic sample. The beat frequency, $f_{AOM} - \Delta f_{SP}$, was found to be exactly the same as the driving frequency of the AOM in all measured spectra shown in Fig. 3b, which implies that the absolute frequency position is well maintained in the plasmonic mode. The ambient temperature and vibration on EOT sample were not intentionally controlled so as to evaluate the performance in normal laboratory environment conditions. Our results show that the frequency comb structures are well maintained under environmental disturbances, for example, temperature variation, mechanical vibration and air fluctuation. This will enable us to develop high-sensitivity frequency-comb-referenced SP sensors working in harsh environments. The phase noise spectra in Fig. 4a also shows a number of minor peaks at 0.2, 1.5, 300 and 600 kHz other than the low-frequency spectral peaks at 12, 17 and 21 Hz observed in Fig. 3b. At higher frequency than 10 kHz, there is a flat noise floor without other spectral peaks or broad pedestals. In S/N ratio measurement, the S/N ratio theoretically could reach 68~75 dB in a 100 kHz RBW because there are $10^4 \sim 10^5$ frequency comb modes in the pass-band of the optical

filter transmittance. The experimental S/N ratio with the plasmonic sample on-resonance position was ~60 dB; this minor deviation could come from imperfect intensity balancing, polarization matching and spatial beam mode-matching. The S/N ratio at 900 nm was 54 dB, relatively lower than that at 800 nm because the quantum efficiency of the avalanche photodetector at 900 nm is ~20% lower than that at 800 nm and the filter bandwidth at 900 nm is 25% of that at 800 nm.

In this article, we have studied SP resonance effects on frequency comb structure in the plasmonic EOT of light through a subwavelength metallic nanohole array. The frequency comb was transduced to plasmonic mode in the sample and reverted to photonic mode without significant changes in linewidth, frequency shift, S/N ratio, phase noise and Allan deviation. The linewidth broadening was <1 Hz (instrument limited), frequency inaccuracy was 6.51×10^{-19}, S/N ratio was higher than 60 dB, Allan deviation increased by 2.92×10^{-19} at 100 s averaging time. This outstanding frequency comb performance in plasmonic nanostructures enables a highly sensitive, high accurate and broadband measurement with direct traceability to standards. This inclusion of frequency comb has the potential to accelerate progresses in various plasmonic applications such as bio-chemical spectroscopy or sensing, quantum optics and sub-diffraction-limit biomedical-imaging. With the aid of SP, frequency-comb-referenced high-speed coherent anti-stokes Raman spectroscopy[14] can be implemented in much smaller nanoscopic volume being requested for single-molecule detection, for example, surface-enhanced coherent anti-stokes Raman spectroscopy[29]. A large number of optical modes in a frequency comb as the time and frequency standard can be coupled at the same time with SP for broadband quantum metrology for entangled atomic qubits or information carrier in subwavelength scale[13,30].

Localized field enhancement of SP will enable highly efficient nonlinear optics[31] coupled with high precision of frequency comb, which is prerequisite for novel sub-diffraction-limit nonlinear biomedical imaging and spectroscopy.

Methods

Frequency comb. A Ti:sapphire femtosecond laser delivers 4.8 fs pulses at a repetition rate of 75 MHz over a broad spectral bandwidth from 1.03 to 2.06 eV (Venteon UB, Venteon). For establishing a frequency comb, the pulse repetition frequency (f_r) and carrier-envelope offset frequency (f_{ceo}) were precisely locked to a reference Rb atomic clock (FS725, Stanford Research Systems) with the aid of a f-$2f$ interferometer and phase-locked control loops (AVR32, TEM-Messtechnik & XPS800-E, Menlosystems). One part of the beam was diverted to and transmitted through an AOM for the frequency shift of 40 MHz to construct a reference frequency comb. If the plasmonic frequency comb suffers from the phase or frequency noise during the plasmonic mode conversion, the noise contributions should be observed at $f_{AOM} - \Delta f_{SP}$ in the forms of linewidth broadening, frequency shift, S/N ratio reduction, increased phase noise or higher Allan deviation. The frequency comb excited the plasmonic sample with the whole-broadband spectrum in a loose focusing geometry with an aspheric lens of 100 mm focal length. The focused peak intensity at the plasmonic sample was set to be < 0.1 MW cm^{-2} not to exceed the thermal damage threshold (~ 1 TW cm^{-2} for Au). Input polarization state was set linear and its direction is parallel to the x axis of periodic holes on plasmonic sample as denoted in Fig. 2e.

Plasmonic EOT sample: design and development. We exploited FDTD solution (XFDTD8.3, Lumerical) to solve Maxwell's equation for plasmonic near-field distribution and transmitted spectrum. Through a series of iterative computations, the optimal geometric parameters were determined as $d = 200$ nm, $l = 530$ nm and $t = 100$ nm. The thickness, t was designed to be much thicker than the Au skin depth (~ 20 nm) here to block the direct transmission through the Au film. The designed nanohole array was fabricated using electron-beam lithography (Raith 150) onto 25-nm-thick ITO-coated quartz substrate.

Evaluation of plasmonic frequency comb. For comparative analysis, interference signals were obtained at three different wavelength regimes – one at plasmonic on-resonance (840 nm) and the others at off-resonance positions (800 and 900 nm) – using optical band-pass filters. The resulting interference beat signal was obtained by high-speed avalanche photodiode and analysed using a high-resolution RF spectrum analyzer (N9020A, Agilent) and a RF frequency counter (53230A, Keysight Technologies). An exemplary RF spectrum is shown in Fig. 3b; the repetition rate (f_r) is located at 75 MHz, the beat frequency ($f_{AOM} - \Delta f_{sp}$) between the frequency-shifted reference frequency comb and the plasmonic frequency comb is at ~ 40 MHz, and the beat frequency ($f_r - f_{AOM} + \Delta f_{sp}$) between the other nearby reference frequency comb modes and the plasmonic frequency comb is at ~ 35 MHz. Other minor spurious peaks are due to the imperfect sinusoidal modulation of AOM; their positions match with beat frequencies between the f_{AOM}-harmonics and the reference frequency comb of $2f_{AOM} - f_r$, $2f_r - 3f_{AOM}$, $3f_{AOM} - f_r$ and $2(f_r - f_{AOM})$.

References

1. Hänsch, T. W. Nobel Lecture: passion for precision. *Rev. Mod. Phys.* **78**, 1297–1309 (2006).
2. Hall, J. L. Nobel Lecture: defining and measuring optical frequencies. *Rev. Mod. Phys.* **78**, 1279–1295 (2006).
3. Steinmetz, T. *et al.* Laser frequency combs for astronomical observations. *Science* **321**, 1335–1337 (2008).
4. Wilken, T. *et al.* A spectrograph for exoplanet observations at the centimetre-per-second level. *Nature* **485**, 611–614 (2012).
5. Predehl, K. *et al.* A 920-kilometer optical fiber link for frequency metrology at the 19th decimal place. *Science* **336**, 441–444 (2012).
6. Giorgetta, F. R. *et al.* Optical two-way time and frequency transfer over free space. *Nat. Photon.* **7**, 434–438 (2013).
7. Newbury, N. R. Searching for applications with a fine-tooth comb. *Nat. Photon.* **5**, 186–188 (2011).
8. Minoshima, K. & Matsumoto, H. High-accuracy measurement of 240-m distance in an optical tunnel by use of a compact femtosecond laser. *Appl. Opt.* **39**, 5512–5517 (2000).
9. Coddington, I., Swann, W. C., Nenadovic, L. & Newbury, N. R. Rapid and precise absolute distance measurements at long range. *Nat. Photon.* **3**, 351–356 (2009).
10. Lee, J., Kim, Y.-J., Lee, K., Lee, S. & Kim, S.-W. Time-of-flight measurement with femtosecond light pulses. *Nat. Photon.* **4**, 716–720 (2010).
11. Rosenband, T. *et al.* Frequency ratio of Al+ and Hg+ single-ion optical clocks; metrology at the 17th decimal place. *Science* **319**, 1808–1812 (2008).
12. Reinhardt, S. *et al.* Test of relativistic time dilation with fast optical atomic clocks at different velocities. *Nat. Phys.* **3**, 861–864 (2007).
13. Hayes, D. *et al.* Entanglement of atomic qubits using an optical frequency comb. *Phys. Rev. Lett.* **104**, 140501 (2010).
14. Ideguchi, T. *et al.* Coherent Raman spectro-imaging with laser frequency combs. *Nature* **502**, 355–358 (2013).
15. Stockman, M. Nanoplasmonics: past, present, and glimpse into future. *Opt. Express* **19**, 22029–22106 (2011).
16. Kauranen, M. & Zayats, A. V. Nonlinear plasmonics. *Nat. Photon.* **6**, 737–748 (2012).
17. Willets, K. A. & Van Duyne, R. P. Localized surface plasmon resonance spectroscopy and sensing. *Annu. Rev. Phys. Chem.* **58**, 267–297 (2007).
18. Zhang, Y. *et al.* Coherent anti-Stokes Raman scattering with single-molecule sensitivity using a plasmonic Fano resonance. *Nat. Commun.* **5**, 4424 (2014).
19. Tame, M. S. *et al.* Quantum plasmonics. *Nat. Phys.* **9**, 329–340 (2013).
20. Melikyan, A. *et al.* High-speed plasmonic phase modulators. *Nat. Photon.* **8**, 229–233 (2014).
21. Haffner, C. *et al.* All-plasmonic Mach–Zehnder modulator enabling optical high-speed communication at the microscale. *Nat. Photon.* **9**, 525–528 (2015).
22. Dennis, B. S. *et al.* Compact nanomechanical plasmonic phase modulators. *Nat. Photon.* **9**, 267–273 (2015).
23. Ebbesen, T. W., Lezec, H. J., Chaemi, H. F., Thio, T. & Wolff, P. A. Extraordinary optical transmission through sub-wavelength hole arrays. *Nature* **391**, 667–669 (1998).
24. Barnes, W. L., Dereux, A. & Ebbesen, T. W. Surface plasmon subwavelength optics. *Nature* **424**, 824–830 (2003).
25. Wurtz, G. A. & Zayats, A. V. Nonlinear surface plasmon polaritonic crystals. *Laser Photon. Rev.* **2**, 125–135 (2008).
26. Garcia-Vidal, F. J., Martin-Moreno, L., Ebbesen, T. W. & Kuipers, L. Light passing through subwavelength apertures. *Rev. Mod. Phys.* **82**, 729–787 (2010).
27. Newbury, N. R. & Swann, W. C. Low-noise fiber-laser frequency combs. *J. Opt. Soc. Am. B* **24**, 1756–1770 (2007).
28. Yue, W. *et al.* Enhanced extraordinary optical transmission (EOT) through arrays of bridged nanohole pairs and their sensing applications. *Nanoscale* **6**, 7917–7923 (2014).
29. Steuwe, C., Kaminski, C. F., Baumberg, J. J. & Mahajan, S. Surface enhanced coherent anti-Stokes Raman scattering on nanostructured gold surfaces. *Nano Lett.* **11**, 5339–5343 (2011).
30. Altewischer, E., Van Exter, M. P. & Woerdman, J. P. Plasmon-assisted transmission of entangled photons. *Nature* **418**, 304–306 (2002).
31. Almeida, E. & Prior, Y. Rational design of metallic nanocavities for resonantly enhanced four-wave mixing. *Sci. Rep.* **5**, 10033 (2015).

Acknowledgements

This work was supported by the Basic Science Research Program (NRF-2013R1A1A2004932; NRF-2014R1A1A1004885), by Global Research Laboratory Program (Grant No. 2009-00439), by the Leading Foreign Research Institute Recruitment Program (Grant No. 2010-00471), by the Max Planck POSTECH/KOREA Research Initiative Program (Grant No. 2011-0031558), by the NRF Grant (No. 2010-0021735), by the Leading Foreign Research Institute Recruitment Program (Grant No. 2012K1A4A3053565) through the NRF funded by the MEST. This work was also supported by a Grant (14CTAP-C077584-01) from Infrastructure and Transportation Technology Promotion Research Program funded by Ministry of Land, Infrastructure and Transport of Korean government. This work was also supported by Singapore National Research Foundation (NRF-NRFF2015-02) and Singapore Ministry of Education under its Tier 1 Grant (RG85/15).

Author contributions

The project was planned and overseen by Y.-J.K., D.E.K. and S.K. Plasmonic sample was prepared and characterized by J.H.S., H.Y. and K.S. Frequency comb experiments were performed by X.T.G., B.J.C., Y.-J.K. and S.K. All authors contributed to the manuscript preparation.

Additional information

Competing financial interests: The authors declare no competing financial interests.

6

Amplification and generation of ultra-intense twisted laser pulses via stimulated Raman scattering

J. Vieira[1], R.M.G.M. Trines[2], E.P. Alves[1], R.A. Fonseca[1,3], J.T. Mendonça[1], R. Bingham[2,4], P. Norreys[2,5] & L.O. Silva[1]

Twisted Laguerre-Gaussian lasers, with orbital angular momentum and characterized by doughnut-shaped intensity profiles, provide a transformative set of tools and research directions in a growing range of fields and applications, from super-resolution microcopy and ultra-fast optical communications to quantum computing and astrophysics. The impact of twisted light is widening as recent numerical calculations provided solutions to long-standing challenges in plasma-based acceleration by allowing for high-gradient positron acceleration. The production of ultra-high-intensity twisted laser pulses could then also have a broad influence on relativistic laser–matter interactions. Here we show theoretically and with *ab initio* three-dimensional particle-in-cell simulations that stimulated Raman backscattering can generate and amplify twisted lasers to petawatt intensities in plasmas. This work may open new research directions in nonlinear optics and high–energy-density science, compact plasma-based accelerators and light sources.

[1]GoLP/Instituto de Plasmas e Fusão Nuclear, Instituto Superior Técnico, Universidade de Lisboa, 1049-001 Lisbon, Portugal. [2]Central Laser Facility, STFC Rutherford Appleton Laboratory, Didcot OX11 0QX, UK. [3]DCTI/ISCTE Lisbon University Institute, 1649-026 Lisbon, Portugal. [4]Department of Physics, 107 Rottenrow East, Glasgow G4 0NG, UK. [5]Department of Physics, University of Oxford, Oxford OX1 3PU, UK. Correspondence and requests for materials should be addressed to J.V. (email: jorge.vieira@ist.utl.pt).

The seminal work by Allen *et al.*[1] on lasers with orbital angular momentum (OAM) has initiated a path of significant scientific developments that can potentially offer new technologies in a growing range of fields, including microscopy[2] and imaging[3], atomic[4] and nano-particle manipulation[5], ultra-fast optical communications[6,7], quantum computing[8] and astrophysics[9]. At intensities beyond material breakdown thresholds, it has been recently shown through theory and simulations that intense (with $\gtrsim 10^{18}$ W cm^{-2} intensities) and short (with 10–100 fs durations) twisted laser beams could also excite strongly nonlinear plasma waves suitable for high-gradient positron acceleration in plasma accelerators[10]. As a result of their importance, many techniques have emerged to produce Laguerre–Gaussian lasers over a wide range of frequencies[11]. Common schemes use spiral phase plates or computer-generated holograms to generate visible light with OAM, nonlinear optical media for high-harmonic generation and emission of XUV OAM lasers[12,13] or spiral electron beams in free-electron lasers to produce OAM X-rays[14,15].

Optical elements such as spiral phase plates are designed for the production of laser beams with pre-defined OAM mode contents. Novel and more flexible mechanisms capable of producing and amplifying beams with arbitrary, well-defined OAM states, using a single optical component, would then be interesting from a fundamental point of view, while also benefiting experiments where OAM light is relevant. In addition, the possibility of extending these mechanisms to the production and amplification of laser pulses with relativistic intensities, well above the damage thresholds of optical devices, could also open exciting perspectives for high-energy-density science and applications. The use of a plasma as the optical medium is a potential route towards the production of OAM light with relativistic intensities. Although other routes may be used to produce high-intensity OAM laser pulses, for instance, by placing spiral phase plates either at the start or at the end of a laser amplification chain[16,17], the use of plasmas can potentially lead to the amplification of OAM light to very high powers and intensities. Plasmas also allow for greater flexibility in the level of OAM in the output laser beam than other more conventional techniques.

Here we show that stimulated Raman scattering processes in nonlinear optical media with a Kerr nonlinearity can be used to generate and to amplify OAM light. Plasmas, optical fibres and nonlinear optical crystals are examples of nonlinear optical media with Kerr nonlinearity. Although optical parametric oscillators have also been used to transfer OAM from a pump to down converted beams[18], here we explore the creation of new OAM states absent from the initial configuration, according to simple selection rules. We also demonstrate that stimulated Raman scattering processes can generate and amplify OAM light even in scenarios where no net OAM is initially present. To this end, we use an analytical theory, valid for arbitrary transverse laser field envelope profiles, complemented by the first three-dimensional (3D) *ab initio* particle-in-cell (PIC) simulation of the process using the PIC code OSIRIS[19], considering that the optical medium is a plasma. Starting from recent experimental and theoretical advances[20–22], our simulations and theoretical developments show that stimulated Raman processes could pave the way to generate OAM light in nonlinear optical media and that the nonlinear optics of plasmas[23,24] could provide a path to generate and amplify OAM light to relativistic intensities[25–28].

Results

Theoretical model. We illustrate our findings considering that the nonlinear optical medium is a plasma. Extension to other materials is straightforward. In a plasma, stimulated Raman

backscattering is a three-wave mode coupling mechanism in which a pump pulse (frequency ω_0 and wavenumber k_0), decays into an electrostatic, or Langmuir, plasma wave (frequency ω_p and wavenumber $2k_0 - \omega_p/c$) and into a counter-propagating seed laser (frequency $\omega_1 = \omega_0 - \omega_p$ and wavenumber $k_1 = \omega_p/c - k_0$). The presence of OAM in the pump and/or seed results in additional matching conditions that ensure the conservation of the angular momentum carried by the pump when the pump itself decays into a scattered electromagnetic wave and a Langmuir wave[29]. These additional matching conditions, which are explored in more detail in Supplementary Notes 1–4 and Supplementary Figs 1–3), correspond to selection rules for the angular momentum carried by each laser and plasma wave. Here we illustrate key properties of OAM generation and amplification by exploring different seed and pump configurations.

In order to derive a model capable of predicting stimulated Raman scattering OAM selection rules, we start with the general equations describing stimulated Raman scattering, given by $D_0 \mathbf{A}_0 = \omega_p^2 \delta n \mathbf{A}_1$, $D_1 \mathbf{A}_1 = \omega_p^2 \delta n^* \mathbf{A}_0$ and $D_p \delta n = e^2 k_p^2 / (2m_e^2) \mathbf{A}_0 \cdot \mathbf{A}_1$, where $D_{0,1} = c^2 (\nabla_\perp^2 \pm 2ik_{0,1}\partial_z) + 2i\omega_{0,1}\partial_t$, $D_p = 2i\omega_p\partial_t$, and where the minus $(-)$ sign is used to describe the seed pulse evolution. Moreover, $\mathbf{A}_{0,1}$ is the envelope of the pump/seed laser, with complex amplitude, given by $\mathbf{A}_1(t, \mathbf{r}_\perp)\exp[ik_{0,1}z - i\omega_{0,1}t] + \text{c.c.}$, where t is the time and z the propagation distance. We note that $\mathbf{A}_{0,1}$ are arbitrary functions of the transverse coordinate $\mathbf{r}_\perp$. The complex amplitude of the plasma density perturbations is $\delta n(t, \mathbf{r}_\perp)\exp[ik_p z - i\omega_p t] + \text{c.c.}$, where $k_p = \omega_p/c$ is the plasma wavenumber, $\omega_p = \sqrt{e^2 n_0/\epsilon_0 m_e}$ the plasma frequency, m_e the mass of the electron, ϵ_0 the vacuum electric permittivity and e the elementary charge. Although these general equations can be used to retrieve the selection rules for the OAM that will be explored throughout this paper, it is possible to derive exact solutions in the long pulse limit, where $k_{0,1}\partial_z \ll \omega_{0,1}\partial_t$ and in the limit where the pump laser contains much more energy than the seed laser energy. In this case, since $\partial_t \mathbf{A}_0^2 \sim \partial_t \mathbf{A}_1^2$, and $\mathbf{A}_0 \gg \mathbf{A}_1$ (pump has more energy than seed), then $\partial_t \mathbf{A}_0 \ll \partial_t \mathbf{A}_1$ (this condition is strictly satisfied in our simulations when new modes are created and until their energy becomes comparable to the energy in the pump pulse). In this case, it is possible to show that stimulated Raman scattering of a seed beam $\mathbf{A}_1$ from a pump beam $\mathbf{A}_0$ creates a plasma wave density perturbation, given by:

$$\delta n^*(t, \mathbf{r}_\perp) = i \frac{e^2 k_p^2}{4\omega_p m_e^2} \left[\mathbf{A}_0^* \cdot \mathbf{A}_1(t=0)\right] \frac{\sinh(\Gamma t)}{\Gamma}, \quad (1)$$

$$\Gamma^2(\mathbf{r}_\perp) = \frac{e^2 k_p^2 \omega_p^2}{8\omega_p \omega_1 m_e^2} |\mathbf{A}_0|^2, \quad (2)$$

where Γ is the growth rate at which the plasma amplitude grows as the interaction progresses and $\mathbf{r}_\perp$ is the transverse position. The amplification of the seed is given by:

$$\mathbf{A}_1(t, \mathbf{r}_\perp) = \left(\mathbf{A}_1(t=0) \cdot \frac{\mathbf{A}_0^*}{|\mathbf{A}_0|}\right) \frac{\mathbf{A}_0}{|\mathbf{A}_0|} \cosh(\Gamma t) + C, \quad (3)$$

where (ω_1, k_1) are the frequency and wavenumber of the seed laser pulse, respectively, and C is a constant of integration. The derivation of equations (1)–(3), presented in detail in Supplementary Note 1, assumes that the pump and the seed satisfy the frequency matching conditions stated above, being valid for arbitrary transverse laser envelope profiles as long as the paraxial equation is satisfied. Neglecting pump depletion does not change the selection rules for the OAM, as discussed in the remainder of this work. Unless explicitly stated, the generic expression for the pump vector potential (or electric field) is

$\mathbf{A}_0 = A_{0x} \exp(i\ell_{0x}\phi)\mathbf{e}_x + A_{0y} \exp(i\ell_{0y}\phi)\mathbf{e}_y$, where $(\mathbf{e}_x, \mathbf{e}_y)$ are the unit vectors in the transverse x and y directions, and ϕ the azimutal angle. Similarly, the generic expression for the seed vector potential is $\mathbf{A}_1 = A_{1x}\exp(i\ell_{1x}\phi)\mathbf{e}_x + A_{1y}\exp(i\ell_{1y}\phi)\mathbf{e}_y$. Selection rules can then be generally derived by inserting these expressions into the factor $(\mathbf{A}_1(t=0) \cdot \mathbf{A}_0^*)\mathbf{A}_0$ in equation (3). Although we have assumed that the plasma is the optical medium, other nonlinear optical media with Kerr nonlinearity will also exhibit similar phenomena.

PIC simulations. We will now use equation (3) to explore OAM generation and amplification in three separate classes of initial set-ups, all identified in Fig. 1. We start by studying the case of the amplification of existing OAM modes. Figure 1a illustrates the process in a set-up leading to the amplification of a seed in an arbitrary, single state of OAM ℓ_1 in a plasma using a counter-propagating Gaussian pump laser without OAM. The mechanism is trivially generalized for a pump with arbitrary OAM ℓ_0. We can then assume a pump linearly polarized in the x direction with OAM ℓ_{0x}, which decays into a Langmuir plasma wave with OAM $\ell_p = \ell_{0x} - \ell_{1x}$ and into a seed, also linearly polarized in the x direction with OAM ℓ_{1x}. Making these substitutions into equation (3) confirms that the amplification of A_1 retains the initial seed OAM. For the specific example in Fig. 1a, where $\ell_{0x} = 0$, direct substitution of $\mathbf{A}_0 \sim \exp(i\ell_{0x}\phi)\mathbf{e}_x$ and $\mathbf{A}_1 \sim \exp(i\ell_{1x}\phi)\mathbf{e}_x$ in equation (1) shows that the plasma wave density perturbations $\delta n \sim \exp[i(\ell_{0x} - \ell_{1x})\phi]$ have OAM $\ell_p = \ell_{0x} - \ell_{1x} = -\ell_{1x}$, that is, the plasma wave absorbs the excess OAM that may exist between the pump and the seed (see Supplementary Note 2 and Supplementary Fig. 1 for several examples illustrating angular and linear momentum matching conditions, demonstrating that the plasma wave always absorbs the excess OAM between pump and seed.) The scheme is thus ideally suited to amplify an existing OAM seed using a long Gaussian pump without OAM. The amplification of circularly polarized OAM lasers, with both spin and OAM, obeys similar selection rules. For amplification to occur in this case, and similar to stimulated Raman backscattering of circularly polarized Gaussian lasers, both seed and pump need to be polarized with the same handedness either in $\mathbf{e}_+ = \mathbf{e}_x + i\mathbf{e}_y$ or in $\mathbf{e}_- = \mathbf{e}_x - i\mathbf{e}_y$.

Figure 2a illustrates 3D simulation results showing the amplification of an $\ell_1 = 1$, linearly polarized seed from a linearly polarized Gaussian pump. Simulation parameters are stated in Table 1. Figure 2a shows that the growth rate for the amplification process is nearly indistinguishable from stimulated Raman amplification of Gaussian lasers. In agreement with equation (2), this result also indicates that, in general, the overall amplification process is OAM-independent.

Stimulated Raman scattering also provides a mechanism to create new OAM modes (that is, modes that are absent from the initial pump/seed lasers) and amplify them to very high intensities. Figure 1b illustrates the process schematically. The pump electric fields can have different OAM components in both transverse directions x and y. Each component is represented in blue and orange in Fig. 1b. The pump electric field component in x has OAM ℓ_{0x}. The pump electric field component in y has OAM ℓ_{0y}. The initial seed electric field contains an OAM ℓ_{1x} component in the x direction. After interacting in the plasma, the pump becomes depleted and a new electric field component appears in the seed with OAM, given by $\ell_{1y} = \ell_{1x} + \ell_{0y} - \ell_{0x}$.

The process can be physically understood by examining the couplings between the plasma and light waves in the example considered above. Initially, a plasma wave will be excited due to beating pump and seed components that have their electric fields pointing in the transverse x direction. According to equation (1), the plasma wave OAM is $\ell_p = \ell_{0x} - \ell_{1x}$. This plasma wave ensures OAM conservation for the pump and seed electric field components in the x direction. The (same) plasma wave also couples the pump and seed modes with electric field components pointing in the y direction. Thus, $\ell_p = \ell_{0y} - \ell_{1y}$ must also hold in order to ensure conservation of angular momentum. This implies the generation of a new seed component with electric field polarized in y so that OAM is conserved at all times and in both components. The OAM of the new seed component is thus $\ell_{1y} = \ell_{0y} - \ell_p = \ell_{1x} - \ell_{0x} + \ell_{0y}$.

Alternatively, this selection rule can also be found by examining equation (3). Direct substitution of a pump profile with $\mathbf{A}_0 \sim \exp(i\ell_{0x}\phi)\mathbf{e}_x + \exp(i\ell_{0y}\phi)\mathbf{e}_y$ and of an initial seed profile with $\mathbf{A}_1 \sim \exp(i\ell_{1x}\phi)\mathbf{e}_x$ then leads to the generation of a new seed with $\exp[i(\ell_{1x} + \ell_{0y} - \ell_{0x})\phi]\mathbf{e}_y$. The same selection rules would also hold if the pump consists of combination of a right- and left-handed circularly polarized modes, each with different OAM, and the seed initially contains only a left- or right-handed circularly polarized mode. In this case, a new seed component would appear with right- or left-handed circular polarization. The new mode is created to ensure conservation of OAM. The selection rules are identical as long as polarizations

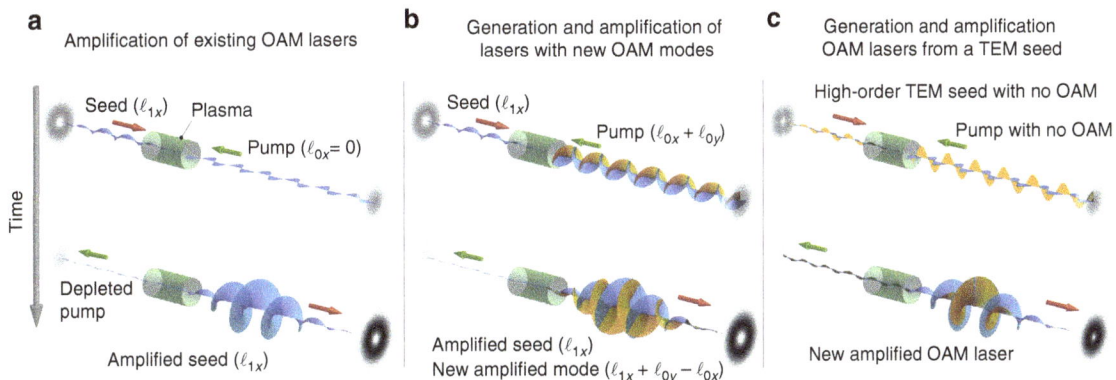

Figure 1 | Generation and amplification of OAM lasers via stimulated Raman backscattering. In Raman amplification, a long pump laser transfers its energy to a short, counter-propagating seed laser in a plasma. The process depletes the pump laser pulse energy and enhances the intensity of the seed laser. The seed/pump lasers propagate in the direction of the red/green arrow, respectively. Polarization in the x/y direction is represented by blue/orange lasers, respectively. The position of the plasma, relative to the seed and pump lasers, is shown by the green cylinders. The back/front projections show the intensity profile of the closest laser. (**a**) A set-up leading to the amplification of a seed with OAM. (**b**) The generation and amplification of new OAM modes. (**c**) The generation and amplification of a new OAM laser in a configuration with no initial OAM.

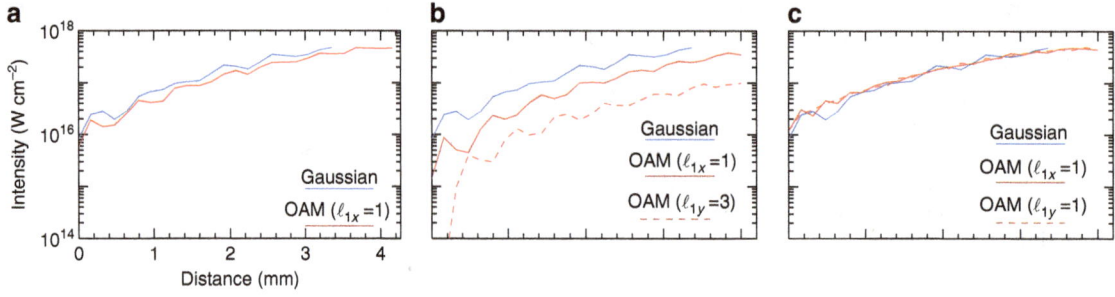

Figure 2 | Simulation results showing the generation and amplification of OAM lasers. Blue refers to the amplification of a Gaussian seed by a Gaussian pump. The initial laser configuration in each panel (**a–c**) corresponds to the initial set-up illustrated by each corresponding panel (**a–c**) in Fig. 1. (**a**) The amplification of a seed with $\ell_{1x} = 1$ using a long Gaussian pump (red). (**b**) The generation and amplification of a new OAM mode with $\ell_{1y} = 3$ (dashed red) and of an existing mode with $\ell_{1x} = 1$ (solid red) from an OAM pump polarized in two directions with $\ell_{0y} = 2$ and $\ell_{0x} = 0$. (**c**) The amplification of a new OAM mode with $\ell_1 = 1$ from a TEM seed with no net initial OAM and from a Gaussian pump. See Table 1 for simulation parameters.

Table 1 | Laser parameters for the different Raman Amplification regimes to generate and amplify OAM lasers.

| | Amplification of existing modes | | Generation and amplification of new OAM modes | | | |
| | Pump | Seed | OAM seed | | TEM seed | |
			Pump	Seed	Pump	Seed
TEM	—	—	—	—	$TEM_{00}\mathbf{e}_x + TEM_{00}\mathbf{e}_y$	$TEM_{01}\mathbf{e}_x + i\,TEM_{10}\mathbf{e}_y$
OAM	$L_{00}\mathbf{e}_x$	$L_{01}\mathbf{e}_x$	$L_{00}\mathbf{e}_x + L_{02}\mathbf{e}_y$	$L_{01}\mathbf{e}_x$	—	—
a_0 (peak)	0.02	0.06	0.02	0.03	0.02	0.08
Spot (µm)	718	435	718	435	718	435
Duration (fs)	25×10^3	25	25×10^3	25	25×10^3	25
w_0/w_p	20	19	20	19	20	19

Simulation parameters are close to ideal Raman amplification regimes determined in ref. 21. In all simulations, the probe has a central wavelength of 1 µm. The background plasma density is $n_0 = 4.3 \times 10^{18}\,cm^{-3}$ for all simulations presented. When the pump/probe initially has components in both the transverse directions, the initial peak a_0, spot size and durations present in the table are identical for every component. L_n^ℓ refers to a Laguerre–Gaussian (n, ℓ) mode, where the OAM corresponds to the index ℓ. TEM_{mn} correspond to Hermite–Gaussian lasers with order (m, n). The table only describes the initial simulation conditions.

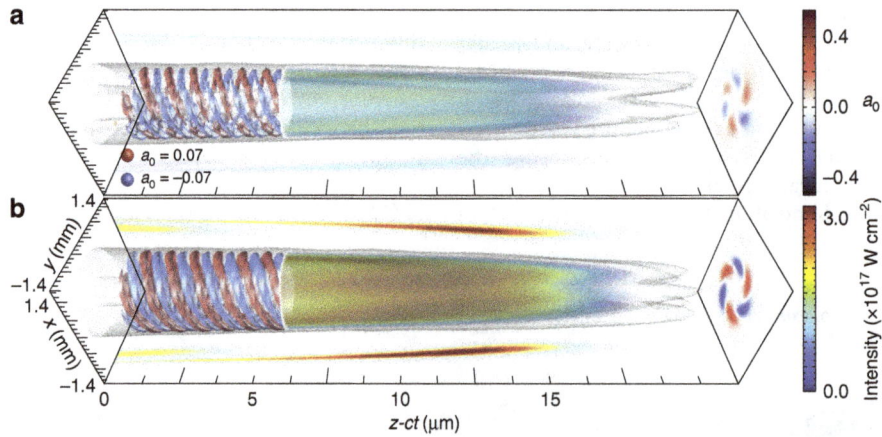

Figure 3 | Simulation results showing the generation and amplification of a new OAM modes. The new mode with $\ell_{1y} = 3$ grow from a seed with $\ell_{1x} = 1$ and a linearly polarized pump with a Gaussian profile in the x direction, and with an OAM $\ell_{0y} = 2$ in the y direction. $z = 2$ mm (**a**) and $z = 6.22$ mm (**b**). The initial set-up is illustrated in Fig. 1b. Projections in the (x,z-ct) and (y,z-ct) planes show intensity profile slices at the mid-plane of the OAM mode (blue–green–red colours). Projections in the (x,y) plane (blue–white–red) show the normalized vector potential (a_0) field envelope of the new OAM mode at the longitudinal slice where the laser intensity is maximum. The envelope of the 3D laser intensity is also shown for $z > 6.25$ mm in blue–green–red colours, and normalized vector potential isosurfaces for $z < 6.25$ mm in blue and red.

in $\mathbf{e}_x/\mathbf{e}_y$ are replaced by polarizations in $\mathbf{e}_+/\mathbf{e}_-$ (where $\mathbf{e}_\pm = \mathbf{e}_x \pm i\mathbf{e}_y$). This set-up provides a robust mechanism for the production and amplification of a new and well-defined OAM mode, absent from the initial set of lasers. The generation of a new OAM mode when a linearly polarized seed interacts with a pump with electric field components in the two orthogonal

directions is also illustrated in Supplementary Note 3 and Supplementary Fig. 2.

Figure 3 shows a result of a 3D PIC simulation illustrating the production of a new seed mode with $\ell_{1y} = 3$, which is initially absent from the simulation, from a pump with $\ell_{0y} = 2$, $\ell_{0x} = 0$ and an initial seed with $\ell_{1x} = 1$ (simulation parameters given in

Table 1). Figure 3 presents several distinct signatures of the new OAM mode with $\ell_{1y} = 3$. The laser vector potential shows helical structures, which indicate that the new mode has OAM. The normalized vector potential forms a pattern that repeats each 3 turns, which turn in the clockwise direction from the front to the back of the pulse, a signature for $\ell_{1y} = 3$. Field projections in the (x,y) plane also show a similar pattern further confirming that the new OAM mode has $|\ell_{1y}| = 3$. The change in colour from blue–green in (a) to (green–red) in (b) is a clear signature for the intensity amplification of the new seed mode. Intensity was calculated using $I(\text{W/cm}^2) = 1.27 \times 10^{18} a_0^2 / \lambda_0^2 (\mu m)$, where $\lambda_0 = 0.8\,\mu m$ is the central laser wavelength. Figure 3 also shows that the amplified laser envelope acquires a bow-shaped profile[21,22,30], a key signature of Raman amplification identified in ref. 21. In agreement with theory (equations (2) and (3)), Fig. 2b shows that the new $\ell_{1y} = 3$ mode and the existing $\ell_{1x} = 1$ OAM mode amplify at nearly coincident growth rates. Still, since it grows from initially higher intensities, the existing $\ell_{1x} = 1$ mode reaches higher final intensities than the new OAM mode with $\ell_{1y} = 3$. The generation of the new modes in Fig. 2b also illustrates the transition from the regime where the depletion of the pump is negligible (small signal and exponential growth) to the regime where the depletion of the pump is not negligible (strong signal and linear growth). Hence, Fig. 2b shows an exponential growth of the new mode up to $z < 0.6$ mm. For $z > 0.6$ mm, the energy in the new seed becomes comparable to the energy contained in the pump pulse. As a result, the growth slows down significantly, becoming linear with the propagation distance[26].

L. Allen et al.[1] showed that particular superpositions of Hermite–Gaussian modes (also called transverse electro-magnetic or TEM modes) are mathematically equivalent to Laguerre–Gaussian modes. Since the transverse amplitude distribution of high-order (transverse) laser modes is usually described by a product of Hermite–Gaussian polynomials, which is also usually associated with TEM modes, this result paved the way for experimental realization of vortex light beams with OAM from existing TEM laser modes. It is thus interesting and important to explore whether and how stimulated Raman backscattering can be used to generate and amplify light with OAM from Hermite–Gaussian laser beams, that is, from initial configurations with no net OAM. Figure 1c illustrates the process. From now on, we refer

to each Hermite–Gaussian beam as a $\text{TEM}_{m,n}$ laser, where (m,n) represents the Hermite–Gaussian mode. The TEM mode electric field is given by equation (5) (see Methods section). We consider first a Gaussian pump linearly polarized at 45°, that is, having similar electric field amplitudes in both transverse directions x and y. The Gaussian pump can then also be written as $\mathbf{A}_0 \sim \text{TEM}_{00}(\mathbf{e}_x + \mathbf{e}_y)$. In addition, we assume a seed with a TEM_{10} mode electric field component in x and with a TEM_{01} electric field component in y. The two seed modes are $\pi/2$ out of phase with respect to each other. The seed is given by $\mathbf{A}_1 \sim \text{TEM}_{10}\mathbf{e}_x + i\text{TEM}_{01}\mathbf{e}_y$. This set-up is represented in Fig. 1c, where blue and orange colours refer to the pump and seed components polarized in the x and y directions, respectively.

Although this set-up has no initial OAM, since both pump and seed have no OAM, it results in the generation and amplification of an OAM mode with $\ell_1 = 1$. In order to understand the OAM generation mechanism, we first consider equation (1). According to equation (1), the beating between the TEM_{10} seed with the Gaussian pump in the x direction will drive a TEM_{10} daughter plasma wave component. The beating between the $i\text{TEM}_{01}$ seed and Gaussian pump in the y direction will drive a $i\text{TEM}_{01}$ daughter plasma wave component. These two plasma wave components are $\pi/2$ out of phase with respect to each other. Hence, the resulting plasma wave will be a combination of TEM modes given by $\delta n \sim \text{TEM}_{10} - i\text{TEM}_{01}$, where the i denotes the phase difference between modes. According to ref. 1, this mode combination is equivalent to a Laguerre–Gaussian mode with $\ell_p = -1$. In order to conserve angular momentum, a new seed component with $\ell_1 = 1$ will then have to be generated and amplified in the direction of polarization of the pump. This process is also illustrated in Supplementary Note 4 and Supplementary Fig. 3.

It is also possible to reach this conclusion by substituting in equation (3) the expressions for initial Gaussian pump transverse profile $[\mathbf{A}_0 \sim \text{TEM}_{00}(\mathbf{e}_x + \mathbf{e}_y)]$ and initial TEM seed profile $(\mathbf{A}_1 \sim \text{TEM}_{10}\mathbf{e}_x + i\text{TEM}_{01}\mathbf{e}_y)$. This substitution yields a new seed transverse profile given by $\mathbf{A}_1 \sim (\text{TEM}_{10} + i\text{TEM}_{01})(\mathbf{e}_x + \mathbf{e}_y)$, corresponding to a Laguerre–Gaussian mode with $\ell_1 = 1$ (ref. 1). Similarly, a circularly polarized Gaussian pump and a $\mathbf{A}_1 \sim \text{TEM}_{10}\mathbf{e}_x + \text{TEM}_{01}\mathbf{e}_y$ seed (that is, without phase difference between the TEM_{10} and TEM_{01} modes) would also lead to a new

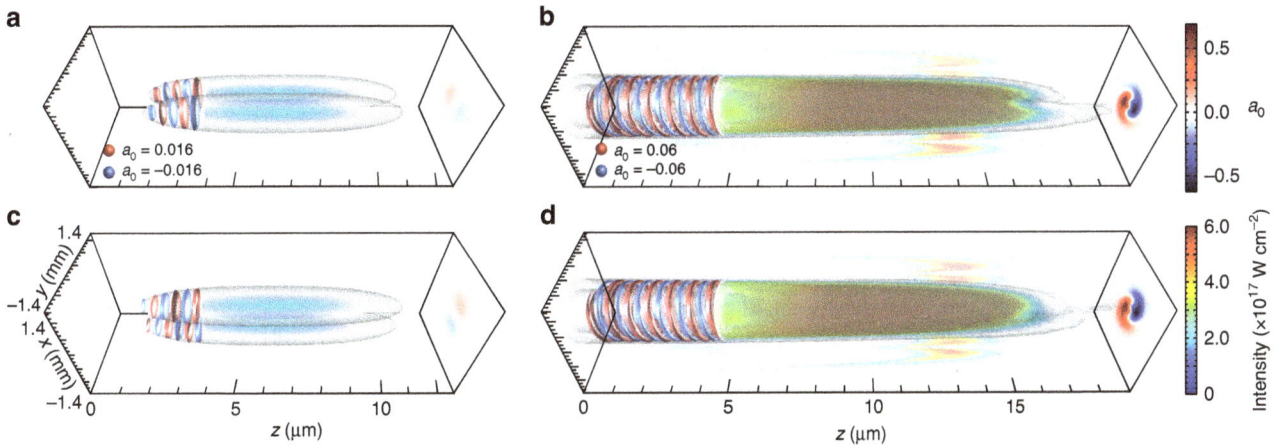

Figure 4 | Simulation result showing the generation and amplification of a new OAM mode from initial configurations with no net OAM. The new mode is linearly polarized in x and in y with $\ell_{1x} = \ell_{1y} = -1$ from an initial seed polarized in the x direction with a TEM_{01} mode and in the y direction with a TEM_{10} mode that is $\pi/2$ out of phase with respect to the TEM_{01} mode polarized in x. The pump is a Gaussian laser linearly polarized at 45°. The initial laser set-up corresponds to Fig. 1c. The meaning of the colour scales and physical quantities plotted in all panels are identical to Fig. 3. The values of the laser vector potential illustrated by the isosurfaces of **a,c** are shown by the spheres in **a**. Those iso-surface values for **b,d** are indicated in **b**. (**a,c**) The initial seed TEM modes in x- and y directions, respectively. (**b,d**) The new OAM mode electric field components at $z = 3.5$ mm in x (**b**) and y (**d**).

seed component with $|\ell_1| = 1$. The plasma can then be viewed as a high-intensity mode converter.

Figure 4 shows results from a 3D simulation that confirms these predictions (see Table 1 for simulation parameters). The simulation set-up follows the example of Fig. 1c described earlier. Simulations show that stimulated Raman scattering leads to a new OAM mode with $\ell_1 = 1$ linearly polarized at $45°$. Figure 2c shows that the amplification rates are comparable to the other typical scenarios shown in Fig. 2a,b. The change on the field topology of the seed normalized vector potential shown in Fig. 4, from plane isosurfaces to helical isosurfaces, indicates the generation of a laser with OAM from a configuration with no net OAM. Normalized vector potential isosurfaces, and projection in the yz direction, form a pattern that repeats each turn and that rotates clockwise from the front to the back of the pulse, thereby indicating an OAM with $\ell_{1y} = 1$.

Discussion

We have so far assumed that the lasers are perfectly aligned. In experiments, however, the beams can only be aligned within a certain precision. In the presence of misalignments, our results will still hold as long as Raman side-scattering can be neglected, that is, when the angle between the two pulses is much smaller than $90°$. The k-matching conditions are then still satisfied in the presence of small misalignments because the nonlinear medium, a plasma in the case of our simulations, absorbs any additional transverse wave vector component. Thus, momentum is still locally conserved, thereby allowing for Raman backscatter processes (interestingly, we note that when using OAM beams, the wave vectors of seed and pump are already locally misaligned). Despite lowering the total interaction time, and possibly the final amplification level, a small angle between the seed and pump will not change the OAM selection rules and the overall physics of stimulated Raman scattering.

We note that our seed laser pulse final intensity, on the order of $10^{17}\,\mathrm{W\,cm^{-2}}$, and seed laser spot size, on the order of 1 mm, indicate the production and amplification of Petawatt class twisted lasers with OAM. Additional simulations (not shown) revealed the generation and amplification of circularly polarized OAM modes using a scheme similar to that in Fig. 1b. Moreover, simulations showed that Raman amplification can also operate in the absence of exact frequency/wavenumber matching between seed and pump as long as the seed is short so that its Fourier components can still satisfy k- and ω-matching conditions.

Finally, we note that our results could be extended to other nonlinear optical media with Kerr nonlinearities. In a plasma, the coupling between seed and pump is through an electron Langmuir wave, which also ensures frequency, wavenumber and OAM matching conditions will hold. In other nonlinear optical media, molecular vibrations, for instance, would play the role of the plasma Langmuir wave. We note that the possibility of OAM transfer has been explored in solids[18]. Similar phenomenology as illustrated in this work could also be obtained in three-wave mixing processes, where an idler wave could play the role of the plasma Langmuir wave. One advantage of testing these set-ups in nonlinear Kerr optical media such as a crystal is that lasers with much lower intensities could be used (see Supplementary Note 5 for a discussion in nonlinear optical media with Kerr nonlinearity admitting three-wave interaction processes). The plasma, however, offers the possibility to amplify these lasers to very high intensities. This scheme could also be used in combination with optical pulse chirped pulse amplification to pre-generate and pre-amplify new OAM modes via stimulated Raman scattering before they enter the plasma to be further amplified. Similar configurations (for example,

stimulated Brillouin backscattering[31]) can also be envisaged to produce intense OAM light.

Methods

Set-up of numerical simulations and simulation parameters. Simulations have been performed using the massively parallel, fully relativistic, electro-magnetic PIC code OSIRIS[19]. In the PIC algorithm, spatial dimensions are discretized by a numerical grid. Electric and magnetic fields are defined in each grid cell and advanced through a finite difference solver for the full set of Maxwell's equations. Each cell contains macro-particles representing an ensemble of real charged particles. Macro-particles are advanced according to the Lorentz force. Since background plasma ion motion is negligible for our conditions, ions have been treated as a positively charged immobile background. The plasma was initialized at the front of the simulation box that moves at the speed of light c. Note that although the simulation are performed in a frame that moves at c, the moving window corresponds to a Galilean transformation of coordinates where all computations are still performed in the laboratory frame. The simulation box dimensions were $50 \times 2{,}870 \times 2{,}870\,\mu m$, it has been divided into $650 \times 2{,}400 \times 2{,}400$ cells and each cell contains $1 \times 1 \times 1$ particles (3.7×10^9 simulation particles in total). Additional simulations with $1 \times 2 \times 2$ particles per cell showed no influence on our conclusions and simulation results. The pump laser was injected backwards from the leading edge of the moving window[32,33]. In order to conserve canonical momentum, the momentum of each plasma electron macro-particle has been set to match the normalized laser vector potential. The particles are initialized with no thermal spread.

The initial OAM seed and pump laser electric field is given by:

$$\mathbf{E} = \frac{1}{2}\frac{\mathbf{E_0}w_0}{w(z)}\left(\frac{r\sqrt{2}}{w(z)}\right)^{|\ell|}L_p^{|\ell|}\left(\frac{2r^2}{w^2(z)}\right)\exp\left(-\frac{r^2}{w^2(z)}\right)$$
$$\times \exp\left[ik(z-z_0)+\frac{ikz}{1+z^2/z_R^2}\frac{r^2}{z_R^2}-i(2p+|\ell|+1)\arctan\left(\frac{z}{Z_r}\right)+i\theta_0+i\ell\phi\right] \quad (4)$$
$$+ \text{c.c.},$$

where c.c. denotes complex conjugate, $\mathbf{E_0} = (E_{0x}, E_{0y})$ is the laser electric field at the focus, with (E_{0x}, E_{0y}) being the electric field amplitudes in the transverse x and y directions, respectively. For a linearly polarized laser, there is no phase difference between E_{0x} and E_{0y}. For circularly polarized light, both components are $\pi/2$ out of phase, that is, $E_{0x} = \pm iE_{0y}$. In addition $w^2(z) = w_0^2\left(1+z^2/Z_r^2\right)$ is the waist of the beam as a function of the propagation distance z in vacuum, w_0 the waist at the focal plane, $Z_r = \pi w_0^2/\lambda$ the Rayleigh length, and $\lambda = 2\pi c/\omega = 2\pi/k$ the central wavelength of the laser, ω and k are its central frequency and wavenumber, respectively. In addition, $L_p^{|\ell|}$ is a generalized Laguerre polynomial with order (p,ℓ), with ℓ being the index that gives rise to the OAM, $r = \sqrt{x^2+y^2}$ the radial distance to the axis, θ_0 an initial phase and z_0 the centre of the laser. We note that all simulations involving Laguerre–Gaussian modes have $p = 0$. The initial electric field of an Hermite–Gaussian (TEM) laser is given by:

$$\mathbf{E} = \frac{1}{2}\frac{\mathbf{E_0}w_0}{w(z)}H_m(x)H_n(y)\exp\left(-\frac{r^2}{w^2(z)}\right)$$
$$\times \exp\left[ik(z-z_0)+i\frac{kz}{1+z^2/z_R^2}\frac{r^2}{z_R^2}-i(m+n+1)\arctan\left(\frac{z}{Z_r}\right)+i\theta_0\right]+\text{c.c.,}$$
$$(5)$$

where H_m is an Hermite polynomial of order m. Moreover, the wavenumber of the pump laser (which travels in the plasma) in all simulations presented in Figs 2–4 is set according to the linear plasma dispersion relation $k^2c^2 = \omega^2 - \omega_p^2$, where $\omega_p = \sqrt{4\pi n_0 e^2/m_e}$ is the plasma frequency associated with a background plasma density n_0, and where e and m_e are, respectively, the elementary charge and electron mass. The seed frequency and wavenumber are set according to the matching conditions for Raman amplification (Table 1).

References

1. Allen, L. et al. Orbital angular momentum of light and the transformation of Laguerre-Gaussian laser modes. Phys. Rev. A **45**, 8185 (1992).
2. Jesacher, A. et al. Shadow effects in spiral phase contrast microscopy. Phys. Rev. Lett. **94**, 233902 (2005).
3. Jack, B. et al. Holographic ghost imaging and the violation of a Bell inequality. Phys. Rev. Lett. **103**, 083602 (2009).
4. Andersen, M. F. et al. Quantized rotation of atoms from photons with orbital angular momentum. Phys. Rev. Lett. **97**, 170406 (2006).
5. Padgett, M. & Bowman, R. Tweezers with a twist. Nat. Photon. **5**, 343 (2011).
6. Wang, J. et al. Terabit free-space data transmission employing orbital angular momentum multiplexing. Nat. Photon. **6**, 488 (2012).
7. Bozinovic, N. et al. Terabit-scale orbital angular momentum mode division multiplexing in fibers. Science **340**, 1145 (2013).
8. Molina-Terriza, G., Torres, J. P. & Torner, L. Twisted photons. Nat. Phys. **3**, 305 (2007).

9. Tamburini, F. *et al.* Twisting of light around rotating black holes. *Nat. Phys.* **7**, 195 (2011).

10. Vieira, J. & Mendonça, J. T. Nonlinear laser driven donut wakefields for positron and electron acceleration. *Phys. Rev. Lett.* **112**, 215001 (2014).

11. Padgett, M., Courtial, J. & Allen, L. Lights orbital angular momentum. *Phys. Today* **57**, 35–40 (2004).

12. Gariepy, G. *et al.* Creating high-harmonic beams with controlled orbital angular momentum. *Phys. Rev. Lett.* **113**, 153901 (2014).

13. Shao, G.-h. *et al.* Nonlinear frequency conversion of fields with orbital angular momentum using quasi-phase-matching. *Phys. Rev. A* **88**, 063827 (2013).

14. Hemsing, E. & Marinelli, A. Echo-enabled X-ray vortex generation. *Phys. Rev. Lett.* **109**, 224801 (2012).

15. Hemsing, E. *et al.* Coherent optical vortices from relativistic electron beams. *Nat. Phys.* **9**, 549 (2013).

16. Shi, Y. *et al.* Light fan driven by relativistic laser pulse. *Phys. Rev. Lett.* **112**, 235001 (2014).

17. Brabetz, C. *et al.* Laser-driven ion acceleration with hollow laser beams. *Phys. Plasmas* **22**, 013102 (2015).

18. Martinelli, M. *et al.* Orbital angular momentum exchange in an optical parametric oscillator. *Phys. Rev. A* **70**, 013812 (2004).

19. Fonseca, R. A. *et al.* OSIRIS: a three-dimensional, fully relativistic particle in cell code for modeling plasma based accelerators. *Lect. Notes Comp. Sci.* vol. 2331/2002 (Springer, 2002).

20. Ren, J. *et al.* A new method for generating ultraintense and ultrashort laser pulses. *Nat. Phys.* **3**, 732 (2007).

21. Trines, R. M. G. M. *et al.* Simulations of efficient Raman amplification into the multipetawatt regime. *Nat. Phys.* **7**, 87 (2011).

22. Trines, R. M. G. M. *et al.* Production of picosecond, kilojoule, and petawatt laser pulses via raman amplification of nanosecond pulses. *Phys. Rev. Lett.* **107**, 105002 (2011).

23. Michel, P., Divol, L., Turnbull, D. & Moody, J. D. Dynamic control of the polarization of intense laser beams via optical wave mixing in plasmas. *Phys. Rev. Lett.* **113**, 205001 (2014).

24. Mori, W. B. The physics of the nonlinear optics of plasmas at relativistic intensities for short-pulse lasers. *IEEE Trans. Plasma Sci.* **33**, 1942 (1997).

25. Shvets, G. *et al.* Superradiant amplification of an ultrashort laser pulse in a plasma by a counterpropagating pump. *Phys. Rev. Lett.* **81**, 4879 (1998).

26. Malkin, V. M. *et al.* Fast compression of laser beams to highly overcritical powers. *Phys. Rev. Lett.* **82**, 4448 (1999).

27. Forslund, D. W., Kindel, J. M. & Lindman, E. L. Theory of stimulated scattering processes in laser-irradiated plasmas. *Phys. Fluids* **18**, 1002 (1975).

28. Malkin, V. M. *et al.* Ultra-powerful compact amplifiers for short laser pulses. *Phys. Plasmas* **7**, 2232 (2000).

29. Mendonça, J. T., Thidé, B. & Then, H. Stimulated Raman and Brillouin backscattering of collimated beams carrying orbital angular momentum. *Phys. Rev. Lett.* **102**, 185005 (2009).

30. Fraiman, G. M. *et al.* Robustness of laser phase fronts in backward Raman amplifiers. *Phys. Plasmas* **9**, 3617 (2002).

31. Alves, E. P. *et al.* A robust plasma-based laser amplifier via stimulated Brillouin scattering. Preprint at http://arxiv.org/abs/1311.2034 (2014).

32. Mardahl, P. *et al.* 43rd APS/DPP meeting, 'XOOPIC simulations of Raman backscattering', Paper KP1.108. *Bull. Am. Phys. Soc* **46**, 202 (2001).

33. Mardahl, P. *PIC Code Charge Conservation, Numerical Heating, and Parallelization: Application of XOOPIC to Laser Amplification via Raman Backscatter.* PhD thesis, Univ. California, Berkeley (2001).

Acknowledgements

This work was supported by the European Research Council through the Accelerates European Research Council project (contract ERC-2010-AdG-267841), Fundação para a Ciência e para a Tecnologia, Portugal (contract EXPL/FIZ-PLA/0834/1012) and the European Union (EUPRAXIA grant agreement 653782). We acknowledge PRACE for access to resources on SuperMUC (Leibniz Research Center).

Author contributions

All authors contributed to all aspects of this work.

Additional information

Competing financial interests: The authors declare no competing financial interests.

Topological phase transitions and chiral inelastic transport induced by the squeezing of light

Vittorio Peano[1], Martin Houde[2], Christian Brendel[1], Florian Marquardt[1,3] & Aashish A. Clerk[2]

There is enormous interest in engineering topological photonic systems. Despite intense activity, most works on topological photonic states (and more generally bosonic states) amount in the end to replicating a well-known fermionic single-particle Hamiltonian. Here we show how the squeezing of light can lead to the formation of qualitatively new kinds of topological states. Such states are characterized by non-trivial Chern numbers, and exhibit protected edge modes, which give rise to chiral elastic and inelastic photon transport. These topological bosonic states are not equivalent to their fermionic (topological superconductor) counterparts and, in addition, cannot be mapped by a local transformation onto topological states found in particle-conserving models. They thus represent a new type of topological system. We study this physics in detail in the case of a kagome lattice model, and discuss possible realizations using nonlinear photonic crystals or superconducting circuits.

[1] Institute for Theoretical Physics, University of Erlangen-Nürnberg, Staudtstr. 7, 91058 Erlangen, Germany. [2] Department of Physics, McGill University, 3600 rue University, Montreal, Quebec, Canada H3A 2T8. [3] Max Planck Institute for the Science of Light, Günther-Scharowsky-Straße 1/Bau 24, 91058 Erlangen, Germany. Correspondence and requests for materials should be addressed to V.P. (email: Vittorio.Peano@fau.de).

Waves are not only ubiquitous in physics, but the behaviour of linear waves is also known to be very generic, with many features that are independent of the specific physical realization. This has traditionally allowed us to transfer insights gained in one system (for example, sound waves) to other systems (for example, matter waves). That strategy has even been successful for more advanced concepts in the field of wave transport. One important recent example of this kind is the physics of topological wave transport, where waves can propagate along the boundaries of a sample, in a one-way chiral manner that is robust against disorder scattering. While first discovered for electron waves, this phenomenon has by now also been explored for a variety of other waves in a diverse set of systems, including cold atoms[1], photonic systems[2] and more recently phononic systems[3-9].

In the case of topological wave transport, the connection between waves in different physical implementations can actually be so close that the calculations turn out to be the same. In particular, if we are dealing with matter waves moving in a periodic potential, the results do not depend on whether they are bosons or fermions, as long as interactions do not matter. The single-particle wave equation to be solved happens to be exactly the same. This has allowed to envision and realize photonic analogues of quantum-Hall effect[10-18], the spin Hall effect[19-22], Floquet topological insulators[23,24] and even Majorana-like modes[25]. More generally, the well-known classification of electronic band structures based on the dimensionality and certain generalized symmetries[26] directly applies to photonic systems provided that the particle number is conserved. As we now discuss, this simple correspondence will fail in the presence of squeezing.

Consider the most general quadratic Hamiltonian describing photons in a periodic potential in the presence of parametric driving:

$$\hat{H} = \sum_{\mathbf{k},n} \varepsilon_n[\mathbf{k}]\hat{b}^\dagger_{\mathbf{k},n}\hat{b}_{\mathbf{k},n} + \sum_{\mathbf{k},n,n'} \left(\lambda_{nn'}[\mathbf{k}]\hat{b}^\dagger_{\mathbf{k},n}\hat{b}^\dagger_{-\mathbf{k},n'} + \text{h.c.} \right). \quad (1)$$

The first term describes a non-interacting photonic band structure, where $\hat{b}_{\mathbf{k},n}$ annihilates a photon with quasimomentum $\mathbf{k}$ in the n-th band. The remaining two-mode squeezing terms are induced by parametric driving and do not conserve the excitation number. As we discuss below, they can be controllably realized in a number of different photonic settings. While superficially similar to pairing terms in a superconductor, these two-mode squeezing terms have a profoundly different effect in a bosonic system, as there is no limit to the occupancy of a particular single-particle state. They can give rise to highly entangled ground states, and even to instabilities.

Given these differences, it is natural to ask how anomalous pairing terms can directly lead to topological phases of light. In this work, we study the topological properties of two-dimensional photonic systems described by Equation (1), in the case where the underlying particle-conserving band structure has no topological structure, and where the parametric driving terms do not make the system unstable. We show that the introduction of particle non-conserving terms can break time-reversal symmetry (TRS) in a manner that is distinct from having introduced a synthetic gauge field, and can lead to the formation of bands having a non-trivial pattern of (suitably defined) quantized Chern numbers. This in turn leads to the formation of protected chiral edge modes: unlike the particle-conserving case, these modes can mediate a protected inelastic (but still coherent) scattering mechanism along the edge (that is, a probe field injected into the edge of the sample will travel along the edge, but emerge at a different frequency). In general, the topological phases we find here are distinct both from those obtained in the particle-conserving case, and from those found in topological super-conductors. We also discuss possible realizations of this model using a nonlinear photonic crystal or superconducting microwave circuits. Finally, we discuss the formal analogies and crucial differences between the topological phases of light investigated here and those recently proposed for other kinds of Bogoliubov quasiparicles[27-31] (see Discussion section).

Results

Kagome lattice model. For concreteness, we start with a system of bosons on a kagome lattice (Fig.1),

$$\hat{H}_0 = \sum_{\mathbf{j}} \omega_0\hat{a}^\dagger_{\mathbf{j}}\hat{a}_{\mathbf{j}} - J \sum_{\langle \mathbf{j},\mathbf{j}'\rangle} \hat{a}^\dagger_{\mathbf{j}}\hat{a}_{\mathbf{j}'} \quad (2)$$

(we set $\hbar = 1$). Here we denote by $\hat{a}_{\mathbf{j}}$ the photon annihilation operator associated with lattice site $\mathbf{j}$, where the vector site index has the form $\mathbf{j} = (j_1,j_2,s)$. $j_1,j_2 \in Z$ labels a particular unit cell of the lattice, while the index $s = A,B,C$ labels the element of the sublattice. $\langle \mathbf{j},\mathbf{j}'\rangle$ indicates the sum over nearest neighbours, and J is the (real valued) nearest-neighbour hopping rate; ω_0 plays the role of an onsite energy. As there are no phases associated with the hopping terms, this Hamiltonian is time-reversal symmetric and topologically trivial. We chose the kagome lattice because it is directly realizable both in quantum optomechanics[5] and in arrays of super-conducting cavity arrays[13,16]; it is also the simplest model where purely local parametric driving can result in a topological phase.

We next introduce quadratic squeezing terms to this Hamiltonian that preserve the translational symmetry of the lattice and that are no more non-local than our original,

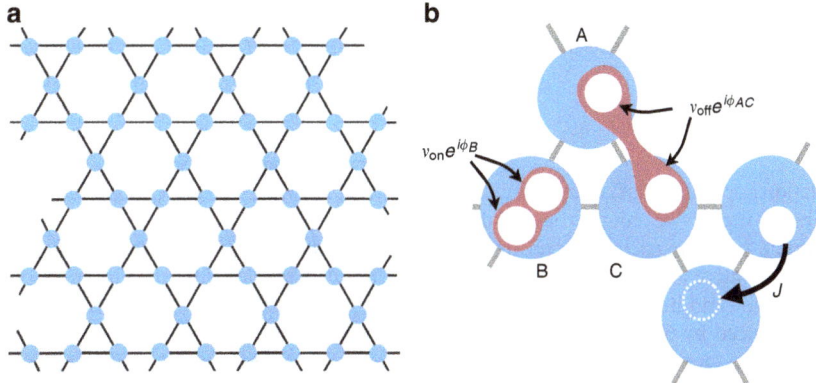

Figure 1 | Setup figure. (a) An array of nonlinear cavities forming a kagome lattice. (b) Photons hop between nearest-neighbour sites with rate J. Each cavity is driven parametrically leading to the creation of photon pairs on the same lattice site (rate v_{on}) and on nearest-neighbour sites (rate v_{off}). A spatial pattern of the driving phase is imprinted on the parametric interactions, breaking the time-reversal symmetry (but preserving the $C3$ rotational symmetry).

nearest-neighbour hopping Hamiltonian:

$$\hat{H}_{\mathrm{L}} = -\frac{1}{2}\left[v_{\mathrm{on}} \sum_{\mathbf{j}} e^{i\phi_s} \hat{a}_{\mathbf{j}}^{\dagger} \hat{a}_{\mathbf{j}}^{\dagger} + v_{\mathrm{off}} \sum_{\langle \mathbf{j}, \mathbf{j}' \rangle} e^{i\phi_{ss'}} \hat{a}_{\mathbf{j}}^{\dagger} \hat{a}_{\mathbf{j}'}^{\dagger} \right] + \mathrm{h.c.} \quad (3)$$

Such terms generically arise from having a nonlinear interaction with a driven auxiliary pump mode (which can be treated classically) on each site, see for example, ref. 32. As we discuss below, the variation in phases in $\hat{H}_{\mathrm{L}}$ from site to site could be achieved by a corresponding variation of the driving phase of the pump. Note that we are working in a rotating frame where this interaction is time independent, and thus ω_0 should be interpreted as the detuning between the parametric driving and the true onsite (cavity) frequency ω_{cav} (that is, $\omega_0 = \omega_{\mathrm{cav}} - \omega_{\mathrm{L}}/2$, where the parametric driving is at a frequency ω_{L}). The parametric driving can cause the system to become unstable; we will thus require that the onsite energy (that is, parametric drive detuning) ω_0 be sufficiently large that each parametric driving term is non-resonant enough to ensure stability. If one keeps ω_0 fixed, this means that the parametric driving amplitudes v_{on}, v_{off} will be limited to some fraction of ω_0 (the particular value of which depends on J, Supplementary Note 1).

For a generic choice of phases in the parametric driving Hamiltonian of Equation (3), it is no longer possible to find a gauge where $\hat{H} = \hat{H}_0 + \hat{H}_{\mathrm{L}}$ is purely real when expressed in terms of real-space annihilation operators: hence, even though the hopping Hamiltonian $\hat{H}_0$ corresponds to strictly zero flux, the parametric driving can itself break TRS. In what follows, we will focus for simplicity on situations where time reversal and particle conservation are the only symmetries broken by the parametric driving: they will maintain the inversion and $\mathcal{C}3$ rotational symmetry of the kagome lattice. We will also make a global gauge transformation so that v_{off} is purely real, while $v_{\mathrm{on}} = |v_{\mathrm{on}}|e^{i\phi_v}$. In this case, the only possible choices for the ϕ phases have the form $(\phi_A, \phi_B, \phi_C) = (\phi_{AB}, \phi_{BC}, \phi_{CA}) = \pm(0, \delta, 2\delta)$ with $\delta = 2\pi m_v/3$, where m_v is an integer and is the vorticity of the parametric driving phases. We stress that these phases (and hence the sign of the TRS breaking) are determined by the phases of the pump modes used to generate the parametric interaction.

Gap opening and non-trivial topology. $\hat{H}_0$ is the standard tight-binding kagome Hamiltonian for zero magnetic field, and does not have band gaps: the upper and middle bands touch at the symmetry point $\mathbf{\Gamma} \equiv (0,0)$, whereas the middle and lower bands touch at the symmetry points $K = (2\pi/3, 0)$ and $K' = (\pi/3, \pi/(3)^{1/2})$ where they form Dirac cones (Fig. 2a).

Turning on the pairing terms, the Hamiltonian $\hat{H} = \hat{H}_0 + \hat{H}_{\mathrm{L}}$ can be diagonalized in the standard manner as $\hat{H} = \sum_{n,\mathbf{k}} E_n[\mathbf{k}] \hat{\beta}_{n,\mathbf{k}}^{\dagger} \hat{\beta}_{n,\mathbf{k}}$, where the $\hat{\beta}_{n,\mathbf{k}}$ are canonical bosonic annihilation operators determined by a Bogoliubov transformation of the form (see Methods section):

$$\hat{\beta}_{n,\mathbf{k}}^{\dagger} = \sum_{s=A,B,C} u_{n,\mathbf{k}}[s]\hat{a}_{\mathbf{k},s}^{\dagger} - v_{n,\mathbf{k}}[s]\hat{a}_{-\mathbf{k},s}. \quad (4)$$

Here $\hat{a}_{\mathbf{k},s}$ are the annihilation operators in quasimomentum space, and $n = 1,2,3$ is a band index; we count the bands by increasing energy. The photonic single-particle spectral function now shows resonances at both positive and negative frequencies, $\pm E_n[k]$, corresponding to particle- and hole-type bands, Fig. 2d. Because of the TRS breaking induced by the squeezing terms, the band structure described by $E_n[\mathbf{k}]$ now exhibits gaps, Fig. 2b; furthermore, for a finite sized system, one also finds edge modes in the gap, Fig. 2d.

The above behaviour suggests that the parametric terms have induced a non-trivial topological structure in the wavefunctions

of the band eigenstates. To quantify this, we first need to properly identify the Berry phase associated with a bosonic band eigenstate in the presence of particle non-conserving terms. For each $\mathbf{k}$, the Bloch Hamiltonian $\hat{H}_{\mathbf{k}}$ corresponds to the Hamiltonian of a multi-mode parametric amplifier. Unlike the particle-conserving case, the ground state of such a Hamiltonian is a multi-mode squeezed state with non-zero photon number; it can thus have a non-trivial Berry's phase associated with it when $\mathbf{k}$ is varied, Supplementary Note 2. The Berry phase of interest for us will be the difference of this ground state Berry phase and that associated with a single quasiparticle excitation. One finds that the resulting Berry connection takes the form

$$\mathcal{A}_n = i\langle \mathbf{k}, n|\hat{\sigma}_z \mathbf{\nabla}_k|\mathbf{k}, n\rangle. \quad (5)$$

Here the six vector of Bogoliubov coefficients $|\mathbf{k}, n\rangle \equiv (u_{n,\mathbf{k}}[A], u_{n,\mathbf{k}}[B], u_{n,\mathbf{k}}[C], v_{n,\mathbf{k}}[A], v_{n,\mathbf{k}}[B], v_{n,\mathbf{k}}[C])$ plays the role of a singe-particle wavefunction, and $\hat{\sigma}_z$ acts in the particle-hole space, associating $+1$ to the u components and -1 to the v components, see Methods section. These effective wavefunctions obey the symplectic normalization condition

$$\langle \mathbf{k}, n|\hat{\sigma}_z|\mathbf{k}, n'\rangle = \sum_s u_{n,\mathbf{k}}^*[s]u_{n',\mathbf{k}}[s] - v_{n,\mathbf{k}}^*[s]v_{n',\mathbf{k}}[s] = \delta_{n,n'}. \quad (6)$$

Having identified the appropriate Berry connection for a band eigenstate, the Chern number for a band n is then defined in the usual manner:

$$C_n = \frac{1}{2\pi}\int_{\mathrm{BZ}} (\mathbf{\nabla} \times \mathcal{A}_n) \cdot \hat{z}. \quad (7)$$

The definition in Eq. (5) agrees with that presented in ref. 27 and (in one-dimension) ref. 29; standard arguments[27] show that the C_n are integers with the usual properties. We note that, as for superconductors, breaking the $U(1)$ (particle-conservation) symmetry remains compatible with a first-quantized picture after doubling the number of bands. The additional hole bands are connected to the standard particle bands by a particle–hole symmetry; see Methods section. In bosonic systems, the requirement of stability generally implies that particle and hole bands can not touch; this is true for our system. Thus, the sum of the Chern numbers over the particle bands (with $E > 0$) must be zero, and there cannot be any edge states with energies below the lowest particle bulk band (or in particular, at zero energy); Supplementary Note 1.

In the special case where we only have onsite parametric driving (that is, $v_{\mathrm{off}} = 0, v_{\mathrm{on}} \neq 0$), the Chern numbers can be calculated analytically (Supplementary Note 3). They are uniquely fixed by the pump vorticity. If $m_v = 0$, we have TRS and the band structure is gapless, while for $m_v = \pm 1$, $\mathbf{C} = (\mp 1, 0, \pm 1)$. This set of topological phases also occurs in a particle-number conserving model on the kagome lattice with a staggered magnetic field, that is, the Oghushi–Murakami–Nagaosa (OMN) model of the anomalous quantum-Hall effect[33,34].

In the general case, where we include offsite parametric driving, entirely new phases appear. We have computed the Chern numbers here numerically, using the approach of ref. 35. In Fig. 3a, we show the topological phase diagram of our system, where J/ω_0 and m_v are held fixed, while the parametric drive strengths $v_{\mathrm{on}}, v_{\mathrm{off}}$ are varied. Different colours correspond to different triplets $\mathbf{C} \equiv (C_1, C_2, C_3)$ of the band Chern numbers, with grey and dark-grey corresponding to the two phases already present in the OMN model. Strikingly, a finite off-diagonal coupling v_{off} generates a large variety of phases which are not present in the OMN model, including phases having bands with $|C_n| > 1$. The border between different topological phases represent topological phase transitions, and correspond to parameter values where a pair of bands touch at a particular symmetry

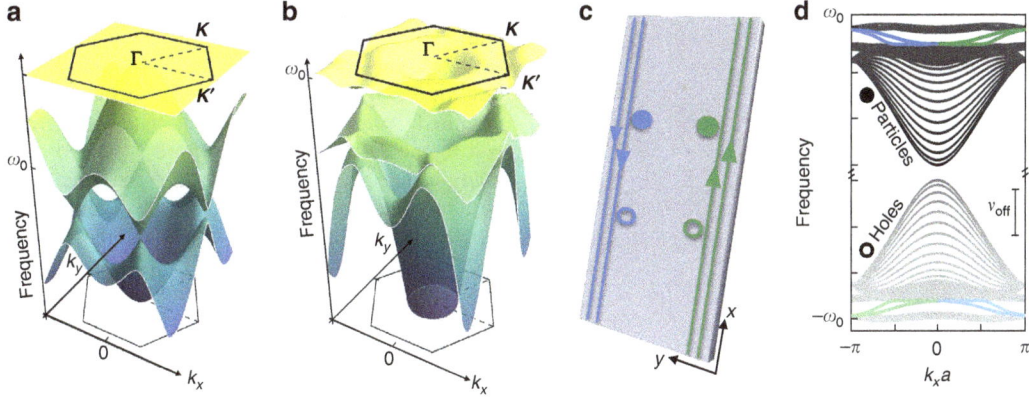

Figure 2 | Topological Band structure. (a,b) 3D plots of the bulk band structure. The hexagonal Brillouin zone is also shown. **(a)** In the absence of parametric driving, neighbouring bands touch at the rotational symmetry points **K**, **K′** and **Γ**. **(b)** The parametric driving opens a gap between subsequent bands. For the chosen parameters, there is a global band gap between the second and third band. **(d)** Hole and particle bands, $\pm E_m[k_x]$, in a strip geometry (sketched in **c**). The line intensity is proportional to the weight of the corresponding resonance in the photon spectral function, Supplementary Note 1. The edge states localized on the right (left) edge, plotted in green (blue), have positive (negative) velocity. Parameters: Hopping rate $J = 0.02\omega_0$ (ω_0 is the onsite frequency); **(b,d)**, the parametric couplings are $v_{\mathrm{on}} = -0.085\omega_0$ and $v_{\mathrm{off}} = 0.22\omega_0$.

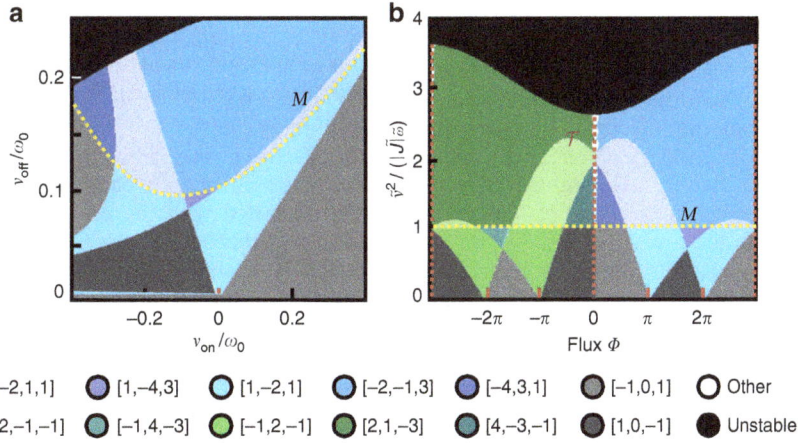

| ⬤ [−2,1,1] | ⬤ [1,−4,3] | ⬤ [1,−2,1] | ⬤ [−2,−1,3] | ⬤ [−4,3,1] | ⬤ [−1,0,1] | ◯ Other |
| ⬤ [2,−1,−1] | ⬤ [−1,4,−3] | ⬤ [−1,2,−1] | ⬤ [2,1,−3] | ⬤ [4,−3,−1] | ⬤ [1,0,−1] | ⬛ Unstable |

Figure 3 | Symplectic Topological phase diagrams. (a) Topological phase diagram for the parametrically driven kagome lattice model. The y (x) axis corresponds to the strength of the onsite parametric drive v_{on} (offsite parametric drive v_{off}), and different colours correspond to different triplets $\mathbf{C} = (C_1, C_2, C_3)$ of Chern numbers for the three bands of the model. Note that only the grey and dark-grey phases are found in the particle-conserving version of our model with a staggered field. We have fixed the hopping rate $J/\omega_0 = 0.02$, and the vorticity of the pump $m_v = 1$. **(b)** Same phase diagram, but now plotted in terms of the effective flux Φ and effective parametric drive $\tilde{v}$ experienced by α quasiparticles.

point; we discuss this further below. Via a standard bulk-boundary correspondence (Supplementary Note 4), the band Chern numbers for a particular phase determine the number of protected edge states that will be present in a system with a boundary; as usual, the number of edge states in a particular bandgap is obtained by summing the Chern numbers of lower-lying bands. We discuss these edge states in greater detail in a following subsection. Finally, the black regions in the phase diagram indicate regimes of instability, which occur when the parametric driving strength becomes too strong.

Dressed-state picture. To gain further insight into the structure of the topological phases found above, it is useful to work in a dressed-state basis that eliminates the local parametric driving terms from our Hamiltonian. We thus first diagonalize the purely local terms in the Hamiltonian; for each lattice site **j** we have

$$\hat{H}_j = \omega_0 \hat{a}_j^\dagger \hat{a}_j \ - \frac{1}{2}\left[v_{\mathrm{on}} e^{i\phi_j} \hat{a}_j^\dagger \hat{a}_j^\dagger + \text{h.c.} \right] = \tilde{\omega}\hat{\alpha}_j^\dagger \hat{\alpha}_j \ . \quad (8)$$

Here $\tilde{\omega} = \sqrt{\omega_0^2 - v_{\mathrm{on}}^2}$, and the annihilation operators $\hat{\alpha}_j$ are given by a local Bogoliubov (squeezing) transformation $\hat{\alpha}_j = e^{i\phi_j} e^{-i\varphi_v/2}(\cosh(r)a_j - e^{i\phi_j}e^{i\varphi_v}\sinh(r)\hat{a}_j^\dagger)$, where the squeezing factor r is

$$r = \frac{1}{4}\ln\left[\frac{\omega_0 + v_{\mathrm{on}}}{\omega_0 - v_{\mathrm{on}}}\right]. \quad (9)$$

On a physical level, the local parametric driving terms attempt to drive each site into a squeezed vacuum state with squeeze parameter r; the $\hat{\alpha}_j$ quasiparticles correspond to excitations above this reference state. Note that we have included an overall phase factor in the definition of the $\hat{\alpha}_j$, which will simplify the final form of the full Hamiltonian.

In this new basis of local quasiparticles, our full Hamiltonian takes the form

$$\hat{H} = \sum_j \tilde{\omega}\hat{\alpha}_j^\dagger \hat{\alpha}_j - \sum_{\langle j,l \rangle} \tilde{J}_{jl}\hat{\alpha}_j^\dagger \hat{\alpha}_l - \left(\frac{\tilde{v}}{2}\sum_{\langle j,l \rangle} \hat{\alpha}_j^\dagger \hat{\alpha}_l^\dagger + \text{h.c.} \right). \quad (10)$$

The transformation has mixed the hopping terms with the non-local parametric terms: The effective counter-clockwise hopping matrix element is

$$\tilde{J}_{jl} = Je^{i\delta} + e^{3i\delta/2}\left[2J\cos\left(\frac{\delta}{2}\right)\sinh^2 r + v_{\text{off}}\sinh 2r\cos\left(\frac{\delta}{2}+\varphi_v\right)\right],\tag{11}$$

and the magnitude of the effective non-local parametric driving is

$$|\tilde{v}| = |\ v_{\text{off}}e^{-i(\delta/2+\varphi_v)} + 2v_{\text{off}}\cos(\delta/2+\varphi_v)\sinh^2 r$$
$$+ J\sinh 2r\ \cos(\delta/2)\ |.\tag{12}$$

Note that the phase of $\tilde{v}$ can be eliminated by a global gauge transformation, and hence it plays no role; we thus take $\tilde{v}$ to be real in what follows.

Our model takes on a much simpler form in the new basis: the onsite parametric driving is gone, and the non-local parametric driving is real. Most crucially, the effective hoppings can now have spatially varying phases, which depend both on the vorticity of the parametric driving in $\hat{H}_L$ (through δ), and the magnitude of the onsite squeezing (through r). In this transformed basis, the effective hopping phases are the only route to breaking TRS. Our model has thus been mapped onto the standard OMN model for the anomalous quantum-Hall effect, with an additional (purely real) nearest-neighbour two-mode squeezing interaction. In the regime where the parametric interactions between the $\hat{\alpha}$ quasiparticles are negligible (Supplementary Note 3), the complex phases correspond in the usual manner to a synthetic gauge field (that is, the effective flux Φ piercing a triangular plaquette would be $\Phi = 3\arg\tilde{J}$). In other words, the squeezing creates a synthetic gauge field for Bogoliubov quasiparticles. However, in the presence of substantial parametric interaction between $\hat{\alpha}$ quasiparticles, the parameter Φ can not be interpreted anymore as a flux: a flux of 2π can not be eliminated by a gauge transformation because the complex phases reappear in the parametric terms. In that case, only a periodicity of 6π in Φ is retained, since that corresponds to having trivial hopping phases of 2π.

Understanding the topological structure of this transformed Hamiltonian is completely sufficient for our purposes: one can easily show that the Chern number of a band is invariant under any local Bogoliubov transformation, hence the Chern numbers obtained from the transformed Hamiltonian in Equation (8) will coincide exactly with those obtained from the original Hamiltonian in Equation (3). We thus see that the topological structure of our system is controlled completely by only three dimensionless parameters: the flux Φ (associated with the hopping phases), the ratio $|\tilde{v}/\tilde{J}|$, and the ratio $\tilde{\omega}/|\tilde{J}|$.

The topological phase diagram for the effective model is shown in Fig. 3b. Again, one sees that as soon as the effective non-local parametric drive $\tilde{v}$ is non-zero, topological phases distinct from the standard (particle-conserving) OMN model are possible. The sign of the parametric pump vorticity m_v determines the sign of the effective flux Φ, c.f. Equation (11). As such, the right half of Fig. 3b (corresponding to $\Phi > 0$) is a deformed version of the phase diagram of the original model for pump vorticity $m_v = 1$, as plotted in Fig. 3a. Changing the sign of m_v (and hence Φ) simply flips the sign of all Chern numbers, Supplementary Note 3.

Our effective model provides a more direct means for understanding the boundaries between different topological phases. Most of these are associated with the crossing of bands at one or more high-symmetry points in the Brillouin zone; this allows an analytic calculation of the phase boundary (Supplementary Note 3). Perhaps most striking in Fig. 3b is the horizontal boundary (labelled $\mathcal{M}$), occurring at a finite value of the effective offsite parametric drive, $\tilde{v} \approx \sqrt{\tilde{J}\tilde{\omega}}$. This boundary is set by the closing of a band gap at the M points; as these points are associated with the decoupling of one sublattice from the other two, this boundary is insensitive to the flux Φ. Similarly, the vertical line labelled $\mathcal{T}$ denotes a line where the system has TRS, and all bands cross at the symmetry points K, K' and Γ. The case of zero pump vorticity $m_v = 0$ (not shown) is also interesting. Here the effective flux Φ depends on the strength of the parametric drivings, but is always constrained to be 0 or 3π. This implies that the effective Hamiltonian has TRS, even though the original Hamiltonian may not (that is, if Im $v_{\text{off}} \neq 0$, the original Hamiltonian does not have TRS). For $m_v = 0$, the parametric

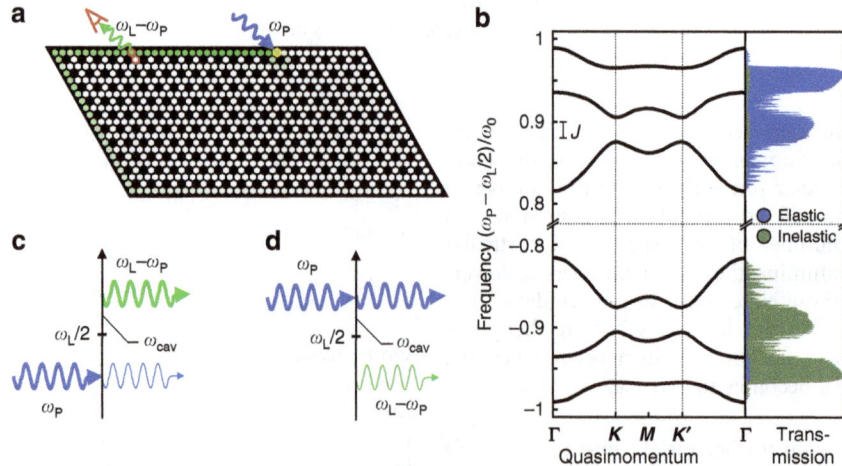

Figure 4 | Topologically protected transport in a finite system. (a) A probe beam at frequency ω_p inside the bulk band gap is focused on a site (marked in yellow) at the edge of a finite sample. The probability map of the light transmitted inelastically at frequency $\omega_L - \omega_p$ (where ω_L is the frequency of the drive tone applied to the auxiliary pump modes) clearly shows that the transport is chiral. (b) The elastic and inelastic transmission probability to a pair of sites along the edges (indicated in red in **a**) is plotted in blue and green, respectively. A cut through the bulk bands is shown to the left. (**c,d**) Sketch of the relevant scattering processes and energy scales. The inelastic (elastic) transmission has a larger rate when the light is injected in the hole (particle) band gap. Parameters: Hopping rate $J = 0.02\omega_0$ (ω_0 is the onsite frequency), parametric couplings $v_{\text{on}} = 0.4\omega_0$ and $v_{\text{off}} = 0.02\omega_0$, optical decay rate $\kappa = 0.001\omega_0$. (**a**) $\omega_p - \omega_L/2 = 0.95\omega_0$.

drivings do not open any band gap and the Chern numbers are not well defined.

Edge states and transport. Despite their modified definition, the Chern numbers associated with our Bogoliubov bands still guarantee the existence of protected chiral edge modes in a system with boundaries via a standard bulk-boundary correspondence, see Supplementary Note 4. These states can be used to transport photons by exciting them with an auxiliary probe laser beam, which is focused on an edge site and at the correct frequency. The lack of particle-number conservation manifests itself directly in the properties of the edge states: along with the standard elastic transmission they can also mediate inelastic scattering processes. In terms of the original lab frame, light injected at a frequency ω_p can emerge on the edge at frequency $\omega_L - \omega_p$, where ω_L is the frequency of the laser parametrically driving the system. This is analogous to the idler output of a parametric amplifier. Here both signal and idler have a topologically protected chirality.

Shown in Fig. 4 are the results of a linear response calculation describing such an experiment, applied to a finite system with corners. We incorporate a finite photon decay rate κ in the standard input–output formalism, see Methods section. Narrow-band probe light inside a topological band gap is applied to a site on the edge, and the resulting inelastic transmission probabilities to each site on the lattice are plotted, Fig. 4a. One clearly sees that the probe light is transmitted in a unidirectional way along the edge of the sample, and is even able to turn the corner without significant backscatter. The corresponding elastic transmission (not shown) is also chiral and shows the same spatial dependence. In Fig. 4b we show the elastic and inelastic transmissions to the sites indicated in red (rescaled by the overall transmission, $1 - R$ where R is the reflection probability at the injection site) as a function of the probe frequency ω_p. By scanning the laser probe frequency, one can separately address particle and hole band gaps. The relative intensity of the inelastic scattering component is highly enhanced when the probe beam is inside a hole band gap, see also the sketches in Fig. 4c,d. When the parametric interaction between the $\hat{\alpha}$ quasiparticles is negligible, the ratio of elastic and inelastic transmissions depends only on the squeezing factor r, (c.f. Equation (9)), see Methods section.

Physical realization. Systems of this type could be implemented in photonic crystal coupled cavity arrays[36] fabricated from nonlinear optical χ^2 materials[37–39]. The array of optical modes participating in the transport would be supplemented by pump modes (resonant with the pump laser at twice the frequency). One type of pump mode could be engineered to be spatially co-localized with the transport modes (v_{on} processes), while others could be located in-between (v_{off}). The required periodic phase pattern of the pump laser can be implemented using spatial light modulators or a suitable superposition of several laser beams impinging on the plane of the crystal. One method for realizing the required kagome lattice of defect cavities was discussed in ref. 5. Optomechanical systems offer another route towards generating optical squeezing terms[40,41], via the mechanically induced Kerr interaction, and this could be exploited to create an optomechanical array with a photon Hamiltonian of the type discussed here. Alternatively, these systems can be driven by two laser beams to create phononic squeezing terms[42]. A fourth alternative consists in superconducting microwave circuits of coupled resonators, where Josephson junctions can be embedded to introduce χ^2 and higher order nonlinearities, as demonstrated in refs 43,44. Kagome lattices of superconducting resonators have recently been implemented[45].

Discussion
Before concluding, it is worthwhile to discuss the connections between our work and other recent studies. A Hamiltonian of the general form of Equation (1) arises naturally in the mean-field description of a Bose-condensed phase. In this setting, the anomalous pairing terms describe the interactions with the condensate treated at the mean-field level. A few recent studies have proposed to take advantage of these interactions to selectively populate topological edge states[28,30] or, closer to our study, to induce novel topological phases. These include a study of a magnonic crystal[27], as well as general Bose–Einstein condensates in one-dimension[29] and in two-dimensions[31].

There are some crucial differences between the above studies and our work. In our case, Equation (1) describes the real particles of our system, not quasiparticles defined above some background. This difference is not just a question of semantics: in our case, topological effects can directly be seen by detecting photons, whereas in refs 29,31, one would need to isolate the contribution of a small number of Bogoliubov quasiparticles sitting on a much larger background of condensed particles. In addition, in our work the pairing terms in Equation (1) are achieved by driving the system, implying that negative and positive frequencies are clearly physically distinguished (that is, they are defined relative to a non-zero pump frequency). This is at the heart of the topologically protected inelastic scattering mechanism we describe, and is something that is not present in previous studies.

Our work opens the door to a number of interesting new directions. On the more practical side, one could attempt to exploit the unique edge states in our system to facilitate directional, quantum-limited amplification. On the more fundamental level, one could use insights from the corresponding disorder problem[46] and attempt to develop a full characterization of particle non-conserving bosonic topological states that are described by quadratic Hamiltonians. This would then be a counterpart to the classification already developed for fermionic systems[26].

Methods
Bogoliubov transformation and first-quantized picture. We find the normal mode decompositions leading to the band structures in Fig. 2 and the topological phase diagrams in Fig. 3 by introducing a first-quantized picture. Since the relevant Hamiltonians do not conserve the excitation number, this is only possible after doubling the degrees of freedom. This is achieved by grouping all annihilation operators with quasimomentum $\mathbf{k}$ and the creation operators with quasimomentum $-\mathbf{k}$ in the $2N$ vector of operators $\hat{\Psi}_\mathbf{k} = (\hat{a}_{k1}, \ldots, \hat{a}_{kN}, \hat{a}^\dagger_{-k1}, \ldots, \hat{a}^\dagger_{-kN})$ (where N is the unit cell dimension), and by casting the second quantized Hamiltonian $\hat{H}$ in the form

$$\hat{H} = \frac{1}{2}\sum_\mathbf{k} \hat{\Psi}^\dagger_\mathbf{k} \hat{h}(\mathbf{k}) \hat{\Psi}_\mathbf{k}. \quad (13)$$

The $2N \times 2N$ hermitian matrix $\hat{h}(\mathbf{k})$ plays the role of a single-particle Hamiltonian and is referred to as the Bogoliubov de Gennes Hamiltonian. By definition of the normal modes $\hat{H} = \sum_{\mathbf{k},n} E_n[\mathbf{k}]\hat{\beta}^\dagger_{n,\mathbf{k}}\hat{\beta}_{n,\mathbf{k}}$, we have $[\hat{H}, \hat{\beta}^\dagger_{n,\mathbf{k}}] = E_n[\mathbf{k}]\hat{\beta}^\dagger_{n,\mathbf{k}}$. By plugging into the above equation the Bogoliubov ansatz Equation (4) one immediately finds

$$\hat{h}(\mathbf{k})|\mathbf{k}_n\rangle = E_n[\mathbf{k}]\hat{\sigma}_z|\mathbf{k}_n\rangle. \quad (14)$$

Likewise, from $[\hat{H}, \beta_{n,-\mathbf{k}}] = -E_n[-\mathbf{k}]\hat{\beta}_{n,-\mathbf{k}}$ one finds

$$\hat{h}(\mathbf{k})(\mathcal{K}\hat{\sigma}_x|-\mathbf{k}_n\rangle) = -E_n[-\mathbf{k}]\hat{\sigma}_z(\mathcal{K}\hat{\sigma}_x|-\mathbf{k}_n\rangle). \quad (15)$$

Here $\mathcal{K}$ denotes the complex conjugation and the matrix $\hat{\sigma}_x$ exchanges the u's and the v's Bogoliubov coefficients,

$$\hat{\sigma}_x = \begin{pmatrix} 0 & \mathbb{1}_N \\ \mathbb{1}_N & 0 \end{pmatrix}.$$

Thus, the spectrum of the $2N$ matrix $\hat{\sigma}_z\hat{h}(\mathbf{k})$ is formed by the set of $2N$ eigenenergies $E_n[\mathbf{k}]$ (belonging to the particle bands) and $-E_n[-\mathbf{k}]$ (belonging to the hole bands). *Vice versa*, to calculate the eigenenergies $E_n[\mathbf{k}]$ and $-E_n[-\mathbf{k}]$ and the vector of Bogoliubov coefficients in Equation (4), we have to solve the eigenvalue problem

$$\hat{\sigma}_z \hat{h}(\mathbf{k}) |m\rangle = \lambda_m |m\rangle. \tag{16}$$

The solutions we are interested in should also display the symplectic orthonormality relations Equation (6).

We note in passing that so far we have implicitly assumed that the normal mode decomposition is possible. However, this is not always the case. When the matrix $\hat{\sigma}_z \hat{h}(\mathbf{k})$ has any complex eigenvalue, the Hamiltonian is unstable. Moreover, at the border of the stable and unstable parameter regions, the matrix $\hat{\sigma}_z \hat{h}(\mathbf{k})$ is not diagonalizable. The Supplementary Note 1 contains a stability analysis of our specific model.

In the stable regime of interest here the matrix $\hat{\sigma}_z \hat{h}(\mathbf{k})$ is diagonalizable and all its eigenvalues are real. In this case, its eigenvectors $|m\rangle$ can be chosen to be mutually $\hat{\sigma}_z$ orthogonal. In addition, there are exactly N positive (negative) norm eigenvectors. Thus, it is always possible to enforce the symplectic orthonormality relations Equation (6) by identifying the (appropriately normalized) positive and negative norm solutions with $|\mathbf{k}_n\rangle$ and $\mathcal{K}\hat{\sigma}_x|-\mathbf{k}_n\rangle$, respectively. The corresponding eigenvalues are then to be identified with $E_n[\mathbf{k}]$ (particle band structure) and $-E_n[-\mathbf{k}]$ (hole band structure), respectively.

Particle–hole symmetry. The Bogoliubov de Gennes Hamiltonian has the generalized symmetry $\mathcal{C}^\dagger \hat{h}(\mathbf{k}) \mathcal{C} = -\hat{h}(-\mathbf{k})$ where the charge conjugation operator C is anti-unitary and $C^2 = \mathbb{1}_{2N}$. Thus, our system represents the bosonic analogue of a superconductor in the Class D of the standard topological classification. This is a simple consequence of the doubling of the degrees of freedom in the single-particle picture. It simply reflects that the set of ladder operators $\hat{\beta}^\dagger_{n,\mathbf{k}}$ and $\hat{\beta}_{n,-\mathbf{k}}$ calculated from $\hat{h}(\mathbf{k})$ are the adjoint of the set of operators $\hat{\beta}_{n,\mathbf{k}}$ and $\hat{\beta}^\dagger_{n,-\mathbf{k}}$ calculated from $\hat{h}(-\mathbf{k})$.

Details of the transport calculations. In our transport calculations we have included photon decay. We adopt the standard description of the dissipative dynamics of photonic systems in terms of the Langevin equation and the input–output theory[47], for each site:

$$\dot{\hat{a}}_j = i[\hat{H}, \hat{a}_j] - \kappa \hat{a}_j/2 + \sqrt{\kappa}\hat{a}_j^{(\text{in})}. \tag{17}$$

In practice, we consider an array of detuned parametric amplifiers with intensity decay rate κ and add to the standard description of each parametric amplifier the inter-cell coherent coupling described in the main text. The last term describes the influence of the input field $\hat{a}_j^{(\text{in})}$ injected by an additional probe drive including also the environment vacuum fluctuations. The field $\hat{a}_j^{(\text{out})}$ leaking out of each cavity at site j is given by the input–output relations

$$\hat{a}_j^{(\text{out})} = \hat{a}_j^{(\text{in})} - \sqrt{\kappa}\hat{a}_j. \tag{18}$$

The above formulas give an accurate description of a photonic system where the intrinsic losses during injection and inside the system are negligible. Intrinsic photon absorption can be incorporated by adding another decay channel to the equation for the light field. It reduces the propagation length but does not change qualitatively the dynamics.

In Fig. 3, we show the probabilities $T_E(\omega, l, j)$ and $T_I(\omega, l, j)$ that a photon injected on site j with frequency $\omega_{\text{in}} = \omega + \omega_L/2$ is transmitted elastically (at frequency $\omega + \omega_L/2$) or inelastically (at frequency $\omega_L/2 - \omega$) to site l where it is detected. From the Kubo formula and the input–output relations we find

$$T_E(\omega, l, j) = \left| \delta_{lj} - i\kappa \tilde{G}_E(\omega, l, j) \right|^2, \tag{19}$$

$$T_I(\omega, l, j) = \kappa^2 \left| \tilde{G}_I(\omega, l, j) \right|^2. \tag{20}$$

Here $\tilde{G}_{E/I}(\omega, l, j)$ are the elastic and inelastic components of the Green's function in frequency space,

$$\tilde{G}_E(\omega, l, j) = -i \int_{-\infty}^{\infty} dt \Theta(t) \left\langle \left[\hat{a}_l(t), \hat{a}_j^\dagger(0) \right] \right\rangle e^{i\omega t}, \tag{21}$$

$$\tilde{G}_I(\omega, l, j) = -i \int_{-\infty}^{\infty} dt \Theta(t) \left\langle \left[\hat{a}_l^\dagger(t), \hat{a}_j^\dagger(0) \right] \right\rangle e^{i\omega t}. \tag{22}$$

In a N site array with single-particle eigenstates $|n\rangle = (u_n[1], \ldots, u_n[N], v_n[1], \ldots, v_n[N])^T$, the Green's functions read

$$G_E(\omega, l, j) = \sum_n \frac{u_n[l] u_n^*[j]}{\omega - E[n] + i\kappa/2} - \frac{v_n^*[l] v_n[j]}{\omega + E[n] + i\kappa/2}, \tag{23}$$

$$G_I(\omega, l, j) = \sum_n \frac{v_n[l] u_n^*[j]}{\omega - E[n] + i\kappa/2} - \frac{u_n^*[l] v_n[j]}{\omega + E[n] + i\kappa/2}. \tag{24}$$

We note that for a probe field inside the bandwidth of the particle (hole) sector but far detuned from the hole (particle) sector, only the first (second) term of the

summand in Equations (23) and (24) is resonant. Thus, as expected, the inelastic scattering is comparatively larger when the probe field is in the hole band gap.

It is easy to estimate quantitatively the relative intensities of elastically and inelastically transmitted light when the parametric interaction of the $\hat{\alpha}$ Bogoliubov quasiparticles is small (the regime where Φ can be interpreted as a synthetic gauge field experienced by the Bogoliubov quasiparticles). In this case, it is straightforward to show that $|v_n[j]/u_n[j]| \approx \tanh r$ independent of the eigenstate n and the site j. By putting together Equations (20, 23, 24) and neglecting the off-resonant terms we find that for $\omega, \tilde{\omega} \gg |\tilde{J}|, \kappa, |\omega - \tilde{\omega}|$,

$$T_I(\omega, l, j) \approx (\tanh r)^2 T_E(\omega, l, j) \approx T_I(-\omega, l, j) \approx (\coth r)^2 T_E(-\omega, l, j).$$

These analytical formulas agree quantitatively with the numerical results shown in Fig. 4b (note that in Fig. 4b the transmission at the output sites is rescaled by the overall transmission, $\sum_{l \neq j} T_I(\omega, l, j) + T_E(\omega, l, j)$).

References

1. Goldman, N., Juzeliunas, G., Öhberg, P. & Spielman, I. B. Light-induced gauge fields for ultracold atoms. *Rep. Prog. Phys.* **77**, 126401 (2014).
2. Lu, L., Joannopoulos, J. D. & Soljacic, M. Topological photonics. *Nat. Photon.* **8**, 821–829 (2014).
3. Prodan, E. & Prodan, C. Topological phonon modes and their role in dynamic instability of microtubules. *Phys. Rev. Lett.* **103**, 248101 (2009).
4. Kane, C. L. & Lubensky, T. C. Topological boundary modes in isostatic lattices. *Nat. Phys.* **10**, 39–45 (2013).
5. Peano, V., Brendel, C., Schmidt, M. & Marquardt, F. Topological phases of sound and light. *Phys. Rev. X* **5**, 031011 (2015).
6. Yang, Z. et al. Topological acoustics. *Phys. Rev. Lett.* **114**, 114301 (2015).
7. Süsstrunk, R. & Huber, S. D. Observation of phononic helical edge states in a mechanical topological insulator. *Science* **349**, 47–50 (2015).
8. Paulose, J., Chen, B. G. & Vitelli, V. Topological modes bound to dislocations in mechanical metamaterials. *Nat. Phys.* **11**, 153–156 (2015).
9. Nash, L. M. et al. Topological mechanics of gyroscopic metamaterials. *Proc. Natl Acad. Sci. USA* **112**, 14495–14500 (2015).
10. Haldane, F. D. M. & Raghu, S. Possible realization of directional optical waveguides in photonic crystals with broken time-reversal symmetry. *Phys. Rev. Lett.* **100**, 013904 (2008).
11. Raghu, S. & Haldane, F. D. M. Analogs of quantum-hall-effect edge states in photonic crystals. *Phys. Rev. A* **78**, 033834 (2008).
12. Wang, Z., Chong, Y., Joannopoulos, J. D. & Soljacic, M. Observation of unidirectional backscattering-immune topological electromagnetic states. *Nature* **461**, 772–775 (2009).
13. Koch, J., Houck, A. A., Le Hur, K. & Girvin, S. M. Time-reversal-symmetry breaking in circuit-QED-based photon lattices. *Phys. Rev. A* **82**, 043811 (2010).
14. Umucallar, R. O. & Carusotto, I. Artificial gauge field for photons in coupled cavity arrays. *Phys. Rev. A* **84**, 043804 (2011).
15. Fang, K., Yu, Z. & Fan, S. Realizing effective magnetic field for photons by controlling the phase of dynamic modulation. *Nat. Photon.* **6**, 782–787 (2012).
16. Petrescu, A., Houck, A. A. & Le Hur, K. Anomalous Hall effects of light and chiral edge modes on the Kagomé lattice. *Phys. Rev. A* **86**, 053804 (2012).
17. Tzuang, L. D., Fang, K., Nussenzveig, P., Fan, S. & Lipson, M. Non-reciprocal phase shift induced by an effective magnetic flux for light. *Nat. Photon.* **8**, 701–705 (2014).
18. Schmidt, M., Kessler, S., Peano, V., Painter, O. & Marquardt, F. Optomechanical creation of magnetic fields for photons on a lattice. *Optica* **2**, 635–641 (2015).
19. Hafezi, M., Demler, E. A., Lukin, M. D. & Taylor, J. M. Robust optical delay lines with topological protection. *Nat. Phys.* **7**, 907–912 (2011).
20. Khanikaev, A. B. et al. Photonic topological insulators. *Nat. Mater.* **12**, 233–239 (2012).
21. Hafezi, M., Mittal, S., Fan, J., Migdall, A. & Taylor, J. M. Imaging topological edge states in silicon photonics. *Nat. Photon.* **7**, 1001–1005 (2013).
22. Mittal, S. et al. Topologically robust transport of photons in a synthetic gauge field. *Phys. Rev. Lett.* **113**, 087403 (2014).
23. Kitagawa, T. et al. Observation of topologically protected bound states in photonic quantum walks. *Nat. Commun.* **3**, 882 (2012).
24. Rechtsman, M. C. et al. Topological creation and destruction of edge states in photonic graphene. *Phys. Rev. Lett.* **111**, 103901 (2013).
25. Bardyn, C.-E. & Imamoğlu, A. Majorana-like modes of light in a one-dimensional array of nonlinear cavities. *Phys. Rev. Lett.* **109**, 253606 (2012).
26. Ryu, S., Schnyder, A. P., Furusaki, A. & Ludwig, A. W. W. Topological insulators and superconductors: tenfold way and dimensional hierarchy. *New J. Phys.* **12**, 065010 (2010).
27. Shindou, R., Matsumoto, R., Murakami, S. & Ohe, J. Topological chiral magnonic edge mode in a magnonic crystal. *Phys. Rev. B* **87**, 174427 (2013).
28. Barnett, R. Edge-state instabilities of bosons in a topological band. *Phys. Rev. A* **88**, 063631 (2013).
29. Engelhardt, G. & Brandes, T. Topological bogoliubov excitations in inversion-symmetric systems of interacting bosons. *Phys. Rev. A* **91**, 053621 (2015).

30. Galilo, B., Lee, D. K. K. & Barnett, R. Selective population of edge states in a 2d topological band system. *Phys. Rev. Lett.* **115**, 245302 (2015).

31. Bardyn, C.-E., Karzig, T., Refael, G. & Liew, T. C. H. Chiral bogoliubov excitations in nonlinear bosonic systems. *Phys. Rev. B* **93**, 020502 (2016).

32. Gerry, C. C. & Knight, P. L. *Introductory Quantum Optics* (Cambridge University Press (2005).

33. Ohgushi, K., Murakami, S. & Nagaosa, N. Spin anisotropy and quantum hall effect in the *kagomé* lattice: chiral spin state based on a ferromagnet. *Phys. Rev. B* **62**, R6065–R6068 (2000).

34. Green, D., Santos, L. & Chamon, C. Isolated flat bands and spin-1 conical bands in two-dimensional lattices. *Phys. Rev. B* **82**, 075104 (2010).

35. Fukui, T., Hatsugai, Y. & Suzuki, H. Chern numbers in discretized brillouin zone: Efficient method of computing (spin) hall conductances. *J. Phys. Soc. Jpn.* **74**, 1674–1677 (2005).

36. Notomi, M., Kuramochi, E. & Tanabe, T. Large-scale arrays of ultrahigh-q coupled nanocavities. *Nat. Photon.* **2**, 741–747 (2008).

37. Mookherjea, S. & Yariv, A. Coupled resonator optical waveguides. *IEEE J. Quantum Elec.* **8**, 448 (2002).

38. Eggleton, B. J., Luther-Davies, B. & Richardson, K. Chalcogenide photonics. *Nat. Photon.* **5**, 141–148 (2011).

39. Dahdah, J., Pilar-Bernal, M., Courjal, N., Ulliac, G. & Baida, F. Near-field observations of light confinement in a two dimensional lithium niobate photonic crystal cavity. *J. Appl. Phys.* **110**, 074318 (2011).

40. Safavi-Naeini, A. H. *et al.* Squeezed light from a silicon micromechanical resonator. *Nature* **500**, 185–189 (2013).

41. Purdy, T. P., Yu, P. L., Peterson, R. W., Kampel, N. S. & Regal, C. A. Strong Optomechanical Squeezing of Light. *Phys. Rev. X* **3**, 031012 (2013).

42. Kronwald, A., Marquardt, F. & Clerk, A. A. Arbitrarily large steady-state bosonic squeezing via dissipation. *Phys. Rev. A* **88**, 063833 (2013).

43. Bergeal, N. *et al.* Analog information processing at the quantum limit with a Josephson ring modulator. *Nat. Phys.* **6**, 296–302 (2010).

44. Abdo, B., Kamal, A. & Devoret, M. Nondegenerate three-wave mixing with the Josephson ring modulator. *Phys. Rev. B* **87**, 014508 (2013).

45. Underwood, D. L., Shanks, W. E., Koch, J. & Houck, A. A. Low-disorder microwave cavity lattices for quantum simulation with photons. *Phys. Rev. A* **86**, 023837 (2012).

46. Gurarie, V. & Chalker, J. T. Bosonic excitations in random media. *Phys. Rev. B* **68**, 134207 (2003).

47. Clerk, A. A., Devoret, M. H., Girvin, S. M., Marquardt, F. & Schoelkopf, R. J. Introduction to quantum noise, measurement, and amplification. *Rev. Mod. Phys.* **82**, 1155–1208 (2010).

Acknowledgements
V.P., C.B., and F.M. acknowledge support by an ERC Starting Grant OPTOMECH, by the DARPA project ORCHID, and by the European Marie-Curie ITN network cQOM. M.H. and A.A.C. acknowledge support from NSERC.

Author contributions
V.P., A.A.C. and F.M. contributed to the conceptual development of the project and interpretation of results. Calculations and simulations were done by V.P., M.H. and C.B.

Additional information

8

Super-crystals in composite ferroelectrics

D. Pierangeli[1], M. Ferraro[1], F. Di Mei[1,2], G. Di Domenico[1,2], C.E.M. de Oliveira[3], A.J. Agranat[3] & E. DelRe[1]

As atoms and molecules condense to form solids, a crystalline state can emerge with its highly ordered geometry and subnanometric lattice constant. In some physical systems, such as ferroelectric perovskites, a perfect crystalline structure forms even when the condensing substances are non-stoichiometric. The resulting solids have compositional disorder and complex macroscopic properties, such as giant susceptibilities and non-ergodicity. Here, we observe the spontaneous formation of a cubic structure in composite ferroelectric potassium-lithium-tantalate-niobate with micrometric lattice constant, 10^4 times larger than that of the underlying perovskite lattice. The 3D effect is observed in specifically designed samples in which the substitutional mixture varies periodically along one specific crystal axis. Laser propagation indicates a coherent polarization super-crystal that produces an optical X-ray diffractometry, an ordered mesoscopic state of matter with important implications for critical phenomena and applications in miniaturized 3D optical technologies.

[1] Dipartimento di Fisica, Università di Roma 'La Sapienza', Rome 00185, Italy. [2] Center for Life Nano Science@Sapienza, Istituto Italiano di Tecnologia, Rome 00161, Italy. [3] Department of Applied Physics, Hebrew University of Jerusalem, Jerusalem 91904, Israel. Correspondence and requests for materials should be addressed to E.D. (email: eugenio.delre@uniroma1.it).

Textbook models of global symmetry-breaking include a low-symmetry low-temperature state with a fixed infinitely extended coherence. In contrast, the spontaneous polarization observed as spatial inversion symmetry is broken during a paraelectric–ferroelectric phase transition generally leads to a disordered mosaic of polar domains that permeate the finite samples[1]. Coherent and ordered ferroelectric states with remarkable properties of both fundamental and technological interest[2–5] can emerge when ferroelectricity is influenced by external factors, such as system dimensionality[6], strain gradients[7–9], electrostatic coupling[10,11] and magnetic interaction[12,13].

Here we report the spontaneous formation of an extended coherent three-dimensional (3D) superlattice in the nominal ferroelectric phase of specifically grown potassium–lithium–tantalate–niobate (KLTN) crystals[14–17]. Visible-light propagation reveals a polarization super-crystal with a micrometric lattice constant, a counterintuitive mesoscopic phase that naturally mimics standard solid-state structures but on scales that are thousands of times larger. The phenomenon is achieved using compositionally disordered ferroelectrics[18–27]. At one given temperature, these have the interesting property of manifesting a single perovskite phase whose dielectric properties depend on the specific composition[28–30]. For example, a compositional gradient along the pull axis leads to a position-dependent Curie point $T_C(\mathbf{r})$, so that for a given value of crystal temperature T a phase separation occurs, where regions with $T > T_C$ are paraelectric and those with $T < T_C$ have a spontaneous polarization[31]. Specifically tailored growth schemes are even able to achieve an oscillating T_C along a given direction, say the x axis[32,33]. Under these conditions, we can expect that, at a given T in proximity of the average (macroscopic) T_C, the sample will be in a hybrid state with alternating regions with and without spontaneous polarization. Crossing the Curie point, under conditions in which perovskite polar domains pervade the volume forming 90° configurations to minimize the free energy associated with polarization charge[34], this oscillation can form a full 3D periodic structure.

Results

Observation of a compositionally induced super-crystal.
To investigate the matter, we make use of top-seeded ferroelectric crystals with an oscillating composition along the growth axis achieved using an off-centre growth technique in the furnace[33,35]. We obtain a zero-cut 2.4 mm by 2.0 mm by 1.7 mm, along the x,y,z directions, respectively, optical-quality KLTN sample with a periodically oscillating niobium composition of period $\Lambda = 5.5\,\mu m$ along the x axis, with an average composition $K_{1-\alpha}Li_\alpha Ta_{1-\beta}Nb_\beta O_3$, where $\alpha = 0.04$ and $\beta = 0.38$ (see Methods). When the crystal is allowed to relax at $T = T_C - 2\,K$, that is, in proximity of the spatially averaged room-temperature Curie point $T_C = 294\,K$, laser light propagating through the sample suffers relevant scattering with strongly anisotropic features (Fig. 1a–d). Typical results are reported in Fig. 1b–d, and they appear as an optical analogue of X-ray diffraction in low-temperature solids. This optical diffractometry provides basic evidence of a 3D superlattice at micrometric scales. Probing the principal crystal directions reveals several diffraction orders that map the entire reciprocal space. The large-scale super-crystal, which permeates the whole sample, overlaps—along the x direction—with the built-in compositional oscillating seed (see Methods). The superlattice extends in full three dimensions, with the same periodicity $\Lambda = 5.5\,\mu m$ of the x-oriented compositional oscillation, also along the orthogonal y and z directions. In particular, Fig. 1d indicates

that in the plane perpendicular to the built-in dielectric microstructure Γ vector, that is, where spatial symmetry should be unaffected by the microstructure in composition, the ferroelectric phase transition leads to a spontaneous pattern of transverse scale Λ. The corresponding elementary structure on micrometric spatial scales is reported in Fig. 1e; it can be represented as a face-centred cubic structure in which the occupation of one of the three faces ($z - y$ face) is missing[36]. The structure, which is, to our knowledge, not observed at atomic scales, can be reduced to a simple cubic structure with a threefold basis and lattice parameter $a = \Lambda$.

As the crystal is brought below the average Curie point, it manifests a metastable (supercooled) and a stable (cold) phase, as analysed in Fig. 2 both in the reciprocal (Fourier) and direct (real) space. In the nominal paraelectric phase, at $T = T_C + 2\,K$ (Fig. 2a), we observe the first Bragg diffraction orders (± 1) consistent with the presence of the seed microstructure, a one-dimensional (1D) transverse sinusoidal modulation acting as a diffraction grating; the distance from the central zero order fulfills the Bragg condition, that is, scattered light forms an angle $\theta_B = \lambda/2n_0\Lambda \simeq 7°$ with the incident wavevector $\mathbf{k}$. Crossing the ferroelectric phase-transition temperature T_C (see Methods), we detect a supercooled metastable state that has an apparently analoguous diffraction effect (Fig. 2b) that is dynamically superseded by the stable and coherent cold superlattice phase (Fig. 2c), in which spatial correlations are extended to the whole crystal volume. In real space, transmission microscopy (see Methods) shows unscattered optical propagation through the paraelectric sample at $T = T_C + 2\,K$ (Fig. 2d), which turns into critical opalescence and scattering from oblique random domains at the structural phase transition (Fig. 2e,f), and into unscattered transmission in the metastable ferroelectric phase at $T = T_C - 2\,K$ (Fig. 2g). After dipolar relaxation has taken place, the cold super-crystal appears in this case as a periodic intensity distribution on micrometric scales, as shown in Fig. 2h.

Spontaneous polarization underlying the ferroelectric superlattice.
To further analyse these supercooled and cold phases, we inspect the supercooled 1D phase (Fig. 2b) that is accessible through linear (unbiased) and electro-optic (biased) polarization-resolved Bragg diffraction measurements. In particular, referring to the set-up illustrated in Fig. 3a, we measure the diffraction efficiency $\eta = P_B/(P_B + P_0)$, where P_B and P_0 are, respectively, the diffracted and non-diffracted powers, in the first Bragg resonance condition, that is, with the incident wavevector $\mathbf{k}$ forming the angle θ_B with respect to the z axis. The diffraction efficiency η is reported in Fig. 3b for different input light polarization and temperature across the average Curie point. Diffraction strongly depends both on the nominal crystal phase and on the polarization of the incident wave: a large increase in η is found for light polarized in the x,z plane (H-polarized). For $T > T_C$, the dependence on light polarization is consistent with what expected in standard periodically index-modulated media (wave-coupled theory), that is, a weak temperature dependence and a maximum η for light polarized normal to the grating vector (V-polarized). In this case, the difference in η_H (Δ) and η_V ($\square$) can be related to the different Fresnel coefficients governing interlayer reflections and is congruently $\eta_V > \eta_H$ by an amount that decreases for larger θ_B (refs 37,38). Consistently, the (H + V)-polarized curve ($\bigcirc$), that is, when the input linear polarization is at 45° with respect to the H and V polarizations, falls between these two curves. Standard behaviour is violated for $T < T_C$, where a large enhancement in η_H rapidly leads to a regime with $\eta_V < \eta_H$.

The physical underpinnings of the super-crystal can be grasped considering the simple model illustrated in Fig. 3c. Here we

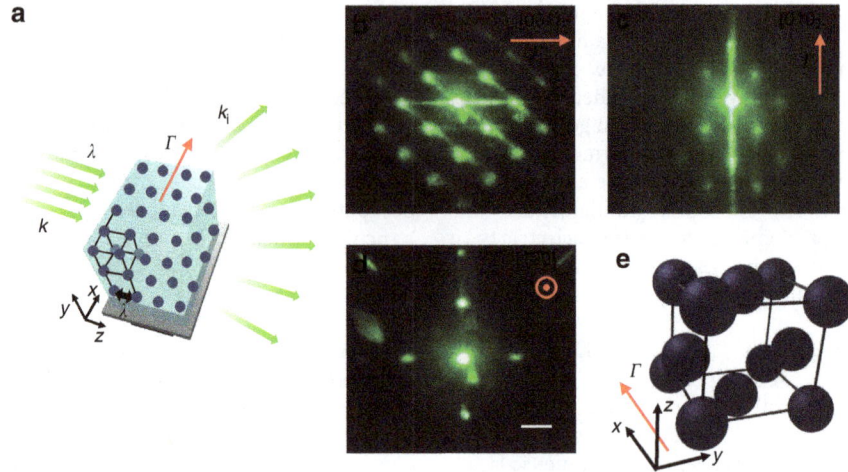

Figure 1 | Super-crystal in the ferroelectric phase. (a) Sketch of visible-light diffraction from micrometric structures through a transparent crystal and **(b-d)** 3D superlattice probed at $T = T_C - 2$ K along the principal symmetry direction of the crystal, respectively, with the incident wavevector **k** parallel to **(b)** the z direction, **(c)** y direction and **(d)** x direction. Crystallographic analysis reveals the elementary cubic structure of lattice constant Λ shown in **e**. Scale bar, 1.2 cm.

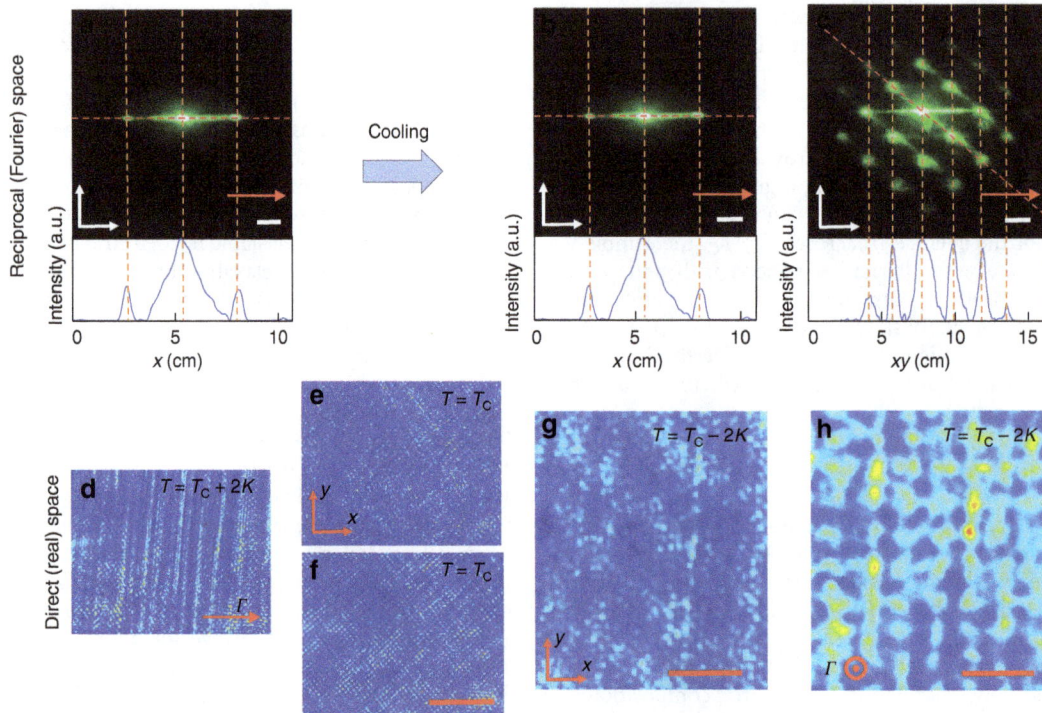

Figure 2 | Light diffraction above and below T_c. (a) Reciprocal space probed at $T = T_C + 2$ K (hot paraelectric phase), showing the first diffraction orders due to the one-dimensional sinusoidal compositional modulation. Cooling below the critical point results at $T = T_C - 2$ K (super-crystal ferroelectric phase) in **(b)** a supercooled (metastable) 1D superlattice with the same diffraction orders that relaxes at the steady state into **(c)** the cold (stable) super-crystals. In both **b,c** the direction of incident light is othogonal to Γ, as in **a**. **(d-h)** Corresponding transmission microscopy images revealing **(d)** unscattered optical propagation, **(e,f)** scattering at the phase transition, **(g)** unscattered optical propagation in the metastable superlattice and **(h)** periodic intensity distribution underlining the 3D superlattice. Metastable and stable (equilibrium) phases are inspected, respectively, at times $t \approx 1$ min and $t \approx 1$ h after the structural transition at $T = T_C$. Bottom profiles in **a-c** are extracted along the red dotted line. Scale bars **(a-c)**, 1.2 cm, **(d-f)**, 100 µm and **(g,h)**, 10 µm.

consider the metastable 1D superlattice (Fig. 2b) before tensorial effects cause the full 3D superlattice relaxation (Fig. 2c). Specifically, for a given T, regions with a local value of T_C such that $T < T_C$ (dark shading) will manifest a finite spontaneous polarization $P_S \neq 0$, whereas region with $T > T_C$ (light shading) will have a $P_S \simeq 0$. Optical measurements are sensitive to the square of the crystal polarization $\langle \mathbf{P} \cdot \mathbf{P} \rangle \simeq P_S^2$ through the

resulting index pattern modulated via the quadratic elecro-optic response $\delta n(P) = -(1/2)n^3 g P^2$, where n is the unperturbed refraction index and g is the corresponding perovskite elecro-optic coefficient[25,39]. Enhanced Bragg-scattering of light polarized parallel to the seed direction Γ (H in Fig. 3b—super-crystal) indicates that $P_S(x)$ is parallel to the seed direction (x axis), where the elecro-optic coefficient g has its maximum value

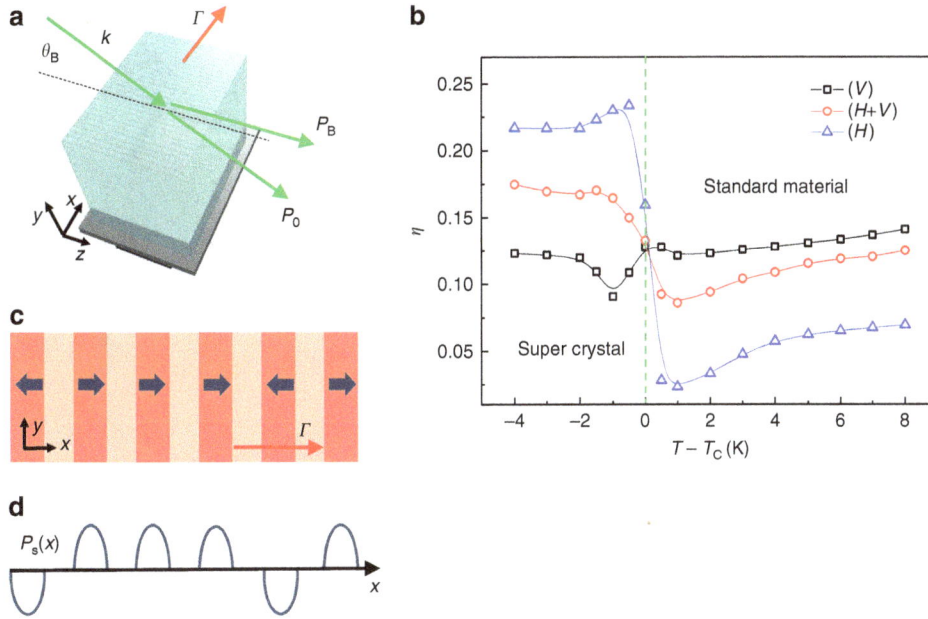

Figure 3 | Diffractive behaviour of the 1D supercooled superlattice. (**a**) Sketch of the experimental geometry and (**b**) detected diffraction efficiency (dots) as a function of temperature in the proximity of ferroelectric transition for different wave polarizations. An anomaly appears crossing T_C for H-polarized light signalling the emergence of the super-crystal. Lines are interpolations serving as guidelines. (**c**) Scheme of the periodically ordered ferroelectric state along the x direction underlying the super-crystal phase for $T < T_C$ and giving the spontaneous polarization $P_S(x)$ sketched in **d**.

$g = 0.16 \, \mathrm{m^4 C^{-2}}$. The resonant response at θ_B and the absence of higher harmonics (Fig. 2b) indicate that this $P_S(x)^2$ distribution is sinusoidal with wavevector Γ. Hence, although in general it may be that macroscopically $\langle \mathbf{P} \rangle \simeq 0$, it turns out that $\langle \mathbf{P}^2 \rangle \simeq P_S^2 \neq \langle \mathbf{P} \rangle^2 \neq 0$ on the micrometric scales, in analogy with optical response in crystals affected by polar nanoregions[25,27,40]. Optical diffraction efficiency reported in Fig. 3b then occurs considering $\eta = \sin^2(\frac{\pi d (\delta n)}{\lambda \cos \theta_B})$, with resonant enhanced diffraction for $T < T_C$ caused by $\delta n = \delta n_0 + \delta n(P)$, where $\delta n_0 \sim 10^{-4}$ is the polarization-independent index change due to the periodic composition variation (Sellmeier's index change).

Electro-optical diffraction analysis. To validate this picture, we perform electro-optic diffractometry experiments, in which a macroscopic polarization activating the nonlinear periodic response is induced via an external static field E applied along x. Results are reported in Fig. 4; in particular, in Fig. 4a the polarization and field dependence of η are shown at $T = T_C + 2 \, \mathrm{K}$. We observe a nearly field-independent behaviour for V-polarized light, which arises from its low electro-optic coupling (bias field and light polarization are orthogonal, $g = -0.02 \, \mathrm{m^4 C^{-2}}$); differently, η_H increases with the field showing a 'discontinuity' at the critical field $E_C = (1.4 \pm 0.1) \, \mathrm{kV \, cm^{-1}}$. The strong similarity between this enhancement and those observed under unbiased conditions at T_C (Fig. 3b) indicates that E_C coincides with the coercive field, and the discontinuity corresponds to the field-induced phase transition[16,26,35]. In fact, in Fig. 4b we repeat this experiment, enhancing the experimental field sensitivity and acquiring data also for decreasing field amplitudes. The result is a partial hysteretic loop for the diffraction efficiency that demontrates the field-induced transition and underlines that, both in the linear and nonlinear (electro-optic) case, the effect of the seeded ferroelectric ordering is to provide a periodic spontaneous polarization along x. We also note a slight asymmetry with respect to positive/negative fields; this is associated with a residual fixed space–charge field that may play an important role in the spontaneous polarization alignment

process and hence in leading to a residual $\langle \mathbf{P} \rangle \neq 0$. The existence of a periodic spontaneous polarization distribution in the superlattice (Fig. 3c) is confirmed in Fig. 4c, where electro-optic Bragg diffraction below T_C is reported. An oscillating full-hysteretic behaviour is observed as a function of the external field, consistently with the prediction $\eta(E) = \sin^2(\frac{\pi d (\delta n (E))}{\lambda \cos \theta_B})$ with $\delta n(E) = \delta n_0 + (1/2) n^3 g \left(P_S^2 + 2 \varepsilon_0 \chi \langle P_S \rangle E + \varepsilon_0^2 \chi^2 E^2 \right)$. The increase in η due to the superlattice polarization allows us to explore its full sinusoidal behaviour, which usually requires extremely large fields in the paraelectric phase and reduces to a parabolic behaviour (Fig. 4d)[41]. From this parabolic behaviour detected at $T = T_C + 5 \, \mathrm{K}$ we estimate that the resulting ampitude in the point-dependent Curie temperature due to the compositional modulation is $\Delta T_C \simeq 2 \, \mathrm{K}$ (ref. 32). Agreement with the periodic polarization model is further stressed by deviations emerging in $\eta(E)$, especially for low and negative increasing fields, where the dependence on $\langle P_S \rangle$ makes observations weakly dependent on the specific experimental realization.

Discussion

An interesting point arising from the experimental results and analysis is how the periodically ordered polarization state along the x direction leads to the super-crystal. Since we pass spontaneously from a metastable to a stable mesoscopic phase, polar-domain dynamics in the presence of the fixed spatial scale Λ play a key role. In fact, we note that the 1D superlattice sketched in Fig. 3c involves the appearence of charge density and associated strains between polar planes, so that the ferroelectric crystal naturally tends to relax into a more stable configuration. In standard perovskites, equilibrium configurations are mainly those involving a 180° and 90° orientation between adjacent polar domains, as schematically shown in Fig. 5a. To explain the 3D polar state and its periodical features underlying the super-crystal, we consider the 90° configuration, which is characterized by 45° domain walls that we observe in a disordered configuration during the ferroelectric phase transition at T_C (Fig. 2f). Owing to the periodic constraint along the x axis, this arrangement has the

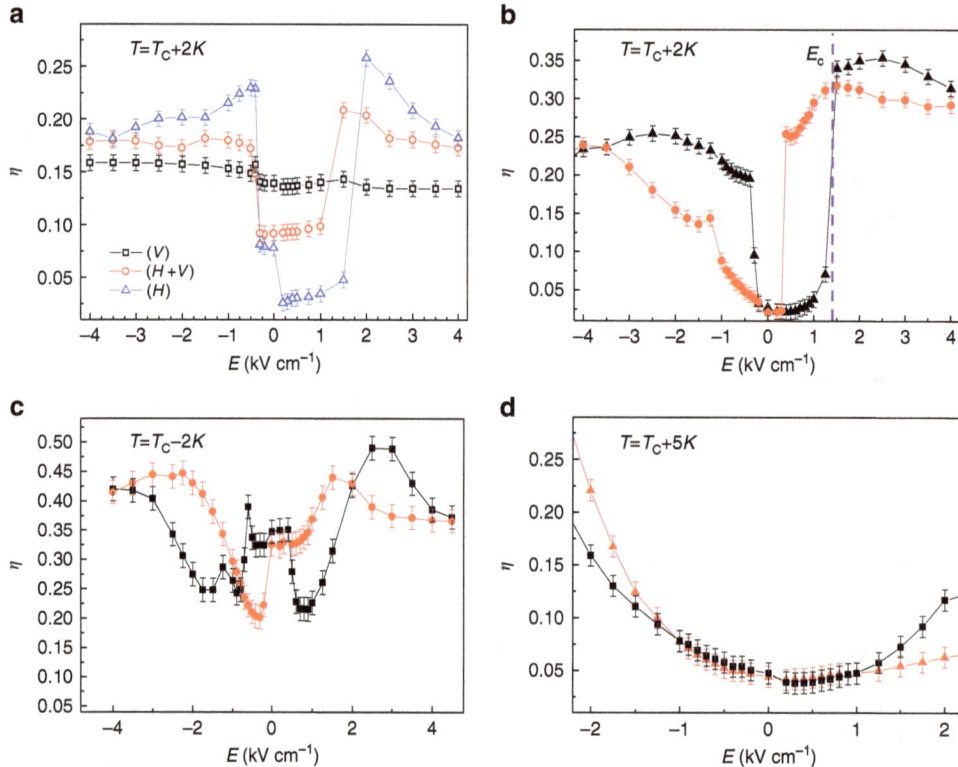

Figure 4 | Electro-optic Bragg diffraction in the critical region. (**a**) Diffraction efficiency as a function of the external applied field for different light polarizations at $T = T_C + 2\,K$; (**b**) hysteresis loop at the same temperature and (**c**) at $T = T_C - 2\,K$ for H-polarization. (**d**) Expected [32] weak-histeretic paraelectric (parabolic) behaviour at $T = T_C + 5\,K$. In **b–d**, black and red dots indicate data obtained, respectively, increasing and decreasing the bias fields. Lines are interpolations serving as guidelines. Error bars are given by the statistics on five experimental realizations.

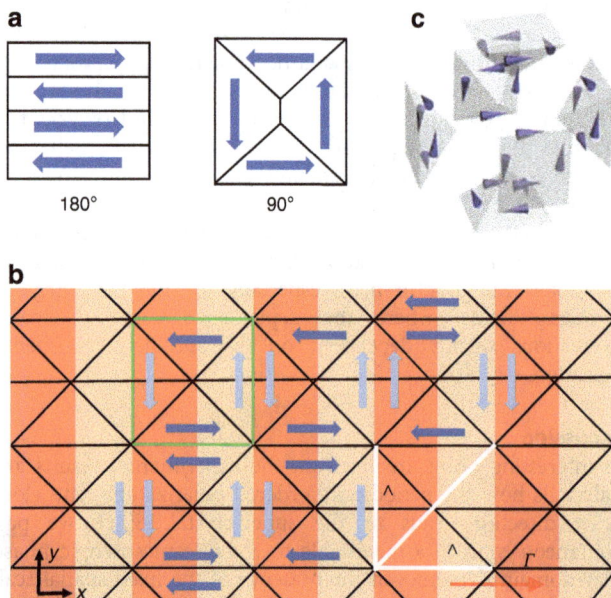

Figure 5 | Polar-domain configuration underlying the 3D superlattice. (**a**) Typical 180° and 90° domain configurations in perovskite ferroelectrics. (**b**) Planar domain arrangement scheme in the stable super-crystal phase obtained with elementary blocks of 90° configurations (green cell). In this periodically ordered ferroelectric state, the compositional modulation (as for Fig. 3c), other domain walls ruling optical diffractometry (black lines), and periods along x, y and xy axes (white bars) are highlighted. Vertical polarizations have a lighter colour to stress their weak optical response in our KLTN sample. (**c**) Extension of the single unit cell (green cell in **b**) in three dimensions.

unique property of reproducing our observations, minimizing energy associated to internal charge density and transferring the built-in 1D order to the whole volume with the same spatial scale Λ. We illustrate the domain pattern in Fig. 5b for the $x - y$ plane, whereas in Fig. 5c the elementary cell is shown in the 3D case, where it maintains its stability features in terms of charge density energy. In particular, in Fig. 5b, domain walls resulting in the diffraction orders of Fig. 1b are marked, as well as the 45° correlation period, which agrees with optical observations of the reciprocal space. We further stress that vertical domains (light blue in Fig. 5b) are optically analoguous to paraelectric regions; moreover, 180° rotations in the polarization direction in each polar region has no effect on the optical response. In view of the symmetry of this arrangement, the observed diffraction anisotropy (Fig. 1d) is then associated to the absence of grating planes in the $y - z$ face.

Further insight on the 3D domain structure requires numerical simulations based on Monte Carlo methods[42,43] and phase-field models[44–47]; they may confirm our picture and reveal new aspects for ferroelectricity, such as polar dynamics, spontaneous long-range ordering and the role of polar strains in composite ferroelectrics with built-in compositional microstructures. In fact, the effect of the composition profile is here crucial in triggering the spontaneous formation of the macroscopic coherent structure, as it sets the typical domain size along the x direction and so rules the whole dynamic towards the equilibrium state. We expect that a different amplitude and period of the modulation may affect the formation, stability, time- and temperature dynamics of the super-crystal; indeed, the parameters of the compositional gradient may be important in determining the interaction between polar regions. Advanced growth techniques[32] can open future perpectives in this direction, as well as towards composite ferroelectrics with different compositional shapes of fundamental and applicative interest.

To conclude, we have reported the formation of a mesoscopic polarization super-crystal in a nanodisordered sample of KLTN. The large-scale coherent state is triggered by a periodically modulated change in composition. Our results show how ferroelectricity can be arranged into new phases, so that in proximity of an average critical temperature a structural order can emerge with a micrometric lattice constant so as to cause light to suffer diffraction as occurs for X-rays in standard crystals. The effect not only opens new avenues in the optical exploration of critical properties and large-scale structures in disordered systems, but also suggests methods to predict and engineer new states of matter. It can also have an impact on the development of innovative technologies, such as nonvolatile electronic and optical structured memories[2-4], microstructured piezo devices and spatially resolved miniaturized electro-optic devices[27,41,48].

Methods

Growth and properties of the microstructured KLTN sample.
We consider a compositionally disordered perovskite of KLTN, $K_{1-x}Li_xTa_{1-\beta}Nb_\beta O_3$ with $\alpha = 0.04$ and $\beta = 0.38$, grown through the top-seeded solution method by extracting a zero-cut 2.4 mm by 2.0 mm by 1.7 mm, along the x, y, z directions, respectively, optical-quality specimen. It shows, through low-frequency dielectric spectroscopy measurements, the spatial-averaged Curie point, which signals the transition from the high-temperature symmetric paraelectric phase to the low-temperature ferroelectric phase, at the room temperature $T_C = 294$ K. A 1D seed microstructure is embedded into the sample as it is grown through the off-centre growth technique so as to manifest a sinusoidal variation in the low-frequency dielectric constant, and thus in the critical temperature T_C, along the growth axis (x direction)[33,35]. This dielectric volume microstructure causes an index of refraction oscillation of period $\Lambda = 5.5\,\mu m$, which is able to diffract light linearly and electro-optically[49]. Details on the technique employed in the sample growth can be found in ref. 33. We note here that the composition amplitude of the periodic microstructure can be estimated from $\Delta\beta/\Delta T$, where $\Delta\beta$ is the amplitude variation in niobium composition and ΔT is the change in the growth temperature incurred by the off-centre rotation. At the growth temperature of $\sim 1{,}470$ K, the ratio $\Delta\beta/\Delta T \approx 0.35$ mol K^{-1} has been extracted from the phase diagram of KTN. The temperature variation incurred by the off-centre rotation was measured to be 3 K, from which we obtain $\Delta\beta \approx 1.05\text{‰}$ mol.

Optical diffraction experiments.
The macroscopic linear and electro-optic diffractive properties of the crystal have been investigated launching low-power (mW) plane waves at $\lambda = 532$ nm that propagate normally and parallelly to the grating vector Γ ($\Gamma = 2\pi/\Lambda$), which is along the x direction (Fig. 3a). Light diffracted by the medium is detected using a broad-area CCD (charge-coupled device) camera placed at $d = 0.2$ m from the crystal output facet or collected into Si power meters. In real-space measurements (Fig. 2d–h), the output crystal facet is imaged on the CCD camera and a cross-polarizer set-up[25,27] has been used to highlight contrast due to polarization inhomogeneities. The time needed to obtain a fully correlated state corresponding to the 3D super-crystal depends on the cooling rate τ and on the details of the thermal environment. Considering, for instance, as a thermal protocol a cooling rate $\tau = 0.05$ K s^{-1} and an environment at $T = T_C + 1$ K (weak thermal gradients), we have found that the metastable 1D lattice state at $T = T_C - 2$ K (Fig. 2b), in which correlations involve mainly in the direction including the Γ vector, lasts ~ 1 h. In this stage, although no macroscopic order occurs in the other directions[50], we observe optimal optical transmission of the sample (Fig. 2g); output light is not affected by scattering related to the existence of random domains and this undelines the presence of a mesoscopic ordering process in which the typical domain size is set. As regards the inspected temperature range, we have found that the super-crystal forms for temperatures till $T = 288$ K, although correlations are weaker at the lower temperatures. This is consistent with the fact that at these temperatures also the regions with a lower local T_C are well below the transition point.

References

1. Rabe, K. M., Ahn, C. H. & Triscone, J. M. *Physics of Ferroelectrics: a Modern Perspective*, Vol. 105 (Springer Science & Business Media, 2007).
2. Choi, K. J. *et al.* Enhancement of ferroelectricity in strained BaTiO3 thin films. *Science* **306**, 1005–1009 (2004).
3. Lee, H. N., Christen, H. M., Chisholm, M. F., Rouleau, C. M. & Lowndes, D. H. Strong polarization enhancement in asymmetric three-component ferroelectric superlattices. *Nature* **433**, 395–399 (2005).
4. Garcia, V. *et al.* Giant tunnel electroresistance for non-destructive readout of ferroelectric states. *Nature* **460**, 81–84 (2005).

5. Kim, W.-H., Son, J. Y., Shin, Y.-H. & Jang, H. M. Imprint control of nonvolatile shape memory with asymmetric ferroelectric multilayers. *Chem. Mater.* **26**, 6911–6914 (2014).
6. Dawber, M., Rabe, K. M. & Scott, J. F. Physics of thin-film ferroelectric oxides. *Rev. Mod. Phys.* **77**, 1083 (2005).
7. Catalan, G. *et al.* Polar domains in lead titanate films under tensile strain. *Phys. Rev. Lett.* **96**, 127602 (2006).
8. Catalan, G. *et al.* Flexoelectric rotation of polarization in ferroelectric thin films. *Nat. Mater.* **10**, 963–967 (2011).
9. Biancoli, A., Fancher, C. M., Jones, J. L. & Damjanovic, D. Breaking of macroscopic centric symmetric in paraelectric phases of ferroelectric materials and implications for flexoelectricity. *Nat. Mater.* **14**, 224–229 (2014).
10. Bousquet, E. *et al.* Improper ferroelectricity in perovskite oxide artificial superlattices. *Nature* **452**, 732–736 (2008).
11. Callori, S. J. *et al.* Ferroelectric PbTiO3/SrRuO3 superlattices with broken inversion symmetry. *Phys. Rev. Lett.* **109**, 067601 (2012).
12. Lawes, G. *et al.* Magnetically driven ferroelectric order in Ni3V2O8. *Phys. Rev. Lett.* **95**, 087205 (2005).
13. Baledent, V. *et al.* Evidence for room temperature electric polarization in RMn2O5 multiferroics. *Phys. Rev. Lett.* **114**, 117601 (2015).
14. DelRe, E., Spinozzi, E., Agranat, A. J. & Conti, C. Scale-free optics and diffractionless waves in nanodisordered ferroelectrics. *Nat. Photon.* **5**, 39–42 (2011).
15. DelRe, E. *et al.* Subwavelength anti-diffracting beams propagating over more than 1,000 Rayleigh lengths. *Nat. Photon.* **9**, 228–232 (2015).
16. Pierangeli, D., DiMei, F., Conti, C., Agranat, A. J. & DelRe, E. Spatial rogue waves in photorefractive ferroelectrics. *Phys. Rev. Lett.* **115**, 093901 (2015).
17. Di Mei, F. *et al.* Observation of diffraction cancellation for nonparaxial beams in the scale-free-optics regime. *Phys. Rev. A* **92**, 013835 (2015).
18. Shvartsman, V. V. & Lupascu, D. C. Lead-free relaxor ferroelectrics. *J. Am. Ceram. Soc.* **95**, 1–26 (2012).
19. Kutnjak, Z., Blinc, R. & Petzelt, J. The giant electromechanical response in ferroelectric relaxors as a critical phenomenon. *Nature* **441**, 956–959 (2006).
20. Glinchuk, M. D., Eliseev, E. & Morozovska, A. Superparaelectric phase in the ensemble of noninteracting ferroelectric nanoparticles. *Phys. Rev. B* **78**, 134107 (2008).
21. Pirc, R. & Kutnjak, Z. Electric-field dependent freezing in relaxor ferroelectrics. *Phys. Rev. B* **89**, 184110 (2014).
22. Manley, M. E. *et al.* Phonon localization drives polar nanoregions in a relaxor ferroelectric. *Nat. Commun.* **5**, 3683 (2014).
23. Phelan, D. *et al.* Role of random electric fields in relaxors. *Proc. Natl Acad. Sci. USA* **111**, 1754–1759 (2014).
24. Chang, Y. C., Wang, C., Yin, S., Hoffman, R. C. & Mott, A. G. Giant electro-optic effect in nanodisordered KTN crystals. *Opt. Lett.* **38**, 4574–4577 (2013).
25. Pierangeli, D. *et al.* Observation of an intrinsic nonlinearity in the electro-optic response of freezing relaxors ferroelectrics. *Opt. Mater. Express* **4**, 1487–1493 (2014).
26. Tian, H. *et al.* Double-loop hysteresis in tetragonal KTa0.58Nb0.42O3 correlated to recoverable reorientations of the asymmetric polar domains. *Appl. Phys. Lett.* **106**, 102903 (2015).
27. Tian, H. *et al.* Dynamic response of polar nanoregions under an electric field in a paraelectric KTa0.61Nb0.39O3 single crystal near the para-ferroelectric phase boundary. *Sci. Rep.* **5**, 13751 (2015).
28. Wemple, S. H. & DiDomenico, M. Oxygen-octahedra ferroelectrics. II. electro-optical and nonlinear-optical device applications. *J. Appl. Phys.* **40**, 735–752 (1969).
29. Sakamoto, T., Sasaura, M., Yagi, S., Fujiura, K. & Cho, Y. In-plane distribution of phase transition temperature of KTa1-xNbxO3 measured with single temperature sweep. *Appl. Phys. Express* **1**, 101601 (2008).
30. Li, H., Tian, H., Gong, D., Meng, Q. & Zhou, Z. High dielectric tunability of KTa0.60Nb0.40O3 single crystal. *J. App. Phys.* **114**, 054103 (2013).
31. Tian, H. *et al.* Variable gradient refractive index engineering: design, growth and electro-deflective application of KTa1-xNbxO3. *J. Mater. Chem. C* **3**, 10968–10973 (2015).
32. Agranat, A. J., Kaner, R., Perpelitsa, G. & Garcia, Y. Stable electro-optic striation grating produced by programmed periodic modulation of the growth temperature. *Appl. Phys. Lett.* **90**, 192902 (2007).
33. de Oliveira, C. E. M., Orr, G., Axelrold, N. & Agranat, A. J. Controlled composition modulation in potassium lithium tantalate niobate crystals grown by off centered TSSG method. *J. Cryst. Growth.* **273**, 203–206 (2004).
34. Lines, M. E. & Glass, A. M. *Principles and Applications of Ferroelectrics and Related Materials* (Oxford University Press, 1977).
35. Agranat, A. J., deOliveira, C. E. M. & Orr, G. Dielectric electrooptic gratings in potassium lithium tantalate niobate. *J. Non-Cryst. Sol.* **353**, 4405–4410 (2007).
36. Ramachandran, G. N. *Advanced Methods of Crystallography* (Academic Press, 1964).

37. Weber, M. F., Stover, C. A., Gilbert, L. R., Nevitt, T. J. & Ouderkirk, A. J. Giant birefringent optics in multilayer polymer mirrors. *Science* **287**, 2451–2456 (2000).

38. Kogelnik, H. Coupled wave theory for thick hologram gratings. *Bell Syst. Tech. J.* **48**, 2909–2947 (1969).

39. Bitman, A. *et al.* Electroholographic tunable volume grating in the g44 configuration. *Opt. Lett.* **31**, 2849–2851 (2006).

40. Gumennik, A., Kurzweil-Segev, Y. & Agranat, A. J. Electrooptical effects in glass forming liquids of dipolar nano-clusters embedded in a paraelectric environment. *Opt. Mater. Express* **1**, 332–343 (2011).

41. Wang, L. *et al.* Field-induced enhancement of voltage-controlled diffractive properties in paraelectric iron and manganese co-doped potassium-tantalate-niobate crystal. *Appl. Phys. Express* **7**, 112601 (2014).

42. Potter, Jr B. G., Tikare, V. & Tuttle, B. A. Monte Carlo simulation of ferroelectric domain structure and applied field response in two dimensions. *J. Appl. Phys.* **87**, 4415–4424 (2000).

43. Li, B. L., Liu, X. P., Fang, F., Zhu, J. L. & Liu, J. M. Monte Carlo simulation of ferroelectric domain growth. *Phys. Rev. B* **73**, 014107 (2006).

44. Chen, L. Q. Phase-field method of phase transitions/domain structures in ferroelectric thin films: a review. *J. Am. Ceram. Soc.* **91**, 1835–1844 (2008).

45. Li, Y. L., Hu, S. Y., Liu, Z. K. & Chen, L. Q. Phase-field model of domain structures in ferroelectric thin films. *Appl. Phys. Lett.* **78**, 3878–3880 (2001).

46. Chu, P. *et al.* Kinetics of 90 domain wall motions and high frequency mesoscopic dielectric response in strained ferroelectrics: a phase-field simulation. *Sci. Rep.* **4**, 5007 (2014).

47. Li, Q. *et al.* Giant elastic tunability in strained BiFeO3 near an electrically induced phase transition. *Nat. Commun.* **6**, 8950 (2015).

48. Parravicini, J. *et al.* Volume integrated phase-modulator based on funnel waveguides for reconfigurable miniaturized optical circuits. *Opt. Lett.* **40**, 1386–1389 (2015).

49. Pierangeli, D. *et al.* Continuous solitons in a lattice nonlinearity. *Phys. Rev. Lett.* **90**, 203901 (2015).

50. Kounga, A. B., Granzow, T., Aulbach, E., Hinterstein, M. & Rödel, J. High-temperature poling of ferroelectrics. *J. App. Phys.* **107**, 024116 (2008).

Acknowledgements

The research leading to these results was supported by funding from grants PRIN 2012BFNWZ2 and Sapienza 2014 Projects. A.J.A. acknowledges the support of the Peter Brojde Center for Innovative Engineering.

Author contributions

D.P., M.F. and E.D. conceived and developed experiments and theory and wrote the article; A.J.A. and C.E.M.d.O. designed and fabricated the KLTN samples and participated in the interpretation of results; and F.D.M. and G.D.D. participated in the experiments, data analysis and interpretation of results.

Additional information

Competing financial interests: The authors declare no competing financial interests.

9

Guided post-acceleration of laser-driven ions by a miniature modular structure

Satyabrata Kar[1], Hamad Ahmed[1], Rajendra Prasad[2], Mirela Cerchez[2], Stephanie Brauckmann[2], Bastian Aurand[2], Giada Cantono[3], Prokopis Hadjisolomou[1], Ciaran L.S. Lewis[1], Andrea Macchi[3,4], Gagik Nersisyan[1], Alexander P.L. Robinson[5], Anna M. Schroer[2], Marco Swantusch[2], Matt Zepf[1,6,7], Oswald Willi[2] & Marco Borghesi[1]

All-optical approaches to particle acceleration are currently attracting a significant research effort internationally. Although characterized by exceptional transverse and longitudinal emittance, laser-driven ion beams currently have limitations in terms of peak ion energy, bandwidth of the energy spectrum and beam divergence. Here we introduce the concept of a versatile, miniature linear accelerating module, which, by employing laser-excited electromagnetic pulses directed along a helical path surrounding the laser-accelerated ion beams, addresses these shortcomings simultaneously. In a proof-of-principle experiment on a university-scale system, we demonstrate post-acceleration of laser-driven protons from a flat foil at a rate of $0.5\,\text{GeV}\,\text{m}^{-1}$, already beyond what can be sustained by conventional accelerator technologies, with dynamic beam collimation and energy selection. These results open up new opportunities for the development of extremely compact and cost-effective ion accelerators for both established and innovative applications.

[1]School of Mathematics and Physics, Queen's University Belfast, Belfast BT7 1NN, UK. [2]Institut für Laser-und Plasmaphysik, Heinrich-Heine-Universität, Düsseldorf D-40225, Germany. [3]Department of Physics E. Fermi, Largo B. Pontecorvo 3, Pisa 56127, Italy. [4]Consiglio Nazionale delle Ricerche, Istituto Nazionale di Ottica, Research Unit Adriano Gozzini, via G. Moruzzi 1, Pisa 56124, Italy. [5]Central Laser Facility, Rutherford Appleton Laboratory, Didcot, Oxfordshire OX11 0QX, UK. [6]Helmholtz Institut Jena, 07743 Jena, Germany. [7]Institut für Optik und Quantenelektronik, Universität Jena, 07743 Jena, Germany. Correspondence and requests for materials should be addressed to S.K. (email: s.kar@qub.ac.uk).

In the context of developing compact, high current ion accelerators, the study of laser-driven acceleration mechanisms and the characterization and optimization of the ion beams produced, have been, over the past decade, very active areas of research[1-4]. Most experimental studies so far have dealt with the so-called target normal sheath acceleration (TNSA) mechanism, where ions are accelerated by space charge fields set up by relativistic electrons at the target surfaces. Despite the remarkable beam emittance and brightness, the inherent divergence and broad exponential energy spectrum of the TNSA beams limit their applicability in wide range of sectors[4], such as healthcare, industry, nuclear physics, high-energy density physics. Furthermore, reaching high-energies ($>$100s MeV) as required by important accelerator applications (for instance, clinical proton therapy[5-7]), will require, according to the current understanding of the acceleration mechanism[2,3], significantly larger laser systems than affordable in many cases. Coupling laser-driven ions to conventional RF stages for post-acceleration and beam control[8] is an approach currently being explored, which however, is inherently less attractive than an all-optical approach in terms of both cost and compactness.

In this paper, we present a laser-driven miniature device for simultaneous energy selection, collimation and post-acceleration of ions, which is an uniquely attractive combination compared with the methods explored so far for improving the quality of laser-driven ion beams[9-15]. The scheme exploits ultra-short (10s of picoseconds), high-amplitude, unipolar electromagnetic (EM) pulses generated in the interaction of a high-power laser with a solid target[16,17]. The technique is demonstrated in a proof-of-principle experiment using a 200 TW university-scale laser, achieving post-acceleration of protons to enhance their energy by $\sim$5 MeV over less than a centimetre of propagation, that is, an accelerating gradient $\sim$0.5 GeV m^{-1}. This is already beyond what can be sustained by conventional accelerator technologies[18-20] and can be enhanced significantly at higher laser intensities.

Results

Experimental setup. The experiment was carried out using the ARCTURUS laser system at Heinrich-Heine-Universität, Düsseldorf, Germany. Laser pulses of $\sim$30 fs duration and energy $\sim$3 J were focused on thin metallic foils by an f/2 off-axis parabola to a spot of $\sim$4 μm full-width at half-maximum, delivering peak intensities of a few times 10^{20} W cm^{-2}. A stack of radiochromic film (RCF) detectors was used to diagnose the spatial and spectral distribution of the proton beam. The stacks of RCF were placed at 35 mm from the proton-generating foils. The RCF dose response was absolutely calibrated against batches of RCFs exposed to different known proton doses from a particle accelerator[21].

Data presented in this paper were collected by using two different target geometries. The propagation of the ultra-short, high-amplitude EM pulse along a thin wire connected to the laser-irradiated target was initially characterized by using the target geometry shown by the schematic representation in Fig. 1a. In this case, the TNSA proton beam generated at the rear of the laser-irradiated target was used as a particle probe in a point-projection arrangement to obtain time resolved snapshots (see Methods section for additional information) of the pulse propagation along the wire (0.1 mm diameter Al wire), which was folded to a square wave pattern in front of and parallel to the interaction foil. The folded wire design was chosen to enable us to follow, in the diagnostic field of view, the pulse propagation along the wire over an extended distance (up to $\sim$35 mm). The second part of the experiment, which demonstrates the technique of using the EM pulse for control and optimization of the proton-beam parameters, was carried out by using a target that comprises a helical coil of suitable dimensions connected to the rear side of the laser irradiated foil, as shown in the Fig. 2a. The details of the target geometries and dimensions are discussed in the respective sections and figure captions.

Experimental results.

Generation of ultra-short EM pulse. At relativistic intensities ($>10^{18}$ W cm^{-2}) the interaction produces a large number of 'hot' (MeV) electrons[22,23]. A small fraction of the hot electron population escapes from the target charging it rapidly to MV potential[9,10,24]. The sudden charge separation leads to the generation of a strong EM pulse (with an electric field amplitude of GV m^{-1} order of magnitude, depending on laser intensity), which travels along the surface of the target at approximately the speed of light[16,17]. If the charged-up target is connected to the ground via a support wire, the pulse effectively contributes to driving a neutralizing current—in a situation analogous to a charged capacitor being suddenly connected to a transmission line. The high temporal resolution of the proton radiography technique[16,25-27] employed in our experiment allowed visualizing the propagation of the pulse along a thin metallic wire attached to the target. The proton radiograph of Fig. 1b shows the presence of an outwardly directed, strong

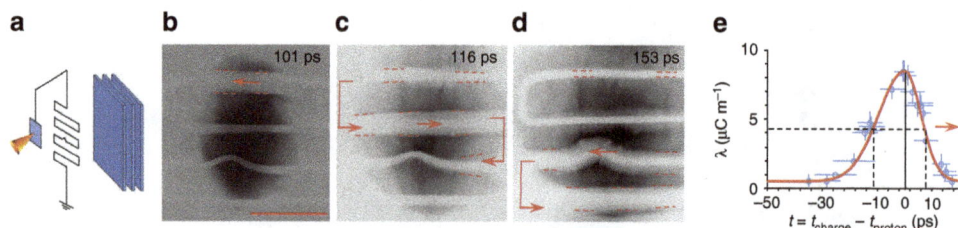

Figure 1 | Proton probing of EM pulse. (**a**) Schematic representation of the setup for diagnosing the EM pulse propagation along a folded wire. A small bent was made in one of the wire segments to act as a fiducial. The corresponding proton radiographs obtained at three different times (as labelled on each image) are shown in **b–d**. In these images, the film darkness is proportional to the proton flux. The red arrows in the images indicate the direction of charge flow in the folded wire pattern. The red dotted lines are eye guides for the width of the proton deflected region. The scale bar (solid red line) shown in **b** refers to 1 mm in the plane of the folded wire pattern. (**e**) Temporal profile of the positive charge pulse travelling along the folded wire. The x axis of the graph corresponds to the relative probing time, $t = t_{charge} - t_{proton}$, where t_{proton} is the probing time of a given point on the folded wire by the protons reaching the Bragg peak in the given RCF layer, and t_{charge} is the time of arrival of the peak of the charge pulse at that point. The experimental uncertainty in time is determined by the transit time of protons through the electric field region and the energy resolution of the active layers of RCF (ref. 25). The uncertainty in the charge density is estimated from the uncertainty in measuring the width of the proton deflection from the RCF data. The red arrow in the graph indicates the direction of propagation of the pulse.

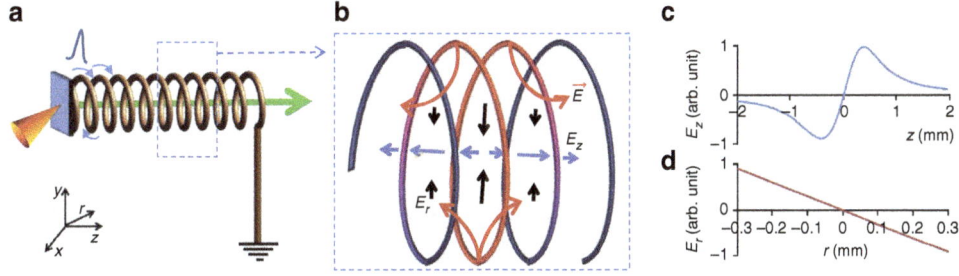

Figure 2 | Helical coil working principle. (a) Schematic representation of the target designed for optimizing the beam parameters of laser-driven protons. In this configuration a helical coil, made of a metallic wire, is attached to the laser-irradiated thin foil at one end and grounded at the other end. The helical coil design guides the EM pulse carrying the neutralizing charge around the proton-beam axis and allows synchronizing its longitudinal propagation (that is, along z) with protons having a given energy within the beam. **(b)** Schematic representation snapshot showing the electric field configuration inside the coil. The red section of the coil represents the segment charged by the travelling pulse at a given moment of time, where the red arrows represent the electric field (**E**) lines originating from the coil, the black and blue arrows represent radial (E_r) and longitudinal (E_z) components of the electric field, respectively. The length of the black and blue arrows represents the relative strength of the field at different locations. **(c,d)** $E_z|_{(r=0)}$ and $E_r|_{(z=0)}$ profiles inside the coil at a given time, where z = 0 corresponds to the location of the peak of charge density along the coil at that time. The field profiles are calculated by the subroutine that defines the input electric field configuration for particle tracing in the PTRACE simulation (see Methods). Dynamics are modelled using an asymmetric Gaussian pulse profile of 5 ps rise and 10 ps decay, as obtained in the experiment shown in Fig. 1e, travelling along a helical coil with the same dimensions as the one used in the experiment illustrated in Fig. 3.

electric field (of the order of GV m^{-1}) around the top segment of the folded wire, causing a clear deflection of the probe protons normal to the wire, while the bottom segments remain electrically neutral. At later times, the electric field region is seen to be moving downwards along the wire, while leaving the top segment electrically neutral (Fig. 1d). This confirms the localization in space of the electrically charged region at any given time. By following the proton deflection across the folded wire geometry as observed in different layers of the RCF stack, a temporal profile of the propagating charge pulse was reconstructed with the help of particle-tracing simulations, employing the PTRACE code (see the Methods section). The temporal profile of the charge pulse is shown in Fig. 1e having rise and decay times of ~5 and ~10 ps, respectively. The velocity of the pulse flowing down the wire is measured as $v = (0.96 \pm 0.04)c$, where c is the speed of light in vacuum. The total charge contained within the pulse is estimated (by integrating the linear charge-density profile $\lambda(t)$ in Fig. 1e) as ~60 nC, which agrees with an order of magnitude estimation of the net charge left in the target because of the escaping hot electrons[9,24,25]. The temporal profile of the neutralizing current pulse can be obtained by multiplying $\lambda(t)$ by its velocity v, which estimates a peak current of the order of kiloamp for the case shown in Fig. 1e.

Laser-driven travelling-wave ion accelerator. The localized, ultra-high electric field associated with the travelling EM pulse can be employed for controlling and optimizing bursts of laser-accelerated MeV ions on a picosecond timescale. By directing the EM pulse along a helical path around the proton-beam axis (see Fig. 2a), one can synchronize, over an extended distance, the longitudinal propagation of the electric field (associated with the pulse) with the transiting protons within a given energy range. A sketch of the electric field configuration inside the coil geometry is shown schematically in Fig. 2b. The calculated longitudinal (E_z) and radial (E_r) electric field profiles inside the coil due to the pulse are shown in Fig. 2c,d. At a given time, the pulse spreads over approximately two windings of the coil used in Fig. 2b, and therefore the profiles are qualitatively similar to those produced by a uniformly charged ring[28] giving, for example, for the longitudinal component, $E_z(z)|_{(r=0)} = 2\pi k\lambda Rz/(R^2 + z^2)^{3/2}$, where k is the Coulomb's constant and R is the radius of the ring. The radial and longitudinal components of the moving electric field act, respectively, towards focussing and acceleration

(or deceleration, depending on the position of protons with respect to the charged segment of the coil) of the protons synchronous with the EM pulse. Conversely, the unsynchronized protons maintain their intrinsic divergence, and can be discarded with an appropriate spatial selection.

Figure 3 shows proof-of-principle results using the helical coil target. In contrast to the typical divergent proton beam produced by a flat foil (Fig. 3a), the helical coil target (Fig. 3b) produced a highly collimated beam of protons of energy significantly higher than from a flat foil, as clearly observed in the RCF stack diagnostic. In the flat-foil proton spectrum shown in Fig. 3c, the proton number after 7 MeV falls below the detection threshold of the RCF—one can estimate 10^7 protons per MeV as an absolute maximum for 8 MeV protons, while assuming a very generous beam diameter (see caption of Fig. 3 for details). On the other hand, a spectral peak at ~9 MeV (with ~10^8 protons per MeV at the peak), with detectable RCF signal up to 10 MeV is observed in the case of Fig. 3b, providing a clear indication of an increase in proton energy resulting from the helical coil. Furthermore, as shown in Fig. 3d, the diameter of the focussed, narrow-band, 9 MeV beam at 35 mm from the target is less than the internal diameter of the coil, which indicates nearly collimated (<0.5° divergence) propagation of the protons after exiting the coil structure.

Particle-tracing simulations (see Methods for more details) carried out using the target parameters and the measured charge pulse characteristics (amplitude, temporal profile and speed) show that in the case of Fig. 3b the pulse is optimally synchronized for input energy 4–5 MeV (see Fig. 4a). While the radial component of the associated field acts on constraining the divergence of these protons, the longitudinal component of the electric field progressively accelerates the leading part of the synchronized proton bunch at a rate close to 0.5 GeV m^{-1}, as shown in Fig. 4c–f. The analogy with a uniformly charged ring model yields an order of magnitude agreement: the longitudinal field has a peak amplitude $E_{z,max} = E_z(z = R/\sqrt{2})|_{(r=0)} = 2kQ/3\sqrt{3}R^2 \sim 0.65$ GeV m^{-1}, while using Q (total charge in the ring) = 30 nC (accounting for the charge spreading over two windings) and $R = 400\,\mu$m. As expected, protons in the trailing part of the synchronized bunch experience deceleration during their travel. The protons which are co-propagating with the central portion of the pulse

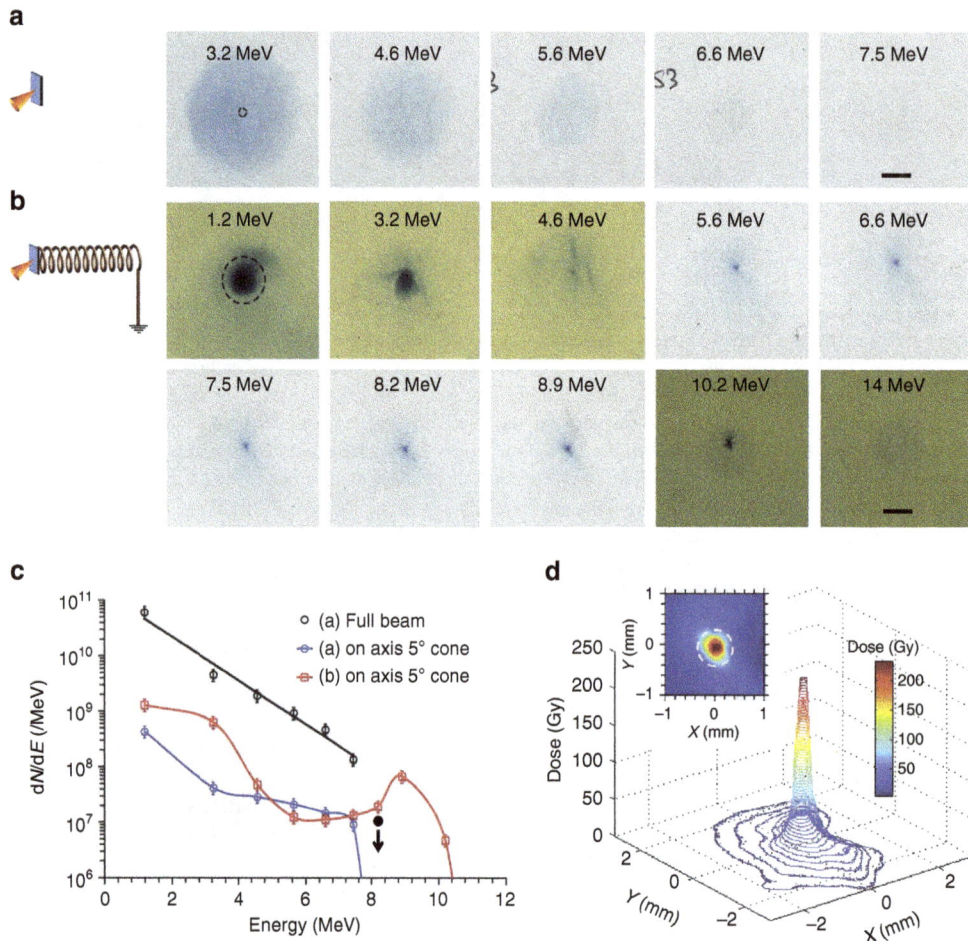

Figure 3 | Effect of helical coil on TNSA beam. In comparison with the typical proton-beam profile obtained from a flat-foil target in the experiment shown in (**a**), (**b**) shows the beam profile obtained from the helical coil target. Au foils (10 μm thick) were used as the laser interaction target in both cases. The coil was made of 100 μm aluminium wire and had internal diameter, pitch and length of 0.7, ~0.28 and 8.7 mm, respectively. The RCF stack was placed at 35 mm from the target. RCF images of 50 mm × 50 mm size are shown in **a**, which are five times larger compared with the RCF images shown in **b** (the black scale bar on the last piece of RCF in **a** and **b** correspond to 10 and 2 mm respectively), in order to account for the large divergence of the TNSA beam produced from the flat foils. The black-dashed circles on the first RCF layer for both **a** and **b** correspond to the projection of the exit ring of the coil on the RCF (~2.8 mm diameter circle corresponding to ~5° full cone angle). (**c**) shows the comparison between proton spectra from the reference flat-foil target and the helical coil target shown in **b** obtained by spectral deconvolution of the RCF signals described in refs 37,38. The error bars were estimated from the possible error in dose conversion[21] and uncertainties in background substraction. Since the flat-foil proton signal at ~8 MeV was below the detection threshold, the solid black circle shows the upper bound for proton signal calculated by considering the detection threshold of the RCF (~10^5 protons per MeV mm^−2) and an overly generous beam size (10 mm diameter on the RCF, which is similar to the beam size at 6.6 MeV shown in **a**) for protons at that energy. (**d**) Three-dimensional profile of the pencil beam of protons obtained in the RCF corresponding to ~9 MeV, where the inset shows the two-dimensional dose map of the central part of the beam. The white dashed circle in the insert represents the internal diameter of the helical coil.

experience virtually no longitudinal field and their energy is unchanged, while they are focused much more strongly, and have already diverged by the time they reach the RCF detector. The simulated proton spectrum agrees well with the experimental data points as shown in Fig. 4b, while using an input proton spectrum mimicking the experimental data from flat foils. The two peaks in the simulated proton spectrum on either side of 5 MeV are due to the simulated energy gain/loss shown in Fig. 4a. Considering that the 10^8 protons per MeV at the spectral peak, observed in the experimental data, are produced by the post-acceleration of ~5 MeV protons, Fig. 3c suggests that gradual focussing of synchronous protons along the coil resulted in an overall collection efficiency of 7%, which is three times better than what would be captured, for an undeflected beam, within the solid angle sustained by the exit ring of the coil.

Discussion

While the experiment shows a proof-of-principle demonstration employing a university-scale laser system, there is significant scope for further development of the technique using higher-intensity lasers and more refined target arrangements. For a given diameter of the helical coil target, the accelerating field inside the coil structure, as discussed above, is directly proportional to the charge ($Q = \int \lambda(l) dl$) contained in the pulse, which is equal to the number of electrons (N_{esc}) escaping from the laser-irradiated target during the interaction times the electron charge (e). The scaling for $eN_{esc}(=Q)$ with incident laser intensity can be obtained by using the simple phenomenological model described by Kar *et al.*[9] Figure 5a shows estimates of Q obtained using this model, for constant laser pulse length (30 fs) and focal spot size on the target (4 μm). A prudent 30% laser-electron conversion efficiency is assumed in this model, which may be an

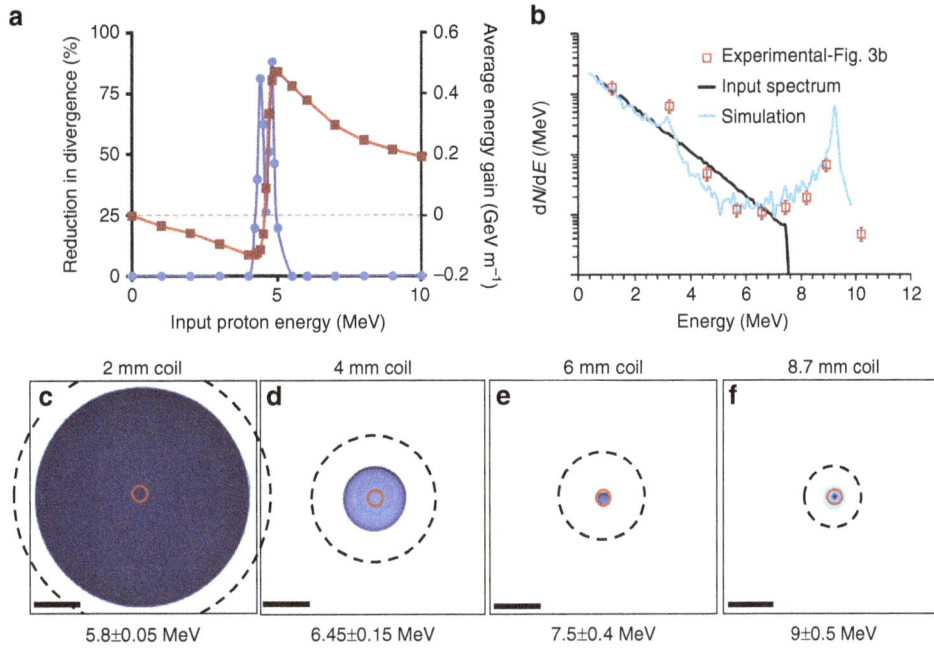

Figure 4 | Simulations of focusing and post-acceleration by the helical coil. (**a**) Reduction in beam divergence (filled blue circles) and the gain in energy (filled red squares) for different input proton energies as obtained from simulations carried out for the case shown in Fig. 3b. (**b**) Comparison between experimental and simulated proton spectra at the detector plane for the case shown in Fig. 3b. An exponential input spectrum for protons (as shown), similar to the one obtained from the reference flat foil, was used in the simulation. (**c-f**) Simulated spatial profiles of the proton beam at the detector plane produced by different lengths of the helical coil target shown in the case of Fig. 3b. The black scale bars at the bottom left of the images correspond to 2 mm on the RCF plane. A divergent beam of 4.8 MeV protons was used as input in the simulations, which experience maximum accelerating field according to the graph shown in **a**. The output beam energy in each case is mentioned on the images. The black-dashed circle represents the projection of the exit ring of the coil on the RCF and the red solid circle shows the internal diameter of the coil.

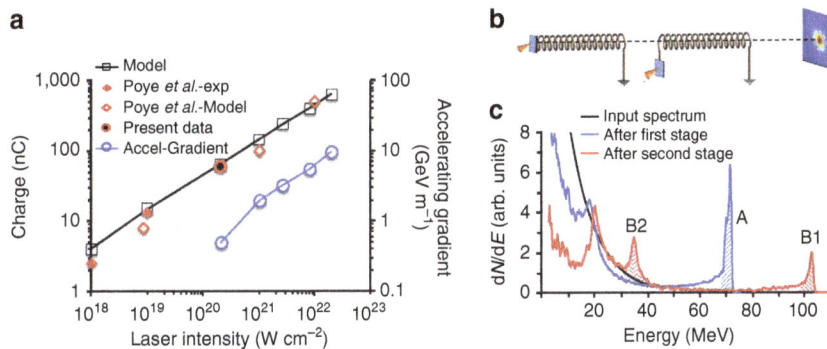

Figure 5 | Scaling to higher laser intensity and multistage acceleration. (**a**) Total charge carried by the pulse moving along the wire connected to the laser-irradiated target, plotted against incident laser intensity on target (black solid line). The blue solid line shows the accelerating gradient inside a helical coil of same diameter and pitch as the one used in our experiment. (**b**) Schematic representation of a double-stage acceleration setup using two helical coils driven by two laser pulses. (**c**) Comparison between the simulated proton spectra, taken at 50 mm from the proton source over a 2 mm × 2 mm area, obtained for: flat-foil proton source (black-input spectrum in the simulation), single-stage (blue) and double-stage (red) coil re-acceleration. Both coils were of 0.5 mm internal diameter and 10 mm long, with variable pitch suited to their input proton energies ($\sim$40 and $\sim$70 MeV, respectively, for the first and second stages). The parameters for the charge pulse in both coils were taken as those expected from the interaction of a PW, 30 fs laser (e.g. as available at GIST, Korea[33]) with a thin target, leading to an acceleration gradient of $\sim$3 GeV m^{-1} (discussed in the text). The shaded areas A, B1 and B2 represent, respectively, the proton bunch accelerated from the first coil and the proton bunches accelerated and decelerated by the second coil. The total number of particles in the bunches B1 and B2 is approximately equal to the number of particles in the bunch A. The time of arrival of the charge pulse at the entrance of the second coil was synchronized with the arrival of the $\sim$70 MeV proton bunch produced by the first coil.

underestimate for intensities above 10^{20} W cm^{-2} (refs 22,29–32). In any case, as can be seen in this figure, the scaling agrees well not only with our experimental data but also with the experimental data and model reported recently by Poye et al.[24] for similar pulse duration and focussing conditions.

The increase in the accelerating gradient with the incident laser intensity is shown in Fig. 5a. This was estimated from PTRACE

simulations while considering the same helical coil geometry as used in our experiment and the peak charge density calculated from the estimated Q for a given laser intensity. Considering for instance the parameters of the PW laser recently commissioned at GIST, Korea[33], the energy gain per unit length for a helical coil (0.5 mm internal diameter and 10 mm long) would be 3 GeV m^{-1}. As shown in Fig. 5b, PTRACE simulations suggest that injecting an exponential spectrum of protons with ~ 40 MeV cutoff (as produced on such a laser system[34]) into the coil, one would expect a narrow energy bandwidth pencil beam of ~ 70 MeV at the coil output. The pitch of the coil employed in this simulation varied along its length in order to keep the charge flow synchronized with the selected proton bunch as it is being accelerated. This approach can, in principle, be extended to maintain an efficient acceleration over coils of several cm length.

Furthermore, since the helical coil device is effectively a separate module with respect to the primary ion acceleration process producing the seed beam which enters the coil, one can envisage a multistage implementation of the process, employing sequential, multiple coils irradiated by separate, appropriately timed laser pulses—a scheme with high promise for the development of a compact linear particle accelerator. As shown in the insert of Fig. 3d, the diameter of the proton beam at 35 mm from the target was smaller than the internal diameter of the coil, which implies that the output beam is quasi-collimated. Therefore, it would be possible to feed the entire output beam from one stage into a second coil of similar (or even smaller) diameter, placed a few mm after the previous coil. As an example, Fig. 5c shows the simulated proton spectrum obtained from a two stage configuration (as shown by a schematic representation in Fig. 5b). The bunch of protons accelerated to ~ 70 MeV by a single-stage coil can be re-accelerated to higher energies by employing suitable coil and charge pulse parameters. It should also be noted that, using a similar arrangement, a separately driven coil could be used for post-acceleration of ion beams generated by any alternative acceleration mechanism[4] on a primary target.

In conclusion, a technique to achieve simultaneous focussing, energy selection and post-acceleration of laser-driven ions is reported, which is based on harnessing, in a travelling wave accelerator arrangement, the fields arising from ultra-short, large-amplitude EM pulses generated by intense lasers. An accelerating gradient of ~ 0.5 GeV m^{-1} is demonstrated by the technique in a proof-of-principle experiment using a 200 TW laser. This development sets a cornerstone for next-generation, extremely compact and cost-effective proton accelerators, which complements the current drive for miniaturization in advanced accelerator technology.

Methods

Proton radiography. In a typical arrangement employing broadband TNSA protons, proton radiography[26,27] provides, in a single-shot, multiple-snapshots (corresponding to sequential time frames) of the probed region. Instead of using a second laser beam to generate the probe protons, in this experiment we used a self-imaging arrangement, where both the probe protons and the EM pulse were generated from the same laser interaction. Protons generated from the rear surface of the target foil were used to probe the propagation of the EM pulse along a folded wire segment, as shown in Fig. 1a. The presence of an electric field in the probed region is detected via the modulation in the proton flux distribution induced by local particle deflections.

PTRACE simulation. The particle-tracing simulations presented in this paper were performed using the PTRACE code (ref. 35). This code simulates in three dimensions the propagation of protons from the source to the detector through the probed region consisting of electric and magnetic fields. At the core of the code, the equation of motion is computed by a Runge–Kutta fourth-order algorithm coupled with an adaptive step size monitoring routine.

The linear charge density at a given time and location on the folded wire segment, shown in Fig. 1, was obtained by matching the experimental proton deflection with the deflection produced by PTRACE for the corresponding probe proton energy. Using the charge-density profile as a template, a set of dynamic PTRACE simulations were carried out in order to evaluate the transit time effect of the probe protons towards the rise time, decay time and peak charge density estimated from the static simulations. We assumed the EM pulse to propagate as a transverse electro-magnetic (TEM) mode[17] and the electric field **E** around the wire was calculated numerically from the linear charge-density profile shown in Fig. 1e. The dynamic simulation was set up by allowing a given positive charge pulse to travel along the folded wire structure with the same dimensions as in the

experiment. The electric field at any given point 'P', experienced by a transiting proton at a given time, was computed by adding electric field vectors at 'P' due to every element of the wire, while considering the position of the pulse in the wire at that probing time. The best match was obtained for a charge pulse profile similar (within the diagnostic's temporal resolution of a few ps) to that obtained from the PTRACE simulations.

The helical coil was modelled in PTRACE using a cylindrical coordinate system. Using the same physical parameters for the coil as used in the experiment, a charge pulse of a given temporal profile was allowed to travel along it. The electric field at any given point at a given time was computed numerically by the method mention above. The proton source in the code was modelled as a point source located at the centre of the entrance plane of the coil, emitting protons towards the coil with a given energy spectrum and divergence. After the transit of the protons through the field region defined by the coil, the PTRACE renderer calculates the response of the stack detector for every incident proton. Energy deposited in each layer of the stack is computed by using the stopping range of protons in the RCF plastic calculated by SRIM (ref. 36) simulation.

References

1. Malka, V. et al. Principles and applications of compact laser-plasma accelerator. Nat. Phys. **4**, 447–453 (2008).
2. Fuchs, J. et al. Laser-driven proton scaling laws and new paths towards energy increase. Nat. Phys. **2**, 48–54 (2006).
3. Robson, L. et al. Scaling of proton acceleration driven by petawatt-laser-plasma interactions. Nat. Phys. **3**, 58–62 (2007).
4. Macchi, A., Borghesi, M. & Passoni, M. Ion acceleration by superintense laser-plasma interaction. Rev. Mod. Phys. **85**, 751–793 (2013).
5. Bulanov, S. V. et al. Oncological hadrontherapy with laser ion accelerators. Phys. Lett. A **299**, 240–247 (2002).
6. Malka, V. et al. Practicality of proton therapy using compact laser systems. Med. Phys. **31**, 1587–1592 (2004).
7. Linz, U. & Alonso, J. What will it take for laser driven proton accelerators to be applied to tumour therapy? Phys. Rev. ST Accel. Beams **10**, 094801 (2007).
8. Antici, P. et al. Post-acceleration of laser generated high energy protons through conventional accelerator linacs. IEEE Trans. Plasma Sci. **36**, 1843–1846 (2008).
9. Kar, S. et al. Dynamic control of laser-produced proton beams. Phys. Rev. Lett. **100**, 105004 (2008).
10. Bartal, T. et al. Focusing of short-pulse high-intensity laser-accelerated proton beams. Nat. Phys. **8**, 139–142 (2012).
11. Kar, S. et al. Ballistic focusing of polyenergetic protons driven by petawatt laser pulses. Phys. Rev. Lett. **106**, 225003 (2011).
12. Patel, P. et al. Isochoric heating of solid-density matter with an ultrafast proton beam. Phys. Rev. Lett. **91**, 125004 (2003).
13. Toncian, T. et al. Ultrafast laser-driven microlens to focus and energy-select mega-electron volt protons. Science **312**, 410–413 (2006).
14. Busold, S. et al. Focusing and transport of high-intensity multi-MeV proton bunches from a compact laser-driven source. Phys. Rev. ST Accel. Beams **16**, 101302 (2013).
15. Schollmeier, M. et al. Controlled transport and focusing of laser-accelerated protons with miniature magnetic devices. Phys. Rev. Lett. **101**, 055004 (2008).
16. Quinn, K. et al. Laser-driven ultrafast field propagation on solid surfaces. Phys. Rev. Lett. **102**, 194801 (2009).
17. Tokita, S. et al. Strong sub-terahertz surface waves generated on a metal wire by high-intensity laser pulses. Sci. Rep. **5**, 8268 (2015).
18. Thwaites, D. I. & Tuohy, J. B. Back to the future: the history and development of the clinical linear accelerator. Phys. Med. Biol. **51**, R343–R362 (2006).
19. Caporaso, G. J. et al. The dielectric wall accelerator. Rev. Accel. Sci. Technol. **2**, 253 (2009).
20. Flanz, J. B. What's new in particle therapy accelerator technology. Nucl. Instrum. Methods Phys. Res. B **261**, 768–772 (2007).
21. Kirby, D. et al. Radiochromic film spectroscopy of laser-accelerated proton beams using the FLUKA code and dosimetry traceable to primary standards. Laser Part. Beams **29**, 231–239 (2011).
22. Wharton, K. B. et al. Experimental measurements of hot electrons generated by ultraintense ($>10^{19}$W cm^{-2}) laser-plasma interactions on solid-density targets. Phys. Rev. Lett. **81**, 822–825 (1998).
23. Kodama, R. et al. Plasma devices to guide and collimate a high density of MeV electrons. Nature **432**, 1005–1008 (2004).
24. Poye, A. et al. Physics of giant electromagnetic pulse generation in short-pulse laser experiments. Phys. Rev. E **91**, 043106 (2015).
25. Borghesi, M. et al. Measurement of highly transient electrical charging following high-intensity lasersolid interaction. Appl. Phys. Lett. **82**, 1529–1531 (2003).
26. Borghesi, M. et al. Electric field detection in laser-plasma interaction experiments via the proton imaging technique. Phys. Plasmas **9**, 2214–2220 (2002).
27. Kar, S. et al. Dynamics of charge-displacement channelling in intense laser-plasma interactions. New J. Phys. **9**, 402 (2007).

28. Pollack, G. L. & Stump, D. R. *Electromagnetism (Addison-Wesly, 2001)*.
29. Wilks, S. C., Kruer, W. L., Tabak, M. & Langdon, A. B. Absorption of ultra-intense laser pulses. *Phys. Rev. Lett.* **69,** 1383–1386 (1992).
30. Wilks, S. C. & Kruer, W. L. Absorption of ultrashort, ultra-intense laser light by solids and overdense plasmas. *IEEE J. Quantum Electron.* **33,** 1954–1968 (1997).
31. Kahaly, S. *et al.* Near-complete absorption of intense, ultrashort laser light by sub-λ grating. *Phys. Rev. Lett.* **101,** 145001 (2008).
32. Levy, M. C., Wilks, S. C., Tabak, M., Libby, S. B. & Baring, M. G. Petawatt laser absorption bounded. *Nat. Commun.* **5,** 4149 (2014).
33. Yu, T. J. *et al.* Generation of high-contrast, 30 fs, 1.5 PW laser pulses from chirped-pulse amplification Ti:sapphire laser. *Opt. Express* **20,** 10807–10815 (2012).
34. Kim, I. J. *et al.* Transition of proton energy scaling using an ultrathin target irradiated by linearly polarized femtosecond laser pulses. *Phys. Rev. Lett.* **111,** 165003 (2013).
35. Schiavi, A. Study of Laser Produced Plasmas by X-ray and Proton Radiography. PhD thesis (Imperial College London, 2003).
36. Ziegler, J. F. SRIM-2013 (Stopping and Range of Ions in Matter). Available at: www.srim.org.
37. Markey, K. Development of Laser Accelerated MeV Ion Sources. PhD thesis, (Queen's University Belfast, 2009).
38. Breschi, E. *et al.* A new algorithm for spectral and spatial reconstruction of proton beams from dosimetric measurements. *Nucl. Instrum. Methods Phys. Res. A* **522,** 190–195 (2004).

Acknowledgements

We acknowledge the use of the TARANIS Laser Facility in the Centre for Plasma Physics at Queens University Belfast for the preliminary work related to the development of the core concept. We thank for the support of D. Doria (QUB), D. Gwynne (QUB), F. Hanton (QUB), K. Naughton (QUB) and A.L. Giesecke (HHU) in carrying out the experiments at TARANIS Laser Facility and T. Wowra (HHU) for the experiment discussed in this paper. We acknowledge funding from EPSRC (EP/J002550/1-Career Acceleration Fellowship held by S.K., EP/L002221/1, EP/K022415/1 and EP/I029206/1), Laserlab-Europe (EC-GA 284464), SFB/TR18, GRK1203 and Invest Northern Ireland (POC-329). We acknowledge Prof. S. Ter-avetisyan (GIST, Korea) and Dr Jason Wiggins (QUB) for useful discussions, and A. Schiavi (Univ. Roma 1, Italy) for the use of the particle tracing code, PTRACE. Data associated with research published in this paper can be accessible at http://dx.doi.org/10.17034/5d88ccf9-31cf-4e13-8023-0693334d27d8.

Author contributions

S.K. developed the concept; the experiment was designed and led by S.K., H.A, O.W. and MB; and supported by S.B., M.C., R.P., P.H., M.S., A.M.S., B.A.. H.A. and S.K. analysed the data. G.C., G.N. and C.L.S.L. contributed towards development of the technique. The manuscript was written by S.K. and M.B. C.L.S.L., M.Z., A.M., A.P.L.R and O.W. were involved in discussions and manuscript preparation.

Additional information

Competing financial interests: The authors declare no competing financial interests.

10

Mapping multidimensional electronic structure and ultrafast dynamics with single-element detection and compressive sensing

Austin P. Spencer[1], Boris Spokoyny[1], Supratim Ray[1], Fahad Sarvari[1] & Elad Harel[1]

Compressive sensing allows signals to be efficiently captured by exploiting their inherent sparsity. Here we implement sparse sampling to capture the electronic structure and ultrafast dynamics of molecular systems using phase-resolved 2D coherent spectroscopy. Until now, 2D spectroscopy has been hampered by its reliance on array detectors that operate in limited spectral regions. Combining spatial encoding of the nonlinear optical response and rapid signal modulation allows retrieval of state-resolved correlation maps in a photosynthetic protein and carbocyanine dye. We report complete Hadamard reconstruction of the signals and compression factors as high as 10, in good agreement with array-detected spectra. Single-point array reconstruction by spatial encoding (SPARSE) Spectroscopy reduces acquisition times by about an order of magnitude, with further speed improvements enabled by fast scanning of a digital micromirror device. We envision unprecedented applications for coherent spectroscopy using frequency combs and super-continua in diverse spectral regions.

[1] Department of Chemistry, Northwestern University, 2145 Sheridan Road, Evanston, Illinois 60208, USA. Correspondence and requests for materials should be addressed to E.H. (email: elharel@northwestern.edu).

The ability to measure quantum correlations in complex systems with high spectral and temporal resolution provides deep physical insights into a wide range of phenomena from intermolecular dynamics in liquids to the electronic and vibrational structures of condensed phase molecular systems[1]. One powerful approach to directly measure intra- and inter-molecular couplings far from equilibrium is two-dimensional photon echo spectroscopy (2D PES)[2–4]. In 2D PES, coherences encoded in multiple time intervals are correlated, providing insight into the quantum states of the system and their interactions with the surroundings. 2D PES has been used to examine the dynamics of energy transfer in photosynthetic proteins[5–7], ultrafast dynamics of solute–solvent species[8] and intraband relaxation in semiconductors[9,10]. However, the advantages of performing 2D PES over one-dimensional spectroscopies come at a cost: increased acquisition time, additional experimental complexity and limited sensitivity in regions of the spectrum outside the visible and infrared. Here we introduce a method to overcome these limitations by combining sparse sampling and spatiotemporal encoding of nonlinear optical signals.

To make this connection, we begin by noting that the signal measured in 2D PES is typically sparse (that is, compressible). In most cases, the phase of the nonlinear signal, which is needed to extract the oscillation frequencies of the system, is measured by spectral interferometry in which an external reference field is coherently mixed with the signal and spectrally dispersed onto a detector[11–14]. Heterodyne detection yields an interferogram that is sparse in one or more Fourier domains, and it is this sparsity that we exploit using compressive sensing (CS) methods.

One consequence of CS theory is effectively to relax the Nyquist–Shannon sampling theorem[15], which enables perfect reconstruction of continuous signals (as a function of space, time, frequency, and so on) from samples acquired at a rate that is greater than the occupied bandwidth of the signal (irrespective of its distribution) rather than its total bandwidth. In effect, the Nyquist–Shannon sampling theorem relates how to sufficiently sample a signal's frequencies while CS relates how to sufficiently sample a signal's information[16]. CS has been shown to greatly reduce the number of measurements needed for signal recovery for a wide range of applications such as magnetic resonance imaging[17], nonlinear optical imaging[18], multidimensional spectroscopy[19,20], holography[21] and super-resolution microscopy[22], to name a few. In an analogous way, CS has been used to reduce the number of costly quantum mechanical calculations needed in computational studies[23–25]. One of the most promising applications of CS is in image reconstruction thanks to the potential to (i) utilize single-element detectors in regions of the spectrum where cameras perform poorly and (ii) bypass uniform sampling requirements imposed by array detectors when such sampling is not optimal.

To apply imaging-based CS to coherent spectroscopy, we first need to physically map the 2D spectrum to an image. In 2D PES, the signal is generated by exciting the sample with a series of ultrafast laser pulses, inducing a time-dependent polarization in the sample. The signal is spectrally dispersed onto an array detector, enabling multi-channel acquisition by detecting all signal frequencies, ω_t, simultaneously. The experiment is repeated for a range of coherence times, τ, and population times, T. Fourier transformation along τ yields the 2D Fourier transform (2DFT) spectrum $S(\omega_\tau, \omega_t; T)$ with parametric dependence on the waiting time, T.

Recently, some of the authors introduced a significantly higher-throughput sampling scheme called GRAPES[14,26,27] (GRadient Assisted Photon Echo Spectroscopy) that breaks with the traditional τ-scan approach. By incorporating a 2D array detector and a spatiotemporal gradient, all the τ delays may be sampled in parallel. The signal, emitted from the narrow focal line spatially overlapped with the excitation pulses in the sample, is spectrally resolved along an orthogonal direction, resulting in a direct image of the signal, $I(\tau, \omega_t)$. Through this imaging arrangement, a 2DFT spectrum is acquired for each laser shot upon Fourier transformation along τ. Combining the spatially encoded signal generated in GRAPES with a programmable spatial mask and a single-element detector enables the single-point array reconstruction by spatial encoding (SPARSE) spectroscopy method described here. This application of CS to 2DFT spectroscopy contrasts with prior approaches[19,20] involving sparse sampling of pulse time delays whereby data acquisition time is reduced by sampling a limited window of time delays.

In this work, we present 2DFT spectra of a carbocyanine dye and a photosynthetic pigment-protein complex measured using SPARSE spectroscopy. The accuracy of SPARSE-detected 2DFT spectra is evaluated based on comparison to conventional camera-detected 2DFT spectra. Reconstruction of spatial spectral interferograms is demonstrated using both CS and Hadamard methods, illustrating the robustness of CS retrieval. We show that in some cases, CS reconstruction requires only one-tenth of the complete set of Hadamard-encoded measurements for accurate interferogram recovery.

Results

Hadamard encoding and CS reconstruction. To verify the experimental methodology of SPARSE spectroscopy, we first implemented a 2D programmable version of Hadamard spectroscopy[28,29]. The Hadamard transform matrix, $\mathbf{H}_n$, is analogous to the discrete Fourier transform matrix, but contains only binary ($+1$ and -1) elements. Hadamard sampling can be expressed as the linear problem $\mathbf{H}_n\mathbf{x} = \mathbf{y}$, where $\mathbf{x}$ is a length-n signal vector and $\mathbf{y}$ is a length-n measurement vector. The unknown signal $\mathbf{x}$ can be reconstructed by performing the inverse Hadamard transform on $\mathbf{y}$. To implement $\mathbf{H}_n$ experimentally, we use a digital micromirror device (DMD)[30], which is a 2D array of electromechanical mirror elements whose surface normal angles can be controlled between two binary states: $+12°$ ('on' state, 1) and $-12°$ ('off' state 0; see Methods for details). In this way, multiplication of the masks (each constructed from a different row (H_i) of the Hadamard matrix $\mathbf{H}_n$) with the unknown image ($\mathbf{x}$) occurs optically by reflecting the image formed at the spectrometer exit off of a spatial mask imprinted on the DMD. The reflected portions (that is, pixels) of the image are summed to form the observation $y_i = \sum H_{ij}x_j$ by focusing the reflected light with a lens onto a small active area photomultiplier tube (PMT) detector (Fig. 1). An example spatial mask (without the random inversions described in Method section) is shown in Fig. 2a.

As with the Fourier transform, Hadamard reconstruction requires Nyquist sampling to recover the signal. However, if the signal is sparse under a suitable unitary transform, then according to CS, an n-element signal may be faithfully reconstructed from fewer than n measurements. CS algorithms (for example, convex optimization and basis pursuit) minimize the L1-norm of the recovered signal subject to the constraint $\mathbf{Ax} = \mathbf{y}$, where $\mathbf{A}$ is an $n \times m$ observation matrix and $\mathbf{y}$ is a length-m measurement vector with $m \le n$. In this work, $\mathbf{A}$ is a pseudo-randomly chosen subset of the rows of a Hadamard matrix $\mathbf{H}_n$, although other forms are possible such as a random matrix of 0 and 1. It is important to note that $\mathbf{x}$ itself does not necessarily have to be sparse, but rather it should be sparse in a suitable basis representation. To solve the constrained L1-norm minimization problem we used l_1-MAGIC, a program that uses standard interior-point methods to solve convex optimization problems[31].

Detection schemes. Spatial spectral interferograms of a carbocyanine dye molecule, IR-144 and photosynthetic pigment-protein complex LH2 were acquired with a GRAPES apparatus using two distinct detection schemes: direct detection with a camera and SPARSE detection using a DMD spatial mask and PMT detector. In the direct case, a 2D array detector at the image plane of the spectrometer recorded 2D interferograms with one spatial dimension (along which τ is encoded) and one spectral dimension ($\lambda \propto 1/\omega_t$). In the second detection scheme, a DMD placed at the image plane selectively reflects light at each pixel towards ('on') or away ('off') from a PMT. A simplified schematic of this apparatus is depicted in Fig. 1. While the camera measures the light intensity incident on each pixel independently, the DMD–PMT detector measures the integrated intensity from all 'on' DMD pixels. Spatial spectral interferograms for IR-144 detected by both of these methods are shown in Fig. 2c. For SPARSE detection, two reconstruction methods are demonstrated, including reconstruction from a Nyquist-sampled set of Hadamard-encoded observations (Fig. 2c, middle panel) as well as CS reconstruction from a $10 \times$ sub-Nyquist sampled (that is, undersampled) subset of the same Hadamard-encoded measurements (Fig. 2c, right panel). All three interferograms are nearly indistinguishable except for minor differences attributable to noise and reconstruction artifacts in the Hadamard and CS interferograms.

Since heterodyne detection by a time-delayed reference field yields a sinusoidal spectral interference pattern, the discrete cosine transform (DCT) was the natural choice for sparsifying transform. To test the validity of this transform, we explicitly compared the DCT of the CS and Hadamard reconstructed interferograms. To perform the 1D DCT, the interferogram is first flattened to a vector according to the order depicted in Supplementary Fig. 1. The DCT of the interferogram, shown in Fig. 2b, consists mostly of isolated groups of non-zero points. The

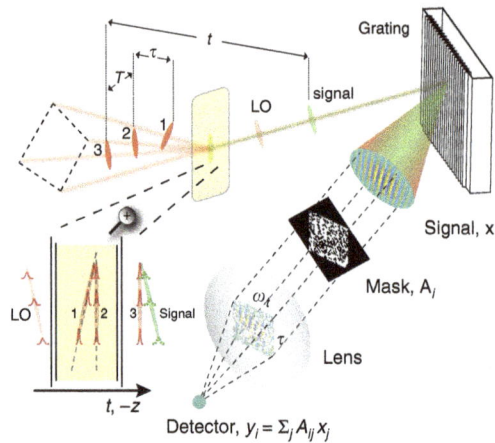

Figure 1 | A simplified schematic of single-shot three-pulse photon echo SPARSE spectroscopy. Three pulses (1, 2 and 3) generate a polarization in the sample which subsequently radiates a signal field that spatially interferes with a reference pulse (or local oscillator, LO) at the exit image plane of a spectrometer. The 2D signal-reference interferogram **x**, which spatially encodes the coherence time τ (inset shows pulse front tilts of each beam at the sample) and detection frequency ω_t, is spatially masked by a DMD (shown as transmissive instead of reflective for simplicity) and then focused by a lens onto a single-element detector. Each mask yields one intensity value on the detector, and by measuring the intensities for a sequence of different masks, the signal-reference interferogram can be retrieved through Hadamard or compressive sensing methods.

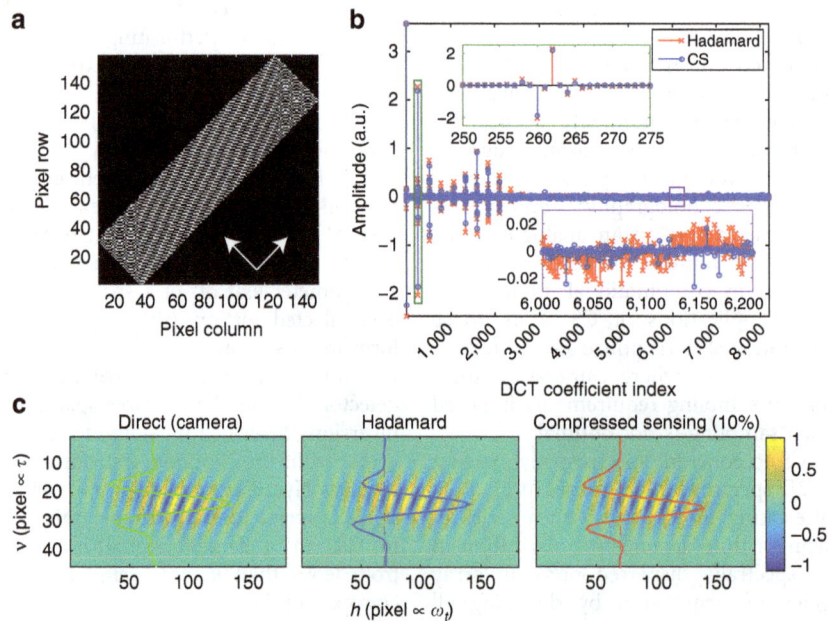

Figure 2 | Hadamard encoding and 1D DCT sparsifying transform for the reconstruction of spatial spectral interferograms. (**a**) A single representative Hadamard spatial mask (without random inversions) tiled into the 'active' region (Methods section) as it appears on the DMD. White arrows indicate horizontal (h) and vertical (v) lab coordinates. (**b**) Comparison of the 1D DCT of a Hadamard-retrieved flattened interferogram to that recovered by compressive sensing using convex optimization. Insets highlight two regions of the DCT interferogram: one at low spatial frequency (green, upper left) and one at high spatial frequency (purple, lower right). (**c**) Comparison of direct (camera), Hadamard and CS ($10 \times$ sub-Nyquist) detected interferograms after Fourier filtering, 45° rotation and cropping. The v axis is proportional to the spatially encoded τ dimension and the h axis is proportional to the detection frequencies, ω_t, in the 2DFT spectrum.

Hadamard measurement confirms that the DCT is effective at making the signal sparse and also demonstrates good agreement with the CS reconstruction. Notable differences appear in the high frequency DCT coefficients where the relatively constant, low amplitude features in the Hadamard reconstruction contrast with the higher, but less frequent spikes in the CS reconstruction. These deviations are expected due to the inability to exactly reconstruct uncorrelated noise from a sub-Nyquist set of measurements.

Comparison of 2DFT spectra. Absolute value 2DFT spectra of IR-144 and LH2 in Fig. 3 were constructed (Methods section) from spatial spectral interferograms collected using each of the three detection methods (camera, Hadamard and CS) described above. The Hadamard and CS (10% of Nyquist for IR-144 and 35% for LH2) reconstructions are faithful to the camera-detected spectra with respect to peak positions, although peak shapes and relative amplitudes do differ somewhat, especially in the case of LH2. The fraction of Hadamard-encoded spatial mask measurements used for CS reconstruction of LH2 and IR-144 interferograms was chosen such that variations between 2DFT spectra generated with different pseudo-randomly chosen measurement subsets were below the 10% level. The higher signal-to-noise ratio of IR-144 measurements enabled greater undersampling than for LH2. Statistical variations in CS reconstruction were explored at a range of sampling percentages to ensure that CS reconstructed 2DFT spectra are reproducible and that they converge to the Hadamard reconstruction as sampling approaches 100% (Supplementary Figs 2–15). Some differences between the SPARSE- and camera-acquired 2DFT spectra are expected based on the larger relative decrease in

responsivity at longer wavelengths of the PMT compared with the camera. This effect at least partially accounts for the differences in peak amplitudes and, perhaps to a lesser extent, peak shapes between 2DFT spectra acquired using a camera versus a PMT.

While IR-144 (Fig. 3a) has one spectrally broad transition centred ~ 800 nm (ref. 32), the spectrum of LH2 contains two absorption bands in the near-infrared, the B800 band at 800 nm ($12,500\ cm^{-1}$) and the B850 band at 850 nm ($11,765\ cm^{-1}$; ref. 33). Since the laser spectrum is sufficiently broad to excite both transitions, two diagonal peaks are expected at early waiting times ($T \approx 0$). As the waiting time progresses, molecules initially excited to the higher energy B800 excited state relax to the B850 excited state, producing an off-diagonal cross peak in the spectrum. By $T = 1$ ps (Fig. 3b), much of the energy transfer has already taken place, leaving a diagonal peak and a cross peak at the B850 emission frequency and a weak diagonal peak at the B800 emission frequency. These features are readily observed in all three detection modalities.

Acquisition time comparison. Although a complete set of 8192 Hadamard-encoded spatial mask measurements is acquired in only 0.82 s, the overall speed of the experiment is limited by the need for signal averaging. Despite this, it takes <3 min to acquire the Hadamard-encoded data needed to generate each 2DFT spectra in the middle column of Fig. 3. Taking into account the ability to undersample when using CS reconstruction, this acquisition time can be reduced by a factor ~ 3–10. Camera-detected 2DFT spectra, on the other hand, required only about 1 s of acquisition time due to the huge multi-channel advantage afforded by the pixel array sensor. In contrast to SPARSE and GRAPES spectroscopy, 2DFT spectroscopy

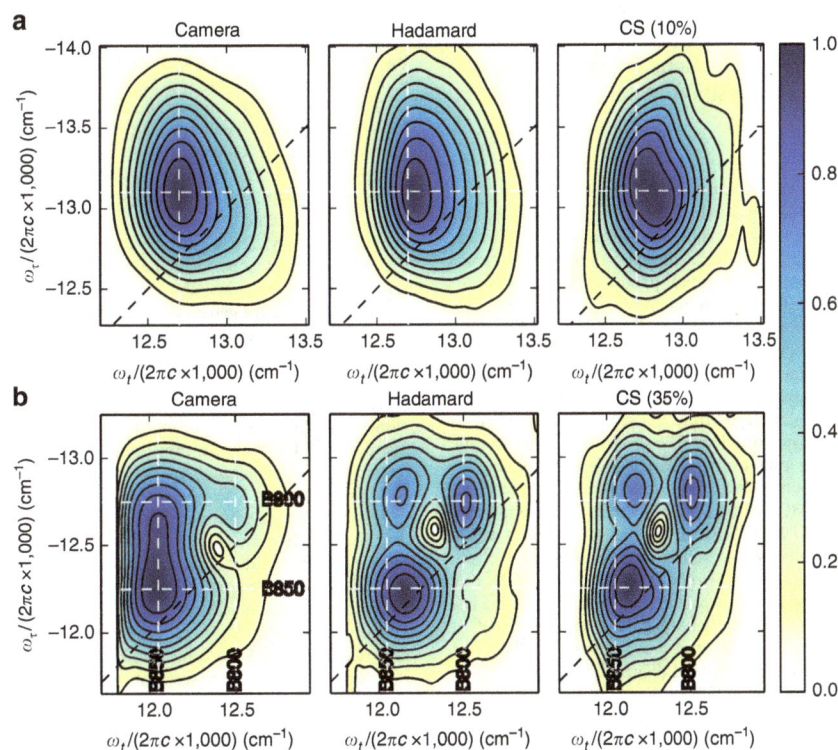

Figure 3 | Comparison of 2DFT spectra. Absolute value 2DFT spectra of (**a**) IR-144 cyanine dye ($T = 20$ fs) and (**b**) LH2 ($T = 1$ ps) detected directly by a camera versus SPARSE (DMD and PMT) detection using either the Hadamard transform (8,192 spatial masks) or compressed sensing with a subset of the Hadamard-encoded measurements (10% (819 spatial masks) for IR-144 and 35% (2,867 spatial masks) for LH2). Diagonal peaks arise from B800 and B850 bands corresponding to the excitation of ring subunits of bacteriochlorophyll pigments in the protein. The upper cross peak results from energy transfer from B800 to B850 in about 1 ps. Approximate locations of band centres are marked with dashed white lines for comparison between figure panels.

techniques that involve scanning pulse time delays can take up to 30 min for one scanned axis[3] or over an hour for two scanned axes[34]. While the most direct comparison for SPARSE spectroscopy would be to phase-sensitive optical detection of 2DFT spectra collected using a single-element detector and scanning both a pulse time delay (τ) and a spectral axis (ω_t), to our knowledge no such study has been reported. The increase in acquisition speed for SPARSE spectroscopy is afforded by the DMD's ability to perform fast 2D scanning of the GRAPES spatial spectral interferogram image instead of relying on slow scanning of pulse time delays.

Discussion

To conclude, we demonstrated the use of SPARSE spectroscopy for acquiring 2DFT spectra with sub-Nyquist sampling, reducing by at least an order of magnitude the number of measurements and, consequently, the data collection time. Array detectors impose uniform sampling even when it greatly oversamples the underlying information content of the image. By contrast, the adaptable nature of SPARSE spectroscopy enables, in principle, use of only the minimum number of measurements necessary to sample the information content of the signal given knowledge of the optimum sparsifying transform. While scientific-grade cameras can currently outperform SPARSE in the visible wavelength region thanks to their substantial multi-channel advantage (proportional to the number of pixels), sensitive cameras with high pixel densities are not available in many other spectral regions. Consequently, one of the most exciting applications of this technique is to expand coherent multidimensional spectroscopic methods to the THz and X-ray regimes where nonlinear spectroscopy is extremely challenging, or for use with frequency combs and supercontinuum sources for high-resolution metrology. Overcoming technical hurdles to detection in these spectral regions will be critically important for elucidating fundamental physical phenomena such as the nature of collective and coherent excitations in liquids and atom-specific electronic coherence in transition metal complexes.

Methods

Experiment. The 1,028 nm wavelength output of a 306 kHz repetition rate Yb:KGW laser system (PHAROS, Light Conversion) with self-contained oscillator and regenerative amplifier is used to pump a visible-near-infrared noncollinear optical parametric amplifier (ORPHEUS-N, Light Conversion), producing tunable pulses with ~ 35 fs duration. The noncollinear optical parametric amplifier output is spatially filtered through a 50-μm diameter pinhole before entering a four-arm interferometer. Four beams, arranged at the corners of a rhombus (or diamond), are focused into the sample cell by a 20-cm focal length cylindrical mirror (refer to Fig. 1 of ref. 14 for details on the beam geometry). The laser centre wavelength was tuned to 775 and 825 nm for measurements on IR-144 and LH2, respectively.

The signal and reference beams are imaged from the sample to the slit of a spectrograph (HR-320, Horiba) by a spherical lens. The spectrograph was modified to create a second output by adding a removable mirror between the second spherical mirror and the standard output. While the standard output houses a CMOS imaging sensor (Zyla 5.5 sCMOS 10-tap, Andor), the added second output is routed to the face of a DMD (DLP9500, Texas Instruments). Since the axes of rotation for the DMD micromirrors lie at a 45° angle relative to the axes of the 1,920 × 1,080 micromirror array, the DMD was rotated by 45° in the plane of its front face so that the reflected light would propagate in a plane parallel to the laser table. The output of the DMD is focused onto a PMT module (H12402, Hamamatsu). The PMT signal is amplified by a fast preamplifier (SR445A, Stanford Research Systems) and then integrated and digitized by a charge-integrating data acquisition (DAQ) system (IQSP518, Vertilon).

The DMD and DAQ are synchronized to the laser pulse train, enabling charge integration of individual laser pulses within a 50 ns window, as well as sub-100 ns control of DMD transition times. The DMD trigger signal, which initiates display of the next available spatial mask in its buffer, was generated by dividing the frequency of the laser clock signal by 32 using a microcontroller (ATmega328, Atmel), ensuring that DMD mask transitions are phase locked to the laser pulse train. In this configuration, each binary mask was held on the DMD for 32 consecutive laser shots, of which 28 laser shots were individually integrated and digitized by the DAQ with the four laser shots closest to the DMD transition being

discarded to ensure the measurements are not influenced by motion of the micromirrors. This results in a DMD frame rate of 9.6 kHz, close to its maximum rate of ~ 10 kHz. Integrated into the DMD control board is a field-programmable gate array, controller chip, driver and high-capacity on-board memory capable of storing ~ 15,000 binary spatial masks.

To perform image reconstruction using a DMD and single-element PMT detector, a set of binary masks was constructed from a 8192nd-order ($(2^{13})^2 = 8192^2$ elements) Hadamard matrix whose columns had been randomly inverted. Such inversions make the mask appear spatially quasi-random, which has the benefit of reducing the influence of undesired diffractive contributions to the measured signal that would otherwise arise from variations in diffraction efficiency between different spatial masks. As opposed to full Hadamard multiplexing which requires balanced detection, an S-matrix-like implementation was used here wherein -1 elements of the Hadamard matrix are replaced by 0 such that only $+1$ elements are detected[29]. To provide adequate spatial resolution while also satisfying the memory limitations of the DMD (see Experimental Limitations and Optimization), the binary mask was limited to an 'active' region (or region of interest) on the DMD composed of 720 by 180 physical pixels, each containing a single micromirror. Groups of 4 by 4 physical pixels were binned (that is, treated as a single unit) to yield a 180 by 45 region containing 8,100 super-pixels. For each binary mask in the 8,192-frame sequence needed for image reconstruction, a row of the Hadamard matrix was tiled into the 8,100 super-pixel 'active' region of the DMD (Supplementary Fig. 1); the remainder of the DMD pixels were set to zero and unused Hadamard row elements were discarded. The 45° rotation of the DMD is compensated for by rotating the rectangular bounds of the active region of the spatial mask such that they lie diagonally in the DMD reference frame. This arrangement makes optimal use of the available 8,100 super-pixels since the signals of interest are typically elongated horizontally (in the lab reference frame) by the angular dispersion induced by the spectrograph grating. This rotation is apparent in the spatial mask shown in a since it is plotted in the DMD reference frame.

The sequence of $2^{13} = 8192$ unique DMD frames was repeatedly cycled through until 10–50 million laser shots had been measured, yielding 43–217 complete data sets that can be averaged and reconstructed into an image of the light intensity at the DMD face.

Data processing. The raw data, each element representing the integrated light intensity of a single laser shot with a given DMD mask, is first divided into complete sequences. Sequences are then averaged together to produce a single 8192-element sequence **y**. To reconstruct the light intensity for each super pixel, the averaged sequence **y** is used to solve the equation $\mathbf{Ax} = \mathbf{y}$, where **A** is the randomly inverted Hadamard matrix used to construct the binary masks and **x** is a vector of super-pixel intensities. Finally, the image at the face of the DMD is obtained by tiling **x** into the active region of the DMD frame exactly as when done to produce the initial DMD masks. An example of an image reconstructed in this way is shown in Fig. 2c (middle panel).

To perform CS reconstruction, first a subset of the Hadamard-encoded measurements described above is pseudo-randomly chosen. The corresponding observation matrix for these measurements is transformed into a sparsifying basis using the 1D DCT along each row. This transformed observation matrix **A** is passed, along with its corresponding intensity measurements **y**, to a convex optimization routine that minimizes the L1-norm of the solution **x** subject to the constraint $\mathbf{Ax} = \mathbf{y}$. The solution is inverse transformed using the 1D DCT to yield the pixel intensity values, which are subsequently tiled into the appropriate spatial arrangement as described above (Supplementary Fig. 1).

2DFT spectra are generated in the same way for both Hadamard and CS reconstructed interferograms. Interferograms first undergo a coordinate transform (involving a 45° rotation with cubic interpolation) from the DMD coordinate space to the camera coordinate space such that a pixel at a given wavelength and τ on the DMD maps to the equivalent wavelength and τ pixel position on the camera. This allows SPARSE-detected and camera-detected interferograms to be processed identically and to share the same pulse delay calibration information. Transformed interferograms are subsequently cropped to the extent of the active area of the DMD spatial mask. The τ axis is calibrated by measuring the spatial spectral interference between beams 1 and 2 using the spectrograph camera, as previously described[14]. The signal-beam 4 interferograms are (piecewise cubic) interpolated from a grid of equidistant wavelengths to a grid of equidistant frequencies. The interpolated interferograms are 2D Fourier transformed and filtered in the conjugate domain to select one of the two AC interference peaks. A 1D inverse Fourier transform along the t dimension (that is, the detection axis) then yields a distorted 2DFT spectrum which must be modified to correct for the crossing angle-induced wavefront tilt between the reference beam (beam 4) and beam 3, as previously reported[14]. Finally, the corrected 2DFT spectra are linearly interpolated to equalize the frequency grid spacing along both the ω_τ and ω_t dimensions. Camera-acquired interferograms are processed identically save for the initial coordinate transform. The camera-acquired interferograms are cropped to the extent of the DMD active area for comparison to SPARSE detection.

Experimental limitations and optimization. There are some properties of the experimental apparatus and methods presented here that are thought to limit the accuracy of 2DFT spectra collected using the implementation of SPARSE detection described here. First, collecting signal from a limited spatial area of the DMD causes the very low-lying wings of the signal to be lost (Fig. 2c). This limitation arises from the finite size of the DMD's on-board memory which, in turn, limits the number of unique DMD spatial masks that can be loaded at a given time. Since the number of pixels in a recovered interferogram image is related to the number of DMD spatial mask measurements, for a fixed number of measurements, there exists a tradeoff between an image's spatial resolution and its size. The spatial size of features in the interferogram image (for example, interference fringes, spectral lineshapes and transient dynamics) sets the minimum required spatial resolution. This spatial resolution requirement, combined with the limit in number of measurements, sets the maximum spatial extent of the region able to be sampled on the DMD. For Hadamard reconstruction, the on-board memory limits the number of recovered image pixel to $\sim 15,000$ since reconstruction of an n pixel image requires measurement of n unique DMD spatial masks. The 720 by 180 physical pixel region used corresponds to a 7.8 mm by 1.9 mm area, less than a quarter of the 16.6 mm by 4.6 mm area collected by the camera. This restriction could be alleviated either by increasing the on-board memory or by using CS reconstruction with $\leq 15,000$ DMD spatial mask chosen randomly from a larger Hadamard matrix (that is, $\mathbf{H}_n$ where $n > 15,000$).

Second, the small active area of the PMT (3 mm by 1 mm) requires careful alignment to ensure that all light reflected from the DMD is collected to avoid attenuating the edges of the interferogram. In addition, the small active area limits the maximum permissible energy per pulse incident on the detector to maintain detector linearity, which is critical for accurate interferogram reconstruction. To avoid detector saturation, likely due to space charge effects[35], lower than optimal reference beam pulse energies were used, limiting the heterodyne advantage of spectral interferometry.

Third, the measured signal is weighted by the spectral response of the PMT, which decreases at longer wavelengths. Although the same is true for the CMOS camera used here, the change in response is much larger for the PMT. Moving from 850 to 800 nm, the spectral response of the PMT increases by a factor of ~ 3 compared with a factor of ~ 1.5 increase for the camera. Overall, the PMT has a low quantum efficiency ($\sim 0.5\%$, compared with $\sim 28\%$ for the camera) which somewhat limits the signal-to-noise ratio of SPARSE detection. Signal-to-noise ratio could be improved by using a higher quantum efficiency detector such as an avalanche photodiode.

References

1. Mukamel, S. *Principles of nonlinear optical spectroscopy* (Oxford University Press, 1995).
2. Cho, M. H. Coherent two-dimensional optical spectroscopy. *Chem. Rev.* **108**, 1331–1418 (2008).
3. Hybl, J. D., Ferro, A. A. & Jonas, D. M. Two-dimensional Fourier transform electronic spectroscopy. *J. Chem. Phys.* **115**, 6606–6622 (2001).
4. Hamm, P., Lim, M., DeGrado, W. F. & Hochstrasser, R. M. The two-dimensional IR nonlinear spectroscopy of a cyclic penta-peptide in relation to its three-dimensional structure. *Proc. Natl Acad. Sci. USA* **96**, 2036–2041 (1999).
5. Harel, E., Long, P. D. & Engel, G. S. Single-shot ultrabroadband two-dimensional electronic spectroscopy of the light-harvesting complex LH2. *Opt. Lett.* **36**, 1665–1667 (2011).
6. Panitchayangkoon, G. *et al.* Long-lived quantum coherence in photosynthetic complexes at physiological temperature. *Proc. Natl Acad. Sci. USA* **107**, 12766–12770 (2010).
7. Collini, E., Wong, C. Y., Wilk, K. E., Curmi, P. M. G., Brumer, P. & Scholes, G. D. Coherently wired light-harvesting in photosynthetic marine algae at ambient temperature. *Nature* **463**, 644–647 (2010).
8. Zheng, J. & Fayer, M. D. Solute-solvent complex kinetics and thermodynamics probed by 2D-IR vibrational echo chemical exchange Spectroscopy. *J. Phys. Chem. B* **112**, 10221–10227 (2008).
9. Scholes, G. D. & Wong, C. Y. Biexcitonic fine structure of cdse nanocrystals probed by polarization-dependent two-dimensional photon echo spectroscopy. *J. Phys. Chem. A* **115**, 3797–3806 (2011).
10. Velizhanin, K. & Piryatinski, A. Probing Interband coulomb interactions in semiconductor nanostructures with 2d double-quantum coherence spectroscopy. *J. Phys. Chem. B* **115**, 5372–5382 (2011).
11. Lepetit, L., Cheriaux, G. & Joffre, M. Linear techniques of phase measurement by femtosecond spectral interferometry for applications in spectroscopy. *J. Opt. Soc. Am. B: Opt. Phys* **12**, 2467–2474 (1995).
12. Brixner, T., Mancal, T., Stiopkin, I. & Fleming, G. Phase-stabilized two-dimensional electronic spectroscopy. *J. Chem. Phys.* **121**, 4221–4236 (2004).
13. Shim, S. H., Strasfeld, D. B., Fulmer, E. C. & Zanni, M. T. Femtosecond pulse shaping directly in the mid-IR using acousto-optic modulation. *Opt. Lett.* **31**, 838–840 (2006).
14. Spencer, A. P., Spokoyny, B., Harel, E. & Enhanced-Resolution Single-Shot, 2DFT Spectroscopy by spatial spectral interferometry. *J. Phys. Chem. Lett.* 945–950 (2015).
15. Bracewell, R. N. *The Fourier Transform and Its Applications* 3rd edn. (McGraw Hill, 2000).
16. Donoho, D. L. Compressed sensing. *ITIT* **52**, 1289–1306 (2006).
17. Lustig, M., Donoho, D., Pauly, J. & Sparse, M. R. I. The application of compressed sensing for rapid MR imaging. *Magn. Reson. Med.* **58**, 1182–1195 (2007).
18. Cai, X. J., Hu, B., Sun, T., Kelly, K. F. & Baldelli, S. Sum frequency generation-compressive sensing microscopy. *J. Chem. Phys.* **135** **(2011)**.
19. Sanders, J. N. *et al.* Compressed sensing for multidimensional spectroscopy experiments. *J. Phys. Chem. Lett.* **3**, 2697–2702 (2012).
20. Dunbar, J. A., Osborne, D. G., Anna, J. M. & Kubarych, K. J. Accelerated 2D-IR using compressed sensing. *J. Phys. Chem. Lett.* **4**, 2489–2492 (2013).
21. Rivenson, Y., Stern, A. & Javidi, B. Overview of compressive sensing techniques applied in holography. *Appl. Opt.* **52**, A423–A432 (2013).
22. Zhu, L., Zhang, W., Elnatan, D. & Huang, B. Faster STORM using compressed sensing. *Nat. Methods* **9**, 721–723 (2012).
23. Andrade, X., Sanders, J. N. & Aspuru-Guzik, A. Application of compressed sensing to the simulation of atomic systems. *Proc. Natl Acad. Sci. USA* **109**, 13938–13933 (2012).
24. McClean, J. R. & Aspuru-Guzik, A. Compact wavefunctions from compressed imaginary time evolution. Preprint at http://arxiv.org/abs/1409.7358 (2014).
25. Sanders, J. N., Andrade, X. & Aspuru-Guzik, A. Compressed sensing for the fast computation of matrices: application to molecular vibrations. *ACS Cent. Sci.* **1**, 24–32 (2015).
26. Harel, E., Fidler, A. & Engel, G. Real-time mapping of electronic structure with single-shot two-dimensional electronic spectroscopy. *Proc. Natl Acad. Sci. USA* **107**, 16444–16447 (2010).
27. Spokoyny, B. & Harel, E. Mapping the Vibronic Structure of a Molecule by Few-Cycle Continuum Two-Dimensional Spectroscopy in a Single Pulse. *J. Phys. Chem. Lett.* **5**, 2808–2814 (2014).
28. Nelson, E. D. & Fredman, M. L. Hadamard Spectroscopy. *J. Opt. Soc. Am.* **60**, 1664 (1970).
29. Graff, D. K. Fourier and Hadamard: transforms in spectroscopy. *J. Chem. Educ.* **72**, 304 (1995).
30. Dudley, D., Duncan, W. & Slaughter, J. in SPIE Proceedings SPIE **4985**, 14–25 (MOEMS Display and Imaging Systems) (2003).
31. Candès, E. & Romberg, J. *l1-magic* v. 1.11California Institute of Technology, 2005).
32. Jonas, D. M. Two-dimensional femtosecond spectroscopy. *Annu. Rev. Phys. Chem* **54**, 425–463 (2003).
33. Cogdell, R., Gall, A. & Kohler, J. The architecture and function of the light-harvesting apparatus of purple bacteria: from single molecules to in vivo membranes. *Q. Rev. Biophys.* **39**, 227–324 (2006).
34. Tekavec, P. F., Lott, G. A. & Marcus, A. H. Fluorescence-detected two-dimensional electronic coherence spectroscopy by acousto-optic phase modulation. *J. Chem. Phys.* **127**, 214307 (2007).
35. HP., K. K. *Photomultiplier Tubes: Basics and Applications* 3rd edn (Hamamatsu Photonics K.K., 2006).

Acknowledgements

The work was supported by the Army Research Office (W911NF-13-1-0290), Air Force Office of Scientific Research (FA9550-14-1-0005) and the Packard Foundation (2013-39272) in part.

Author contributions

E.H. conceived the experiments. A.P.S. and B.S. conducted the experiments. S.R. developed the computer interface to the DMD. E.H., F.S. and B.S. developed CS implementation. A.P.S., B.S. and E.H. performed data analysis. E.H. and A.P.S. contributed to the preparation of manuscript.

Additional information

11

Shaping the nonlinear near field

Daniela Wolf[1,2], Thorsten Schumacher[1] & Markus Lippitz[1]

Light scattering at plasmonic nanoparticles and their assemblies has led to a wealth of applications in metamaterials and nano-optics. Although shaping of fields around nanostructures is widely studied, the influence of the field inside the nanostructures is often overlooked. The linear field distribution inside the structure taken to the third power causes third-harmonic generation, a nonlinear optical response of matter. Here we demonstrate by a far field Fourier imaging method how this simple fact can be used to shape complex fields around a single particle alone. We employ this scheme to switch the third-harmonic emission from a single point source to two spatially separated but coherent sources, as in Young's double-slit assembly. We envision applications as diverse as coherently feeding antenna arrays and optical spectroscopy of spatially extended electronic states.

[1] Experimental Physics III, University of Bayreuth, Universitätsstrasse 30, D-95440 Bayreuth, Germany. [2] Max Planck Institute for Solid State Research, Heisenbergstrasse 1, D-70569 Stuttgart, Germany. Correspondence and requests for materials should be addressed to M.L. (email: markus.lippitz@uni-bayreuth.de).

When a noble metal nanostructure is excited by ultrafast laser pulses, nonlinear optical effects such as higher harmonics generation or multiphoton-induced luminescence can be observed[1-4]. The correlation between linear and nonlinear response[5,6], the spectral dependence of the nonlinear signal[7,8] and the use of nanoantennas to boost optical nonlinearities[9,10] has been investigated in detail. Although the linear response of plasmonic nanostructures is well understood and almost any desired field distribution can be realized by well-chosen arrangements of multiple particles[11-13], the spatial origin of nonlinear signals such as second-harmonic[14,15] and third-harmonic generation (THG)[16-18] is still under debate. Spatial features that are below the diffraction limit of the fundamental wavelength can still be resolved at the shorter wavelength of the higher harmonic signals[15,19]. However, real-space imaging does not give access to the coherence properties of different emitting spots. In Young's double-slit experiment[20], the coherent emission from two slits interferes due to the wave nature of light and a characteristic pattern of dark and bright stripes can be observed in the far field. The emerging Fraunhofer diffraction pattern corresponds to the Fourier transform of the apertures[21]. The shape and intensity distribution of the interference pattern in the far field is thus uniquely related to both width and separation of the slits, as well as their relative phase.

We set out to control the spatial distribution of third-harmonic emission in a plasmonic nanostructure. A schematic overview of the experiment is given in the introductory Fig. 1. In analogy to Young's experiment, the emission properties manifest in the far field interference pattern, allowing a reconstruction of the nonlinear near fields. Using this method, we prove that it is the field inside the structure and not the external hotspots that cause THG. As illustrated in Fig. 1, we show that the emission of

a gold rod can be switched between a configuration with one single and two separated sources when the excitation wavelength is tuned over a higher-order plasmon resonance. Our results are supported by calculations of the emission patterns using single dipoles, as well as simulations of the linear and nonlinear fields using a finite element method.

Results

Nonlinear plasmonic analogue of a single slit. To introduce our experimental technique, we first discuss the nonlinear plasmonic analogue of a single slit as illustrated in Fig. 2a. A single 270-nm-long gold nanorod on a glass substrate is illuminated by a focused beam of short near-infrared laser pulses, leading to THG. Other nonlinear signals are filtered spectrally (see Supplementary Fig. 1). At 1,170 nm wavelength, the fundamental dipolar plasmon mode is excited resonantly in this rod, as can be seen in Fig. 2a from the field distribution calculated with a finite element method. Regarding the THG, the field inside the particle is the crucial parameter[17]. The internal field is highest right in the centre of the rod, which is in stark contrast to the field outside the particle, peaking at the ends of the rod. As the nonlinear material polarization is proportional to the third power of this internal field[1], the calculation predicts that the third-harmonic signal is emitted from the centre of the structure for this geometry. In free space, the angular radiation pattern would be that of a free dipole, that is, of toroidal shape with the nanorod being on the symmetry axis. The air–glass interface refracts and reflects part of this emission so that a typical two-lobed angular pattern remains in the direction of the glass substrate[22]. We detect this angular emission pattern by imaging the back focal plane of the microscope objective that collects the light emitted into the substrate (see Methods section). The field distribution in the back focal plane is the Fourier transform of the sample plane[21], corresponding thus to the Fraunhofer pattern in the classical diffraction experiments. The finite acceptance angle α of the microscope objective given by the numerical aperture $NA = n \sin(\alpha)$ limits the observable region in reciprocal space to $|k/k_0| < NA$, with n being the refractive index of the substrate and $k_0 = 2\pi/\lambda$ the wave vector in free space. As shown in Fig. 2b, we find good agreement between the measured emission pattern of the nanorod and the calculated pattern of a single dipole oriented parallel to the interface, corresponding to emission from the centre of the particle. Assuming two dipoles oscillating in phase as in Fig. 2c, corresponding to emission from the ends of the rod, clearly yields a different radiation pattern (for full emission patterns, see Supplementary Fig. 2 and for quantitative analysis see Supplementary Fig. 3). We therefore conclude that it is the high field inside the gold structure that causes THG. The emission properties of a single resonantly excited nanorod can simply be described by a single dipole oriented parallel to an interface (see Supplementary Fig. 4).

Nonlinear plasmonic analogue of Young's double slit. We now turn to the nonlinear plasmonic analogue of Young's double-slit experiment. In the classical experiment, a double-slit assembly is illuminated by spatially coherent light. In our plasmonic analogue, we exploit the coherence of the third-harmonic emission from two spatially separated nanorods excited by the same laser focus (Fig. 3a). Each nanorod emits similar to a single dipole, as confirmed by numerical simulations. For a pair of identical rods, an interference pattern is superimposed on the dipolar radiation pattern of the individual rods. As shown in Fig. 3b for a distance of 930 nm, again we find good agreement between the patterns from experiment and dipole model, regarding both the position of the extrema and the intensity

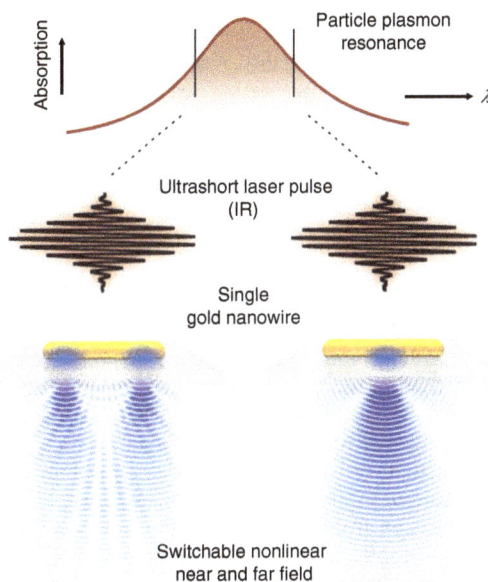

Figure 1 | Nonlinear near field and emission control. Slightly tuning the excitation wavelength over a plasmon resonance drastically changes the local fields and thus the far field response of a simple plasmonic structure. In our experiment, we switch between a configuration with one single and two separated emitting centres. In analogy to classical diffraction experiments, the coherent emission from different sources leads to characteristic interference patterns in the far field. We exploit this effect to determine the position and relative phase of emitting centres in a single plasmonic nanoparticle and the concomitant nonlinear near fields.

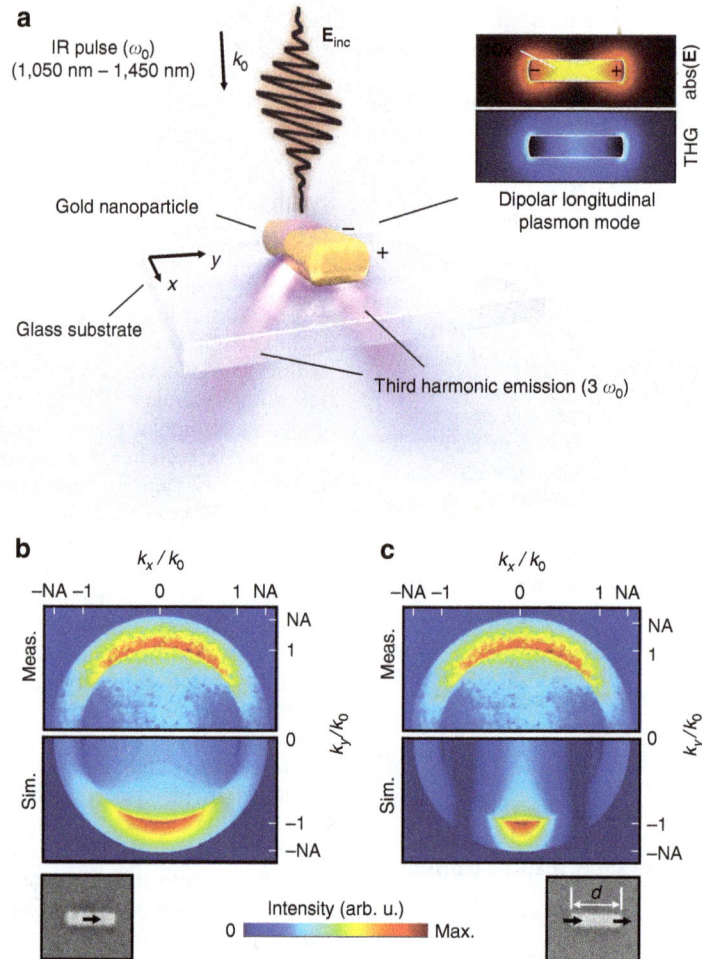

Figure 2 | Localizing the emission from a single nanorod. (**a**) Schematic representation of the measurement method. A single gold nanoparticle is excited by infrared light from the air side and the generated third harmonic is detected through the glass substrate. Inset: electric field and resulting third-harmonic field of a nanorod excited at the fundamental plasmon resonance calculated with a finite element method. The linear field inside the particle is scaled up by a factor of 10 for better visibility. (**b,c**) Measured back focal plane image of a 270-nm-long rod compared with two hypotheses: (**b**) the emission stems from one spot in the centre, modelled as one dipole or (**c**) the emission stems from the end surfaces, modelled by two equal, in phase dipoles, separated by $d = 270$ nm. Shown is always one half of the symmetric emission pattern. Excitation wavelength is 1,170 nm.

distribution as well. To further analyse the data, we project the back focal plane images onto the k_x axis, which retains all relevant details on the interference process (for the full data set, see Supplementary Fig. 5). The resulting intensity profiles for two different rod distances are shown in Fig. 3c, corresponding to the typical intensity patterns in the classical experiments. The number and especially the exact position of the minima in Fig. 3d depend strongly on the separation of the emitting centres but agree well between measurements and simulations for rod pairs with distances between 330 and 930 nm. It is noteworthy that the gap between the two particles is always larger than 60 nm so that plasmonic coupling can be excluded. The careful analysis of the interference pattern thus allows us to accurately measure the emitter distance. That way, this experiment further confirms the centres of the rods as sources of the nonlinear signal (see also Supplementary Figs 2 and 3).

Nonlinear emission properties of a long rod. This fact becomes even more apparent when instead of two short rods a single long rod is investigated. We demonstrate that the nonlinear emission and near field can be switched between one and two emitting centres by slightly tuning the fundamental wavelength.

The experiment is summarized in Fig. 4. The angular emission pattern of a 925-nm-long gold rod at an excitation wavelength of 1,320 nm clearly displays the characteristic interference pattern of a double slit where both apertures emit in phase (see Fig. 4a). With the method presented above, the separation of the emitting centres can be determined to approximately 600 nm, as indicated in the scanning electron microscopic image. When the fundamental wavelength is tuned to 1,420 nm, we obtain an angular pattern that deviates only slightly from a single dipole, implying a dominant emission from the centre of the rod (see Fig. 4c). Our conclusions are confirmed by numerical simulations of equivalent rod structures as shown in Fig. 4a,c.

We take the emission patterns at 490 and 425 nm TH wavelength as templates for the states $|1\rangle$ and $|2\rangle$, respectively, and fit the patterns at all other wavelengths by a linear superposition $(1 - a)|1\rangle + a|2\rangle$ (see Supplementary Note 1 for explanation of the fitting method). As shown in Fig. 4b, the weight a obtained in this way displays a steep transition within 25 nm at the third-harmonic wavelength between a single emitting spot in the centre (state $|1\rangle$) and two in-phase spots with well-defined separation (state $|2\rangle$).

To explain the switching of the emission pattern, we need to consider in more detail the modes of the fundamental field. The

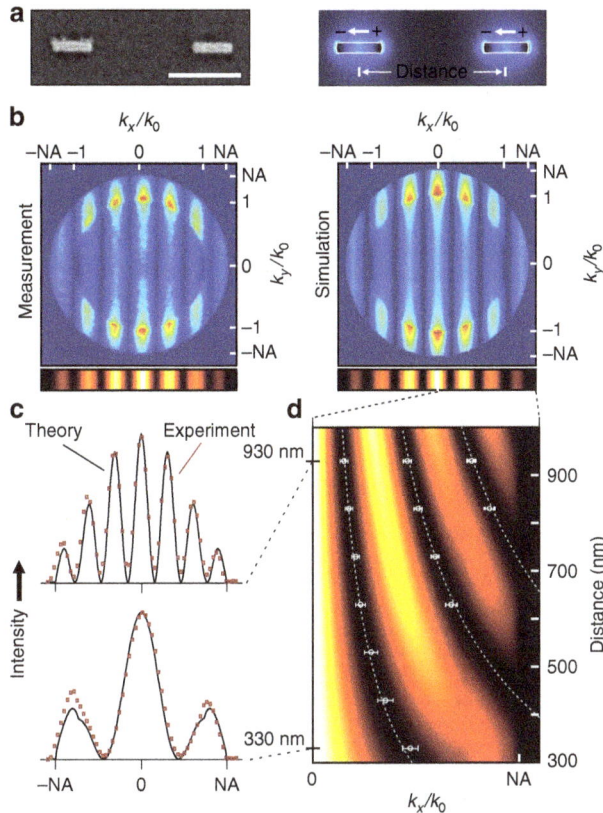

Figure 3 | Interference in the emission of two rods. (**a**) Scanning electron microscopic image of a pair of 270-nm-long nanorods; scale bar, 500 nm. The centre distance is varied between 330 and 930 nm. According to the finite element method calculations, the particles act as two separated dipoles oscillating in phase. (**b**) Measured and calculated radiation patterns for 930 nm distance with intensity projection onto the k_x axis. (**c**) Measured (red squares) and calculated (black lines) intensity profiles for 330 and 930 nm distance, corresponding to cuts through the distance-dependent intensity projection shown in **d**. The squares and dashed lines indicate the positions of the minima from measurement and calculation, respectively. The error bars correspond to an increase to three times the noise level above the respective minimum.

modes of long nanorods resemble standing waves, where only odd modes can be excited optically in our configuration[23]. Here, the dipolar mode is shifted far into the infrared, while the third-order mode shows a resonance in the wavelength regime where the experiments are carried out (see Fig. 4d). When the excitation wavelength is tuned over a plasmon resonance, the phase changes by π. In the vicinity of the third-order resonance, the phase of the third-order mode undergoes this change, whereas the phase of the dipolar mode is unaffected. Tuning the fundamental wavelength over the third-order resonance thus changes the relative phase of the two modes by π. It is noteworthy that the third-order resonance of the investigated structure is red-shifted compared with the calculation due to fabrication inaccuracies. The field distribution of the dipolar and the third-order mode can be described by a single dipole and by three counter-oscillating dipoles, respectively. We always excite both modes with varying efficiency and observe their superposition. For wavelengths below the resonance, the overall phase between the modes vanishes. In the dipole picture, the single dipole and the inner dipole of the third-order mode oscillate against each other and cancel. As only the outer dipoles remain, this corresponds to the double slit behaviour. Above the resonance, the overall phase between the modes is π. Hence, the single dipole and the inner dipole of the

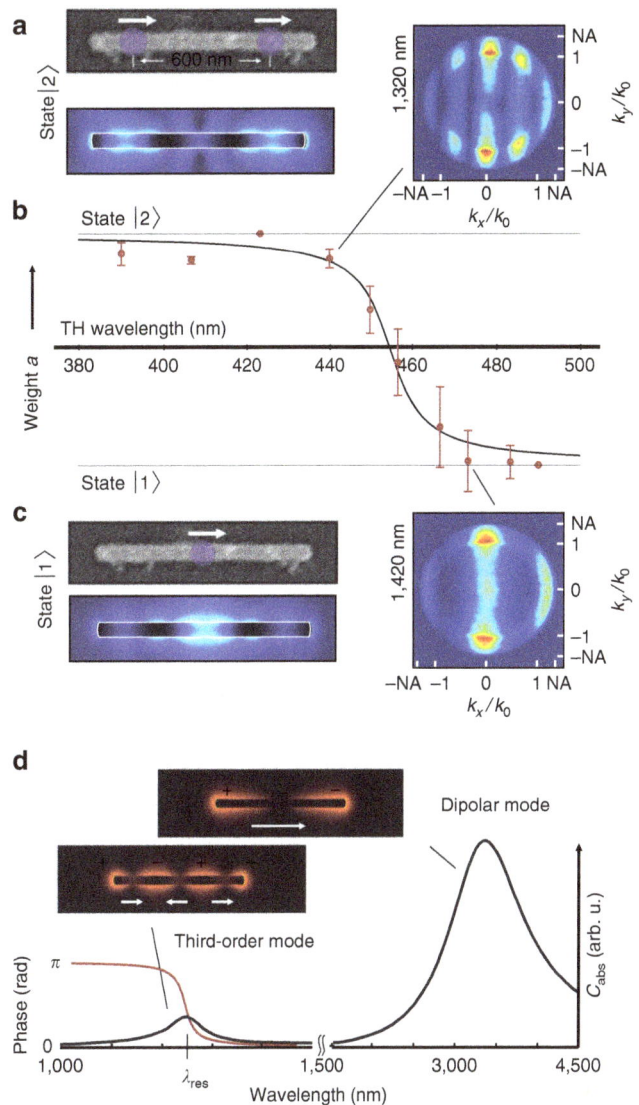

Figure 4 | Switching of nonlinear emission and near fields. (**a,c**) The emission pattern of a 925-nm-long rod depends on the excitation wavelength. The two states $|1\rangle$ and $|2\rangle$ differ in the number of emitting spots, as indicated in the scanning electron microscopic images and the calculated TH fields. (**b**) We observe a transition within a wavelength range of about 25 nm for the weight a of state $|1\rangle$ (red dots). The error bars are three times the s.d. of the fit. The black line is a guide to the eye. (**d**) Calculated absorption cross-section (black line) and phase of the third-order mode (red line) of a 925-nm-long rod. Insets: linear field distributions at the fundamental and third-order resonance.

third-order mode add up so that the centre dipole dominates. The nonlinear third-order process of THG amplifies amplitude and phase differences of the fundamental field inside the rod and thus drastically changes the third-harmonic emission properties (see Supplementary Fig. 6). Evidently, the generated near field switches accordingly as demonstrated by the numerical simulations shown in Fig. 4, whereas only slight changes are observed in the linear near field. The same switching behaviour is observed when tuning the plasmon resonance via the length of the nanorod and keeping the excitation wavelength fixed (see also Supplementary Fig. 6).

Discussion

In summary, we have localized spatially the third-harmonic emission centres of a single plasmonic nanorod by analysing interference in the far field. The experiment unambiguously shows that it is not the high field around the tips of the nanorod, but the standing waves inside the rod that cause the emission at the third-harmonic wavelength. This allows us to control and tune the optical near field patterns at the third harmonic solely by slight variations of the incoming field. Although optical near fields at the fundamental wavelength are affected as well, the optical nonlinearity drastically amplifies the effect. We demonstrated a switching of the emission centres by a wavelength shift of only 25 nm.

Our experiments open up a new direction for nanophotonics[24]. The local field around plasmonic nanostructures can not only be sculptured by well-chosen arrangements of many nanoparticles, but with a similar efficiency also by an engineered distribution of the fundamental field inside a continuous piece of metal. Owing to the high optical nonlinearities of noble metals[16], a conceivable field strength at the third harmonic is obtained, resulting in unprecedented control over the placement of light sources on the nanoscale. We envision applications as diverse as coherently feeding antenna arrays[25] or optical circuits[26,27] and optical spectroscopy of spatially extended electronic states[28].

Methods

Experimental set-up and sample fabrication. The signal output of a Ti:Sapphire pumped optical parametric oscillator (76 MHz, 150 fs, 1,050–1,450 nm) is focused onto the sample with an infrared lens (NA 0.55), leading to a spot size of $\sim 1.5\,\mu m$ on the sample. The gold nanostructures are fabricated on 170-μm-thick glass coverslips using electron beam lithography followed by metal evaporation and lift-off. All structures are 60 nm wide and 30 nm high. Their length is 270 nm for the short rods and 925 nm for the long rod. The polarization of the excitation is chosen parallel to the long axis of the nanorods. Excitation powers are 2.5 and 10 mW for the short and long rods, respectively.

The transmitted light and the generated higher harmonic is collected by an oil-immersion objective (NA 1.35). The near-infrared excitation light is eliminated with a Schott KG5 filter. A back focal plane image is acquired on the charge-coupled device camera by placing an additional lens (focal length 150 mm), acting as a Bertrand lens, in front of the spectrograph. For the measurements at 1,170 nm excitation wavelength, a narrow bandpass filter transmits only the third harmonic at 390 nm and the back focal plane images are acquired using the spectrograph with a mirror instead of a grating. For the wavelength-dependent measurements, the bandpass filter is removed and a grating disperses the light. The convolution of spatial pattern and emission spectrum has little influence, as the emission spectrum is very peaked. To locate the structures, the sample is scanned with a piezo stage and the emitted light is detected by an avalanche photodiode. In this case, the Bertrand lens is removed.

Simulation methods. To calculate the radiation patterns, we assume a dipole in air ($n=1$) at a height of 15 nm above the glass interface ($n=1.5$). We calculate the fields from the Fresnel coefficients and project the resulting intensity distribution onto the back focal plane[22,29]. When two to four dipoles are considered, all dipoles emit with the same amplitude and phase. The fields from the individual dipoles are superimposed to obtain the total field, which is again projected onto the back focal plane.

For the finite element calculations, we use the commercial software package 'Comsol Multiphysics'. The simulations for the linear (at ω_0) and nonlinear (at $3\omega_0$) response are performed in frequency domain and separated into two models. In both, the dimensions of the structures are identical and matched to those in the experiment. As dielectric function of gold, we use the data reported by Johnson and Christy[30]. For the linear response, we assume a plane wave excitation and solve Maxwell's equations for the given boundary value problem. The shown linear field distributions are the absolute values of the electric field $\mathbf{E}(\omega_0)$ in a plane 15 nm above the substrate ($n=1.5$). The third-harmonic polarization is calculated as described in previous work[17] from $\mathbf{P}_{THG,loc} \sim \chi^3_{Au}\mathbf{E}^3_{loc} \sim \mathbf{P}^3_{loc}$. To obtain the third-harmonic fields around the structure, we use the third-harmonic polarization as surface boundary value at the gold structures and solve the system at $3\omega_0$. The third-harmonic field plots show the absolute values of the nonlinear electric field $\mathbf{E}(3\omega_0)$ in the same plane as the linear fields before.

In Supplementary Table 1, we provide information about which simulation method has been used in the individual graphs in this paper.

References

1. Boyd, R. W. *Nonlinear Optics* 3rd edn (Academic Press, 2008).
2. Lippitz, M., Dijk, M. A. & Orrit, M. Third-harmonic generation from single gold nanoparticles. *Nano Lett.* **5**, 799–802 (2005).
3. Renger, J., Quidant, R., Van Hulst, N. & Novotny, L. Surface-enhanced nonlinear four-wave mixing. *Phys. Rev. Lett.* **104**, 046803 (2010).
4. Kauranen, M. & Zayats, A. V. Nonlinear plasmonics. *Nat. Photonics* **6**, 737–748 (2012).
5. Hentschel, M., Utikal, T., Giessen, H. & Lippitz, M. Quantitative modeling of the third harmonic emission spectrum of plasmonic nanoantennas. *Nano Lett.* **12**, 3778–3782 (2012).
6. O'Brien, K. *et al.* Predicting nonlinear properties of metamaterials from the linear response. *Nat. Mater.* **14**, 379–383 (2015).
7. Hanke, T. *et al.* Efficient nonlinear light emission of single gold optical antennas driven by few-cycle near-infrared pulses. *Phys. Rev. Lett.* **103**, 257404 (2009).
8. Metzger, B., Hentschel, M., Lippitz, M. & Giessen, H. Third-harmonic spectroscopy and modeling of the nonlinear response of plasmonic nanoantennas. *Opt. Lett.* **37**, 4741–4743 (2012).
9. Harutyunyan, H., Volpe, G., Quidant, R. & Novotny, L. Enhancing the nonlinear optical response using multifrequency gold-nanowire antennas. *Phys. Rev. Lett.* **108**, 217403 (2012).
10. Aouani, H., Rahmani, M., Navarro-Ca, M. & Maier, S. A. Third-harmonic-upconversion enhancement from a single semiconductor nanoparticle coupled to a plasmonic antenna. *Nat. Nanotechnol.* **9**, 1–5 (2014).
11. Prodan, E., Radloff, C., Halas, N. J. & Nordlander, P. A hybridization model for the plasmon resonance of complex nanostructures. *Science* **302**, 419–422 (2003).
12. Schuller, J. A. *et al.* Plasmonics for extreme light concentration and manipulation. *Nat. Mater.* **9**, 193–204 (2010).
13. Luk'yanchuk, B. *et al.* The Fano resonance in plasmonic nanostructures and metamaterials. *Nat. Mater.* **9**, 707–715 (2010).
14. Valev, V. K. *et al.* Asymmetrical optical second-harmonic generation from chiral G-shaped gold nanostructures. *Phys. Rev. Lett.* **104**, 127401 (2010).
15. Mascheck, M. *et al.* Observing the localization of light in space and time by ultrafast second-harmonic microscopy. *Nat. Photonics* **6**, 293–298 (2012).
16. Utikal, T. *et al.* Towards the origin of the nonlinear response in hybrid plasmonic systems. *Phys. Rev. Lett.* **106**, 133901 (2011).
17. Metzger, B., Schumacher, T., Hentschel, M., Lippitz, M. & Giessen, H. Third harmonic mechanism in complex plasmonic Fano structures. *ACS Photonics* **1**, 471–476 (2014).
18. Liu, X., Larouche, S., Bowen, P. & Smith, D. R. Clarifying the origin of third-harmonic generation from film-coupled nanostripes. *Opt. Express* **23**, 19565–19574 (2015).
19. Hanke, T. *et al.* Tailoring spatiotemporal light confinement in single plasmonic nanoantennas. *Nano Lett.* **12**, 992–996 (2012).
20. Young, T. *A Course of Lectures on Natural Philosophy and the Mechanical Arts* vol. 1 (Printed for J. Johnson, 1807).
21. Goodman, J. W. *Introduction to Fourier Optics* 3rd edn (MaGraw-Hill, 2005).
22. Novotny, L. & Hecht, B. *Principles of Nano-Optics* 2nd edn (Cambridge University Press, 2012).
23. Dorfmüller, J. *et al.* Plasmonic nanowire antennas: experiment, simulation, and theory. *Nano Lett.* **10**, 3506–3603 (2010).
24. Koenderink, A. F., Alù, A. & Polman, A. Nanophotonics: Shrinking light-based technology. *Science* **348**, 516–521 (2015).
25. Dregely, D. *et al.* Imaging and steering an optical wireless nanoantenna link. *Nat. Commun.* **5**, 4354 (2014).
26. Engheta, N. Circuits with light at nanoscales: optical nanocircuits inspired by metamaterials. *Science* **317**, 1698–1702 (2007).
27. Gramotnev, D. K. & Bozhevolnyi, S. I. Plasmonics beyond the diffraction limit. *Nat. Photonics* **4**, 83–91 (2010).
28. Dubin, F. *et al.* Macroscopic coherence of a single exciton state in a polydiacetylene organic quantum wire. *Nat. Phys.* **2**, 32–35 (2006).
29. Lieb, M. A., Zavislan, J. M. & Novotny, L. Single-molecule orientations determined by direct emission pattern imaging. *J. Opt. Soc. Am. B* **21**, 1210–1215 (2004).
30. Johnson, P. B. & Christy, R. W. Optical constants of noble metals. *Phys. Rev. B* **6**, 4370–4379 (1972).

Acknowledgements

We gratefully acknowledge financial support from the Deutsche Forschungsgemeinschaft through the doctoral training centre GRK1640 and SPP1391 Ultrafast Nanooptics.

Author contributions

M.L. and T.S. designed the experiment. D.W. fabricated the sample and conducted the experiments. T.S. and D.W. analysed the data and performed the simulations. D.W., T.S. and M.L. wrote the paper. All authors contributed through scientific discussions.

Additional information

Competing financial interests: The authors declare no competing financial interests.

12

Optically controlled electroresistance and electrically controlled photovoltage in ferroelectric tunnel junctions

Wei Jin Hu[1], Zhihong Wang[2], Weili Yu[1] & Tom Wu[1]

Ferroelectric tunnel junctions (FTJs) have recently attracted considerable interest as a promising candidate for applications in the next-generation non-volatile memory technology. In this work, using an ultrathin (3 nm) ferroelectric $Sm_{0.1}Bi_{0.9}FeO_3$ layer as the tunnelling barrier and a semiconducting Nb-doped $SrTiO_3$ single crystal as the bottom electrode, we achieve a tunnelling electroresistance as large as 10^5. Furthermore, the FTJ memory states could be modulated by light illumination, which is accompanied by a hysteretic photovoltaic effect. These complimentary effects are attributed to the bias- and light-induced modulation of the tunnel barrier, both in height and width, at the semiconductor/ferroelectric interface. Overall, the highly tunable tunnelling electroresistance and the correlated photovoltaic functionalities provide a new route for producing and non-destructively sensing multiple non-volatile electronic states in such FTJs.

[1] Materials Science and Engineering, King Abdullah University of Science and Technology (KAUST), Thuwal 23955-6900, Saudi Arabia. [2] Advanced Nanofabrication Core Lab, King Abdullah University of Science and Technology (KAUST), Thuwal 23955-6900, Saudi Arabia. Correspondence and requests for materials should be addressed to W.J.H. (email: weijin.hu@kaust.edu.sa) or to T.W. (email: tao.wu@kaust.edu.sa).

Aferroelectric layer sandwiched between two metal electrodes serves as the basic building block in diverse applications, such as ferroelectric capacitors[1-3] and ferroelectric diodes[4,5], which utilize polarization as the switching degree of freedom to perform logic and memory functions. As the thickness of the ferroelectric layer decreases to a few nanometres, electrons can quantum mechanically tunnel through the ferroelectric barrier, which is the basis for the operation of so-called ferroelectric tunnel junctions (FTJs)[6,7]. In fact, the concept of FTJ was formulated by Esaki et al.[8] more than four decades ago, but it has only recently aroused considerable interest following the demonstration of ferroelectricity in ultrathin perovskite films with a thickness of several unit cells[9-11].

A basic feature of FTJs is the modulation of conductance between on and off states with the reversal of ferroelectric polarization, namely, the tunnelling electroresistance (TER)[6,7]. Tsymbal and Kohlstedt[12] theoretically demonstrated that polarization-related physical factors, such as interface charge screening, interface atomic bonding and the inverse piezoelectric effect, can change the interfacial barrier height and/or width, thereby leading to the TER effect. Most FTJs with metallic electrodes exhibit TER values on the order of $\sim 1-10^3$, and the effect is attributed to the modulated ferroelectric barrier height[13-21]. Recently, a giant TER of up to 10^4 was observed in a FTJ of Pt/ BaTiO$_3$ (BTO)/Nb:SrTiO$_3$ (NSTO)[22]. This remarkable improvement in performance originates from the suppression/ enhancement of the space charge layer in the semiconducting NSTO electrode, that is, both the height and width of the ferroelectric tunnel barrier were modulated[22,23]. Enhancement of TER was also reported in La$_{0.7}$Sr$_{0.3}$MnO$_3$/BTO/La$_{0.5}$Ca$_{0.5}$MnO$_3$/ La$_{0.7}$Sr$_{0.3}$MnO$_3$(LSMO) tunnel junctions (TER ~ 100 at 5 K) (ref. 18) and PbZr$_{0.2}$Ti$_{0.8}$O$_3$/La$_{1-x}$Sr$_x$MnO$_3$ heterostructures (TER ~ 300 at 300 K) (ref. 24), which were attributed to the ferroelectric-induced phase modulation of the manganite electrode. Recently, it was recognized that ultrathin ferroelectric films are in poly-domain states with nm-scale domains[25], which is a direct result of the fact that the ferroelectric domain size scales with the square root of the film thickness[26]. This insight offers the opportunity of engineering the domain states and creating multiple resistive states with memristive behaviour in FTJs based on ultrathin BTO (ref. 27) and tetragonal BiFeO$_3$ (BFO; ref. 28) films. FTJ-based memristor was also demonstrated in Co/BTO/LSMO heterostructures, and the operating mechanism was related to the field-induced interface charge redistribution[29]. These ground breaking results provided new routes towards the development of high-performance FTJs, and thus, the underlying mechanism clearly warrants further investigations.

Concurrently, the demands for higher density, low-power data processing/storage devices have motivated the development of oxide heterostructures that possess integrated functionalities and that can be manipulated by multiple external parameters[30-35]. Effective approaches towards integrating novel functionalities into FTJs are, however, sparse[36]. The most notable approach is the multiferroic tunnel junction, in which ferroelectric tunnel layers are combined with ferromagnetic electrodes and in which magnetic properties such as magnetoresistance and spin polarization can be controlled by an electric field in addition to a magnetic field[16-18]. Because some perovskite oxides are known to be good light absorbers[31,37,38], light illumination could be envisioned as an additional tuning parameter for controlling the operation of FTJs.

Herein, we report complimentary electrical and optical control over the resistive switching in FTJs of Pt/Sm$_{0.1}$Bi$_{0.9}$FeO$_3$(SBFO)/ NSTO. A colossal TER of greater than 10^5 at room temperature

has been observed in our FTJs, which could be further modulated by ~ 10-fold through light illumination. Our detailed analysis suggests that the colossal TER originates from the electrically induced transition between the conduction modes of direct tunnelling and Schottky thermionic emission at the ferroelectric/ semiconductor interface. In addition, the light-absorbing depletion layer in NSTO induces a photovoltaic (PV) effect, which is dependent on the resistance state of the device and provides a new route to non-destructive reading in addition to the conventional current-based route.

Results

Device structure and band alignment. Schematics of the FTJ and the corresponding energy band diagrams are shown in Fig. 1, illustrating the complimentary effects of light illumination and electric field on the device transport. The device has a pushpin-type structure (Fig. 1a) with an effective junction area of $5\,\mu m \times 5\,\mu m$ (Fig. 1b) (see Methods for details of device fabrication). The unique feature of the ITO/semi-transparent Pt top electrode allows light penetration, which enables the modulation of the junction transport by light illumination in addition to the polarization states as shown by the schematic band alignments in Fig. 1c,d. When the polarization is pointing to the left (right), NSTO interface is accumulated (depleted) due to the charge screening effect, which yields substantial changes of the barrier width in addition to the barrier height and hence leads to a giant TER effect. More importantly, the light illumination could change the band bending of the NSTO interface by exciting photo carriers, leading to additional tuning of the TER effect. Simultaneously, the isolation of excited electrons and holes due to the band bending of NSTO results in a PV effect, providing us a new route to read the non-volatile electronic state of FTJs.

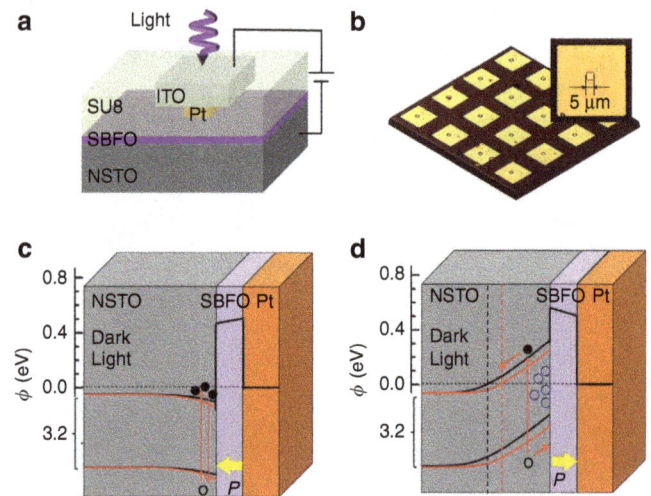

Figure 1 | Mechanism of complementary electroresistance and photovoltaic effects. (a) Schematic representation of structure of the NSTO/SBFO/Pt FTJs. **(b)** Optical image of the as-prepared FTJs with an effective junction area of $5\,\mu m \times 5\,\mu m$ (shown in the inset). Band alignment in the dark and under light illumination are shown for ferroelectric polarization (yellow arrow) pointing to either **(c)** the semiconducting NSTO electrode or **(d)** the metal electrode Pt. Polarization-modified carrier accumulation/depletion at the NSTO/SBFO interface contributes to the complementary electroresistance and PV effects. Electric-field-induced drift of positively charged oxygen vacancies (blue circles) from the NSTO bulk towards the interface is also illustrated in **d**. Black filled (empty) circles represent photon-excited electrons (holes).

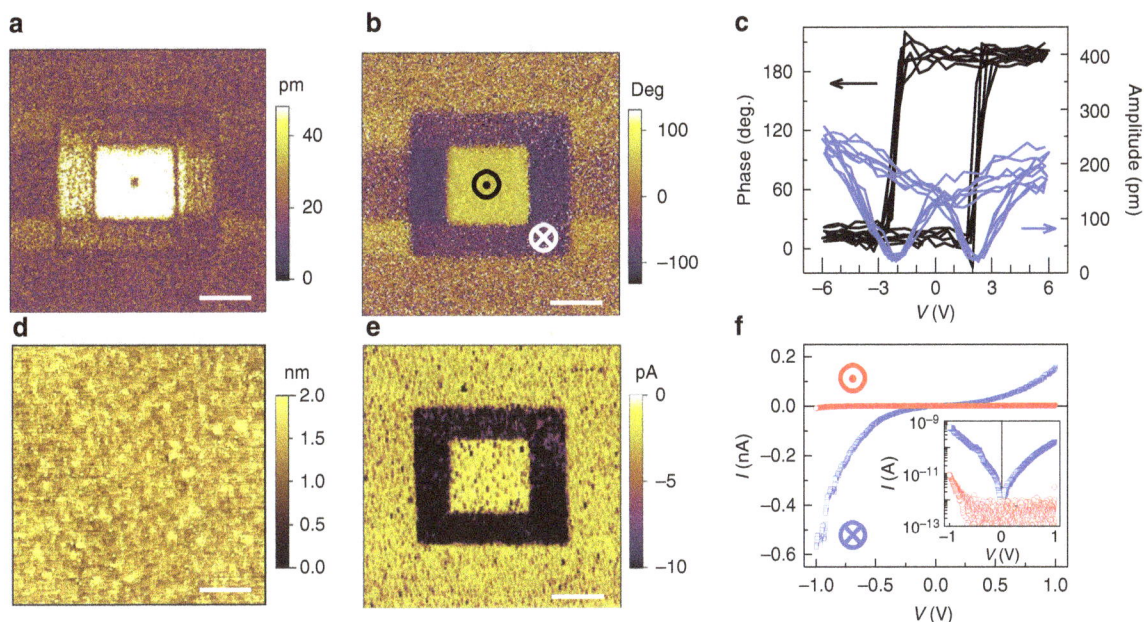

Figure 2 | PFM and CAFM measurements on the ultrathin (3 nm) SBFO film. (**a**) PFM amplitude image and (**b**) phase image of square domains with opposite polarization directions written using an AFM tip with ±4 V bias. (**c**) Local PFM hysteresis curves. (**d**) AFM image of the film surface. The RMS roughness is 0.4 nm. (**e**) Current mapping taken at a low bias of − 0.5 V after writing opposite domain patterns on the same film. (**f**) Local $I − V$ curves for two opposite polarization domains (blue, down; red, up) by CAFM. Inset, $I − V$ data on a log scale. Note that DC voltages were applied on the sample for current mapping and $I − V$ curve measurements. Scale bars, 1 µm.

Characterization and ferroelectricity of SBFO films. We first characterized the ferroelectricity and the polarization controlled resistive switching of the bare ferroelectric film. Epitaxial SBFO films of $3 − 9$ nm were grown using pulsed-laser deposition (PLD) on (001) 0.7 wt.% NSTO single crystal substrates. Atomic force microscopy measurement on the 3 nm bare film shows smooth surface with roughness of 0.4 nm (shown in Fig. 2d). SBFO was chosen as the barrier due to its smaller leakage current compared with BFO (Supplementary Fig. 1). NSTO is a degenerate semiconductor and is used as both the growth substrate and the bottom electrode (see Methods for details). The basic structural properties of the films are presented in Supplementary Fig. 2.

The SBFO films were in-plane compressively strained, ensuring stabilization of ferroelectricity in the ultrathin films at room temperature, which was confirmed using piezoresponse force microscopy (PFM). As shown in Fig. 2a,b, clear amplitude and a phase contrast of ∼180° were observed in the bare 3 nm SBFO film after writing square patterns with alternative + 4 and − 4 V biases applied on the PFM tip. These results clearly indicate the ferroelectric nature of the 3 nm SBFO film, which was further supported by the butterfly-like amplitude and hysteresis behaviour of the phase signal in the local PFM measurement, as shown in Fig. 2c. Similar ferroelectric properties were also observed in 6 and 9 nm films (Supplementary Figs 3,4). The acquired coercive electric field of ∼730 MV m^{-1} for 3 nm SBFO film is comparable with those reported for 5 nm BFO films ($∼200 − 600$ MV m^{-1}) (ref. 25), whereas they are one or two orders of magnitude larger than those of thick BFO films ($∼10$ MV m^{-1}) (ref. 39) and BFO single crystals ($∼2$ MV m^{-1}) (ref. 5), indicating the important role of the interface in the switching behaviour of ultrathin ferroelectric films.

In conjunction with the PFM, we also investigated the local current switching properties of the as-grown SBFO film using conductive atomic force microscopy (CAFM). Current mapping was acquired by scanning a polarization-patterned area with a dc bias of − 0.5 V applied on the sample (Fig. 2e). The current contrast clearly suggests that the polarization reversal modifies

the local conduction of the 3 nm SBFO film. A larger current was observed for the downward-polarized domain, which is qualitatively consistent with the band alignment shown in Fig. 1c,d. $I − V$ curves shown in Fig. 2f were acquired in local areas with opposite polarizations. The $I − V$ curve of low-resistance state (LRS) shows a parabolic behaviour, consistent with the direct tunnelling across the ultrathin SBFO layer, whereas the $I − V$ curve of high-resistance state (HRS) shows as an asymmetric Schottky-emission-type behaviour. A HR/LR ratio of up to ∼1,000 (inset of Fig. 2f) was obtained. Current mapping (Supplementary Figs 3b,e) and local $I − V$ curves (Supplementary Fig. 5) were also performed on SBFO films with thicknesses of 6 and 9 nm. Different from the 3 nm ultrathin film, both HRS and LRS of thicker films showed typical Schottky-emission behaviour, indicating that the conduction of heterostructures with thick SBFO barriers is no longer governed by direct tunnelling.

Transport mechanisms and voltage/light-induced TER of FTJs. Now we turn to the electric properties of FTJs with the ultrathin SBFO films being sandwiched between NSTO and Pt electrodes (Fig. 1a). For electric characterizations, applied voltage is defined as positive when the top Pt electrode is positively biased. Indium was soldered on the NSTO substrate to form good Ohmic contacts. Figure 3a shows the $I − V$ characteristics of a typical junction by sweeping the voltage from 0 to 3 V, then to − 16 V, and finally back to 0 V. Clear resistive switching was observed with the positive (negative) voltage leading to LRS (HRS). The rectification ratio at 3 V is $5.6 × 10^4$ for HRS, whereas it decreases to 32 for LRS. The prototypical diode behaviour observed in the high bias regime indicates that the Schottky barrier, rather than the tunnel barrier, dominates the resistance of the junction (Supplementary Fig. 6).

Figure 3b presents the $I − V$ curves of the FTJ in the low-bias regime (− 0.5 to + 0.5 V) of LRS and HRS after applying voltage pulses of + 3 and − 16 V both in the dark and under UV light illumination. It was found that the nonlinear LRS $I − V$ curve can

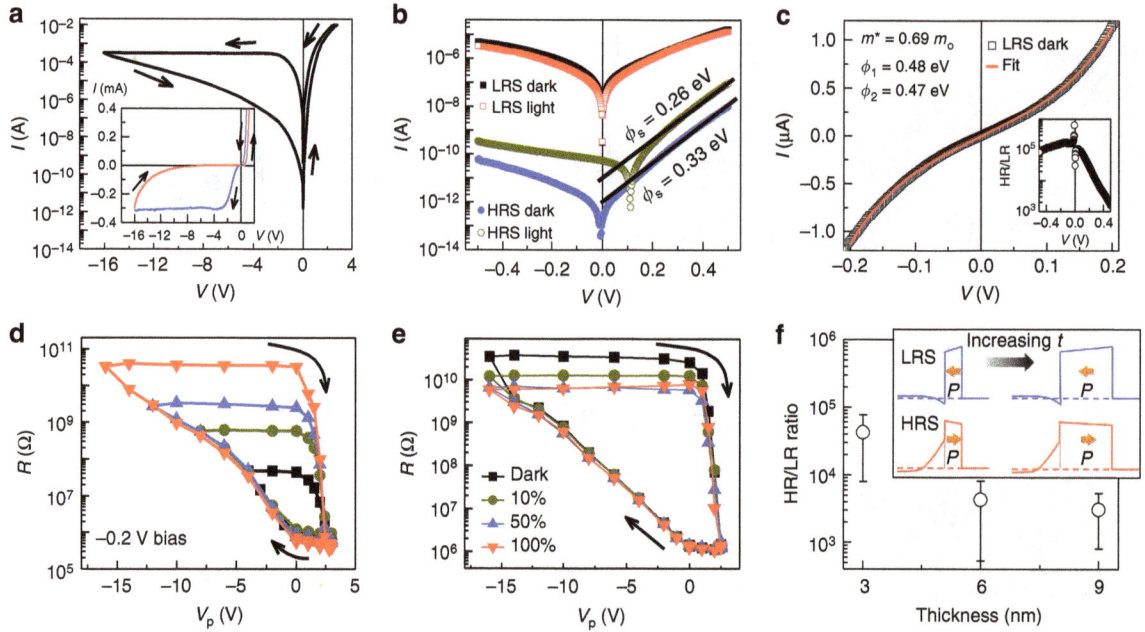

Figure 3 | Tunable electroresistance controlled by voltage and light in NSTO/SBFO (3 nm)/Pt FTJs. (a) $I - V$ switching characteristics of the FTJ as the voltage sweeps from 0 to $+3$ V, to -16 V and then back to 0 V. Inset is the same set of data shown on a linear scale. **(b)** Typical $I - V$ data of LRS and HRS in the low-bias region after writing with voltages of $+3$ and -16 V, both in the dark and under UV light illumination. **(c)** Tunnelling behaviour of the LRS. Inset, HR/LR ratio in the dark calculated from the $I - V$ curves shown in **b**. $R - V$ hysteresis loops measured with **(d)** increasing voltage pulses in the dark and **(e)** increasing light intensities. **(f)** HR/LR ratio as a function of the SBFO thickness. The error bars represent the s.d. measured in 18 junctions for each barrier thickness. Inset is the schematics of corresponding band alignment.

be well described by the direct tunnelling theory based on the WKB approximation[19],

$$J(V) = C \frac{\exp\left\{\alpha(V)[(\phi_2 - eV/2)^{3/2} - (\phi_1 + eV/2)^{3/2}]\right\}}{\alpha^2(V)[(\phi_2 - eV/2)^{1/2} - (\phi_1 + eV/2)^{1/2}]^2}$$
$$\times \sinh\left\{\frac{3}{2}\alpha(V)[(\phi_2 - eV/2)^{1/2} - (\phi_1 + eV/2)^{1/2}]\frac{eV}{2}\right\},$$

(1)

Where $C = -(4em^*)/(9\pi^2\hbar^3)$ and $\alpha(V) = [4d(2m^*)^{1/2}]/[3\hbar(\phi_1 - \phi_2 + eV)]$. In this equations, $\phi_{1,2}$ are the tunnelling barrier heights at the two interfaces and m^* is the effective mass of electrons in the SBFO tunnel barrier. Figure 3c shows the fitting to the LRS curve using the following parameters: $\phi_{1,2} = 0.48$ (0.47) eV for the NSTO/SBFO (Pt/SBFO) barrier and m^* of $0.69m_o$. Similar fitting results were achieved on nearly 20 junctions (Supplementary Fig. 7).

In contrast, for the $I - V$ data of HRS, no reasonable fit to the direct tunnelling mechanism could be obtained (Supplementary Fig. 8). However, as shown in Fig. 3b, the conduction behaviour of HRS can be described by Schottky thermionic emission, for which the forward current is governed by the following equation[40]:

$$I(V) = A^*AT^2 \exp(-\phi_s/k_BT) \exp\left(-\alpha_T d\sqrt{\phi_T}\right)$$
$$\exp(qV/nk_BT - 1)$$
$$= A^{**}AT^2 \exp(-\phi_s/k_BT) \exp(qV/nk_BT - 1),$$

(2)

where A is the junction area, A^* is the standard Richardson constant (600 A cm^{-2} K^{-2} by assuming an effective mass of $5m_0$ for NSTO),[41] ϕ_s is the Schottky barrier height, T is the temperature and n is the ideality factor. The term $\exp\left(-\alpha_T d\sqrt{\phi_T}\right)$ denotes the current-reducing effect due to the tunnelling barrier, where $\alpha_T = 2(2qm^*)^{1/2}/\hbar$, d is the thickness of

SBFO and ϕ_T is the average tunnelling barrier height. The impact of the tunnelling barrier is reflected by the reduced Richardson constant $A^{**}(\sim 0.2$ A m^{-2} K^{-2}) by assuming a ϕ_T of 0.475 eV for simplicity, the same as that of LRS. The Schottky barrier height ϕ_s of HRS can then be estimated from fitting the forward current to equation 2, which provides 0.33 and 0.26 eV for ϕ_s in the dark and under illumination, respectively. Furthermore, the depletion layer thickness W_d could be estimated using the following relationship: $\phi_s = (qN_d/2\varepsilon_0\varepsilon_s)W_d^2$ (ref. 40), where N_d ($\sim 10^{20}$ cm^{-3}) and ε_s (~ 200) are the free carrier density and the static dielectric constant of NSTO[22], respectively. This estimation provides a W_d of ~ 8.5 nm for the FTJ in the dark, which decreases to ~ 7.6 nm under light illumination. In addition, note that the derived ideality factor n of 1.9 deviates from 1, which is consistent with the simultaneous existence of tunnelling electrons[40,42]. The direct tunnelling character of LRS and the contribution of electron tunnelling to the Schottky thermionic emission of HRS were further confirmed by temperature-dependent $I - V$ measurements (Supplementary Fig. 9 and Supplementary Table 1).

Based on the above analysis, we conclude that in the low-bias regime (<0.2 V), the LRS is dominated by the direct tunnelling, whereas the HRS is dominated by the Schottky thermionic emission. This conclusion is consistent with the band structure of the Pt/SBFO/NSTO tunnel junction shown in Fig. 1c. In the LRS, the polarization of SBFO points to the NSTO electrode, which leads to the accumulation of charges at the NSTO/SBFO interface, and electrons can directly tunnel through the ultrathin SBFO barrier. In contrast, in the HRS, the polarization of SBFO points away from NSTO, the positive screening charges lead to upward bending of the NSTO conducting band, and the Schottky barrier is thus significantly enhanced. This increase in the depletion region in NSTO results in the onset of HRS, which is accompanied by a change in the transport mechanism from direct tunnelling to Schottky emission.

Figure 3d presents typical non-volatile resistive memory loops controlled by voltage pulses. The junction was first set to LRS by a pulse voltage of $+3$ V, followed by sweeping the voltage to -4 (or -8, -12 and -16) V and back to $+3$ V. The resistance of FTJ was read at a bias of -0.2 V. A series of intermediate resistance states can be achieved by adjusting the magnitude of the applied maximum negative voltage; thus, the device behaves as a memristor[43,44]. A giant HR/LR ratio of approximately 10^5 was finally achieved, which is consistent with that calculated directly from the $I-V$ curves, as shown in the inset of Fig. 3c. This colossal TER is considerably larger than that previously reported for a BFO-based FTJ (~ 0.8 at 80 K) (ref. 45) and 10-times larger than FTJs based on tetragonal-BFO[28]. To the best of our knowledge, the TER achieved here is the highest ever reported for FTJs[36]. Furthermore, this colossal TER exhibits only slight decrease on increasing negative bias (see the inset of Fig. 3c), allowing to increase the readout current levels without losing the performance.

Interestingly, the $R-V$ loops in Fig. 3d are highly asymmetric, that is, the transition from HRS to LRS is abrupt, whereas the transition from LRS to HRS is gradual. This is different from the symmetric ferroelectric hysteresis loops (Fig. 2c), where full polarization-reversal occurs at approximately 2–3 V and no strong imprint effect was observed. Thus, we believe that in addition to the polarization reversal, other factors such as drifting of oxygen vacancies under the intense electric field could contribute to the observed giant ER effect[29,46]. As shown in Fig. 1d, the accumulation of oxygen vacancies at the NSTO/SBFO interface under the negative electric field enhances the depletion layer and consequently increases the electroresistance. In addition, we found that the ER magnitude decreases with increasing SBFO thickness (see Fig. 3f and Supplementary Fig. 10), which is consistent with the fact that the effect of tuning NSTO barrier on ER becomes weaker when the thicker SBFO barrier dominates the junction transport (inset of Fig. 3f). Because of the same reason, in FTJs with thicker SBFO (6 and 9 nm), Schottky diode effect, instead of electron tunnelling, dominates the transport behaviour of both LRS and HRS, which is similar to the observations made on thick BFO films[47].

More interestingly, we found that the electroresistance of FTJs could also be modulated by UV-light illumination. As shown in Fig. 3b, UV-light illumination produced almost no effect on LRS, but notable tuning was observed for HRS. The modulation of electroresistance by illumination was also confirmed by the resistance memory loops (Fig. 3e) measured under illumination with different light intensities. Furthermore, switch cycling tests were conducted under both dark and light conditions (Supplementary Fig. 11). In general, the HR/LR ratio was reduced by ~ 10-fold when the FTJs were exposed to UV light. This result is consistent with the Schottky barrier reduction of ~ 0.07 eV obtained from fitting the HRS $I-V$ curves. Because this effect was observed only for HRS and because NSTO is the main light absorber in the devices, we believe that this effect is correlated with the tuning of the NSTO depletion layer near the SBFO/NSTO interface. The light excited free electrons will diffuse towards the NSTO interface and reduce the depletion width and barrier height for HRS (Fig. 1d). This was further confirmed by the light-tuning effect observed in capacitance measurements (Supplementary Fig. 12). Such complementary modulations of the resistance states by electric field and light illumination produce additional resistance states and enhance the memory density of the FTJs.

PV effect and non-destructive reading of the memory states. For a metal – insulator – semiconductor heterostructure, when the illuminated wavelength matches the optical bandgap of the active material, photo-excitation of charge carriers occurs. The generated electrons and holes will be separated by the internal electric field, which leads to the PV effect[40]. In our case, the main light-absorbing layer is NSTO rather than SBFO because of the considerably thinner character of the latter, which was confirmed by investigating the wavelength dependence of the PV effect (Supplementary Fig. 13). In other words, the short-circuit current I_{sc} (current at zero bias) and open-circuit voltage V_{oc} (voltage at zero current) are mainly determined by the depletion layer near the NSTO/SBFO interface. Consequently, the PV effect is expected to be dependent on the memory states of the FTJ and thus tunable.

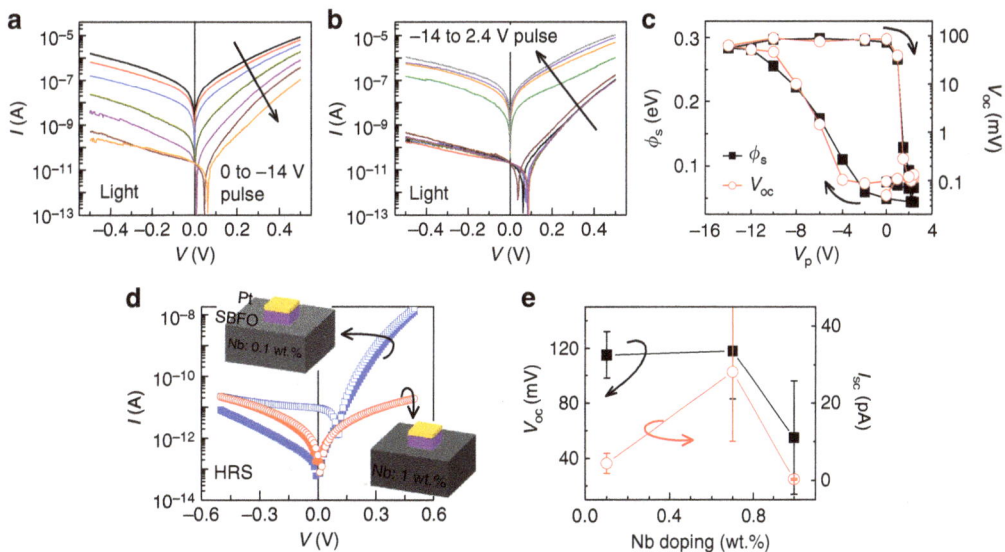

Figure 4 | Electrically tuned PV effect in NSTO/SBFO/Pt FTJs. Typical $I-V$ curves of FTJ measured under light illumination (**a,b**). The successive voltage pulses are 0, -2, -4, -6, -8, -10 and -14 V in **a**; -14, -10, -6, -2, 0, 1, 1.5, 2, 2.2 and 2.4 V in **b**. The device was first set to LRS by a $+2.4$ V pulse before the measurement. (**c**) Schottky barrier ϕ_s (left) and V_{oc} (right) as a function of the voltage pulse. ϕ_s is obtained from the thermionic emission fitting to the forward $I-V$ curves measured in the dark. (**d**) Typical HRS $I-V$ curves measured on FTJs with Nb doping levels of 0.1 and 1 wt.% (solid, dark; empty, light). (**e**) V_{oc} and I_{sc} of HRS measured on FTJs using NSTO substrate with different Nb doping levels. The error bars represent the s.d. measured in 18 junctions for each doping level.

Figure 4a and 4b show the $I - V$ curves measured under light illumination after switching with a series of voltage pulses (V_p) from 0 to $- 14$ V (LRS to HRS) and from $- 14$ to 2.4 V (HRS to LRS), respectively. We found that the HRS $I - V$ curves exhibit notable shifts along the voltage axis, and V_{oc} is as high as ~ 100 mV. In contrast, V_{oc} is negligible small for LRS (< 0.1 mV). Whereas no shift has been observed for $I - V$ curves measured in the dark (Supplementary Fig. 14). This result unambiguously suggests that the magnitude of V_{oc} depends on the band bending in NSTO near the NSTO/BFO interface (Fig. 1c,d). The strong upward band bending in the HRS (Fig. 1d) drives the photo-generated electrons into the NSTO bulk, and the holes tunnel through the SBFO barrier, leading to the PV effect. Both the resistance state and the PV effect exhibit similar dependence on the switching voltage bias, indicating the shared mechanism related to the depletion/accumulation of NSTO near the interface, revealing the importance of the electric field induced polarization reversal and the interface charge redistribution (Fig. 1c,d). Consequently, the evolution of V_{oc} as a function of the applied voltage pulses (Fig. 4c, right) is very similar to that of the Schottky barrier (Fig. 4c left, obtained from the fitting of $I - V$ curves in the dark) and the resistance memory loops (Fig. 3e). Furthermore, the dependence of I_{sc} on the applied voltage pulses also shows hysteresis behaviour (Supplementary Fig. 15). In general, such a resistance-state-dependent PV effect provides a new route by using V_{oc} (or I_{sc}) for sensing the memory states of FTJs in addition to the conventional resistive reading. This new route is non-destructive and reliable because light illumination will not change the polarization state of the devices. Because V_{oc} is in principle independent of the lateral size of the junctions, this V_{oc}-based reading scheme is amenable to the miniaturization of data storage devices and promising to facilitate the high-density integration.

Note that the PV effect observed in our metal/ferroelectric/metal FTJs is different from that recently reported for FE capacitors in which the ferroelectric layer itself is the active light-sensing component[31]. The present device scheme allows us to engineer the PV effect by adjusting the Nb doping level of the NSTO substrates. Figure 4d shows the typical HRS $I - V$ curves of FTJs fabricated on NSTO substrates with Nb doping levels of 0.1 and 1 wt.%. Although the NSTO substrate with 1 wt.% Nb doping has the strongest light absorption (Supplementary Fig. 16), its higher carrier concentration leads to thinner depletion layer and weaker PV effect[40]. In addition, different with the highly asymmetric HRS $I - V$ curves of FTJs based on NSTO (0.1 and 0.7 wt.%), the HRS $I - V$ curves of FTJ based on NSTO (1 wt.%) show parabolic direct tunnelling behaviour, in line with the much reduced Schottky barrier formed at the NSTO (1 wt.%)/SBFO interface. The dependence of the PV effect on the Nb doping level is summarized in Fig. 4e. Clearly, FTJs fabricated on 0.7 wt.% doped NSTO substrate presents the best PV performance.

Device performance and reliability. For practical non-volatile memory applications, a large off/on resistance ratio, fast writing speed, long data retention and stable fatigue properties are the most important figures of merit. Fig. 5a shows the HRS and LRS reading at a $- 0.2$ V bias for 18 randomly selected devices with the SBFO barrier thickness of 3 nm. The resistance ratios fall in the range of 10^4–10^5, indicating the high yield and good uniformity of the FTJs. Note that the writing current density of our FTJs is only $\sim 2 \times 10^4$ A cm^{-2}, which is comparable with the previously reported FTJ based on tetragonal BFO thin films[11], and two to three orders smaller than those of spin-based devices such as magnetic random access memories (MRAMs)[48], spin-transfer-torque-based MRAM[33], and spin–orbit-torque-based MRAM[49].

To investigate the operation speed of resistive switching, we measured the influence of the poling pulse width on the resistance switching from HRS (LRS) to LRS (HRS). Before the measurements, the device was first set to HRS (LRS) by applying a voltage pulse of $- 10$ V ($+ 2.6$ V) with duration of 1 s. Then, square pulses with varied pulse widths of 16 ns to 1 s were applied on the device, and the resistance of FTJ was subsequently recorded at a bias of $- 0.2$ V. As shown in Fig. 5b, resistance switching emerges when the pulse width exceeds ~ 10 µs. This result was also confirmed by the characteristics of the resistive switching data measured using different pulse widths, as shown in Fig. 5c. The switching of the FTJs is faster than that of flash memory[50] but slower than resistive random access memory[51] and MRAM[52]. A writing speed of up to 10 ns has recently been realized in LSMO/BTO/Co FTJs with sizes on the nanometre scale[13,27], suggesting that our FTJs still have the potential for improvement.

Generally speaking, there are two factors that limit the writing speed of FTJs: the switching time of the polarization and the field-induced drift of charged species at the interface. Considering the switching of the polarization could be as fast as 1 ns for ferroelectrics[31], we believe that the switching speed of our FTJs is mainly limited by the dynamics of the drifting charges. Semiconducting NSTO has a depletion layer thickness of ~ 8 nm, which is much larger than that of LSMO (~ 0.2–1.9 nm) (ref. 16) and may lead to a longer switching time. Since the writing speed determines the ultimate operating bandwidth of integrated FTJ memories, more in-depth studies are warranted in future.

Figure 5d shows the resistance retention at a reading bias of $- 0.2$ V. The values of HRS and LRS slightly decrease and increase with time, respectively, but an off/on ratio of more than 10^3 is still retained after 10^5 s. This value is ~ 260 after 10 years through a simple extrapolation, suggesting the excellent non-volatility of our device[22,53]. Furthermore, the off and on states are stable under repeated bipolar switching cycles, as shown in Fig. 5e. Voltage pulses of $- 10$ and $+ 2.6$ V with a pulse width of 10 ms were applied in this endurance test. Slight resistance decreases were observed for both states with increasing cycle numbers, and a large off/on ratio of ~ 200 was preserved after $\sim 10^7$ writing cycles. This fatigue property of our FTJs is comparable or better than those in previous reports, for example on BTO based FTJs (off/on ratio of ~ 100 over 3,000 writing cycles)[22], and BFO-based FTJs (off/on ratio of ~ 100 up to 4×10^6 cycles)[53]. In addition, the HR and LR states of our FTJs appear to be more stable without much resistive fluctuation compared with the previously reported FTJs based on BFO barrier[53]. The substantial improvements in both the fatigue and the stability of the non-volatile states could be attributed to the Sm doping in the ultrathin ferroelectric BFO layers. As previously reported, rare earth doping could reduce the migration of oxygen vacancies[54], which has been widely believed to be one of the microscopic origins of fatigue in ferroelectric oxides[55]. We also monitored the retention and fatigue behaviours of V_{oc} for HRS and LRS under light illumination (Fig. 5f,g), no deterioration in retention up to 20 h and in fatigue up to 10^4 cycles was observed, suggesting the feasibility and robustness of the PV route for reading the non-volatile states of FTJs.

Discussion

In this work, we demonstrated that using an ultrathin (3 nm) SBFO as the tunnel barrier, in conjunction with a semiconducting Nb:STO as the bottom electrode, can significantly improve the performance of FTJs. In particular, the off/on ratio reaches a giant value of 10^5, which, to the best of knowledge, is the highest reported to date in the literature. One important factor underlying the achieved improvements is the excellent ferroelectric

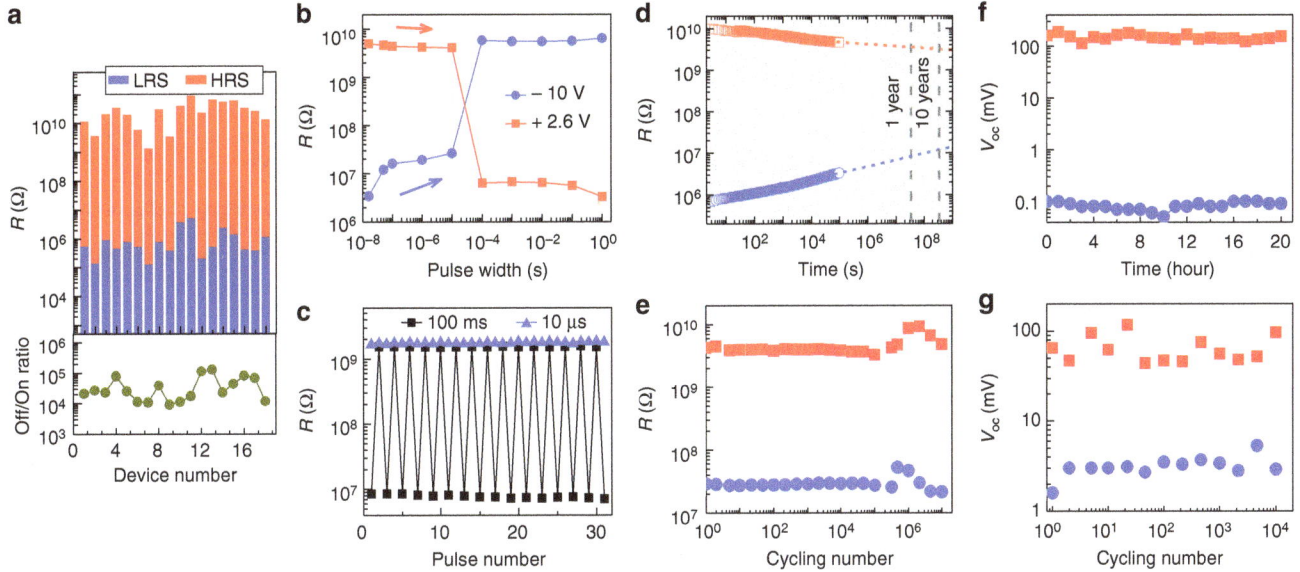

Figure 5 | Retention and fatigue properties of NSTO/SBFO (3 nm)/Pt FTJs. (a) Off and on resistance states read at a -0.2 V bias and the corresponding off/on ratios measured on 18 junctions. **(b)** Resistance evolution of the FTJ after writing with increasing voltage pulse widths from 16 ns to 1s. Writing voltages of -10 and $+2.6$ V were used. **(c)** Consecutive resistance switching cycles induced by voltage pulses with durations of 10 μs and 100 ms. **(d)** Retention properties of the HRS and the LRS after setting by -16 and $+3$ V with pulse width of 1s. **(e)** Fatigue characteristics up to 10^7 cycles of a typical FTJ device. Each cycle consists of writing at $+2.6$ V, reading at -0.2 V, writing at -10 V and reading at -0.2 V. The pulse width is 10 ms. **(f)** Retention and **(g)** fatigue behaviours of V_{oc} of typical devices.

properties and the ultra-low leakage current of the 3 nm SBFO layer. BFO and its doped variants are known for their high spontaneous polarization; for example, the polarization of bulk BFO reaches $\sim 70\,\mu C\,cm^{-2}$ along [001] (ref. 39), whereas it is $\sim 26\,\mu C\,cm^{-2}$ for BTO[56]. Ultralow leakage current also plays a very important role in increasing the performance of FTJs because the undesired leakage current would suppress the switching effect of ferroelectric polarization and reduce the off/on ratio of FTJs. Doping Sm to a level of 10% was found to significantly reduce the leakage current without losing the polarization (Supplementary Fig. 1). In addition, the piezoelectric coefficient could be significantly enhanced after Sm doping[57]. All these factors are potentially beneficial to achieving TER with even larger magnitudes[7,12]. Overall, our results highlight ultrathin SBFO layers as an excellent choice as the tunnel barrier in high-performance FTJs.

The semiconducting NSTO electrode is another important factor behind the excellent performance of our FTJs. In previous reports, the memristive behaviour observed in FTJs such as LSMO/BTO/Co was attributed to the electric modulation of ferroelectric domains[27] or the interface-charge redistribution[29]. In the present work, we demonstrated that the colossal TER originates from the bias-induced transition between the direct tunnelling across the SBFO barrier and the thermionic emission over the Schottky barrier at the NSTO/SBFO interface. We should note that strictly speaking, our Pt/SBFO/NSTO heterostructures are no longer FTJs in the HRS because the junction transport is predominantly limited by the Schottky barrier instead of the tunnel barrier. This is intrinsically different from the transport mechanism reported previously for most other FTJs[13,14,16,18–21,58], where the direct tunnelling was claimed to be responsible for both HRS and LRS. There have been only a few reports on the thermionic emission effect in FTJs[36,59]. In fact, our findings are qualitatively consistent with the results from Pantel and Alexe[60]. Their calculations showed that switching between different transport mechanisms by polarization reversal could yield large electroresistance. In our case, the effective barrier widths were estimated as ~ 3 and ~ 11.5 nm for LRS and HRS,

respectively. Such a large modulation in the barrier width naturally leads to the transition of transport mechanism between direct tunnelling and Schottky emission, which is the main mechanism underlying the colossal TER achieved in our FTJs.

Another benefit of using such a semiconducting NSTO electrode is the light-enabled tunability observed in the FTJ transport. On the one hand, light illumination can be used as an additional parameter to control the resistance state of the FTJs (Fig. 3e and Supplementary Fig. 11), which allows the realization of multiple non-volatile states and enhances the data storage capability. On the other hand, the observed PV effect provides an alternative route for sensing the electronic states of FTJs non-destructively.

In summary, by doping the ultrathin ferroelectric layer and using the light-absorbing semiconducting substrate, we realized a colossal TER of up to 10^5 in NSTO/SBFO/Pt FTJs, which, to the best of our knowledge, is the largest ever reported in such devices. We demonstrated the complementary modulation of the TER by light illumination and electric field, as well as the PV effect for the first time in FTJs. The results reported here establish a strong connection between the giant TER/PV effects and the modulation of the NSTO interface triggered by applied electric field and light illumination. The good device performance of these FTJs in terms of retention, endurance, and reproducibility underscores their potential for future device applications.

Methods

Device preparation. The PLD technique was used to prepare epitaxial $Sm_{0.1}Bi_{0.9}FeO_3$ films on a (001) 0.7 wt% Nb-doped $SrTiO_3$ substrate. Similar to our previous reports on growing oxide heterostructures, KrF excimer laser (248 nm) was used with an energy density of $1\,J\,cm^{-2}$ and a repetition rate of 3 Hz. Films with thicknesses from 3 to 9 nm were grown using a $Sm_{0.1}Bi_{0.9}FeO_3$ target at a substrate temperature of 600 °C and an oxygen pressure of 10 mTorr. Pushpin-shaped FTJs with top Pt (15 nm)/ITO (350 nm) electrodes were then prepared on the films using a two-step photolithography technique, and the FTJs had a working size of 5 μm × 5 μm. In detail, the sample was first spin coated with negative photoresist SU8-2000.5 (MicroChem) at 3,500 r.p.m. for 30 s and baked at 100 °C for 60 s. The sample was then attached to a photomask for UV light exposure with an intensity of 70 mJ cm^{-2}, and the patterns were obtained by developing in SU8 developer (MicroChem) for 2 min. The top openings (50 μm × 50 μm) were then fabricated by spin coating positive photoresist AZ1505 (MicroChemicals) at 3,000 r.p.m., exposed under UV light of 20 mJ cm^{-2} and followed by developing in AZ726 (MicroChemicals) for 20 s. After photolithography, Pt/ITO top electrodes were prepared by magnetic sputtering and lift-off.

Electrical characterizations. PFM and conducting atomic force microscopy measurements were performed using a commercial atomic force microscope (Asylum Research MFP-3D) with Pt/Ir-coated Si cantilever tips and diamond-coated Si cantilever tips, respectively. In typical PFM measurements, an ac voltage of amplitude of 600 mV and at a contact resonance frequency of ∼ 300 kHz was applied on the tip. For the local electric measurements, the bias voltage was applied on the sample. $I-V$ measurements on FTJs were performed on a probe station (Cascade MPS150) equipped with a multi-SourceMeter (Keithley 2635 A). UV light was provided by a halogen lamp (Asahi Max-303) with wavelength from 250–385 nm and illumination energy density of up to 60 mW cm^{-2}. Voltage pulses were supplied by an arbitrary waveform generator (Agilent 33522 A). For the fatigue measurements, the voltage stress pulses were applied using a Multiferroic tester (Radiant Technologies). Unless specified otherwise, the setting/resetting voltage pulses have a fixed length of 1 s. Capacitance was measured using a LCR meter (Agilent E4980A) with a frequency ranging from 20 Hz to 20 MHz. In all measurements, the bottom electrodes were grounded and voltages were applied on the top electrodes.

Optical characterizations. To check the absorption of Nb doped SrTiO$_3$, 100 nm thin films of different doping levels were grown by PLD on double-side-polished quartz substrate at a temperature of 750 °C and in vacuum (∼ 2 × 10^{-7} torr) followed by annealing at 600 °C for 30 min under oxygen pressure of 10 mtorr. Then the transmittance was characterized by photospectrometry (Cary 5000, Agilent technologies) using quartz substrate as the reference.

References

1. Scott, J. F. Applications of modern ferroelectrics. *Science* **315**, 954–959 (2007).
2. Hu, W. J. *et al.* Universal ferroelectric switching dynamics of vinylidene fluoride-trifluoroethylene copolymer films. *Sci. Rep.* **4**, 4772 (2014).
3. Hu, W. J., Wang, Z. H., Du, Y. M., Zhang, X. X. & Wu, T. Space-charge-mediated anomalous ferroelectric switching in P(VDF-TrEE) polymer films. *ACS Appl. Mater. Interfaces* **6**, 19057–19063 (2014).
4. Blom, P. W. M., Wolf, R. M., Cillessen, J. F. M. & Krijn, M. P. C. M. Ferroelectric Schottky diode. *Phys. Rev. Lett.* **73**, 2107–2110 (1994).
5. Choi, T., Lee, S., Choi, Y. J., Kiryukhin, V. & Cheong, S. W. Switchable ferroelectric diode and photovoltaic effect in BiFeO$_3$. *Science* **324**, 63–66 (2009).
6. Zhuravlev, M. Y., Sabirianov, R. F., Jaswal, S. S & Tsymbal, E. Y. Giant electroresistance in ferroelectric tunnel junctions. *Phys. Rev. Lett.* **94**, 246802 (2005).
7. Kohlstedt, H., Pertsev, N. A., Contreras, J. R. & Waser, R. Theoretical current-voltage characteristics of ferroelectric tunnel junctions. *Phys. Rev. B* **72**, 125341 (2005).
8. Esaki, L., Laibowitz, R. B. & Stiles, P. J. Polar switch. *IBM Tech. Discl. Bull.* **13**, 2161–2162 (1971).
9. Fong, D. D. *et al.* Ferroelectricity in ultrathin perovskite films. *Science* **304**, 1650–1653 (2004).
10. Lichtensteiger, C., Triscone, J. M., Junquera, J. & Ghosez, P. Ferroelectricity and tetragonality in ultrathin PbTiO$_3$ films. *Phys. Rev. Lett.* **94**, 047603 (2005).
11. Tenne, D. A. *et al.* Probing nanoscale ferroelectricity by ultraviolet raman spectroscopy. *Science* **313**, 1614–1616 (2006).
12. Tsymbal, E. Y. & Kohlstedt, H. Tunneling across a ferroelectric. *Science* **313**, 181–183 (2006).
13. Chanthbouala, A. *et al.* Solid-state memories based on ferroelectric tunnel junctions. *Nat. Nanotechnol.* **7**, 101–104 (2012).
14. Pantel, D. *et al.* Tunnel electroresistance in junctions with ultrathin ferroelectric Pb(Zr$_{0.2}$Ti$_{0.8}$)O$_3$ barriers. *Appl. Phys. Lett.* **100**, 232902 (2012).
15. Garcia, V. *et al.* Ferroelectric control of spin polarization. *Science* **327**, 1106–1110 (2010).
16. Gajek, M. *et al.* Tunnel junctions with multiferroic barriers. *Nat. Mater.* **6**, 296–302 (2007).
17. Pantel, D., Goetze, S., Hesse, D. & Alexe, M. Reversible electrical switching of spin polarization in multiferroic tunnel junctions. *Nat. Mater.* **11**, 289–293 (2012).
18. Yin, Y. W. *et al.* Enhanced tunneling electroresistance effect due to a ferroelectrically induced phase transition at a magnetic complex oxide interface. *Nat. Mater.* **12**, 397–402 (2013).
19. Gruverman, A. *et al.* Tunneling electroresistance effect in ferroelectric tunnel junctions at the nanoscale. *Nano Lett.* **9**, 3539–3543 (2009).
20. Lu, H. *et al.* Ferroelectric tunnel junctions with graphene electrodes. *Nat. Commun.* **5**, 5518 (2014).
21. Li, Z. *et al.* An epitaxial ferroelectric tunnel junction on silicon. *Adv. Mater.* **26**, 7185–7189 (2014).

22. Wen, Z., Li, C., Wu, D., Li, A. & Ming, N. Ferroelectric-field-effect-enhanced electroresistance in metal/ferroelectric/semiconductor tunnel junctions. *Nat. Mater.* **12**, 617–621 (2013).
23. Tsymbal, E. Y. & Gruverman, A. Ferroelectric tunnel junctions – beyond the barrier. *Nat. Mater.* **12**, 602–604 (2013).
24. Jiang, L. *et al.* Tunneling electroresistance induced by interfacial phase transitions in ultrathin oxide heterostructures. *Nano Lett.* **13**, 5837–5843 (2013).
25. Rana, A. *et al.* Scaling behavior of resistive switching in epitaxial bismuth ferrite heterostructures. *Adv. Funct. Mater.* **24**, 3962–3969 (2014).
26. Catalan, G., Seidel, J., Ramesh, R. & Scott, J. F. Domain wall nanoelectronics. *Rev. Mod. Phys.* **84**, 119–156 (2012).
27. Chanthbouala, A. *et al.* A ferroelectric memristor. *Nat. Mater.* **11**, 860–864 (2012).
28. Yamada, H. *et al.* Giant electroresistance of super-tetragonal BiFeO$_3$-based ferroelectric tunnel junctions. *ACS Nano* **7**, 5385–5390 (2013).
29. Kim, D. J. *et al.* Ferroelectric tunnel memristor. *Nano. Lett.* **12**, 5697–5702 (2012).
30. Ma, J., Hu, J., Li, Z. & Nan, C. W. Recent progress in multiferroic magnetoelectric composites: from bulk to thin films. *Adv. Mater.* **23**, 1062–1087 (2011).
31. Guo, R. *et al.* Non-volatile memory based on the ferroelectric photovoltaic effect. *Nat. Commun.* **4**, 1990 (2013).
32. Kundys, B., Viret, M., Colson, D. & Kundys, D. O. Light-induced size changes in BiFeO$_3$ crystals. *Nat. Mater.* **9**, 803–805 (2010).
33. Wang, W. G., Li, M., Hageman, H. & Chien, C. L. Electric-field-assisted switching in magnetic tunnel junctions. *Nat. Mater.* **11**, 64–68 (2012).
34. Lu, C. L., Hu, W. J., Tian, Y. F. & Wu, T. Multiferroic oxide thin films and heterostructures. *Appl. Phys. Rev.* **2**, 021304 (2015).
35. Lin, W. N. *et al.* Electrostatic modulation of LaAlO$_3$/SrTiO$_3$ interface transport in an electric double-layer transistor. *Adv. Mater. Interfaces* **1**, 1300001 (2014).
36. Garcia, V. & Bibes, M. Ferroelectric tunnel junctions for information storage and processing. *Nat. Commun.* **5**, 4289 (2014).
37. Lee, S., Apgar, B. A. & Martin, L. W. Strong visible-light absorption and hot-carrier injection in TiO$_2$/SrRuO$_3$ heterostructures. *Adv. Energy Mater.* **3**, 1084–1090 (2013).
38. Sheng, Z. G. *et al.* Magneto-tunable photocurrent in manganite-based heterojunctions. *Nat. Commun.* **5**, 4584 (2014).
39. Wang, J. *et al.* Epitaxial BiFeO$_3$ multiferroic thin film heterostructures. *Science* **299**, 1719–1722 (2003).
40. Sze, S. M. & Ng, K. K. *Physics of Semiconductor Devices* 3rd edn (Wiley, 2007).
41. Wunderlich, W., Ohta, H. & Koumoto, K. Enhanced effective mass in doped SrTiO$_3$ and related perovskites. *Phys. B* **404**, 2202–2212 (2009).
42. Bera, A. *et al.* A versatile light-switchable nanorod memory: Wurtzite ZnO on perovskite SrTiO$_3$. *Adv. Funct. Mater.* **23**, 4977–4984 (2013).
43. Chua, L. O. Memristor–the missing circuit element. *IEEE Trans. Circuit Theory* **18**, 507–519 (1971).
44. Strukov, D. B., Snider, G. S., Stewart, D. R. & Williams, R. S. The missing memristor found. *Nature* **453**, 80–83 (2008).
45. Hambe, M. *et al.* Crossing an interface: ferroelectric control of tunnel currents in magnetic complex oxide heterostructures. *Adv. Funct. Mater.* **20**, 2436–2441 (2010).
46. Wu, S. X. *et al.* Nonvolatile resistive switching in Pt/LaAlO$_3$/SrTiO$_3$ heterostructures. *Phys. Rev. X* **3**, 041027 (2013).
47. Jiang, A. Q. *et al.* A resistive memory in semiconducting BiFeO$_3$ thin-film capacitors. *Adv. Mater.* **23**, 1277–1281 (2011).
48. Ikeda, S. *et al.* Magnetic tunnel junctions for spintronic memories and beyond. *IEEE Trans. Electron. Dev.* **54**, 991–1002 (2007).
49. Yu, G. Q. *et al.* Switching of perpendicular magnetization by spin-orbit torques in the absence of external magnetic fields. *Nat. Nanotechnol.* **9**, 548–554 (2014).
50. Shim, S. I., Yeh, F. C., Wang, X. W. & Ma, T. P. SONOS-type flash memory cell with metal/Al$_2$O$_3$/SiN/Si$_3$N$_4$/Si structure for low-voltage high-speed program/erase operation. *IEEE Electron. Dev. Lett.* **29**, 512–514 (2008).
51. Waser, R. & Aono, M. Nanoionics-based resistive switching memories. *Nat. Mater.* **6**, 833–840 (2007).
52. Chappert, C., Fert, A. & Van Dau, F. N. The emergence of spin electronics in data storage. *Nat. Mater.* **6**, 813–823 (2007).
53. Boyn, S. *et al.* High-performance ferroelectric memory based on fully patterned tunnel junctions. *Appl. Phys. Lett.* **104**, 052909 (2014).
54. Lee, D., Kim, M. G., Ryu, S., Jang, H. M. & Lee, S. G. Epitaxially grown La-modified BiFeO$_3$ magnetoferroelectric thin films. *Appl. Phys. Lett.* **86**, 222903 (2005).
55. Zou, X. *et al.* Mechanism of polarization fatigue in BiFeO$_3$. *ACS Nano* **6**, 8997–9004 (2012).
56. Jaffe, B., Cook, Jr W. R. & Jaffe, H. *Piezoelectric Ceramics* (Academic Press, 1971).
57. Kan, D. *et al.* Universal behavior and electric-field-induced structural transition in rare-earth-substituted BiFeO$_3$. *Adv. Funct. Mater.* **20**, 1108–1115 (2010).

58. Soni, R. *et al.* Giant electrode effect on tunneling electroresistance in ferroelectric tunnel junctions. *Nat. Commun.* **5,** 5414 (2014).

59. Pantel, D., Goetze, S., Hesse, D. & Alexe, M. Room-temperature ferroelectric resistive switching in ultrathin $Pb(Zr_{0.2}Ti_{0.8})O_3$ films. *ACS Nano* **5,** 6032–6038 (2011).

60. Pantel, D. & Alexe, M. Electroresistance effects in ferroelectric tunnel barriers. *Phys. Rev. B* **82,** 134105 (2010).

Acknowledgements

This work was supported by King Abdullah University of Science and Technology (KAUST).

Author contributions

W.J.H. and T.W. conceived the idea and designed the experiment. W.J.H prepared the films and devices and conducted the electrical and PV characterizations. Z.H.W. and W.J.H. performed the PFM and CAFM measurements. W.L.Y. assisted with the UV light correlated measurements. W.J.H. and T.W. wrote the manuscript. All authors discussed the results and commented on the manuscript.

Additional information

Competing financial interests: The authors declare no competing financial interests.

13

Suppression law of quantum states in a 3D photonic fast Fourier transform chip

Andrea Crespi[1,2], Roberto Osellame[1,2], Roberta Ramponi[1,2], Marco Bentivegna[3], Fulvio Flamini[3], Nicolò Spagnolo[3], Niko Viggianiello[3], Luca Innocenti[3,4], Paolo Mataloni[3] & Fabio Sciarrino[3]

The identification of phenomena able to pinpoint quantum interference is attracting large interest. Indeed, a generalization of the Hong–Ou–Mandel effect valid for any number of photons and optical modes would represent an important leap ahead both from a fundamental perspective and for practical applications, such as certification of photonic quantum devices, whose computational speedup is expected to depend critically on multi-particle interference. Quantum distinctive features have been predicted for many particles injected into multimode interferometers implementing the Fourier transform over the optical modes. Here we develop a scalable approach for the implementation of the fast Fourier transform algorithm using three-dimensional photonic integrated interferometers, fabricated via femtosecond laser writing technique. We observe the suppression law for a large number of output states with four- and eight-mode optical circuits: the experimental results demonstrate genuine quantum interference between the injected photons, thus offering a powerful tool for diagnostic of photonic platforms.

[1] Istituto di Fotonica e Nanotecnologie, Consiglio Nazionale delle Ricerche (IFN-CNR), Piazza Leonardo da Vinci, 32, I-20133 Milano, Italy. [2] Dipartimento di Fisica, Politecnico di Milano, Piazza Leonardo da Vinci, 32, I-20133 Milano, Italy. [3] Dipartimento di Fisica, Sapienza Università di Roma, Piazzale Aldo Moro 5, I-00185 Roma, Italy. [4] Università di Roma Tor Vergata, Via della ricerca scientifica 1, I-00133 Roma, Italy. Correspondence and requests for materials should be addressed to R.O. (email: roberto.osellame@polimi.it) or to F.S. (email: fabio.sciarrino@uniroma1.it).

The amplitude interference between wavefunctions corresponding to indistinguishable particles lies at the very heart of quantum mechanics. Right after the introduction of laser amplification, the availability of strong coherent pulses allowed to test interference between different light pulses[1,2], while generation of pairs of identical photons through parametric fluorescence[3] led subsequently to the milestone experiment of Hong et al.[4-8]. Later on, photonic platforms have been demonstrated to be in principle capable to perform universal quantum computing[9].

Recently, multi-particle interference effects of many photons in large interferometers are attracting a strong interest, as they should be able to show unprecedented evidences of the superior quantum computational power compared with that of classical devices[10-12]. The main example is given by the boson sampling[13] computational problem, which consists in sampling from the probability distribution given by the permanents of the $n \times n$ submatrices of a given Haar random unitary. The problem is computationally hard (in n) for a classical computer, since calculating the permanent of a complex-valued matrix is a #P-hard problem. However, sampling from the output distribution can be efficiently achieved by letting n indistinguishable photons evolve through an optical interferometer implementing the unitary transformation in the Fock space, and by detecting output states with an array of single-photon detectors. The chance to provide evidences of a post-classical computation with this relatively simple set-up has triggered a large experimental effort, leading to small-scale implementations[14-20], as well as theoretical analyses on the effects of experimental imperfections[21,22] and on possible implementations including alternative schemes[23,24].

In the context of searching for experimental evidences against the extended Church–Turing thesis, a boson sampling experiment poses a problem of certification of the result's correctness in the computationally hard regime[25]. The very complexity of the boson-sampling computational problem precludes the use of a brute-force approach, that is, calculating the expected probability distribution at the output and comparing it with the collected data. Efficient statistical techniques able to rule out trivial alternative distributions have been proposed[26] and tested[18,19], but the need for more stringent tests able to rule out less trivial distributions has led, and continues to encourage, additional research efforts in this direction.

In particular, an efficient test able to confirm true n-photon interference in a multimode device has been recently proposed[27]. The protocol is based on the use of an interferometer implementing the transformation described by the n^p-dimensional Fourier matrix, with p being any integer. When feeding this device with multi-photon states of a specific symmetry, suppression of many output configurations is observed[28], due to granular[27] many-particle interference. This effect is able to rule out alternative models requiring only coarse-grained features like the ones present in Bose–Einstein condensates[29-31]. Indeed, the implications of this effect go well beyond the certification of boson sampling devices. As a generalization of the two-photon/two-modes Hong–Ou–Mandel (HOM) effect, the suppression law, also named Zero-Transmission law[28], is important at a fundamental level, while at the practical level it could be used as a diagnostic tool for a wide range of photonic platforms[27,32,33]. During the review process of this work, the implementation of a discrete Fourier transform circuit in a fully reconfigurable chip has been reported[34]. The Zero-Transmission law for three-photon no-bunching events has been demonstrated in this planar six-mode interferometer.

In this article, we report the experimental observation of the recent theoretically proposed[27] suppression law for Fourier matrices, and its use to validate quantum many-body interference against alternative non-trivial hypotheses resulting in similar output probability distributions. The Fourier matrices have been implemented with an efficient and reliable approach by exploiting the quantum version of the fast Fourier transform (qFFT), an algorithm developed by Barak and Ben-Aryeh[35] to optimize the number of optical elements required to build the Fourier transform over the optical modes. Here we implement the qFFT on photonic integrated interferometers by exploiting the three-dimensional (3D) capabilities of femtosecond laser writing[36,37], which makes it possible to fabricate waveguides arranged in 3D structures with arbitrary layouts[38-40], by adopting an architecture scalable to a larger number of modes. The observations have been carried out with two-photon Fock states injected into four-mode and eight-mode qFFT interferometers. The peculiar behaviour of Fock states compared with other kinds of states is investigated, showing in principle the validity of the certification protocol for the identification of true granular n-particle interference, which is the source of a rich landscape of quantum effects such as the computational complexity of boson sampling.

Results

Suppression law in Fourier transform matrices. As a generalization of the HOM effect, it has been pointed out that quantum interference effects in multimode interferometers may determine suppression of a large fraction of the output configurations[28,31,41], depending on the specific unitary transformation being implemented and on the symmetry of the input state. In particular[28,31], let us consider a cyclic input, that is, an n-photon Fock state over $m = n^p$ modes (for some integer p) where the occupied modes j_r^s are determined by the rule $j_r^s = s + (r-1)n^{p-1}$, with $r = 1, \ldots, n$ and $s = 1, \ldots, n^{p-1}$. The index s takes into account the fact that there are n^{p-1} possible n-photon arrangements with periodicity n^{p-1}, which simply differ by a translation of the occupational mode labels. For example, for $n = 2$ and $m = 4$ there are $2^1 = 2$ possible cyclic states, $(1,0,1,0)$ and $(0,1,0,1)$, while for $n = 2$ and $m = 8$ there are $2^2 = 4$ possible (collision-free) cyclic inputs, that is, the states $(1,0,0,0,1,0,0,0)$, $(0,1,0,0,0,1,0,0)$, $(0,0,1,0,0,0,1,0)$ and $(0,0,0,1,0,0,0,1)$.

We consider the evolution of such states through an interferometer implementing the transformation described by the Fourier matrix

$$(U_m^F)_{l,q} = \frac{1}{\sqrt{m}} e^{i\frac{2\pi l q}{m}}. \tag{1}$$

Such evolution results in the suppression of all output configurations not fulfilling the equation

$$\mathrm{mod}\left(\sum_{l=1}^{n} k_l, n\right) = 0, \tag{2}$$

where k_l is the output mode of the lth photon. An interesting application of suppression laws is to certify the presence of true many-body granular interference during the evolution in the interferometer, ruling out alternative hypotheses which would result in similar output probability distributions. In particular, in the case of Fourier matrices, the observation of the suppression law (2) allows to certify that the sampled output distribution is not produced by either distinguishable particles or a mean field state (MF)[27]. The latter is defined as a single-particle state $|\psi^s\rangle$ of the form

$$|\psi^s\rangle = \frac{1}{\sqrt{n}} \sum_{r=1}^{n} e^{i\theta_r} |j_r^s\rangle, \tag{3}$$

with a random set of phases θ_r for each state, being $|j_r^s\rangle$ a single-particle state occupying input mode j_r^s. This state reproduces macroscopic interference effects, such as bunching or bosonic clouding[19], and cannot be distinguished from true multi-particle interference with criteria based on these features. Since it is possible to efficiently simulate the evolution of a MF with a classical algorithm[27], it is of fundamental importance to assess the ability of a validation scheme to discriminate such a state in the context of an untrusted party claiming to perform a boson sampling experiment. Hence, the MF represents an optimal test bed for the certification protocol based on the suppression law (see Fig. 1).

It is possible to quantify the degree of violation $\mathcal{D} = N_{forbidden}/N_{events}$ of the suppression law as the number of observed events in forbidden output states divided by the total number of events[27]. If a Fock state is injected in a Fourier interferometer, a violation $\mathcal{D} = 0$ would be observed. In the case of distinguishable photons there is no suppression law, and the violation would be simply the fraction of suppressed outputs, each one weighted with the number of possible arrangements of the n distinguishable particles in that output combination. In the case of two-photon states, the weighting factor is 2 for collision-free outputs and 1 otherwise, and a degree of violation of 1/2 is expected (see Section 'Observation of the suppression law'). In contrast, in the case of two-photon MF, bunching effects occur leading to an expected degree of violation of half the weighted fraction of suppressed outputs (1/4 for two-photon MF). It has been shown that the fraction of forbidden outputs is always large[31]. Hence, a comparison of the observed value of $\mathcal{D}$ with the expected one represents an efficient way, in terms of necessary experimental runs, to discriminate between Fock states, distinguishable particles states and MFs.

Realization of 3D qFFT interferometers. Let us now introduce our experimental implementation of the qFFT. The general method to realize an arbitrary unitary transformation using linear optics was introduced by Reck et al.[42], who provided a

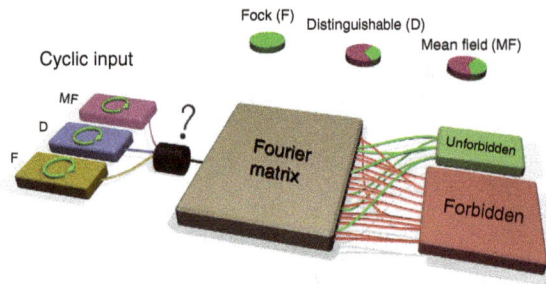

Figure 1 | Suppression law for Fock states in a Fourier interferometer.
Conceptual scheme of the protocol: the possible configurations of n photons at the output of an m-mode interferometer can be divided into two categories, unforbidden and forbidden, depending on whether they satisfy or not the suppression condition (2), respectively. The pie charts show the expected output statistics with different classes of particles, where green and red areas represent events with unforbidden and forbidden outputs, respectively. The injection of a cyclic Fock state (beige box) in an m-mode Fourier interferometer results in total suppression of forbidden output states. Cyclic states with distinguishable particles (blue box) show no suppression, so that each output combination is equally likely to occur. A mean field state (purple box), which reproduces some of the features of bosonic statistics, shows suppression with highly reduced contrast.
Therefore, with a cyclic input the m-mode Fourier interferometer is able to discriminate, through the measurement of degree of violation
$\mathcal{D} = N_{forbidden}/N_{events}$, which of these three hypotheses the input state belongs to.

decomposition of a unitary of dimension m as a sequence of $m(m-1)/2$ beam splitters and phase shifters. However, in the special case of Fourier matrices a more efficient method has been proposed[35,43], which takes advantage of their symmetries to significantly reduce the number of linear optical elements required. On the basis of the classical algorithm of Cooley and Tukey[44], who first introduced the fast Fourier transform algorithm as a more efficient way to calculate the discrete Fourier transform, Barak and Ben-Aryeh developed a quantum analogue in the linear optics domain, leading to the concept of qFFT. This approach, valid for 2^p-dimensional Fourier matrices, requires only $(m/2)\log m$ beam splitters and phase shifters, to be compared with the $O(m^2)$ elements needed for the more general Reck decomposition, thus enhancing the compactness and scalability of the platform for a more reliable experimental realization. The overall linear transformation on the optical modes implemented by the qFFT circuit is naturally equivalent to the transformation described by the Fourier matrix, hence $U_m^{qFFT} = U_m^F$.

Here we introduce a new methodology for an integrated implementation of the qFFT, which exploits the 3D capabilities of the femtosecond laser writing technique. The sequential structure arising from the decomposition of the m-dimensional Fourier matrix using the Barak and Ben-Aryeh algorithm is reproduced by the consecutive layers shown in Fig. 2. The complex arrangement of pairwise interactions necessary for the qFFT method cannot be easily implemented using a planar architecture. However, femtosecond laser writing technique allows to overcome this issue exploiting the third dimension, arranging the waveguides along the bidimensional sections of the integrated chip.

The strategy can be outlined as follows (see also Supplementary Note 1): the 2^p modes are ideally placed on the vertices of a p-dimensional hypercube; in each step of the algorithm the vertices connected by parallel edges having one specific direction are made to interact by a two-mode Hadamard transformation, with proper phase terms. An optical interferometer implementing this procedure is thus composed of $\log_2 m = p$ sections, each employing $m/2$ balanced beam splitters and phase shifters.

We fabricated waveguide interferometers realizing the Fourier matrix for $m = 4$ and 8 modes in borosilicate glass chips using femtosecond laser micromachining[36,37]. A schematic representation of these two interferometers is given in Fig. 2. According to the scheme outlined above and by exploiting the 3D capabilities of the fabrication technique, the waveguides are placed, for what concerns the cross-section of the device, on the vertices of a two-dimensional projection of the p-dimensional hypercube (see also Supplementary Fig. 1). 3D directional couplers, with proper interaction length and distance to achieve a balanced splitting, connect in each step the required vertices. The insets of Fig. 2 show, at each step i, which modes are connected by directional couplers (L_i) and the amount of phase shift that needs to be introduced in specific modes (P_i). Phase shifters, where needed, are implemented by geometrical deformation of the connecting S-bends. Fan-in and fan-out sections at the input and output of the devices allows interfacing with 127 µm spaced single-mode fibre arrays. Note that in our device geometry, in each step, the vertices to be connected are all at the same relative distance. This means that, unless geometric deformations are designed where needed, light travelling in different modes does not acquire undesired phase delays. It is worth noting that the geometric construction here developed is scalable to an arbitrary number of modes with a number of elements increasing as $m\log_2 m$.

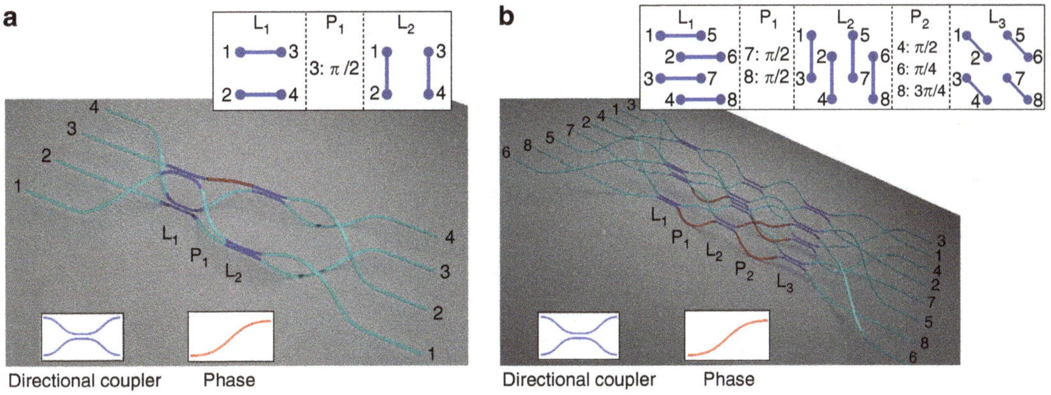

Figure 2 | Schematic representation of the structure of the integrated devices. Internal structure of the four-mode (**a**) and eight-mode (**b**) integrated interferometers implementing the qFFT over the optical modes. In the eight-mode case, the Barak and Ben-Aryeh algorithm requires an additional relabelling of the output modes (not shown in the figure), namely 2↔5 and 4↔7, to obtain the effective Fourier transformation. The mode arrangement has been chosen in a way to minimize bending losses. The insets show the actual disposition of the waveguides in the cross-section of the devices. The modes coupled together in each step (L_i) of the interferometer are joined by segments. The implemented phase shifts in each step (P_i) are also indicated.

One- and two-photon measurements. The two implemented interferometers of $m = 4$ and 8 modes are fed with single- and two-photon states. The experimental set-up, preparing a biphoton wave packet to be injected into the devices, is shown in Fig. 3. Further details on the photon generation and detection scheme are described in the Methods section. To test the validity of the suppression law, we measured the number of coincidences at each forbidden output combination injecting cyclic inputs with two indistinguishable photons. The degree of violation $\mathcal{D}$ of the suppression law could simply be evaluated with a counting experiment. Alternatively, the same quantity $\mathcal{D}$ can be expressed as a function of single-photon input–output probabilities and of the HOM visibilities, defined as

$$V_{i,j} = \frac{N_{i,j}^D - N_{i,j}^Q}{N_{i,j}^D} \tag{4}$$

where $N_{i,j}^D$ is the number of detected coincidences for distinguishable photons and $N_{i,j}^Q$ for indistinguishable photons. The subscripts (i,j) are the indexes of the two output modes, for a given input state. The degree of violation can therefore be expressed as

$$\mathcal{D} = \frac{N_{\text{forbidden}}}{N_{\text{events}}} = P_{\text{forbidden}} =$$
$$= \sum_{(i,j)_{\text{forbidden}}} P_{i,j}^Q = \sum_{(i,j)_{\text{forbidden}}} P_{i,j}^D (1 - V_{i,j}) \tag{5}$$

where $P_{i,j}^Q$ ($P_{i,j}^D$) are the probabilities of having photons in the outputs i,j in the case of indistinguishable (distinguishable) particles. Here $P_{i,j}^D$ can be obtained from single-particle probabilities. The visibilities are measured by recording the number of coincidences for each output combination as a function of the temporal delay between the two injected photons.

For the four-mode device, we measured the full set of $\binom{4}{2}^2 = 36$ collision-free input–output combinations, that is, where the two photons exit from different output ports. These contributions have been measured by recording the number of coincidences for each combination of two outputs as a function of the temporal delay between the two input photons. Because of the law given by equation (2), we expect to observe four suppressed outcomes (over six possible output combinations) for the two cyclic input states (1,3) and (2,4). Since distinguishable photons exhibit no interference, HOM dips in the coincidence patterns are expected for the suppressed output states. Conversely, peaks are

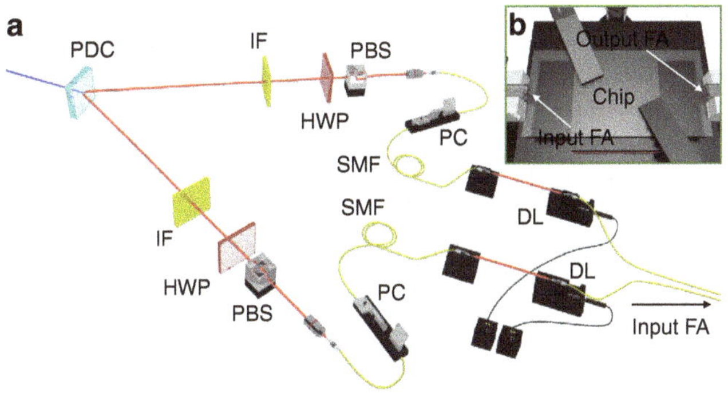

Figure 3 | Experimental apparatus for input state preparation. (a) The photon source (IF, HWP, PBS, PC, PDC, DL and SMF). (**b**) Photon injection (extraction) before (after) the evolution through the interferometer. DL, delay lines with motorized stages; FA, fibre array; HWP, half-wave plate; IF, interferential filter; PBS, polarizing beam splitter; PC, polarization compensator; PDC, parametric downconversion; SMF, single-mode fibre.

expected in the non-suppressed output combinations. The experimental results are shown in Fig. 4a, where the expected pattern of four suppressions and two enhancements is reported, with average visibilities of $\overline{V}_{\text{supp}} = 0.899 \pm 0.001$ and $\overline{V}_{\text{enh}} = -0.951 \pm 0.004$ for suppression and enhancement, respectively.

For the cyclic inputs, we also measured the interference patterns for the output contributions where the two photons exit from the same mode. These terms have been measured by inserting an additional symmetric beam splitter on each output mode, and by connecting each of its two outputs to a single-photon detector. These cases correspond to a full-bunching scenario with $n = 2$, and a HOM peak with $V = -1$ visibility is expected independently from the input state and from the unitary operation[45]. This feature has been observed for the tested inputs, where an average visibility of $\overline{V}_{\text{bunch}} = -0.969 \pm 0.024$ has been obtained over all full-bunching combinations. Note that the measured two-mode correlation matrix is not compatible with classical light (see Supplementary Note 4).

The existence of a general rule for the prediction of suppressed output combinations when injecting a cyclic Fock state in a Fourier interferometer is due to the intrinsic symmetry of the problem, as opposed to the general boson sampling scenario[13]. Suppressed outputs for non-cyclic inputs can be predicted by

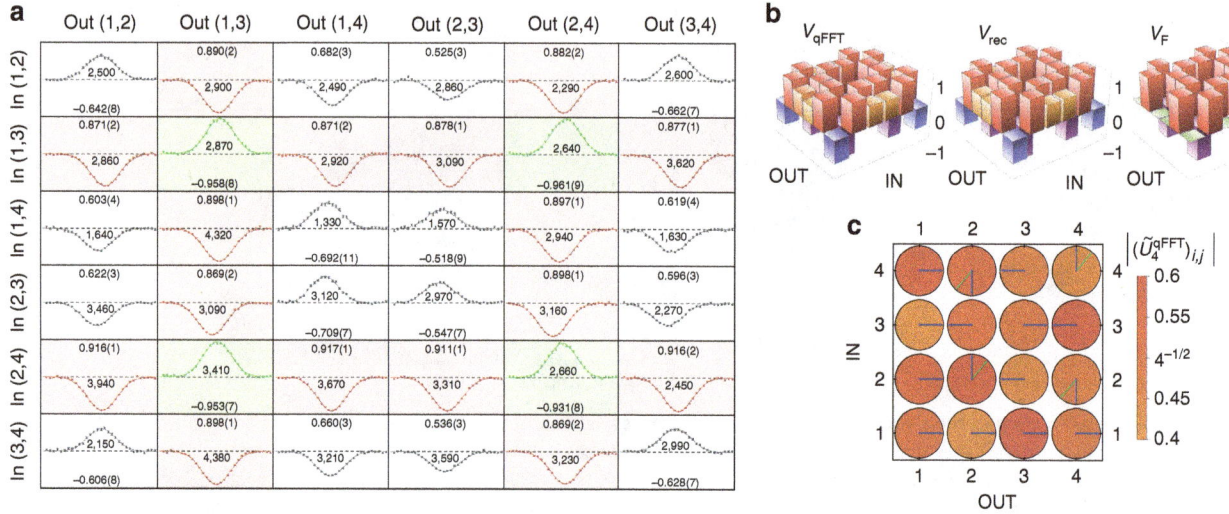

Figure 4 | Suppression law in a four-mode qFFT integrated chip. (**a**) Complete set of 36 measured coincidence patterns (raw experimental data) for all input–output combinations in the four-mode chip. For each input–output combination, the measured coincidence pattern as a function of the time delay is shown (points: experimental data, lines: best-fit curves). Cyclic inputs (1,3) and (2,4) exhibit enhancement (green) and suppression (red) on cyclic and non-cyclic outputs, respectively. For all points, error bars are due to the Poissonian statistics of the events. In each subplot the measured visibility with corresponding error and the sample size are reported. For each visibility, the error is obtained through a Monte Carlo simulation by averaging over 3,000 simulated data sets. In each subplot the zero level coincides with the baseline, while a dashed line represents the number of coincidence events in the distinguishable limit. (**b**) HOM visibilities for all 36 input–output configurations. (left to right) Experimental measured visibilities (V_{qFFT}, obtained from raw experimental data), visibilities calculated from the reconstructed unitary (V_{rec}), and visibilities calculated from the theoretical unitary (V_F). (**c**) Representation of the reconstructed experimental transformation $\tilde{U}_4^{qFFT}$, and comparison with U_4^F. Coloured disks represent the moduli of the reconstructed matrix elements (all equal to $4^{-1/2}$ for U_4^F). Arrows represent the phases of the unitary matrix elements (green: reconstructed unitary, blue: Fourier matrix).

calculating the permanent of the submatrix given by the intersection of the columns and rows of U^F corresponding to the occupied input and output modes, respectively. The complete set of measured dips and peaks is shown in Fig. 4a, highlighting the symmetry in the Fourier transform interference pattern. The injection of the non-cyclic input states has been employed for the complete reconstruction of the chip action $\tilde{U}_4^{qFFT}$, using a data set statistically independent from the one adopted to observe the suppression law. The adopted reconstruction algorithm, which exploits knowledge on the internal structure of the interferometers (specified in Fig. 2), works in two steps. In a first step, the power-splitting ratios measured with classical light are employed to extrapolate the transmissivities of the directional couplers. In a second step, the two-photon visibilities for the non-cyclic inputs are used to retrieve the values of the fabrication phases. In both steps the parameters are obtained by minimizing a suitable χ^2 function. The results are shown in Fig. 4c. The fidelity between the reconstructed unitary $\tilde{U}_4^{qFFT}$ and the theoretical Fourier transform U_4^F is $\mathcal{F} = 0.9822 \pm 0.0001$, thus confirming the high quality of the fabrication process. The error in the estimation of the fidelity is obtained through a Monte Carlo simulation, properly accounting for the degree of distinguishability of the photons with a rescaling factor in the visibilities.

For the eight-mode chip we recorded all the $\binom{8}{2} = 28$ two-photon coincidence patterns, as a function of the relative delay between the input photons, for each of the four collision-free cyclic inputs and for one non-cyclic input. The reconstruction of the actual unitary transformation $\tilde{U}_8^{qFFT}$ implemented has been performed with the same algorithm of the four-modes, by using the power-splitting ratios measured with classical light and the two-photon visibilities for one non-cyclic input. The latter has been chosen in a way to maximize the sensitivity

of the measurements with respect to the five fabrication phases. The results are shown in Fig. 5. The fidelity between the reconstructed unitary $\tilde{U}_8^{qFFT}$ and the ideal eight-mode Fourier transform U_8^F is $\mathcal{F} = 0.9527 \pm 0.0006$. More details on the reconstruction algorithm can be found in the Supplementary Note 3.

Observation of the suppression law. The suppression of events which do not satisfy equation (2) is fulfilled only when two perfectly indistinguishable photons are injected in a cyclic input of a perfect Fourier interferometer. In such a case, we would have the suppression of all output states whose sum of the indexes corresponding to the occupied modes is odd. For the four-mode (eight-mode) interferometer, this corresponds to four (16) suppressed and two (12) non-suppressed collision-free outputs (each one given by two possible arrangements of the two distinguishable photons), plus four (8) terms with two photons in the same output, each one corresponding to a single possible two-photon path.

The expected violation for distinguishable particles can be obtained from classical considerations. Let us consider the case with $n = 2$. The two distinguishable photons evolve independently from each other, and the output distribution is obtained by classically mixing single-particle probabilities. All collision-free terms are equally likely to occur with probability $q = 2/m^2$, while full-bunching events occur with probability $q' = q/2 = 1/m^2$. The degree of violation $\mathcal{D}_D$ can then be obtained by multiplying the probability q by the number of forbidden output combinations. As a result, we expect a violation degree of $\mathcal{D}_D = 0.5$ for distinguishable two-photon states. The evaluation of the expected value for a MF state, which is due to single-particle bosonic statistic effects, requires different calculations[27]. It can be shown that for $n = 2$ the degree of violation is $\mathcal{D}_{MF} = 0.25$.

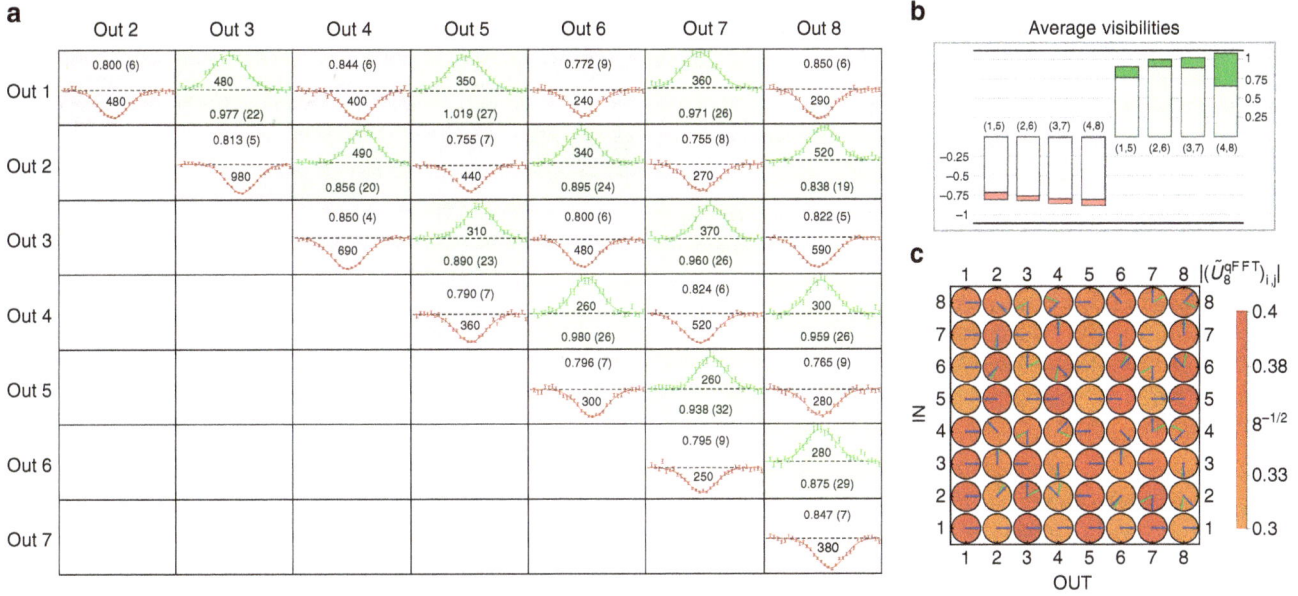

Figure 5 | Suppression law in a eight-mode qFFT integrated chip. (**a**) Set of 28 measured coincidence patterns (raw experimental data), corresponding to all collision-free output combinations for the input (2,6) of the eight-mode interferometer. For each output combination, the measured coincidence pattern as a function of the time delay is shown (points: experimental data, lines: best-fit curves). Red or green backgrounds correspond to dips and peaks, respectively. For all points, error bars are due to the Poissonian statistics of the events. In each subplot the measured visibility with corresponding error and the sample size are reported. For each visibility, the error is obtained through a Monte Carlo simulation by averaging over 3,000 simulated data sets. In each subplot the zero level coincides with the baseline, while a dashed line represents the number of coincidence events in the distinguishable limit. (**b**) Average visibilities of dips (red bars) and peaks (green bars) observed for the four collision-free cyclic inputs [(1,5), (2,6), (3,7), (4,8)]. Darker regions correspond to error bars of ±1 s.d. (**c**) Representation of the reconstructed experimental transformation $\tilde{U}_8^{qFFT}$, and comparison with U_8^F. Coloured disks represent the moduli of the matrix elements (all equal to $8^{-1/2}$ for U_8^F). Arrows represent the phases of the unitary matrix elements (green: reconstructed unitary, blue: Fourier matrix).

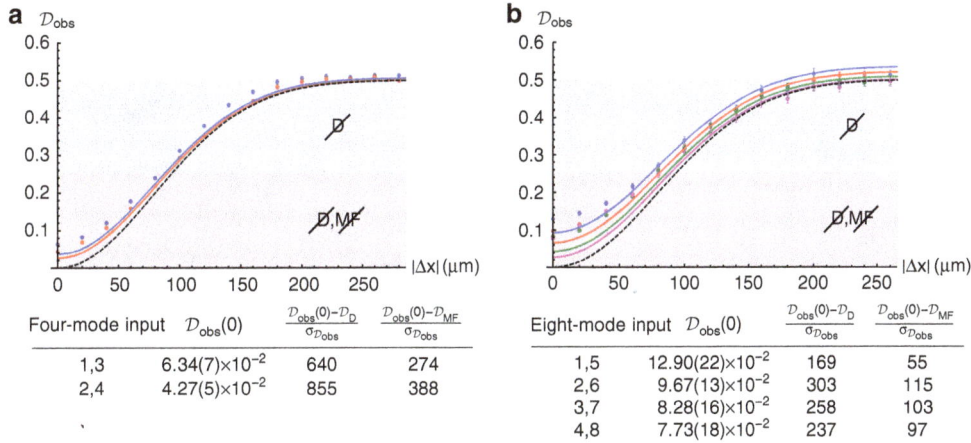

Four-mode input	$\mathcal{D}_{obs}(0)$	$\dfrac{\mathcal{D}_{obs}(0)-\mathcal{D}_{D}}{\sigma_{\mathcal{D}_{obs}}}$	$\dfrac{\mathcal{D}_{obs}(0)-\mathcal{D}_{MF}}{\sigma_{\mathcal{D}_{obs}}}$
1,3	$6.34(7)\times10^{-2}$	640	274
2,4	$4.27(5)\times10^{-2}$	855	388

Eight-mode input	$\mathcal{D}_{obs}(0)$	$\dfrac{\mathcal{D}_{obs}(0)-\mathcal{D}_{D}}{\sigma_{\mathcal{D}_{obs}}}$	$\dfrac{\mathcal{D}_{obs}(0)-\mathcal{D}_{MF}}{\sigma_{\mathcal{D}_{obs}}}$
1,5	$12.90(22)\times10^{-2}$	169	55
2,6	$9.67(13)\times10^{-2}$	303	115
3,7	$8.28(16)\times10^{-2}$	258	103
4,8	$7.73(18)\times10^{-2}$	237	97

Figure 6 | Measured violations. Observed violations $\mathcal{D}_{obs}$ as a function of the path difference $|\Delta x| = c|\Delta\tau|$ between the two photons. Blue shaded regions in the plots correspond to the cases where the hypothesis of distinguishable particles can be ruled out. Red regions correspond to the cases when both the hypotheses of distinguishable particles and mean field state can be ruled out, and true two-particle interference is present. (**a**) Data for the four-mode interferometer. Blue points: input (1,3). Red points: input (2,4). Blue solid line: theoretical prediction for input (1,3). Red solid line: theoretical prediction for input (2,4). Black dashed line: theoretical prediction for a Fourier matrix. (**b**) Data for the eight-mode interferometer. Blue points: input (1,5). Red points: input (2,6). Green points: input (3,7). Magenta points: input (4,8). Coloured solid lines: corresponding theoretical predictions for the different inputs. Black dashed line: theoretical prediction for a Fourier matrix. Tables: violations $\mathcal{D}_{obs}(0)$ at $\Delta x = 0$ and discrepancies (in sigmas) with the expected values for distinguishable particles ($\mathcal{D}_{D}$) and MFs ($\mathcal{D}_{MF}$), for the cyclic inputs of the two interferometers. $\mathcal{D}_{obs}(0)$ are calculated following formula (5), while expected values for the other two cases are $\mathcal{D}_{D} = 0.5$ and $\mathcal{D}_{MF} = 0.25$. Error bars in all experimental quantities are due to the Poissonian statistics of measured events. All theoretical predictions in solid lines are calculated from the reconstructed unitaries, obtained from different sets of experimental data to ensure statistical independence. See Supplementary Note 2 for the modelling of imperfect state preparation.

For each of the cyclic input, we have evaluated here the violation degree $\mathcal{D}_{obs}$ resulting from collected data. By measuring the coincidence pattern as a function of the path difference $\Delta x = c\Delta\tau$ between the two photons, and thus by tuning their degree of distinguishability, we could address the transition from distinguishable to indistinguishable particles. The value of $\mathcal{D}_{obs}$ as a function of Δx has been obtained as $\sum_{(i,j)_{\text{forbidden}}} P_{i,j}^D (N_{i,j}^{\Delta x}/N_{i,j}^D)$, where $N_{i,j}^{\Delta x}$ and $N_{i,j}^D$ are the number of measured coincidences for a given value of Δx and for distinguishable particles respectively. Two different regions can be identified. For intermediate values of Δx with respect to the coherence length of the photons, the measured data fall below the threshold $\mathcal{D}_D$, and hence the hypothesis of distinguishable particles can be ruled out. Then, for smaller values of the path difference up to $\Delta x \to 0$, true two-photon interference can be certified since both hypothesis of distinguishable particles and MF state can be ruled out. The maximum violation occurring at $\Delta x = 0$ delay can be evaluated using equation (5). The experimental results retrieved from the protocol are shown in the tables of Fig. 6, in which we compare the values $\mathcal{D}_{\text{obs}}(0)$ with the expected values for distinguishable particles $\mathcal{D}_D$ and for a MF state $\mathcal{D}_{\text{MF}}$. As shown for our implementation, the robustness of the protocol is ensured by the high number of s.d. separating the values in each comparison, thus unambiguously confirming the success of the certification protocol. In conclusion, the alternative hypotheses of distinguishable particles and of a MF state can be ruled out for all experiments.

Discussion

We have reported on the experimental observation of the suppression law on specific output combinations of a Fourier transformation due to quantum interference between photons. The observation of the suppression effect allowed us to rule out alternative hypotheses to the Fock state. The use of a novel implementation architecture, enabled by the 3D capabilities of femtosecond laser micromachining, extends the scalability of this technique to larger systems with lower experimental effort with respect to other techniques. While the presented architecture is designed to implement a Fourier matrix for a number of modes equal to $m = 2^p$, a generalization of the approach can be obtained by adopting a building block different from the beam splitter. For devices of odd dimension, for instance, such a tool can be provided by the tritter transformation[39]. At the same time, the universality of a generalized HOM effect with an arbitrary number of particles and modes is expected to make it a pivotal tool in the diagnostic and certification of quantum photonic platforms. Boson sampling represents a key example, since the scalability of the technique is expected to allow efficient certification of devices outperforming their classical counterparts. An interesting open problem is whether the computational hardness of boson sampling is maintained if the Haar-randomness condition is relaxed[46], and thus which is the minimal interferometer architecture required for an evidence of post-classical computation.

Fourier matrices can find application in different contexts. For instance, multiport beam splitters described by the Fourier matrix can be employed as building blocks for multiarm interferometers, which can be adopted for quantum-enhanced single and multi-phase estimation protocols[47]. This would also allow the measurement of phase gradients with precision lower than the shot-noise limit[48]. Other fields where Fourier matrices are relevant include quantum communication scenarios[49], observation of two-photon correlations as a function of geometric phase[50], fundamental quantum information theory including mutually unbiased bases[51], as well as entanglement generation[52].

Methods

Waveguide device fabrication. Waveguide interferometers are fabricated in EAGLE2000 (Corning Inc.) glass chips. To inscribe the waveguides, laser pulses with 300 fs duration, 220 nJ energy and 1 MHz repetition rate from an Yb:KYW cavity dumped oscillator (wavelength 1,030 nm) are focused in the bulk of the glass,

using a 0.6 NA microscope objective. The average depth of the waveguides, in the 3D interferometric structure, is 170 μm under the sample surface. The fabricated waveguides yield single-mode behaviour at 800 nm wavelength, with about 0.5 dB cm^{-1} propagation losses. The central part of the 3D interferometer, which includes all the relevant couplers, have a cross-section of about 50 μm × 50 μm (95 μm × 95 μm) for a length of 9.0 mm (14.7 mm) in the four-(eight-)modes case. The length of each fan-in and fan-out section, needed to bring the waveguides at 127 μm relative distance, is 7.8 mm (13.2 mm).

Photon generation and manipulation. The generation of two-photon states is performed by pumping a 2-mm long BBO crystal with a 392.5 nm wavelength Ti:Sa pulsed laser, with average power of 750 mW, which generates photons at 785 nm with a type II parametric downconversion process. The two photons are spectrally filtered by means of 3 nm interferential filters, and coupled into single-mode fibres. The indistinguishability of the photons is then ensured by a polarization compensation stage, and by propagation through independent delay lines (used to adjust the degree of temporal distinguishability) before injection within the interferometer via a single-mode fibre array. After the evolution through the integrated devices, photons are collected via a multimode fibre array. The detection system consists of four (8) single-photon avalanche photodiodes used for the four- (eight-) modes chip. An electronic data acquisition system allowed us to detect coincidences between all pairs of output modes. Typical coincidence rates for each collision-free output combination with distinguishable photons amounted to ~70–80 Hz (for the four-mode chip) and ~10–20 Hz (for the eight-mode chip).

References

1. Magyar, G. & Mandel, L. Interference fringes produced by superposition of two independent maser light beams. *Nature* **198**, 255–256 (1963).
2. Pfleegor, R. L. & Mandel, L. Interference of independent photon beams. *Phys. Rev. Lett.* **159**, 1084–1088 (1967).
3. Burnham, D. C. & Weinberg, D. L. Observation of simultaneity in parametric production of optical photon pairs. *Phys. Rev. Lett.* **25**, 84–87 (1970).
4. Hong, C. K., Ou, Z. Y. & Mandel, L. Measurement of subpicosecond time intervals between two photons by interference. *Phys. Rev. Lett.* **59**, 2046–2016 (1987).
5. Cosme, O., Padua, S., Bovino, F., Sciarrino, F. & De Martini, F. Hong-Ou-Mandel interferometer with one and two photon pairs. *Phys. Rev. A* **77**, 053822-1–053822-10 (2008).
6. Ou, Z. Y. & Mandel, L. Violation of Bell's inequality and classical probability in a two-photon correlation experiment. *Phys. Rev. Lett.* **61**, 50 (1988).
7. Walborn, S. P., Oliveira, A. N., Padua, S. & Monken, C. H. Multimode Hong-Ou-Mandel interference. *Phys. Rev. Lett.* **90**, 143601 (2003).
8. Sagioro, M. A., Olindo, C., Monken, C. H. & Padua, S. Time control of two-photon interference. *Phys. Rev. A* **69**, 053817 (2004).
9. Knill, E., Laflamme, R. & Milburn, G. J. A scheme for efficient quantum computation with linear optics. *Nature* **409**, 46–52 (2001).
10. O'Brien, J. L. Optical quantum computing. *Science* **318**, 1567–1570 (2007).
11. Walther, P. et al. Experimental one-way quantum computing. *Nature* **434**, 169–176 (2005).
12. Walmsley, I. A. & Raymer, M. G. Toward quantum-information processing with photons. *Science* **307**, 1733–1734 (2005).
13. Aaronson, S. & Arkhipov, A. in *Proceedings of the 43rd Annual ACM symposium on Theory of Computing* 333–342 (ACM Press, 2011).
14. Broome, M. A. et al. Photonic boson sampling in a tunable circuit. *Science* **339**, 794–798 (2013).
15. Spring, J. B. et al. Boson Sampling on a photonic chip. *Science* **339**, 798–801 (2013).
16. Tillmann, M. et al. Experimental boson sampling. *Nat. Photon.* **7**, 540–544 (2013).
17. Crespi, A. et al. Integrated multimode interferometers with arbitrary designs for photonic boson sampling. *Nat. Photon.* **7**, 545–549 (2013).
18. Spagnolo, N. et al. Experimental validation of photonic boson sampling. *Nat. Photon.* **8**, 615–620 (2014).
19. Carolan, J. et al. On the experimental verification of quantum complexity in linear optics. *Nat. Photon.* **8**, 621–626 (2014).
20. Bentivegna, M. et al. Experimental scattershot boson sampling. *Sci. Adv.* **1**, e1400255 (2015).
21. Rohde, P. P. & Ralph, T. C. Error tolerance of the boson sampling model for linear optics quantum computing. *Phys. Rev. A* **85**, 022332 (2012).
22. Leverrier, A. & Garcia-Patron, R. Analysis of circuit imperfections in boson sampling. *Quantum Inf. Comput.* **15**, 0489–0512 (2015).
23. Motes, K. R., Dowling, J. P. & Rohde, P. P. Spontaneous parametric down-conversion photon sources are scalable in the asymptotic limit for boson-sampling. *Phys. Rev. A* **88**, 063822 (2013).
24. Rohde, P. P., Moten, K. R. & Dowling, J. P. Evidence for the conjecture that sampling generalized cat states with linear optics is hard. *Phys. Rev. A* **91**, 012342 (2015).

25. Gogolin, C., Kliesch, M., Aolita, L. & Eisert, J. Boson-sampling in the light of sample complexity. Preprint at http://lanl.arxiv.org/abs/1306.3995 (2013).
26. Aaronson, S. & Arkhipov, A. Bosonsampling is far from uniform. *Quantum Info. Comput.* **14**, 1383–1423 (2014).
27. Tichy, M. C., Mayer, K., Buchleitner, A. & Molmer, K. Stringent and efficient assessment of boson-sampling devices. *Phys. Rev. Lett.* **113**, 020502 (2014).
28. Tichy, M. C., Tiersch, M., De Melo, F. & Mintert, F. and Buchleitner, A. Zero-transmission law for multiport beam splitters. *Phys. Rev. Lett.* **104**, 220405 (2010).
29. Cennini, G., Geckeler, C., Ritt, G. & Weitz, M. Interference of a variable number of coherent atomic sources. *Phys. Rev. A* **72**, 051601 (2005).
30. Hadzibabic, Z., Stock, S., Battelier, B., Bretin, V. & Dalibard, J. Interference of an array of independent Bose-Einstein condensates. *Phys. Rev. Lett.* **93**, 180403 (2004).
31. Tichy, M. C., Tiersch, M., Mintert, F. & Buchleitner, A. Many-particle interference beyond many-boson and many-fermion statistics. *New J. Phys.* **14**, 093015 (2012).
32. Peruzzo, A. *et al.* Quantum walks of correlated photons. *Science* **329**, 1500–1503 (2010).
33. Crespi, A. *et al.* Anderson localization of entangled photons in an integrated quantum walk. *Nat. Photon.* **7**, 322–328 (2013).
34. Carolan, J. *et al.* Universal linear optics. *Science* **349**, 711–716 (2015).
35. Barak, R. & Ben-Aryeh, Y. Quantum fast Fourier transform and quantum computation by linear optics. *J. Opt. Soc. Am. B* **24**, 231–240 (2007).
36. Osellame, R. *et al.* Femtosecond writing of active optical waveguides with astigmatically shaped beams. *J. Opt. Soc. Am. B* **20**, 1559–1567 (2003).
37. Gattass, R. & Mazur, E. Femtosecond laser micromachining in transparent materials. *Nat. Photon.* **2**, 219–225 (2008).
38. Meany, T. *et al.* Non-classical interference in integrated 3D multiports. *Opt. Express* **20**, 26895–26905 (2012).
39. Spagnolo, N. *et al.* Three-photon bosonic coalescence in an integrated tritter. *Nat. Commun.* **4**, 1606 (2013).
40. Poulios, K. *et al.* Quantum walks of correlated photon pairs in two-dimensional waveguide arrays. *Phys. Rev. Lett.* **112**, 143604 (2013).
41. Crespi, A. Suppression laws for multiparticle interference in Sylvester interferometers. *Phys. Rev. A* **91**, 013811 (2015).
42. Reck, M., Zeilinger, A., Bernstein, H. J. & Bertani, P. Experimental realization of any discrete unitary operator. *Phys. Rev. Lett.* **73**, 58–61 (1994).
43. Törmä, P. Beam splitter realizations of totally symmetric mode couplers. *J. Mod. Opt.* **43**, 245–251 (1996).
44. Cooley, J. W. & Tukey, W. An algorithm for the machine calculation of complex Fourier series. *Math. Comput.* **19**, 297–301 (1965).
45. Spagnolo, N. *et al.* General rules for bosonic bunching in multimode interferometers. *Phys. Rev. Lett.* **113**, 130503 (2013).
46. Matthews, J. C. F., Whittaker, R., O'Brien, J. L. & Turner, P. S. Testing randomness with photons by direct characterization of optical t-designs. *Phys. Rev. A* **91**, 020301(R) (2015).
47. Spagnolo, N. *et al.* Quantum interferometry with three-dimensional geometry. *Sci. Rep.* **2**, 862 (2012).
48. Motes, K. R. *et al.* Linear optical quantum metrology with single photons: exploiting spontaneously generated entanglement to beat the shot-noise limit. *Phys. Rev. Lett.* **114**, 170802 (2015).
49. Guha, S. Structured optical receivers to attain superadditive capacity and the Holevo limit. *Phys. Rev. Lett.* **106**, 240502 (2011).
50. Laing, A., Lawson, T., Martín López, E. & O'Brien, J. L. Observation of quantum interference as a function of Berry's phase in a complex Hadamard optical network. *Phys. Rev. Lett.* **108**, 260505 (2012).
51. Bengtsson, I. *et al.* Mutually unbiased bases and Hadamard matrices of order six. *J. Math. Phys.* **48**, 052106 (2007).
52. Lim, Y. L. & Beige, A. Multiphoton entanglement through a Bell-multiport beam splitter. *Phys. Rev. A* **71**, 062311 (2005).

Acknowledgements

We acknowledge technical support from Sandro Giacomini and Giorgio Milani. This work was supported by the ERC-Starting Grant 3D-QUEST (3D-Quantum Integrated Optical Simulation; grant agreement no. 307783): http://www.3dquest.eu, by the PRIN project Advanced Quantum Simulation and Metrology (AQUASIM) and by the H2020-FETPROACT-2014 Grant QUCHIP (Quantum Simulation on a Photonic Chip; grant agreement no. 641039). F.S. had full access to all the data in the study and takes responsibility for the integrity of the data and the accuracy of the data analysis.

Author contributions

A.C., R.O., M.B., F.F., N.S., N.V. and F.S. conceived the experimental approach for the observation of the suppression law. A.C. and R.O. developed the technique for three-dimensional circuits, and fabricated and characterized the integrated devices using classical optics. M.B., N.V., F.F., N.S., and F.S. carried out the quantum experiments, F.F., L.I., N.S., M.B., N.V. and F.S. elaborated the data. All authors discussed the experimental implementation and results, and contributed to writing the paper.

Additional information

Wavelength-tunable sources of entangled photons interfaced with atomic vapours

Rinaldo Trotta[1], Javier Martín-Sánchez[1], Johannes S. Wildmann[1], Giovanni Piredda[2], Marcus Reindl[1], Christian Schimpf[1], Eugenio Zallo[3,4], Sandra Stroj[2], Johannes Edlinger[2] & Armando Rastelli[1]

The prospect of using the quantum nature of light for secure communication keeps spurring the search and investigation of suitable sources of entangled photons. A single semiconductor quantum dot is one of the most attractive, as it can generate indistinguishable entangled photons deterministically and is compatible with current photonic-integration technologies. However, the lack of control over the energy of the entangled photons is hampering the exploitation of dissimilar quantum dots in protocols requiring the teleportation of quantum entanglement over remote locations. Here we introduce quantum dot-based sources of polarization-entangled photons whose energy can be tuned via three-directional strain engineering without degrading the degree of entanglement of the photon pairs. As a test-bench for quantum communication, we interface quantum dots with clouds of atomic vapours, and we demonstrate slow-entangled photons from a single quantum emitter. These results pave the way towards the implementation of hybrid quantum networks where entanglement is distributed among distant parties using optoelectronic devices.

[1] Institute of Semiconductor and Solid State Physics, Johannes Kepler University Linz, Altenbergerstr. 69, A-4040 Linz, Austria. [2] Forschungszentrum Mikrotechnik, FH Vorarlberg, Hochschulstr. 1, A-6850 Dornbirn, Austria. [3] Institute for Integrative Nanosciences, IFW Dresden, Helmholtzstr. 20, D-01069 Dresden, Germany. [4] Paul-Drude-Institut für Festkörperelektronik, Hausvogteilplatz 5-7, 10117 Berlin, Germany. Correspondence and requests for materials should be addressed to R.T. (email: rinaldo.trotta@jku.at).

The possibility of exploiting quantum effects in the solid state for quantum communication applications is nowadays one of the main driving forces behind current research efforts on semiconductor nanostructures. While the investigation of single nanostructures as hosts or sources of quantum bits is already advanced, using several of them for the envisioned applications requires an important challenge to be overcome: nanostructures are not identical and their properties cannot be predicted with the desired accuracy, a property arising from our limited capability to control their fabrication processes with atomic-scale precision[1]. A prominent example is represented by semiconductor quantum dots (QDs)[2]. During the radiative decay of a confined biexciton XX, these nanostructures can generate entangled photon pairs[3–6] with high efficiency[7–9], high degree of entanglement[9–11], high indistinguishability[12,13], and—in contrast to parametric down-conversion[14] and four-wave mixing[15] sources—they can deliver photons deterministically[12,16]. Compared with other single quantum emitters[17], QDs have the advantage of being compatible with the mature semiconductor technology[18,19]. For these reasons, it has been recently argued that they have the potential to become the 'perfect' sources of entangled photons[20]. In spite of these accomplishments, there are two points that are often overlooked: First, in the presence of unavoidable structural asymmetries the anisotropic electron-hole exchange interaction induces a fine structure splitting (FSS; s) between the intermediate exciton X levels[21] that markedly lowers the degree of entanglement of the source[22], which eventually emits only classically polarization-correlated photons. Even though methods to suppress the FSS exist[6,23], advanced quantum optics experiments[24] are still carried out using single 'hero' QDs that have—for probabilistic reasons[25]—zero FSS. Second, even having at hand a bunch of these special QDs, each of them emits entangled photons at a different random energy, and any attempt to modify these energies via external perturbations restores the FSS, thus spoiling entanglement[10]. This represents a serious hurdle when one aims at using several QDs for quantum communication. In fact, the tunability in energy of the entangled photons is a fundamental prerequisite to teleport entanglement between the distant nodes of a quantum network via quantum interference[26] and, to date, has been demonstrated only for Poissonian sources of entangled photons[27,28]. Control over the photon energy is also an essential requirement for efficient photon storage in quantum memories based on atomic clouds[29] or Bose–Einstein condensates[30], as the existing protocols require precise colour matching with atomic resonances.

Here we demonstrate that it is possible to control the energy of the entangled photons emitted by arbitrarily selected QDs without degrading the degree of entanglement of the quantum source. The core idea of our work is to manipulate the strain state of the QD and surrounding semiconductor matrix so as to achieve full control over the anisotropic electron-hole exchange interaction[21], and to modify the energy levels involved in the generation of entangled photons (X and XX) without opening the FSS. Addressing this task is not trivial. Theory[6,23] and experiments[6] have demonstrated that the X level degeneracy of any arbitrary QD can be restored via the combined action of two independent external perturbations, such as stress and electric fields. Suppression of the FSS occurs, however, for a particular combination of the magnitude of the two fields and, as a consequence, for a rather small and unpredictable spectral range of the emitted photons[10]. Although for QDs with special properties energy tuning at small (albeit non zero) FSS can be achieved using a combination of magnetic and electric fields[31], the demonstration of a truly energy-tunable source of entangled photons—whose entanglement degree is not dependent on the

photon energy—has been till now lacking. Recently, some of us have theoretically shown that an energy-tunable source of entangled photons based on an arbitrary QD requires at least three independent 'tuning knobs'[32], thus explaining the unsuitability of previous approaches[10,31]. Such degrees of freedom are conveniently provided by the three components of the QD in-plane strain tensor. We now present the experimental implementation of this theoretical concept. To benchmark our results we precisely tune a QD ('artificial atom') to emit entangled photons in the spectral region between double absorption resonances of natural atoms, and we demonstrate slow-entangled photons.

Results

Mastering the exciton fine structure splitting. Our device (Fig. 1b,c) is built-up merging piezoelectric and semiconductor technologies[33] and it allows any arbitrary QD (Fig. 1a) to be tuned for the generation of polarization-entangled photons with tunable wavelength (Fig. 1d) in the spectral region of the D_1 lines of a cloud of cesium atoms (Fig. 1e). The self-assembled In(Ga)As QDs studied here are embedded in the free-standing area of a 300-nm-thick GaAs nanomembrane that is bonded onto a micromachined 300-μm-thick $[Pb(Mg_{1/3}Nb_{2/3})O_3]_{0.72}$-$[PbTiO_3]_{0.28}$ (PMN-PT) piezoelectric substrate. The actuator features six trapezoidal areas ('legs') separated by trenches aligned at ∼60° with respect to each other (for details about the device fabrication, see Supplementary Fig. 1 and Supplementary Note 1). The key idea behind this design is that full control of the in-plane strain tensor can be achieved by applying three independent uniaxial stresses in the nanomembrane plane. With our actuator, quasi-uniaxial stresses in the membrane can be obtained by applying three independent voltages (V_1, V_2 and V_3) at the bottom of opposite legs (labelled as Leg 1,2,3) with respect to the top part of the piezo, which is electrically grounded. Figure 1c shows a top-view microscope picture of the central part of the device. The regions of the nanomembrane that are suspended can be clearly distinguished from those that are bonded on the piezo legs.

The first question we address is: how to achieve full control over the FSS of any arbitrary QD? External fields with two different degrees of freedom are required to master two different QD parameters[32]: the magnitude of the FSS (s), and the polarization direction of the exciton emission (θ). The latter parameter gives the in-plane orientation of the exciton dipoles with respect to the crystal axis[34], and it provides information about the QD anisotropy that is fundamental to drive the device during the experiment. A robust approach[6] to achieve zero FSS in any QD with two 'tuning knobs' can be illustrated by picturing the QD anisotropy as an ellipse with axes given by the two in-plane spin–spin coupling constants[21] (Fig. 2a): the FSS can be suppressed every time one external field is used to align θ along the direction of application of the second field (ϕ, see the central panel of Fig. 2a), which is then capable to compensate completely for the asymmetries in the in-plane QD confining potential (see the right panel of Fig. 2a), that is, it is capable to tune the FSS through zero. Following this picture, the first step we performed in the experiment is a polarization-resolved measurement aimed at quantifying s and θ for a randomly chosen QD when no voltages are applied to the piezo legs. These two quantities are encoded in the polar plots of Fig. 2b, where the length and orientation of the 'petals' give the value of the s and θ, respectively (see Methods). The measurement at zero applied voltages (see the black data points in Fig. 2b) reveals $s = 20 \pm 0.3$ μeV and $\theta = 109° \pm 0.4°$ with respect to the [110] direction of the GaAs nanomembrane, which was aligned perpendicularly to Leg 2

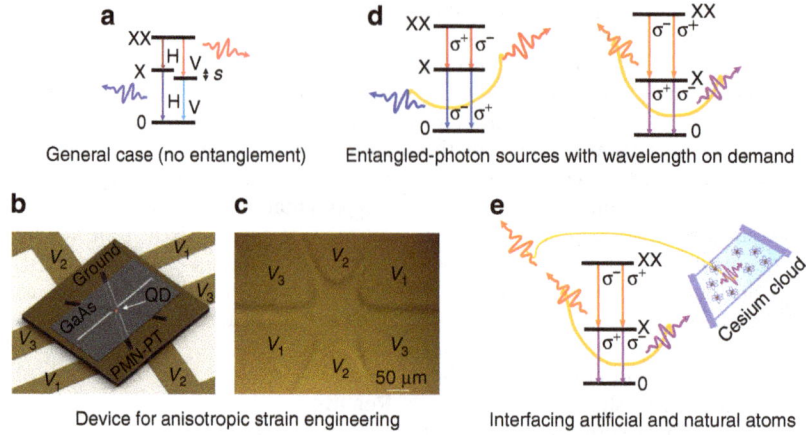

Figure 1 | A six-legged semiconductor-piezoelectric device for quantum optics. (a) Sketch of the radiative decay of a confined biexciton (XX) to the ground state (0) in a generic as-grown QD. In the presence of a FSS (s), the emitted photons are only classically correlated. H (V) indicates horizontally (vertically) polarized photons. **(b)** Sketch of the six-legged device used to engineer the strain status of a nanomembrane (grey region) containing QDs. **(c)** Microscope picture of the central part of the final device. **(d)** Same as in **a** for a QD embedded in the device show in **c**, where anisotropic in-plane strains are first used to restore the exciton (X) degeneracy (left panel), and then to modify the X and XX energies without affecting the FSS (right panel). The yellow lines indicate entanglement. σ^+ (σ^-) indicates right (left) circularly polarized photons. **(e)** Sketch of a QD whose X photon—polarization entangled with the XX photon—is tuned to the middle of the hyperfine levels of cesium (Cs) and is slowed down.

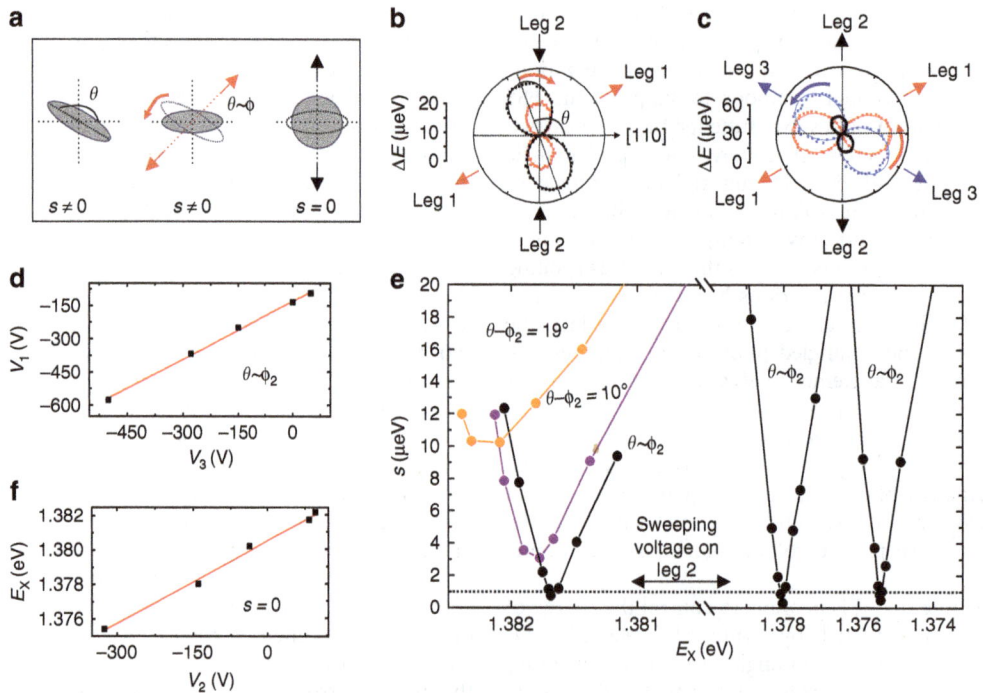

Figure 2 | Tuning the exciton energy at zero fine structure splitting. (a) Sketches of the in-plane QD anisotropy under the effect of external perturbations, see the main text. **(b)** Dependence of ΔE (see Methods) as a function of the angle the linear polarization analyser forms with the [110] crystal axis. The length and orientation of the petals give the value of the s and θ, respectively. The black data points correspond to zero applied voltages, while the red data points show the configuration in which Leg 1 is used to achieve $\theta \sim \phi_2$. The solid lines are sinusoidal fits to the experimental data. **(c)** Same as in **b** when Leg 3 is used to change the QD strain status (see the blue data points) while Leg 1 is again used to achieve $\theta \sim \phi_2$ (see the red data points). The curve for zero applied voltages is also reported for reference (see the black data points). **(d)** Linear dependence of V_1 versus V_3 when $\theta \sim \phi_2$. **(e)** Behaviour of the FSS as a function of E_X, modified as described in the text. The dashed line shows the threshold of $1\,\mu eV$. **(f)** Behaviour of E_X as a function of V_2 at $s = 0$.

during device fabrication. Since θ differs from the direction of application of the stress exerted by Leg 2 (labelled ϕ_2 in the figure) by 19°, it is not possible to suppress the FSS using only this leg. In fact, Fig. 2e shows clearly that sweeping the voltage across Leg 2 leads to an anticrossing between the two X states[35], with a lower bound for the FSS of $\sim 10\,\mu eV$. The X degeneracy can

instead be restored by first using Leg 1 to align θ along Leg 2 (see red data points in Fig. 2b), and then employing Leg 2 to cancel the FSS. Figure 2e shows that when $\theta \sim \phi_2$ the FSS can be tuned well below the threshold of $1\,\mu eV$ usually required to observe entangled photon emission (see the black data points on the left-hand side). These results represent the first experimental evidence that the FSS can be fully controlled with two independent stresses, in line with earlier theoretical predictions[23]. We observe this behaviour in all eight QDs we selected randomly in two different devices, thus proving the general relevance of our findings. It is important to note that, for each QD, there is only one combination of V_1 and V_2 that allow the condition $s = 0$ to be reached, similarly to the case of strain and electric field[10].

The second question is: how to control the X energy (E_x) of an arbitrary QD without re-opening the FSS? In other words: how to use any QD as energy-tunable source of entangled photons? As recently proposed, this can be achieved by adding an isotropic biaxial strain in the plane of the QD[32,36], which does not affect its in-plane anisotropy (Fig. 2a). To accomplish this task we make use of the third pair of piezo legs: By applying a voltage V_3 across it we modify the strain configuration and thus the initial values of s and θ (see the blue data points in Fig. 2c). We then follow the two-step procedure described above to erase the FSS: (1) We use Leg 1 to rotate θ until the exciton aligns along Leg 2 or the perpendicular direction, that is, until we reach the condition $\theta \sim \phi_2$ (see the Supplementary Fig. 2 and the Supplementary Note 2). The specific direction is determined by the handedness of θ under the stress exerted by Leg 1, being clockwise (anticlockwise) when it is larger (smaller) than 120°. (2) Finally, we suppress the FSS by sweeping again the voltage on Leg 2 (Fig. 2e). Having modified the QD strain status with Leg 3, the condition $s = 0$ is now obtained for a different combination of V_1 and V_2 and, as a consequence, for a different X energy (see the black data points on the right-hand side of Fig. 2e). It is worth mentioning that the role of the legs can be exchanged without affecting the final results, as there exists only one combination of the three in-plane components of the stress tensor which leads to $s = 0$ at each specific X energy[32]. To change E_X, it is then sufficient to repeat the three step procedure described for a different combination of V_1, V_2 and V_3 (see the other black points on the right-hand side of Fig. 2e). This is made easier considering that the voltages V_1 and V_3 to be applied to maintain the condition $\theta \sim \phi_2$ scale linearly (Fig. 2d). Finally, Fig. 2f shows that at $s = 0$ E_X changes linearly with V_2 and, most importantly, that it can be tuned across a spectral range of $\sim$7 meV. The experimental results shown so far are fully in line with a theoretical model based on $\mathbf{k} \cdot \mathbf{p}$ theory that describes the behaviour of the X states under the influence of in-plane strains with variable magnitude and anisotropy[32]. This model further confirms that our six-legged device is capable of delivering three independent stresses to single QDs (Supplementary Fig. 2), the key ingredient to reach the results shown in this work.

Entangled photon sources with wavelength on demand. Having demonstrated for the first time that it is possible to tune the energy of the X at $s = 0$, we now demonstrate that our device can be used as a truly energy-tunable source of entangled photons, that is, a source where the level of entanglement does not depend on the energy of the emitted photons. For this experiment, we choose a different QD and we repeat the procedure detailed above by carefully tuning the FSS down to zero with a resolution of $\sim$0.2 μeV (Fig. 3a). For two different X energies (highlighted with E_{X1} and E_{X2} in Fig. 3a) we performed polarization-resolved cross-correlation measurements between the X and XX photons (see Methods). In the case $s = 0$, the two-photon state can be expressed by the maximally entangled Bell state $\psi = (|R_{XX}L_X\rangle + |L_{XX}R_X\rangle) / \sqrt{2}$, which can be equivalently rewritten as $\psi = (|H_{XX}H_X\rangle + |V_{XX}V_X\rangle) / \sqrt{2}$ or $\psi = (|D_{XX}D_X\rangle + |A_{XX}A_X\rangle) / \sqrt{2}$, where H (V), D (A) and R (L) indicate horizontally (vertically) polarized, diagonally

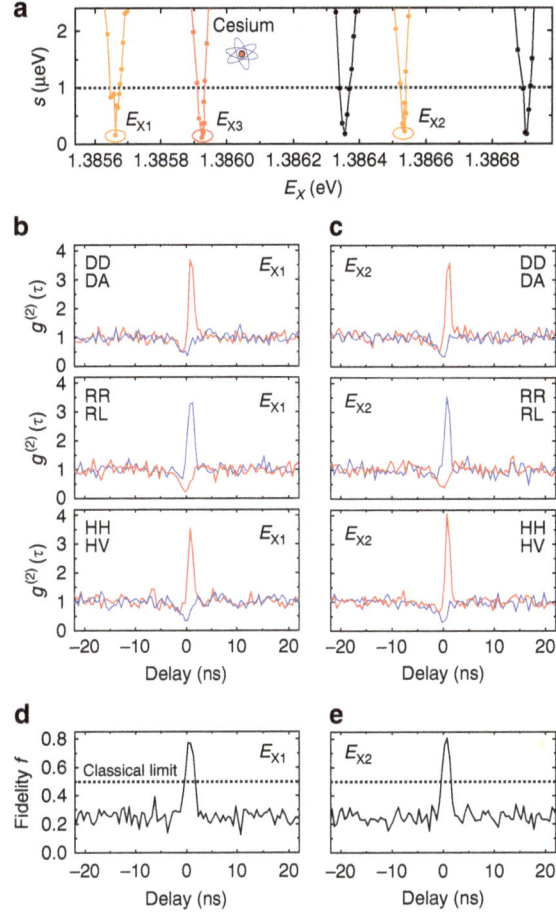

Figure 3 | Entangled photon pairs at different photon energies. (a) Same as in Fig. 2e for a different QD whose FSS is fine-tuned to $s < 0.2$ μeV. The circles indicate the X energies ($E_{X1,2,3}$) where cross-correlation measurements have been performed. In particular, E_{X3} correspond to the position in between the hyperfine lines of Cs. **(b)** XX-X cross-correlation measurements for diagonal (top panel), circular (central panel) and linear (bottom panel) polarization basis when the X energy is set to E_{X1}. **(c)** Same as in **b** for E_{X2}. **(d)** Fidelity as a function of the time delay when the X energy is set to E_{X1}. The dashed line indicates the threshold for entanglement, that is, the classical limit. **(e)** Same as in **d** for E_{X2}.

(antidiagonally) polarized, and right (left)-circularly polarized photons, respectively. Therefore, when performing polarization-resolved cross-correlation measurements, one expects a strong bunching peak for co-linear, co-diagonal, and cross-circular polarization and antibunching for the opposite polarization settings. This is exactly the behaviour we observe in our measurements (Fig. 3b and 3c), which represent a clear signature of entanglement. To estimate the degree of entanglement, we calculated the fidelity f to the Bell state ψ (see Methods) and we found—by integrating over 0.5 ns around the central peak—$f_1 = 0.80 \pm 0.05$ and $f_2 = 0.78 \pm 0.06$ for E_{X1} and E_{X2}, respectively (Fig. 3d,e), similarly to what we measured in different QDs by full reconstruction of the two-photon density matrix[10]. These values are larger than the classical limit (0.5) for a source emitting polarization-correlated photons, and prove for the first time that QDs can produce entangled states of light at different energies. The fact that the two values of the fidelity are identical within the error bounds demonstrates that the degree of entanglement of the source does not depend on the photon energy, an important requirement for applications[37]. Despite the fidelity we measure is among the highest ever reported so far for QDs[9,10], it is not yet perfect (~ 1) due to depolarization of X states[38] and recapture processes[7]. However, we strongly believe that the level of entanglement can be even improved further using faster photon detectors and resonant two-photon excitation schemes[12,16].

Slowing-down entangled photons from QDs. To prove that we are able to control the energy of the entangled photons with the precision required for advanced quantum optics experiments, we show that it is possible to interface entangled photons emitted by a QD with clouds of natural atoms operated as a slow-light medium. Pioneering works on the field[39,40] have demonstrated that single photons emitted by GaAs QDs can be slowed down using warm rubidium vapours, which can be also exploited as absolute energy reference to interconnect the different nodes of a quantum network. The concept has been recently adopted to interface single photons emitted by single molecules with vapours of alkali atoms[41]. However, it has never been applied to slow down entangled photon pairs from single quantum dots. The reason is that the energy of one photon of the entangled pair must match the spectral window of double absorption resonances in warm atomic vapours, and this possibility was out of reach before our work.

The emission spectrum of our QDs is centred at ~ 900 nm, close to the D_1 lines of cesium (Cs). Therefore, we insert a temperature-stabilized quartz cell containing Cs vapour[42,43] in the optical path of the exciton photon (see Methods). Figure 4a shows a sketch of the Cs D_1 lines of relevance for this experiment, that is, the transitions involving the $6^2P_{1/2}$ level and the hyperfine-split $6^2S_{1/2}$ doublet[44]. To observe slow-entangled photons, one has to tune the energy of the X (or XX) transition in between the hyperfine lines[39] while s is kept at 0 (see purple arrows in Fig. 4a). Experimentally, this condition is identified by recording the intensity of the X emission while scanning its energy across the Cs lines (Fig. 4b). This allows us to observe two absorption resonances resulting from a convolution between the inhomogeneously broadened X transition and the Doppler-broadened Cs doublet (Supplementary Note 3). When the X energy is tuned exactly in the middle of the hyperfine lines (E_{X3}, Figs 3a and 4b), the QD photons probe a strongly dispersive medium and are slowed down[39]. Taking into account the temperature of the Cs cell used in the experiment ($T_{Cs} = 140\,°C$), we estimate a differential delay of up to ~ 2 ns on a 7.5 cm long path. To observe this delay, it is sufficient to perform X-XX cross-correlation measurements with and without the Cs cell. Figure 4c shows the result of such measurements when both X and XX photons are collected with the same diagonal polarization. The bunching peak clearly shifts by >1.8 ns when the Cs cell is inserted into the optical path. However, the peak appears less pronounced and broader compared with the measurements performed without Cs cell, which we attribute to dispersion combined with the spectral broadening of the X emission line ($\sim 35\,\mu eV$, see Fig. 4b, Supplementary Note 3 and ref. 43). It is therefore interesting to investigate whether the degree of entanglement of photon pairs is affected by the presence of the atomic cloud. Figure 4d shows the fidelity calculated after performing XX-X cross-correlation measurements with the same polarization settings used in Fig. 3b and 3c. The measurements without Cs cell show a peak fidelity of $f_3^{\text{off}} = 0.8 \pm 0.02$, which nicely matches the values obtained for the other E_{X1} and E_{X2} energies (Fig. 3) and further confirms that the degree of entanglement of our source does not change with the X energy. The measurements with the Cs cell show instead a peak fidelity of $f_3^{\text{on}} = 0.65 \pm 0.08$. Despite this value is above the classical limit of 0.5, it may suggest a degradation of the degree of entanglement. However, careful inspection of the

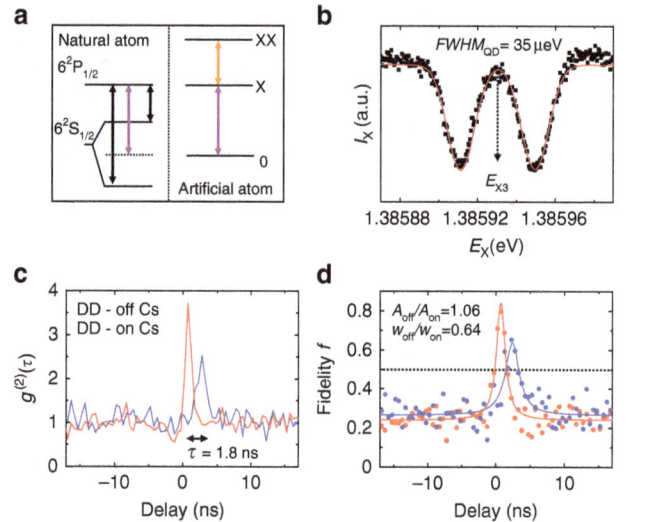

Figure 4 | Entangled photons from artificial atoms interfaced with natural atoms. (**a**) Sketch of the energy levels of a Cs atom (left panel) and of a QD fine-tuned for entanglement (right panel). The energy axis (not to scale) has been shifted so that E_X (purple arrow) matches the middle of the two D_1 lines of Cs (black arrows). (**b**) Intensity of the X when its energy is swept across the D_1 lines of Cs. The solid line is a fit to the theoretical model (see Supplementary Note 3) used to extract the linewidth of the X transition (indicated with $FWHM_{QD}$). The dashed arrow indicates the E_{X3}. (**c**) XX-X cross-correlation measurements with the same diagonal polarization with (blue line) and without (red line) the Cs cell in the X optical path. The temporal delay introduced by the Cs cell (τ) is also indicated. (**d**) Fidelity with (blue data points) and without (red data points) the Cs cell in the X optical path. The solid lines are Lorentzian fits to the data. The ratio between the full-width-half maxima (FWHM; w) of the curves and the integrated area (A) are also indicated.

data shows that this is not the case and that the peak fidelity is not the most accurate parameter to estimate entanglement. The solid lines in Fig. 4d show lorentzian fits to the experimental data. In the presence of the Cs cloud there is a considerable temporal broadening (a factor almost 2), while the integrated area (A) remains practically unchanged ($A_{\text{off}}/A_{\text{on}} = 1.06$). These results clearly imply that the degree of entanglement of the photons is not affected by the presence of the Cs cell, thus opening new venues for advanced quantum optics experiments with the hybrid artificial–natural atomic interface.

Discussion

In conclusion, we have demonstrated that it is possible to modify the energy of the polarization-entangled photons emitted by arbitrary QDs without affecting the degree of entanglement of the photons. These results have been achieved by developing a novel class of electrically controlled semiconductor-piezoelectric devices that allow the optical and electronic properties of every single QD to be arbitrarily re-shaped via anisotropic strain engineering. The tunability of the photon energy has opened up the unprecedented possibility to interface entangled photons from QDs with clouds of natural atoms and, in turn, to demonstrate for the first time slow-entangled photons from a single QD. In light of our results, it is possible to envisage a new era for QDs in the field of large distance quantum communication. In fact, dissimilar QDs can now be used for entanglement teleportation (or swapping), as the entangled photons can be colour matched for quantum interference at a beam splitter[26]. Furthermore, the hybrid artificial-atomic interface we have built-up can be further developed up to the limit where entangled states are stored and retrieved at the single-photon level[29]. This could allow the whole concept of a quantum repeater[45] to be implemented. Addressing these applications most likely requires the use of our energy-tunable entangled-photon source in combination with broadband photonic structures featuring GaAs microlenses that are capable to boost the flux of QD photons[46,47] and with resonant excitation schemes for the deterministic generation of entangled photons with high degree of indistinguishability[12,16]. However, the implementation of the perfect source of entangled photons is worth the efforts, as the realization of a solid state-based quantum network for the distribution of quantum entanglement among distant parties[48] will be revolutionary.

Methods

Sample growth and device fabrication. In(Ga)As QDs were grown by molecular beam epitaxy. Following oxide desorption and buffer growth, a 100-nm-thick $Al_{0.75}Ga_{0.25}As$ sacrificial layer was deposited before a 300-nm-thick GaAs layer containing the QDs. The QDs were grown at $500\,°C$ and capped by an indium flush technique. Using standard epoxy-based photoresist, the sample (coated with a 100-nm-thick Cr/Au layer) was integrated via a flip-chip process onto a Cr/Au-coated micro-machined PMN-PT actuator. The GaAs substrate, buffer layer and AlGaAs sacrificial layer were removed using wet chemical etching (Supplementary Fig. 1), thus leaving the GaAs nanomembrane tightly bound on the piezoelectric actuator. The PMN-PT actuator was processed in the six-leg design shown in Fig. 1 with a femtosecond laser having central wavelength of 520 nm, pulse duration of 350 fs, repetition rate of 25 kHz and a $\sim 5\,\mu m$ spot size. Six metal contacts were fabricated at the bottom of the piezo legs so that independent voltages can be applied with respect to the top contact, which is set to ground. The same voltage is applied to opposite legs to limit displacements of the central structure. The voltage applied on each pair of legs leads to a well-controlled deformation of the GaAs nanomembrane suspended in the central part (Supplementary Note 4 and Supplementary Fig. 3). The piezoelectric actuator was poled so that a voltage $V<0$ ($V>0$) applied to each pairs of aligned legs result in an out-of-plane electric field that leads to an in-plane contraction (expansion) of the piezo leg. Therefore, an in-plane tensile (compressive) strain is transferred to the central part of the nanomembrane (Supplementary material). In the tensile regime ($V<0$), the X emission line of the QD of Fig. 2 shifts approximately linearly with the applied voltages, with a slope that depends on the used leg. In particular, we measure slopes of 10, 7, 3 $\mu eV/V$ for Legs 1, 2 and 3, respectively. This difference may arise from the fact that the QD under investigation is not sitting in the very central part of the device (Supplementary Note 4).

Micro-photoluminescence and photon-correlation spectroscopy. Conventional micro-photoluminescence spectroscopy is used for the optical characterization of the devices. The measurements are performed at low temperatures (typically 4–10 K) in a helium flow cryostat. The QDs are excited non-resonantly at 840 nm with a femtosecond Ti:Sapphire laser and focused by a microscope objective with 0.42 numerical aperture. The measurements in Figs 3 and 4 appear to be performed with a continuous wave laser due to long lifetime of the X transition. The same objective is used for the collection of the photoluminescence signal, which is spectrally analysed by a spectrometer and detected by a nitrogen-cooled silicon charge-coupled device. Polarization-resolved micro-photoluminescence experiments are performed combining a rotating half-wave plate and a fixed linear polarizer placed after the microscope objective. The transmission axis of the polarizer is set parallel to the [110] direction of the GaAs crystal (within 2°) and perpendicular to the entrance slit of the spectrometer, which defines the laboratory reference for vertical polarization. The FSS and the polarization angle of the X emission are evaluated using the same procedure reported in ref. 10, which ensures sub-microelectronvolt resolution. In particular, Fig. 2 shows ΔE as a function of the angle the linear polarization analyser forms with the [110] crystal axis, where ΔE is half of the difference between the XX and X energy minus its minimum value.

For photon-correlation measurements, the signal is split into two parts after the microscope objective using a non-polarizing 50/50 beam splitter, spectrally filtered with two independent spectrometers tuned to the XX and X energies, and finally sent to two Hanbury Brown and Twiss setups. Each Hanbury Brown and Twiss setups consists of a polarizing 50/50 beam splitter placed in front of two avalanche photodiodes (APDs), whose output is connected to a four-channel correlation electronics for reconstructing the second-order cross-correlation function between the XX and X photons, $g^{(2)}(\tau)$. The temporal resolution of the set-up is $\sim 500\,ps$, mainly limited by the time jitter of the APDs. Properly oriented half-wave plates and quarter-wave plates are placed right after the non-polarizing beamsplitter to select the desired polarization. With a single measurement, this experimental set-up allows $g^{(2)}_{XX,X}(\tau)$ to be reconstructed in four different polarization settings, so as to minimize the effect of possible sample drifts. The count rate at each APD is typically 10^3–10^4 count per seconds and the integration time used for each measurement is $\sim 1\,h$.

For cross-correlation measurements in the presence of the atomic cloud we use a 7.5-cm-long quartz cell filled with Cs vapours and wounded up with heating foils for precise temperature tuning. The temperature chosen for the experiment is $140 \pm 0.1\,°C$. The Cs cell is inserted in the X optical path, after the non-polarizing beam splitter and the retarder wave plates, right before the entrance slit of the spectrometer. When the X is tuned in between the hyperfine lines of Cs, the count rate at the APDs is reduced by a factor of ~ 2 due to the optical absorption in the Cs vapour.

Entanglement analysis. Raw data were used to evaluate the second-order correlation functions, without any background light subtraction. As mentioned in the previous section, the experimental set-up allows the second order XX-X cross-correlation function $g^{(2)}_{AB}(\tau)$ to be evaluated in four different polarization settings (AB) with a single measurement, that is, $g^{(2)}_{AB}(\tau)$, $g^{(2)}_{BA}(\tau)$, $g^{(2)}_{AA}(\tau)$, $g^{(2)}_{BB}(\tau)$ are measured simultaneously. Having four measurements for each polarization basis has two main advantages: it compensates for measurement drifts and it allows us to calculate the degree of correlation without any assumption on the polarization state of the photons. In particular, we do not need to assume, for example, that $g^{(2)}_{AA}(\tau) = g^{(2)}_{BB}(\tau)$, and we can simply average the observed results, that is, $g^{*(2)}_{AA} = (g^{(2)}_{AA} + g^{(2)}_{BB}) / 2$. The degree of correlation is then calculated using the following formula $C_{AB} = (g^{*(2)}_{AA} - g^{*(2)}_{AB}) / (g^{*(2)}_{AA} + g^{*(2)}_{AB})$. Finally, the fidelity is calculated via[5] $f = (1 + |C_{HV}| + |C_{DA}| + |C_{RL}|) / 4$.

References

1. Weitering, H. H. Quantum dots: one atom at a time. *Nat. Nanotechnol.* **9**, 499–500 (2014).
2. Michler, P. (ed.). in *Single Quantum Dots* (Springer, 2009).
3. Benson, O., Santori, C., Pelton, M. & Yamamoto, Y. Regulated and entangled photons from a single quantum dot. *Phys. Rev. Lett.* **84**, 2513–2516 (2000).
4. Akopian, N. *et al.* Entangled photon pairs from semiconductor quantum dots. *Phys. Rev. Lett.* **96**, 130501–130505 (2006).
5. Young, R. J. *et al.* Improved fidelity of triggered entangled photons from single quantum dots. *New J. Phys.* **8**, 29 (2006).
6. Trotta, R. *et al.* Universal recovery of the energy-level degeneracy of bright excitons in InGaAs quantum dots without a structure symmetry. *Phys. Rev. Lett.* **109**, 147401–147405 (2012).
7. Dousse, A. *et al.* Ultrabright source of entangled photon pairs. *Nature* **466**, 217–220 (2010).
8. Huber, T. *et al.* Polarization entangled photons from quantum dots embedded in nanowires. *Nano. Lett.* **14**, 7107–7114 (2014).
9. Versteegh, M. A. M. *et al.* Observation of strongly entangled photon pairs from a nanowire quantum dot. *Nat. Commun.* **5**, 5298 (2014).
10. Trotta, R., Wildmann, J. S., Zallo, E., Schmidt, O. G. & Rastelli, A. Highly entangled photons from hybrid piezoelectric-semiconductor quantum dot devices. *Nano Lett.* **14**, 3439–3444 (2014).
11. Kuroda, T. *et al.* Symmetric quantum dots as efficient sources of highly entangled photons: Violation of Bell's inequality without spectral and temporal filtering. *Phys. Rev. B* **88**, 041306 R (2013).
12. Müller, M., Bounouar, S., Jöns, K. D., Glässl, M. & Michler, P. On-demand generation of indistinguishable polarization-entangled photon pairs. *Nat. Photon.* **8**, 224–228 (2014).
13. Stevenson, R. M. *et al.* Indistinguishable entangled photons generated by a light-emitting diode. *Phys. Rev. Lett.* **108**, 040503 (2012).
14. Kwiat, P. G. *et al.* New high-intensity source of polarization entangled photon pairs. *Phys. Rev. Lett.* **75**, 4337 (1995).
15. Fulconis, J., Alibart, O., O'Brien, J. L., Wadsworth, W. J. & Rarity, J. G. Nonclassical interference and entanglement generation using a photonic crystal fiber pair photon source. *Phys. Rev. Lett.* **99**, 120501 (2007).
16. Jayakumar, H. *et al.* Deterministic photon pairs and coherent optical control of a single quantum dot. *Phys. Rev. Lett.* **110**, 135505 (2013).
17. Aspect, A., Grangier, P. & Roger, G. Experimental tests of realistic local theories via Bell's theorem. *Phys. Rev. Lett.* **47**, 460–463 (1981).
18. Salter, C. L. *et al.* An entangled-light-emitting diode. *Nature* **465**, 594–597 (2010).
19. Juska, G., Dimastrodonato, V., Mereni, L. O., Gocalinska, A. & Pelucchi, E. Towards quantum-dot arrays of entangled photon emitters. *Nat. Photon.* **7**, 527 (2013).
20. Lu, C.-Y. & Pan, J.-W. Quantum optics: push-button photon entanglement. *Nat. Photon.* **8**, 174–176 (2014).
21. Bayer, M. *et al.* Electron and hole g factors and exchange interaction from studies of the exciton fine structure in In0.60Ga0.40As quantum dots. *Phys. Rev. Lett.* **82**, 1748–1751 (1999).
22. Santori, C., Fattal, D., Pelton, M., Solomon, G. S. & Yamamoto, Y. Polarization-correlated photon pairs from a single quantum dot. *Phys. Rev. B* **66**, 045308 (2002).

23. Wang, J., Gong, M., Guo, G.-C. & He, L. Eliminating the fine structure splitting of excitons in self-assembled InAs/GaAs quantum dots via combined stresses. *Appl. Phys. Lett.* **101,** 063114 (2012).
24. Nilsson, J. *et al.* Quantum teleportation using a light-emitting diode. *Nat. Photon.* **7,** 311–315 (2013).
25. Gong, M. *et al.* Statistical properties of exciton fine structure splitting and polarization angles in quantum dot ensembles. *Phys. Rev. B* **89,** 205312 (2014).
26. Pan, J.-W., Bouwmeester, D., Weinfurter, H. & Zeilinger, A. Experimental entanglement swapping: entangling photons that never interacted. *Phys. Rev. Lett.* **80,** 3891–3894 (1998).
27. Predojević, A., Grabher, S. & Weihs, G. Pulsed Sagnac source of polarization entangled photon pairs. *Opt. Express* **20,** 25022–25029 (2012).
28. Fedrizzi, A., Herbst, T., Poppe, A., Jennewein, T. & Zeilinger, A. A wavelength-tunable fiber-coupled source of narrowband entangled photons. *Opt. Express* **15,** 15377 (2007).
29. Choi, K. S., Deng, H., Laurat, J. & Kimble, H. J. Mapping photonic entanglement into and out of a quantum memory. *Nature* **452,** 67–71 (2008).
30. Lettner, M. *et al.* Remote entanglement between a single atom and a Bose-Einstein condensate. *Phys. Rev. Lett.* **106,** 210503 (2011).
31. Pooley, M. A. *et al.* Energy-tunable quantum dot with minimal fine structure created by using aimultaneous electric and magnetic fields. *Phys. Rev. Applied* **1,** 024002 (2014).
32. Trotta, R., Martín-Sánchez, J., Daruka, I., Ortix, C. & Rastelli, A. Energy-tunable sources of entangled photons: a viable concept for solid-state-based quantum relays. *Phys. Rev. Lett.* **114,** 150502 (2015).
33. Trotta, R. *et al.* Nanomembrane quantum-light-emitting diodes integrated onto piezoelectric actuators. *Adv. Mater.* **24,** 2668 (2012).
34. Gong, M., Zhang, W., Guo, G.-C. & He, L. Exciton polarization, fine-structure splitting, and the asymmetry of quantum dots under uniaxial stress. *Phys. Rev. Lett.* **106,** 227401 (2011).
35. Bennett, A. J. *et al.* Electric-field-induced coherent coupling of the exciton states in a single quantum dot. *Nat. Phys.* **6,** 947–950 (2010).
36. Wang, J., Gong, M., Guo, G.-C. & He, L. Towards scalable entangled photon sources with self-assembled InAs/GaAs quantum dots. *Phys. Rev. Lett.* **115,** 067401 (2015).
37. Ekert, A. K. Quantum cryptography based on Bell's theorem. *Phys. Rev. Lett.* **67,** 661–663 (1991).
38. Stevenson, R. M. *et al.* Coherent entangled light generated by quantum dots in the presence of nuclear magnetic fields. Preprint at http://arxiv.org/abs/1103.2969 (2011).
39. Akopian, N., Wang, L., Rastelli, A., Schmidt, O. G. & Zwiller, V. Hybrid semiconductor-atomic interface: slowing down single photons from a quantum dot. *Nat. Photon.* **5,** 230–233 (2011).
40. Akopian, N. *et al.* An artificial atom locked to natural atoms. Preprint at http://arxiv.org/abs/1302.2005 (2013).
41. Siyushev, P., Stein, G., Wrachtrup, J. & Gerhardt, I. Molecular photons interfaced with alkali atoms. *Nature* **509,** 66–70 (2014).
42. Ulrich, S. M. *et al.* Spectroscopy of the D1 transition of cesium by dressed-state resonance fluorescence from a single (In,Ga)As/GaAs quantum dot. *Phys. Rev. B* **90,** 125310 (2014).
43. Wildmann, J. S. *et al.* Atomic clouds as spectrally-selective and tunable delay lines for single photons from quantum dots. *Phys. Rev. B* **92,** 235306 (2015).
44. Steck, D. A. Cesium D Line Data. Available at http://steck.us/alkalidata (revision 2.1.4, 23 December 2010).
45. Duan, L.-M., Lukin, M. D., Cirac, J. I. & Zoller, P. Long-distance quantum communication with atomic ensembles and linear optics. *Nature* **414,** 413 (2001).
46. Gschrey, M. *et al.* Highly indistinguishable photons from deterministic quantum-dot microlenses utilizing three-dimensional in situ electron-beam lithography. *Nat. Commun.* **6,** 7662 (2015).
47. Ma, Y., Ballesteros, G., Zajac, J. M., Sun, J. & Gerardot, B. D. Highly directional emission from a quantum emitter embedded in a hemispherical cavity. *Opt. Lett.* **40,** 2373–2376 (2015).
48. Kimble, H. J. The quantum internet. *Nature* **453,** 1023–1030 (2008).

Acknowledgements

We thank G. Katsaros, G. Bauer, A. Predojecvic, C. Ortix and F. Schäffler for fruitful discussions and T. Lettner and D. Huber for help. We acknowledge O.G. Schmidt for support in the sample growth. The work was supported financially by the European Union Seventh Framework Programme 209 (FP7/2007-2013) under Grant Agreement No. 601126 210 (HANAS), and the AWS Austria Wirtschaftsservice, PRIZE Programme, under Grant No. P1308457.

Author contributions

R.T. conceived the experiment. R.T., with help from J.M.-S., J.S.W., M.R. and A.R., performed measurements and data analysis. J.M.-S., with help from C.S., R.T. and A.R., designed the actuator and processed the device. G.P., S.S. and J.E. processed the piezoelectric actuators. E.Z. performed MBE growth. R.T. wrote the manuscript with help from all the authors. R.T. and A.R. coordinated the project.

Additional information

15

Selectively enhanced photocurrent generation in twisted bilayer graphene with van Hove singularity

Jianbo Yin[1,*], Huan Wang[1,*], Han Peng[2,*], Zhenjun Tan[1,3], Lei Liao[1], Li Lin[1], Xiao Sun[1,3], Ai Leen Koh[4], Yulin Chen[2], Hailin Peng[1] & Zhongfan Liu[1]

Graphene with ultra-high carrier mobility and ultra-short photoresponse time has shown remarkable potential in ultrafast photodetection. However, the broad and weak optical absorption (~2.3%) of monolayer graphene hinders its practical application in photodetectors with high responsivity and selectivity. Here we demonstrate that twisted bilayer graphene, a stack of two graphene monolayers with an interlayer twist angle, exhibits a strong light–matter interaction and selectively enhanced photocurrent generation. Such enhancement is attributed to the emergence of unique twist-angle-dependent van Hove singularities, which are directly revealed by spatially resolved angle-resolved photoemission spectroscopy. When the energy interval between the van Hove singularities of the conduction and valance bands matches the energy of incident photons, the photocurrent generated can be significantly enhanced (up to ~80 times with the integration of plasmonic structures in our devices). These results provide valuable insight for designing graphene photodetectors with enhanced sensitivity for variable wavelength.

[1] Center for Nanochemistry, Beijing Science and Engineering Center for Nanocarbons, Beijing National Laboratory for Molecular Sciences, College of Chemistry and Molecular Engineering, Peking University, 202 Chengfu Road, Haidian District, Beijing 100871, China. [2] Clarendon Laboratory, Department of Physics, University of Oxford, Parks Road, Oxford OX1 3PU, UK. [3] Academy for Advanced Interdisciplinary Studies, Peking University, Beijing 100871, China. [4] Stanford Nano Shared Facilities, Stanford University, Stanford, California 94305, USA. * These authors contributed equally to this work. Correspondence and requests for materials should be addressed to Z.L. (email: zfliu@pku.edu.cn) or to H.P. (email: hlpeng@pku.edu.cn) or to Y.C. (email: yulin.chen@physics.ox.ac.uk).

The unique Dirac-cone band structure makes graphene a promising material for photodetection. Its linearly dispersive band structure near Fermi level results in massless Dirac fermion type of carriers, large Fermi velocity ($\sim 1/300$ of the speed of light) and surprisingly high carrier mobility[1–4]. In graphene device, the photovoltage generation time is shorter than 50 fs, which is associated with the carrier heating time[5]. In addition, the rapid cooling process of photoexcited carriers ($\sim$ picoseconds) in the monolayer graphene results in a quick annihilation of photoelectrical signal in the electric circuit[5–12]. These advantages of the monolayer graphene facilitate its applications associated with ultrafast photodetection, such as high-speed optical communications[13–17] and terahertz oscillators[18]. However, it remains a great challenge to achieve high photoresponsivity and selectivity in the monolayer-graphene-based detectors due to the weak and broadband absorption (only 2.3%, from the ultraviolet to the infrared)[19] and the short photocarrier cooling time ($\sim$ picoseconds)[5–12].

On the other hand, twisted bilayer graphene (tBLG) is non-AB stacked bilayer graphene in which one graphene monolayer sheet rotates by a certain angle (θ) relative to the other (Fig. 1a). Recent theoretical studies of tBLG have shown that the Dirac band dispersions change dramatically and become strongly warped with small twist angles ($\theta \leq 5°$)[20–23]. Even at relatively large twist angles, the electronic coupling between the two monolayers, albeit weak, can still introduce new band structures[20–27]. Unlike the parabolic band structure in AB-stacked bilayer graphene[28–34], the band structure of tBLG with large twist angle (typically larger than 5°)[20,35] maintains linear near the Dirac point and thus it inherits some unique properties of monolayer graphene[36]. Away from the Dirac point, Dirac cones of the two individual

monolayers intersect and form saddle points in reciprocal space of tBLG[24], leading to the formation of van Hove singularities (VHSs) in the density of state (DOS)[25,26,35,37,38], which then gives rise to some interesting phenomena such as enhanced optical absorption, Raman G-band resonance and enhanced chemical reactivity of tBLG[27,37,39–45].

In this study, to address the problem of low photoresponsivity and selectivity in the monolayer graphene photodetection, we explore the high-performance photodetector based on tBLG with VHSs. For the first time, we report that the VHSs in tBLG leads to a prominent photocurrent enhancement of tBLG photodetectors with a wavelength selectivity under incident light irradiation.

Results

Structure and Raman spectra. tBLG samples were grown on copper foil via chemical vapour deposition (CVD) method and then transferred to heavily doped Si substrate, which was capped with 90 nm SiO_2. As shown in typical optical image and scanning electron microscopy images (Fig. 1b,c), both the overlayer and underlayer in tBLG exhibit hexagonal shapes with sharp edges, which implies highly crystalline qualities of tBLG domains[46–48]. The interlayer twist angle can be measured from the relative misalignment of the straight edges, which is consistent with the observation by transmission electron microscopy (TEM) (Supplementary Fig. 1 and Supplementary Note 1). tBLG domains with different twist angles can be readily obtained in our samples (Fig. 1d), which provide a platform for the study of θ-dependent light–matter interactions. The highly crystalline quality and clean interface between two monolayers of our CVD sample are evidenced by the moiré pattern in high-resolution

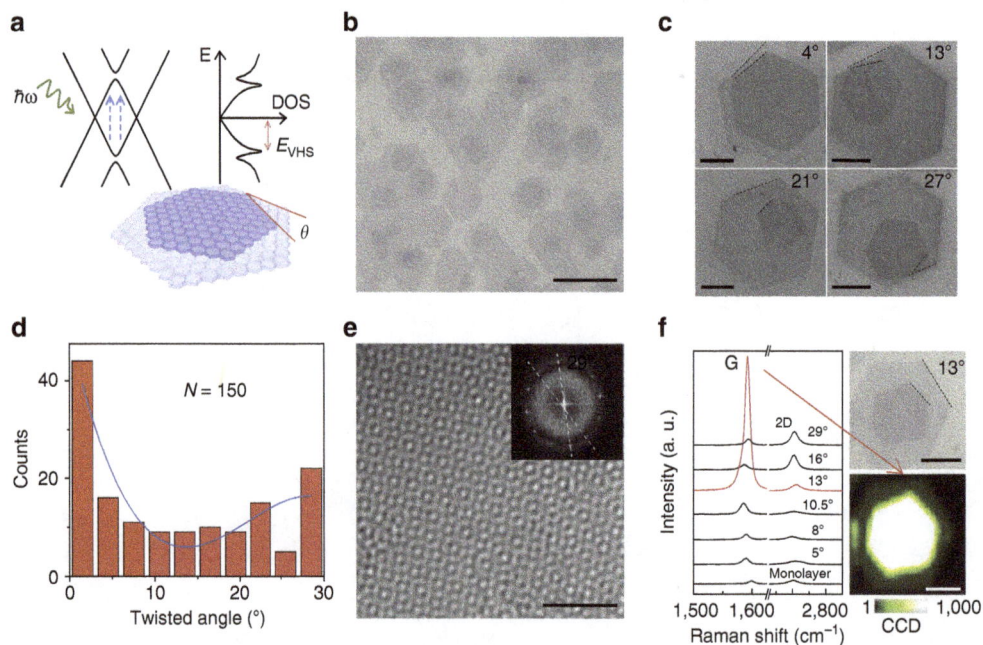

Figure 1 | Structures and Raman spectra of tBLG with different twist angles. (a) Schematics for band structure with minigaps (top left) and the corresponding DOS with VHSs (top right) in tBLG (bottom). Blue arrows describe the photoexcitation process as the energy interval of two VHSs ($2E_{VHS}$) matches the energy of incident photon. (b) The optical image of tBLG domains grown by CVD on Cu and then transferred onto SiO_2 (90 nm)/Si substrate. Scale bar, 30 μm. (c) Scanning electron microscopy (SEM) images of tBLG domains with different twist angles on SiO_2/Si. The twist angles are measured from the edges of over- and underlayer of tBLG domains. Scale bars, 5 μm. (d) Histogram of twist angles measured from tBLG domains in the CVD sample as shown in b. (e) Typical high-resolution TEM (HRTEM) image of tBLG. The periodicity of the moiré pattern is ~ 0.455 nm. The inset is the fast Fourier transform (FFT) of the image, showing that the twist angle is 29°. Scale bar, 2 μm. (f) Left column, Raman spectra of monolayer graphene and tBLG domains with twist angle of 5°, 8°, 10.5°, 13°, 16° and 29°, respectively. The incident laser wavelength is 532 nm (2.33 eV). Top right: the optical image of 13° tBLG domain on SiO_2/Si. Bottom right: G-band intensity mapping image of the 13° tBLG domain shows uniformity of the intensity enhancement of Raman G-band. Scale bars, 10 μm.

TEM image (Fig. 1e). This clean interface guarantees the interaction and coupling of electronic states from the over- and underlayer of tBLG. This interlayer electronic coupling is also proved by the enhanced G-band peak in Raman spectra (Fig. 1f). Taking 13° tBLG domain as an example, the Raman G-band intensity displays a tremendous enhancement of ~20 folds under 532 nm laser (2.33 eV), which is consistent with the previously reported results[27,28,39-43]. This Raman G-band enhancement implies that an interlayer coupling introduces new band structures in tBLG. In addition, the enhanced G-band intensity of 13° tBLG domain was found to be uniform across the whole domain as shown in the G-band mapping image (Fig. 1f), which further confirms the high quality of our CVD tBLG samples. The Raman G-band enhancement is believed to correlate with the formation of VHSs in tBLG[27,37,39-42,49].

Micro-ARPES spectra of tBLG. To unravel the nature of VHSs, we directly investigate the band structures of CVD-grown tBLG domains using spatially resolved angle-resolved photoemission spectroscopy with submicrometre spatial resolution (micro-ARPES). Owing to the twist angle (θ) between over- and underlayer of tBLG, the two sets of (six) Dirac points originated from each layer are rotated relatively by the angle θ as well (see Fig. 2a), which we mark as the $\mathbf{k}$ (left cones) and $\mathbf{k}_\theta$ (right cones) points, respectively. The band structures of a tBLG domain are shown in Fig. 2, where the constant energy contours (Fig. 2b), the band dispersions cutting across (Fig. 2c) and perpendicular (Fig. 2d) to the two adjacent Dirac points are presented, respectively. In Fig. 2b, the stacking plots of the band contours at different binding energies clearly depict the typical two Dirac-cone dispersions of tBLG and each preserves the linear dispersion of monolayer graphene. One of the Dirac cone exhibits a weaker intensity and higher electron doping level, indicating its origin from the underlayer graphene, as the photoelectrons from the bottom layer are screened by the top layer (thus leading to a weaker intensity), and being closer to the Cu substrate also increases its charge transfer[50,51].

By measuring the separation between the two Dirac points (Fig. 2b,c), we can determine the twist angle (θ) of this tBLG domain as 19.1° (Supplementary Fig. 2 and Supplementary Note 2). Without interlayer coupling, the two Dirac cones in Fig. 2 shall intersect and cross each other at higher binding energy. Instead, the band structure at Fig. 2b clearly shows fine structures at the intersection (indicated by red arrows in Fig. 2b) and the dispersion in Fig. 2c shows the opening of the gap at the crossing point of the dispersions from the two Dirac cones, which is indicated by the faint intensity in the spectra intensity map (left panel) and the dip in the DOS plot (right panel, indicated by red arrows). This gap opening in the band structure is a typical anticrossing behaviour introduced by interlayer electronic coupling[24], which leads to the formation of the VHS (Fig. 1a). In addition, from Fig. 2d, one can see that the anticrossing affects the hyperbolic curve as well and results in split and parallel dispersions.

With the same method, we further studied tBLG domains with various different twist angles and tracked the positions of VHSs with respect to the twist angles, as can be seen in Fig. 2e. At small angles, the value of E_{VHS} increases almost linearly with θ, in consistence with the theoretical prediction (Supplementary Fig. 2). This dependence also helps explain the Raman G-band enhancement at specific twist angle (Fig. 1f) for a given incident laser frequency. If the energy of incident photon matches the energy interval of the two VHSs of tBLG ($\hbar\omega \approx 2E_{VHS}$, see Fig. 1a), the electrons are excited and transit between the fine band structure, causing the increase of the intensity of Raman G

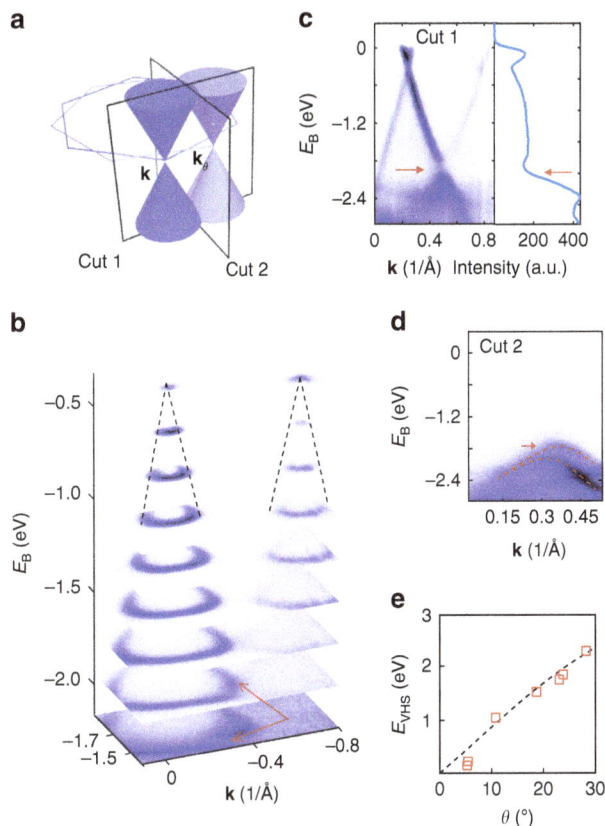

Figure 2 | Micro-ARPES spectra of tBLG. (a) Schematic illustration of the first primitive Brillouin zones (hexagons) and Dirac cones of over- and underlayer of tBLG. **(b)** Stacking plot of constant-energy contours at different binding energies (E_B) of tBLG. **(c)** ARPES spectra along Cut 1 as labelled in **a**. The right curve is energy spectrum density curve (EDC) integrated from the spectrum. **(d)** ARPES spectra along Cut 2 as labelled in **a**. Red arrows in **b,c** and **d** indicate the minigap band topology and the split parallel branches arising from interlayer coupling. **(e)** E_{VHS} versus the twist angle (θ) of tBLG domains. The E_{VHS}, measured from micro-ARPES data of tBLG, is the energy interval between the minigap (VHS) and Dirac point. The E_{VHS} varies almost linearly with twist angle. The black dashed line is a theoretical curve.

band peak (see Supplementary Fig. 3 and Supplementary Note 3 for details).

Selectively enhanced photocurrent generation of tBLG. The strong light–matter interaction of tBLG selectively enhanced by the VHSs can also enhance the generation of photocurrent under illumination. As an example, two adjacent tBLG domains with twist angle of 13° and 7° transferred onto SiO$_2$ (90 nm)/Si were etched into a strip and then embedded into two-terminal devices in parallel (Fig. 3a,b). Raman spectroscopy and two-dimensional maps of the two adjacent tBLG domains were first measured under the 532-nm laser (2.33 eV). As expected, the G-band intensity of whole 13° domain exhibits a uniformly 20-fold enhancement as compared with the 7° domain (see Fig. 3c), as the energy interval of the two VHSs in 13° domain matches the energy of incident photon ($\hbar\omega \approx 2E_{VHS}$). To generate photocurrent selectively, interfacial junctions of tBLG-metal electrodes were used to separate the photoexcited electrons and holes under illumination[52,53]. As shown in current–bias voltage curves, both tBLG domains produce pronounced photocurrent

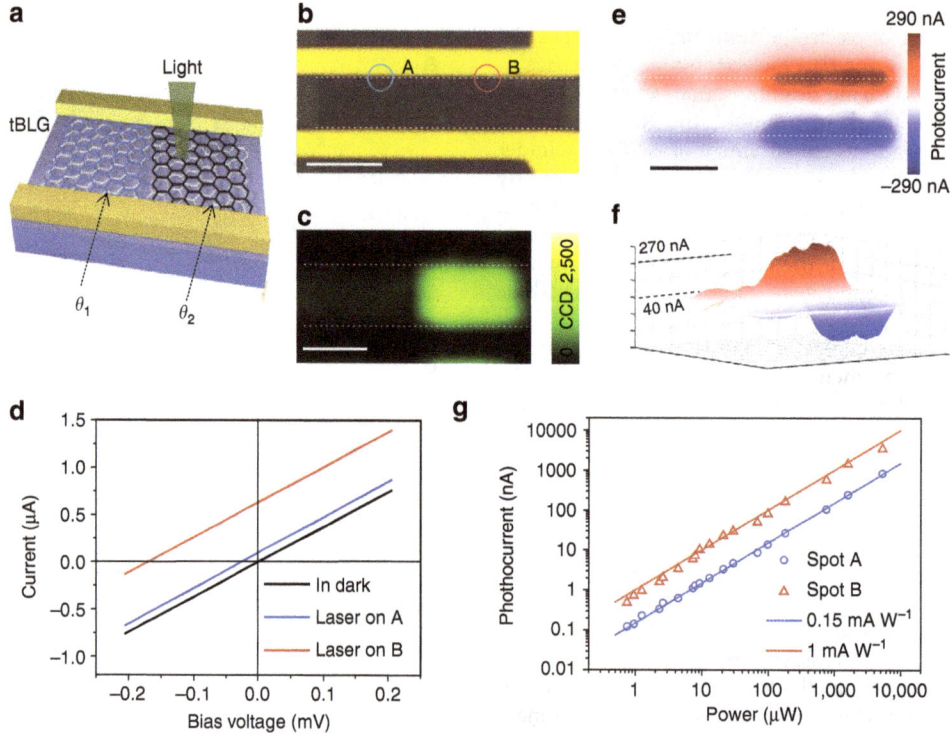

Figure 3 | Selectively enhanced photocurrent generation in tBLG photodetection devices. (**a**) Schematic illustration of a tBLG photodetection device. The channel comprises of two adjacent tBLG domains with different twist angles of θ_1 and θ_2, respectively. (**b**) Optical image of the tBLG photodetection device. The θ_1 and θ_2 are 7° and 13°, respectively. (**c**) Raman G-band intensity mapping image under 532 nm (2.33 eV) laser. 13° tBLG domain exhibits an enhanced G-band intensity. (**d**) Current versus source-drain bias (*I–V*) curve without laser on and with laser focusing on 7° (spot A) and 13° (spot B) tBLG domains, respectively. The intercepts at current axis represent the net photocurrents. (**e**) Scanning photocurrent images of the same tBLG device. A 532-nm laser with power of 200 µW is focused on the device, while the net photocurrent is amplified and then detected by a lock-in amplifier. All the photocurrents here are generated without source-drain bias and gate bias. (**f**) Three-dimensional view of the scanning photocurrent image of the same tBLG device. (**g**) Photocurrents generated from 7° (spot A) and 13° (spot B) tBLG domains as a function of incident power, respectively. The white dashed lines in **c** and **e** show the positions of graphene–metal electrode interfaces, respectively. Scale bars, 5 µm (all).

shifts (Fig. 3d). Remarkably, the 13° tBLG domain generates a much larger net photocurrent (0.63 µA) at zero bias than that of the 7° domain (0.097 µA), originated from selectively enhanced light–matter interaction of 13° tBLG domain with the 532-nm laser.

We further conducted net photocurrent mapping of the device by using scanning photocurrent microscopy, in which the net photocurrent was recorded while scanning a focused 532-nm laser spot with a diameter of ~1 µm over the device (Fig. 3e,f). The photocurrent was observed to exhibit contrary directions at the two graphene–metal electrode interfaces in the device. Significantly, the intensity of photocurrent generated at 13° tBLG domain is ~6.6 times stronger than that at the 7° tBLG domain. This twist angle-related photocurrent enhancement holds great promise in high-selectivity photodetection applications.

To further evaluate the photoresponsivity of tBLG, we performed photocurrent measurements of tBLG devices under different incident power of 532 nm laser illumination, respectively. As shown in Fig. 3g, the photocurrents from 7° and 13° tBLG domains both increase as the incident power rises from ~1 µW to ~5 mW. The photoresponsivity of 7° and 13° tBLG domain is measured as ~0.15 and ~1 mA W^{-1}, respectively, indicating a robust and strong enhancement in 13° tBLG domain under different incident power of 532 nm laser illumination.

From the unravelling of band structures, the energy interval of the two VHSs (2E_{VHS}) of 13° tBLG domain is ~2.34 eV, which matches the energy of incident photon (2.33 eV, λ = 532 nm) and thus leads to a strong light–matter interaction. When we changed the wavelength of incident laser from 532 to 632.8 nm (1.96 eV),

the photocurrent was found to be selectively enhanced in a 10.5° tBLG domain device with 2E_{VHS} of ~1.89 eV (Supplementary Fig. 4 and Supplementary Note 4). To further investigate the correlation of 2E_{VHS} with $\hbar\omega$ in photocurrent generation of 13° and 10.5° tBLG domains, $\hbar\omega$ was gradually changed from 1.77 to 2.48 eV (500 to 700 nm in wavelength), while the power of incident laser was kept unchanged. As shown in Fig. 4a, the photocurrents of 13° and 10.5° tBLG domains exhibit peaks at ~2.30 and ~1.94 eV, agreeing well with 2E_{VHS} values (~2.34 and ~1.89 eV), respectively.

Discussion

The origin of the photocurrent enhancement can be understood qualitatively when taking the unique electronic state of tBLG into account. In the photoexcitation process, the interband transition has to satisfy both momentum and energy conservation. For momentum conservation, the electrons are confined to transit between states with the same **k** value in reciprocal space, owing to the very small momentum of incident photons. As for energy conservation, the energy difference between these two states equals to $\hbar\omega$. When $\hbar\omega \approx 2E_{VHS}$, the initial and final states are both near VHSs (thus with enhanced DOS, see Fig. 1a). Specifically, the effect of VHSs on the photoexcitation process can be evaluated by joint DOS (JDOS), which is defined as:

$$JDOS(\omega) = \frac{1}{4\pi^3}\int \delta[E_c(\mathbf{k}) - E_v(\mathbf{k}) - \hbar\omega]d\mathbf{k} \quad (1)$$

where E_C and E_V represent the energies of the conduction and valence bands, respectively. JDOS is associated with the process in which an electron absorbs a photon with energy $\hbar\omega = E_c - E_v$ and then transits from conduction to valence band. A calculated JDOS shows an abrupt increase associated with the VHSs when $\hbar\omega \approx 2E_{VHS}$ (ref. 40). This leads to an enhanced photo-excitation process, consistent with experimental observations in Raman (Fig. 1f) and absorption spectra[27,45]. As a result, the intensified photoexcitation process may result in the enhanced photocurrent generation.

Besides the efficient photoexcitation, the improvement of separation efficiency of excited carriers can facilitate the photocurrent generation in tBLG. A gate voltage applied on the tBLG photodetection device can manipulate the doping level of

graphene in channel and thus change the value of Seebeck coefficient, which may simultaneously lead to the photocurrent change[54–57]. As shown in Fig. 4b, the photocurrent of 13° tBLG domain has a 2.6-fold increase from ~ 25 to ~ 66 nA when the back-gate voltage decreases from 0 to -20 V. As the back-gate voltage increases from 0 to 20 V, the photocurrent first flips its polarity at 2.5 V and then reaches a value of about -82 nA. The inset in Fig. 4b shows the two band profiles of graphene–metal electrode junctions in the tBLG photodetection device under the applied back-gate voltage. From the transfer curve (Supplementary Fig. 5 and Supplementary Note 5), we believe that the positive gate voltage manipulates the graphene in channel from p- to n-type doping, which gives rise to the change of Seebeck coefficient. In contrast, owing to the Fermi-level pinning of graphene underneath the metal electrodes, its Seebeck coefficient keeps unchanged. Therefore, the difference of these two Seebeck coefficients could be tuned and flipped by gate voltage, which leads to the value and polarity change of photocurrent.

The responsivity of tBLG is measured as ~ 1 mA W^{-1} at the resonance frequency, which is about 20 times enhancement compared with that of mechanically exfoliated monolayer graphene (~ 0.05 mA W^{-1}) with similar device configuration (Supplementary Fig. 6). To further improve the responsivity, we have integrated tBLG with plasmonic electrode structures as shown in Fig. 5a. A tBLG domain and an adjacent monolayer domain were embedded into the same two-terminal electrodes. A finger-patterned plasmonic structure (Ti/Au, 5/45 nm in thickness) with 110 nm finger width and 300 nm pitch[58] were fabricated on the tBLG domain as shown in Fig. 5b,c. The Raman mapping image in Fig. 5d exhibits uniformly enhanced G-band intensity, which confirms that the interval of two VHSs ($2E_{VHS}$) of the tBLG domain matches the energy of incident photon (532 nm and 2.33 eV). The scanning photocurrent results of the device (Fig. 5e,f) show that the photocurrents of tBLG and the adjacent monolayer domain are measured as ~ 10 and ~ 0.95 nA,

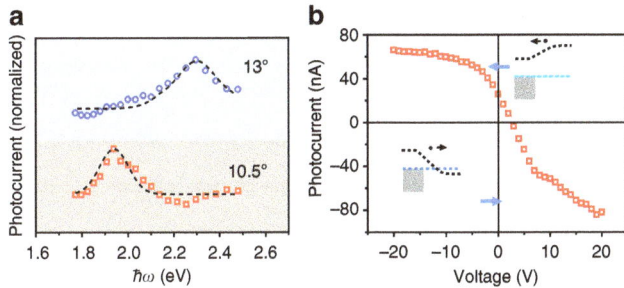

Figure 4 | The variation of photocurrent with photon energy and gate voltage. (**a**) Photocurrent versus energy of incident photon ($\hbar\omega$). tBLG domains with 10.5° and 13° twist angles show different peak positions. Incident photons with energy ω near $2E_{VHS}$ generate an enhanced photocurrent, while photons with energy lower or higher than $2E_{VHS}$ excite ordinary optoelectronic processes. Dotted lines were used to guide the eyes. The plots are normalized with that of AB-stacked bilayer graphene. (**b**) Plot of photocurrent as function of gate voltage. Insets are the corresponding band profiles, where the grey boxes, blue dotted lines and black dotted lines represent Ti electrodes, Fermi levels and positions of Dirac points of tBLG, respectively.

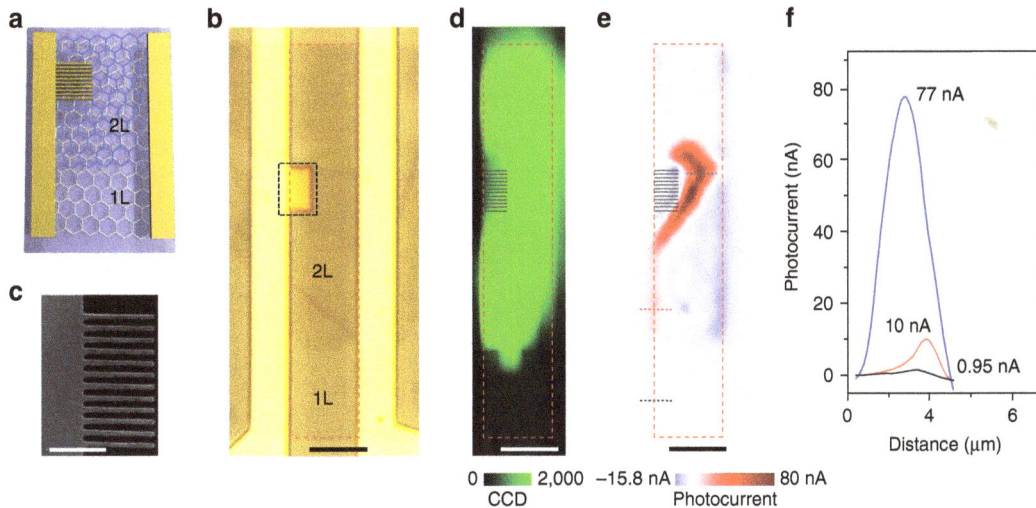

Figure 5 | tBLG photodetector integrated with plasmonic structure. (**a**) Schematic illustration of the detector. The channel comprises graphene monolayer domain and tBLG domain which are labelled by '1L' and '2L', respectively. The electrode is integrated with finger structure. (**b**) Optical image of the tBLG photodetector. (**c**) Scanning electron microscopy (SEM) image of the finger structure labelled by black dashed rectangle in **b**. (**d**) Raman G-band intensity mapping image of the device under 532 nm laser. The tBLG domain exhibits an enhanced G-band intensity. (**e**) Scanning photocurrent image of the photodetector. The exciting light is a focused 532 nm laser with power of 30 μW. (**f**) Line-scanning photocurrent of the photodetector. The blue, red and black curves correspond to photocurrent distributions along the tBLG near figure structure, tBLG and graphene monolayer, which are labelled by blue, red and black dashed lines in **e**. The maximum photocurrent at tBLG with the plasmonic structure is 77 nA, while the maximum photocurrents at the tBLG and monolayer graphene are 10 and 0.95 nA, respectively. All the photocurrents here are generated without source-drain bias and gate bias. The red dashed rectangles in **b,d** and **e** correspond to the channel of the photodetector. Scale bar, 1 μm (finger structure in **c**). Scale bars, 5 μm (others).

respectively. The enhancement of about 10.5 times is achieved. Remarkably, the photocurrent generated on tBLG near the finger structure is further enhanced to the value of ~ 77 nA, which is ~ 80-fold enhancement compared with the photocurrent of adjacent monolayer graphene channel. The mechanism of such strong photoresponsivity enhancements can be ascribed to the combination of the selective resonance enhancement of VHSs in tBLG and plasmonic enhancement of the finger structure.

In summary, we have experimentally demonstrated that the light–matter interaction in tBLG with VHSs is dependent on the interlayer twist angle (θ) and then can lead to a selective enhancement in photocurrent generation. Micro-ARPES was performed to unravel the band structure of CVD-grown tBLG and reveal the emergence of θ-dependent VHSs. The photocurrent of tBLG photodetectors exhibits a ~ 6.6-fold enhancement at a suitable θ, when the energy interval of the VHSs ($2E_{VHS}$) matches the energy of incident photon ($\hbar\omega$). By integrating plasmonic structures, the responsivity of tBLG photodetector can be significantly enhanced by ~ 80 folds compared with the monolayer graphene. Our results open a new route to graphene-based optoelectronic applications.

Methods

tBLG growth and characterization. The tBLG was grown on copper foil in a home-made low-pressure CVD system. The growth was carried out under the flow of H_2 and CH_4 (600:1 in volume) with a pressure of 600 Pa at 1,030 °C for 40 min. The samples were characterized by Olympus BX51 microscope, scanning electron microscopy (Hitachi S-4800 operated at 2 kV), TEM (FEI Tecnai F30 operated at 300 kV for the diffraction image and FEI 80-300 Cs image-corrected Titan operated at 80 kV for the moiré pattern image) and Raman spectroscopy (Horiba HR800). The micro-ARPES measurements were carried out at the spectromicroscopy beamline at Elettra Synchrotron Radiation lab in Italy, with energy resolution of 50 meV, spatial resolution of 0.8 μm and angle resolution of 0.5°. Before the micro-ARPES measurements, in-situ annealing at 350 °C was carried out to clean the sample surface.

Device fabrication and measurement. tBLG samples on copper were transferred to a highly doped Si substrate with 90 nm SiO_2 with the help of poly(methyl methacrylate)[59]. The Ti/Au (20/30 nm) electrodes and Ti/Au (5/45 nm) figure structures were fabricated by electron-beam lithography and the following electron-beam evaporation. The electrical measurements were performed by Keithley SCS-4200. The photoelectrical measurements were performed by a scanning photocurrent microscopy. In the set-up, 532 and 632.8 nm, and Supercontinuum Laser Sources (NKT Photonic) were used as laser sources. The chopper-modulated (~ 1 kHz) laser beams were focused to ~ 1 μm on the device using × 100 objective and the short-circuit photocurrents were then measured by pre-amplifier and lock-in amplifier. When scanning the laser spot over the device, the induced photocurrents and beam positions were recorded and displayed simultaneously with the assistance of a computer, which communicated with lock-in amplifier and motorized stage (with device on it). A voltage source (Keithley 2400) was used to supply the gate voltage. All the electrical and photoelectrical measurements were performed in air at room temperature.

References

1. Novoselov, K. S. et al. Two-dimensional gas of massless Dirac fermions in graphene. Nature **438**, 197–200 (2005).
2. Novoselov, K. S. et al. Electric field effect in atomically thin carbon films. Science **306**, 666–669 (2004).
3. Zhang, Y. B. et al. Experimental observation of the quantum Hall effect and Berry's phase in graphene. Nature **438**, 201–204 (2005).
4. Geim, A. K. & Novoselov, K. S. The rise of graphene. Nat. Mater. **6**, 183–191 (2007).
5. Tielrooij, K. J. et al. Generation of photovoltage in graphene on a femtosecond timescale through efficient carrier heating. Nat. Nanotechnol. **10**, 437–443 (2015).
6. Sun, D. et al. Ultrafast relaxation of excited Dirac fermions in epitaxial graphene using optical differential transmission spectroscopy. Phys. Rev. Lett. **101**, 157402 (2008).
7. George, P. A. et al. Ultrafast optical-pump terahertz-probe spectroscopy of the carrier relaxation and recombination dynamics in epitaxial graphene. Nano Lett. **8**, 4248–4251 (2008).
8. Tani, S. et al. Ultrafast carrier dynamics in graphene under a high electric field. Phys. Rev. Lett. **109**, 166603 (2012).
9. Graham, M. W. et al. Photocurrent measurements of supercollision cooling in graphene. Nat. Phys. **9**, 103–108 (2013).
10. Tielrooij, K. J. et al. Photoexcitation cascade and multiple hot-carrier generation in graphene. Nat. Phys. **9**, 248–252 (2013).
11. Tse, W.-K. & Das Sarma, S. Energy relaxation of hot Dirac fermions in graphene. Phys. Rev. B **79**, 235406 (2009).
12. Bistritzer, R. & MacDonald, A. H. Electronic cooling in graphene. Phys. Rev. Lett. **102**, 206410 (2009).
13. Xia, F. N. et al. Ultrafast graphene photodetector. Nat. Nanotechnol. **4**, 839–843 (2009).
14. Mueller, T. et al. Graphene photodetectors for high-speed optical communications. Nat. Photonics **4**, 297–301 (2010).
15. Gan, X. et al. Chip-integrated ultrafast graphene photodetector with high responsivity. Nat. Photonics **7**, 883–887 (2013).
16. Pospischil, A. et al. CMOS-compatible graphene photodetector covering all optical communication bands. Nat. Photonics **7**, 892–896 (2013).
17. Schall, D. et al. 50 GBit/s photodetectors based on wafer-scale graphene for integrated silicon photonic communication systems. ACS Photonics **1**, 781–784 (2014).
18. Boubanga-Tombet, S. et al. Ultrafast carrier dynamics and terahertz emission in optically pumped graphene at room temperature. Phys. Rev. B **85**, 035443 (2012).
19. Nair, R. R. et al. Fine structure constant defines visual transparency of graphene. Science **320**, 1308–1308 (2008).
20. Shallcross, S. et al. Quantum interference at the twist boundary in graphene. Phys. Rev. Lett. **101**, 056803 (2008).
21. Shallcross, S. et al. Emergent momentum scale, localization, and van Hove singularities in the graphene twist bilayer. Phys. Rev. B **87**, 245403 (2013).
22. Shallcross, S. et al. Electronic structure of turbostratic graphene. Phys. Rev. B **81**, 1 (2010).
23. Landgraf, W. et al. Electronic structure of twisted graphene flakes. Phys. Rev. B **87**, 075433 (2013).
24. Ohta, T. et al. Evidence for interlayer coupling and moire periodic potentials in twisted bilayer graphene. Phys. Rev. Lett. **109**, 186807 (2012).
25. Li, G. H. et al. Observation of Van Hove singularities in twisted graphene layers. Nat. Phys. **6**, 109–113 (2010).
26. Yan, W. et al. Angle-dependent van Hove singularities in a slightly twisted graphene bilayer. Phys. Rev. Lett. **109**, 126801 (2012).
27. Havener, R. W. et al. Angle-resolved Raman imaging of inter layer rotations and interactions in twisted bilayer graphene. Nano Lett. **12**, 3162–3167 (2012).
28. Ohta, T. et al. Controlling the electronic structure of bilayer graphene. Science **313**, 951–954 (2006).
29. McCann, E. Asymmetry gap in the electronic band structure of bilayer graphene. Phys. Rev. B **74**, 161403 (2006).
30. Castro, E. V. et al. Biased bilayer graphene: semiconductor with a gap tunable by the electric field effect. Phys. Rev. Lett. **99**, 216802 (2007).
31. McCann, E. & Fal'ko, V. I. Landau-level degeneracy and quantum hall effect in a graphite bilayer. Phys. Rev. Lett. **96**, 086805 (2006).
32. Guinea, F. et al. Electronic states and Landau levels in graphene stacks. Phys. Rev. B **73**, 245426 (2006).
33. Ohta, T. et al. Interlayer interaction and electronic screening in multilayer graphene investigated with angle-resolved photoemission spectroscopy. Phys. Rev. Lett. **98**, 206802 (2007).
34. Zhang, Y. B. et al. Direct observation of a widely tunable bandgap in bilayer graphene. Nature **459**, 820–823 (2009).
35. Luican, A. et al. Single-layer behavior and its breakdown in twisted graphene layers. Phys. Rev. Lett. **106**, 126802 (2011).
36. dos Santos, J. et al. Graphene bilayer with a twist: electronic structure. Phys. Rev. Lett. **99**, 256802 (2007).
37. Coh, S. et al. Theory of the Raman spectrum of rotated double-layer graphene. Phys. Rev. B **88**, 165431 (2013).
38. Brihuega, I. et al. Unraveling the intrinsic and robust nature of van Hove Singularities in twisted bilayer graphene by scanning tunneling microscopy and theoretical analysis. Phys. Rev. Lett. **109**, 196802 (2012).
39. Kim, K. et al. Raman spectroscopy study of rotated double-layer graphene: misorientation-angle dependence of electronic structure. Phys. Rev. Lett. **108**, 246103 (2012).
40. Sato, K. et al. Zone folding effect in Raman G-band intensity of twisted bilayer graphene. Phys. Rev. B **86**, 125414 (2012).
41. Carozo, V. et al. Resonance effects on the Raman spectra of graphene superlattices. Phys. Rev. B **88**, 085401 (2013).
42. He, R. et al. Observation of low energy Raman modes in twisted bilayer graphene. Nano Lett. **13**, 3594–3601 (2013).
43. Ni, Z. et al. Reduction of Fermi velocity in folded graphene observed by resonance Raman spectroscopy. Phys. Rev. B **77**, 235403 (2008).
44. Liao, L. et al. van Hove singularity enhanced photochemical reactivity of twisted bilayer graphene. Nano Lett. **15**, 5585–5589 (2015).

45. Wang, Y. Y. *et al*. Stacking-dependent optical conductivity of bilayer graphene. *ACS Nano* **4**, 4074–4080 (2010).
46. Zhou, H. L. *et al*. Chemical vapour deposition growth of large single crystals of monolayer and bilayer graphene. *Nat. Commun.* **4**, 2096 (2013).
47. Yan, Z. *et al*. Toward the synthesis of wafer-scale single-crystal graphene on copper foils. *ACS Nano* **6**, 9110–9117 (2012).
48. Ma, T. *et al*. Repeated growth-etching-regrowth for large-area defect-free single-crystal graphene by chemical vapor deposition. *ACS Nano* **8**, 12806–12813 (2014).
49. Ni, Z. H. *et al*. G-band Raman double resonance in twisted bilayer graphene: evidence of band splitting and folding. *Phys. Rev. B* **80**, 125404 (2009).
50. Giovannetti, G. *et al*. Doping graphene with metal contacts. *Phys. Rev. Lett.* **101**, 026803 (2008).
51. Zhou, S. Y. *et al*. Substrate-induced bandgap opening in epitaxial graphene. *Nat. Mater.* **6**, 770–775 (2007).
52. Xia, F. N. *et al*. Photocurrent imaging and efficient photon detection in a graphene transistor. *Nano Lett.* **9**, 1039–1044 (2009).
53. Mueller, T. *et al*. Role of contacts in graphene transistors: a scanning photocurrent study. *Phys. Rev. B* **79**, 245430 (2009).
54. Song, J. C. W. *et al*. Hot carrier transport and photocurrent response in graphene. *Nano Lett.* **11**, 4688–4692 (2011).
55. Gabor, N. M. *et al*. Hot carrier-assisted intrinsic photoresponse in graphene. *Science* **334**, 648–652 (2011).
56. Xu, X. D. *et al*. Photo-thermoelectric effect at a graphene interface junction. *Nano Lett.* **10**, 562–566 (2010).
57. Sun, D. *et al*. Ultrafast hot-carrier-dominated photocurrent in graphene. *Nat. Nanotechnol.* **7**, 114–118 (2012).
58. Echtermeyer, T. J. *et al*. Strong plasmonic enhancement of photovoltage in graphene. *Nat. Commun.* **2**, 458 (2011).
59. Reina, A. *et al*. Transferring and identification of single- and few-layer graphene on arbitrary substrates. *J. Phys. Chem. C* **112**, 17741–17744 (2008).

Acknowledgements

We are grateful to Dr Yao Guo and Mr Chen Peng from Department of Electronics, Peking University, for their suggestions in device fabrication, and Mr Ziwei Li from School of Physics, Peking University, for the calculation regarding to plasmonic structures. We acknowledge financial support from the National Basic Research Program of China (numbers 2014CB932500, 2011CB921904 and 2013CB932603), the National Natural Science Foundation of China (numbers 21173004, 21222303, 51121091 and 51362029), the National Program for Support of Top-Notch Young Professionals and Beijing Municipal Science and Technology Commission (Z131100003213016). Part of this work was performed at the Stanford Nano Shared Facilities.

Author contributions

J.Y. and H.L.P. conceived and designed the experiments. J.Y and H.W. performed the synthesis and structural characterization. J.Y. made the devices and carried out optoelectronic measurements. Z.T., L. Liao, L. Lin and X.S. assisted in experimental work and contributed to the scientific discussions. H.P. and Y.L.C. preformed micro-ARPES. A.L.K. and H.L.P. conducted the TEM, high-resolution TEM and aberration-corrected high-resolution TEM experiments. J.Y., H.L.P. and H.P. wrote the paper. H.L.P., Z.L. and Y.L.C. supervised the project. All the authors discussed the results and commented on the manuscript.

Additional information

Competing financial interests: The authors declare no competing financial interests.

Wavelength-tunable entangled photons from silicon-integrated III–V quantum dots

Yan Chen[1], Jiaxiang Zhang[1], Michael Zopf[1], Kyubong Jung[1], Yang Zhang[1], Robert Keil[1], Fei Ding[1] & Oliver G. Schmidt[1,2]

Many of the quantum information applications rely on indistinguishable sources of polarization-entangled photons. Semiconductor quantum dots are among the leading candidates for a deterministic entangled photon source; however, due to their random growth nature, it is impossible to find different quantum dots emitting entangled photons with identical wavelengths. The wavelength tunability has therefore become a fundamental requirement for a number of envisioned applications, for example, nesting different dots via the entanglement swapping and interfacing dots with cavities/atoms. Here we report the generation of wavelength-tunable entangled photons from on-chip integrated InAs/GaAs quantum dots. With a novel anisotropic strain engineering technique based on PMN-PT/silicon micro-electromechanical system, we can recover the quantum dot electronic symmetry at different exciton emission wavelengths. Together with a footprint of several hundred microns, our device facilitates the scalable integration of indistinguishable entangled photon sources on-chip, and therefore removes a major stumbling block to the quantum-dot-based solid-state quantum information platforms.

[1] Institute for Integrative Nanosciences, IFW Dresden, Helmholtzstraße 20, 01069 Dresden, Germany. [2] Material Systems for Nanoelectronics, Chemnitz University of Technology, Reichenhainer strasse 70, 09107 Chemnitz, Germany. Correspondence and requests for materials should be addressed to F.D. (email: f.ding@ifw-dresden.de).

A topical challenge in quantum information processing (QIP) is the generation and manipulation of polarization-entangled photon pairs[1,2]. Spontaneous parametric-down-conversion (SPDC) and four-wave-mixing (FWM) have served as the main workhorses for these purposes in the past decade, and the implementation of a fully integrated quantum device is within reach by marrying these sources with chip-scale silicon photonics[3–6]. However, the generated photons are characterized by Poissonian statistics, that is, one usually does not know when an entangled photon pair is emitted. This fundamentally limits their applications in complex quantum protocols, for example, an event-ready test of Bell's inequality and high-efficiency entanglement purifications, where deterministic operations are much favoured[1].

The intrinsic limitations of SPDC and FWM processes call for next generation entangled photon sources. III–V semiconductor quantum dots, often referred to as artificial atoms, are among the leading candidates for deterministic quantum light sources. As proposed by Benson *et al.* single quantum dots (QDs) can generate polarization-entangled photon pairs via its biexciton (XX) cascade decay through the intermediate exciton states X, Fig. 1 (ref. 7). In real III–V QDs the anisotropy in strain, composition and shape reduces the QD symmetry to C_{2v} or the even lower C_1, leading to the appearance of an energetic splitting between the two bright X states, the so-called fine structure splitting (FSS)[8]. High fidelity to the entangled state $|\Psi^+\rangle = 1/\sqrt{2}(|H_{XX}H_X\rangle + |V_{XX}V_X\rangle)$, with H and V denoting the horizontal and vertical polarizations, can be observed only with a vanishing FSS (typically, smaller than the radiative linewidth of $\sim 1\,\mu eV$). The probability of finding such QDs in an as-grown sample is $<10^{-2}$. After extensive efforts by many groups, the elimination of FSS can be achieved by applying rapid thermal annealing[9], optical Stark effect[10], magnetic field[11,12],

electric field[13,14], and more recently, anisotropic strain fields[15,16] to the QDs. In the past years we have witnessed considerable progress in this field, and entangled photon emissions can be triggered optically[11,17] from single QDs with high brightness (up to 0.12 pair per excitation pulse)[18] and high indistinguishability (0.86 ± 0.03 for the XX photons)[19]. III–V QDs also possess an important advantage of being compatible with mature semiconductor technology, and electrically triggered entangled photon sources have been successfully demonstrated[16,20].

Armed with these powerful techniques, III–V QDs have the potential to fulfil the 'wish-list' of a perfect entangled photon source[21]. Among the next goals are the miniaturization and scaling up of the technology. Several important issues need to be considered. First, the FSS of each QD can only be eliminated under particular tuning parameters, and any attempt to manipulate the emission wavelength increases the FSS and spoils the entanglement. This fact undoubtedly restricts the entangled photon emissions at arbitrary wavelengths. The inability to tune the emission wavelength without restoring the FSS, which is unfortunately the common disadvantage associated with all FSS tuning technologies to date, has become a major stumbling block to the QIP applications based on scalable QD sources. Second, as being investigated with SPDC and FWM sources[3–5], the integration of wavelength-tunable quantum light sources on silicon is arguably one of the most promising choices for on-chip QIP applications[22].

Here we demonstrate wavelength-tunable entangled photon sources based on III–V QDs integrated on a silicon chip. There are two recent theoretical proposals[23,24] on generating wavelength-tunable entangled photon from QDs, however, the experimental implementations of their proposals are quite challenging. We design and fabricate a device consisting of QD-embedded nanomembranes suspended on a four-legged

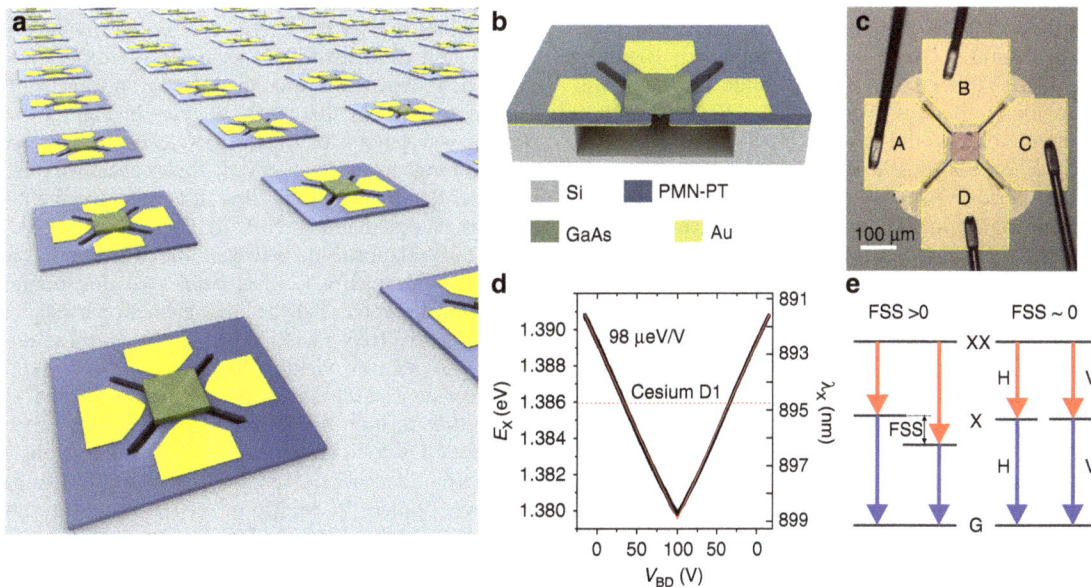

Figure 1 | Wavelength-tunable polarization-entangled photon sources integrated on silicon. (**a**) MEMS devices for anisotropic strain engineering of III–V QD-based quantum light sources. Owing to its small footprint and the compatibility with mature semiconductor technologies, large scale on-chip integration is feasible. (**b**) Schematic of the cross section of a single device. Focused ion beam (FIB) cut is used to define trenches on the PMN-PT thin film, and then wet-chemical undercut is used to form four suspended actuation legs. A thin GaAs nanomembrane containing In(Ga)As QDs is transferred onto the suspended region between the four legs. (**c**) Micrograph showing the zoom-in of a completed device. Electrical contacts are made on the four legs A–D. The centre region is a bonded QD-containing nanomembrane. (**d**) Performance of a typical device. The exciton wavelength of a single QD is recorded when the voltage on legs B&D is scanned. The actuation legs contract under positive voltages, leading to the tensile stresses on the nanomembrane and to the red shift of the QD emission. The red solid lines show the linear fit. The dashed line indicates the caesium D1 absorption line. (**e**) Illustration of the FSS in a QD. Polarization-entangled photons are emitted from the XX cascade emission only when the FSS is tuned to near zero.

thin-film PMN-PT ($[Pb(Mg_{1/3}Nb_{2/3})O_3]_{0.72}[PbTiO_3]_{0.28}$) actuator integrated on a silicon substrate. With the combined uniaxial stresses along two orthogonal directions, we are able to keep the FSS strictly below 1 μeV while shifting the exciton wavelength/energy by more than 3,000 times of the QD radiative linewidth. High-fidelity entangled photon emission is demonstrated when the FSS is tuned to below 1 μeV. Therefore wavelength-tunable entangled photons are generated on chip with a single-device footprint of a few hundred microns.

Results

Concept of device. For the device fabrication we use the industrial transfer printing and die bonding techniques to realize the novel integration of III–V, PMN-PT and Si. Unlike the piezo substrate used in all previous works[15,16,24–27], a 15-μm PMN-PT thin-film bonded on a silicon substrate is employed here to realize novel micro-electromechanical system (MEMS) devices with sophisticated functionalities on chip (Fig. 1, see also Methods section). Arrays of QD-containing GaAs nanomembranes, each $80 \times 80\ \mu m^2$ in size, were then transferred onto the PMN-PT MEMS with four actuation legs (Fig. 1a–c). The crystal axes [1–10] and [110] of the GaAs nanomembrane were carefully aligned along the designed stress axes of the actuators. When applying negative (positive) voltages to the electric contacts, the PMN-PT legs expand (contract) in-plane and therefore exert quasi-uniaxial compressive (tensile) stresses to the QDs.

This new device concept has several advantages. First, a controllable anisotropic strain is achieved by the four-legged configuration, which is not possible with any piezo substrate. Second, the use of piezoelectric film alleviates us from high voltages (typically, up to thousands volts) required for the bulk PMN-PT substrate, which is certainly important for on-chip integration. And third, theoretically proposed scalable sources of strain-controlled quantum light sources[23,24] can be realized on chip with such devices.

By sweeping the voltage on only one pair of opposite legs, the exciton emission is shifted over a large range (up to 10 nm in wavelength or 12 meV in energy, Fig. 1d) due to the quasi-uniaxial stresses along the legs. For suitable QDs the exciton emission can be tuned across the caesium D1 line at ∼ 894.7 nm, which is required for realizing a hybrid quantum memory[28]. To confirm the performance of the device, we have also performed finite-element modelling of our device and two principle stresses (with a magnitude of ∼2 GPa at voltages of 50 V) along orthogonal directions can be identified (not shown here).

Theory[26,27,29] predicts that the uniaxial strain tuning behaviour of FSS is determined by the QD principal axis with respect to the uniaxial stress direction. For a QD whose principal axis is closely aligned with the stress direction, the FSS can be effectively eliminated. For typical self-assembled semiconductor QDs the directions of the QD principal axes have a Gaussian-like distribution, and a significant amount[30] of QDs are closely aligned with the crystallographic directions due to the anisotropic surface adatom diffusion. An example of the FSS tuning behaviour for such an aligned QD is given in Fig. 2a. The uniaxial stress tuning is done by sweeping the voltage V_{AC} from 0 to 100 V on one pair of legs, whereas fixing the voltage V_{BD} at 0 V on the other pair of legs. With increasing V_{AC} the FSS first decreases monotonically to a minimum value and then increases. At V_{AC} of ∼73 V the FSS is completely eliminated. The phase θ, which indicates the angle (see inset of Fig. 2a) between the exciton polarization and the [1–10] crystallographic direction of GaAs, undergoes a sharp phase change of 90°. This result is in agreement with theoretical predictions[27].

How does a QD behave when both voltages V_{AC} and V_{BD} are turned on, that is, under the application of a pair of orthogonal uniaxial stresses? This has never been studied due to the lack of realistic experimental techniques. In Fig. 2b we present the FSS tuning result at a different V_{BD} of − 25 V. Similar to Fig. 2a we also observe a zero FSS and an abrupt change in θ by exactly 90°. The only difference between the two situations ($V_{BD} = 0$ and − 25 V) is the voltage V_{AC} at which the FSS is erased. This is an indication of the recovery of QD electronic symmetry at different experimental conditions. It is then interesting to see how the exciton emission wavelength changes, Fig. 2c. With negative (positive) voltages applied, the legs exert uniaxial compressive (tensile) stresses to the QD which causes a blue (red) shift in exciton emission. This is confirmed by sweeping V_{AC} from 0 to 100 V, from which we observe a red shift of the emission. The device performs remarkably well and we do not observe any hysteresis in the wavelength tuning, see the linear fit. The effect is similar when changing V_{BD} from − 25 to 0 V at a fixed V_{AC}. Therefore we have a high degree of control on the exciton wavelength by using two pairs of actuation legs.

Wavelength-tunable entangled photons from QDs. Two-dimensional scanning on the two pairs of legs by sweeping both V_{BD} and V_{AC} is then performed. In Fig. 2d we show the results in a three-dimensional plot. The astonishing result is that, with this four-legged device providing orthogonal uniaxial stresses, multiple zero FSS points with different exciton wavelength λ_X (energy E_X) can be achieved. At different V_{BD}, the electronic symmetry of quantum dot can be always recovered by sweeping V_{AC} and the FSS is erased. The dashed line on the bottom plane of the plot indicates the combinations of (V_{AC} and V_{BD}) at which the FSS reaches its minimum. A linear relationship is found for the ratio of voltage changes $\Delta V_{AC}/\Delta V_{BD}$. In terms of the applied stresses (X, Y), indeed, an effective two-level model (Supplementary Note 1) for the FSS of QDs with exciton polarization closely aligned to principal stress axes predicts a zero FSS with a linear relationship $\Delta X/\Delta Y$ and confirms this experimental finding.

As the independent tunability of exciton wavelength and FSS is a main concern in this work, we plot in Fig. 3a the FSS versus exciton wavelength for different V_{BD}. For clarity, we show only the FSS range from 0 to 5 μeV. It shows exactly a linear behaviour (represented by the solid line fits) in agreement with a **k.p** analysis, Supplementary Note 1. For self-assembled InAs/GaAs QDs the relationship between the FSS and the attainable entanglement fidelity f^+ has been well documented and it is commonly accepted that the entanglement persists for a FSS even up to 3–4 μeV (refs 15,16,31). As shown in these reports, tuning the FSS to <1 μeV yields a high fidelity f^+ of >0.7. With this novel device, it is clear that the FSS can be 'locked' strictly below 1 μeV, that is, high-fidelity entangled photons can be generated from the QDs, for a large range of exciton emissions. Considering the typical lifetime (500 ps–1 ns) of our QDs and therefore a radiative limited linewidth of ∼1 μeV, the tuning range of 3.7 meV (2.3 nm) shown in Fig. 3a corresponds to >3,000 times of the radiative linewidth. This tunability is >1 order of magnitude larger than what has been demonstrated for SPDC sources[32].

We have performed the polarization cross-correlation spectroscopy[11,14,15,17] on a brighter QD-embedded inside another device on the same chip. The FSS is tuned to around zero (0.21 ± 0.20 μeV) to demonstrate the polarization entanglement, and the data are presented in Fig. 3b. A key criterion for entanglement is the presence of a correlation independent of the chosen polarization basis, that is, $|\Psi^+\rangle = 1/\sqrt{2}(|H_{XX}H_X\rangle + |V_{XX}V_X\rangle) = 1/\sqrt{2}(|D_{XX}D_X\rangle + |A_{XX}A_X\rangle) = 1/\sqrt{2}(|R_{XX}L_X\rangle + |L_{XX}R_X\rangle)$, with

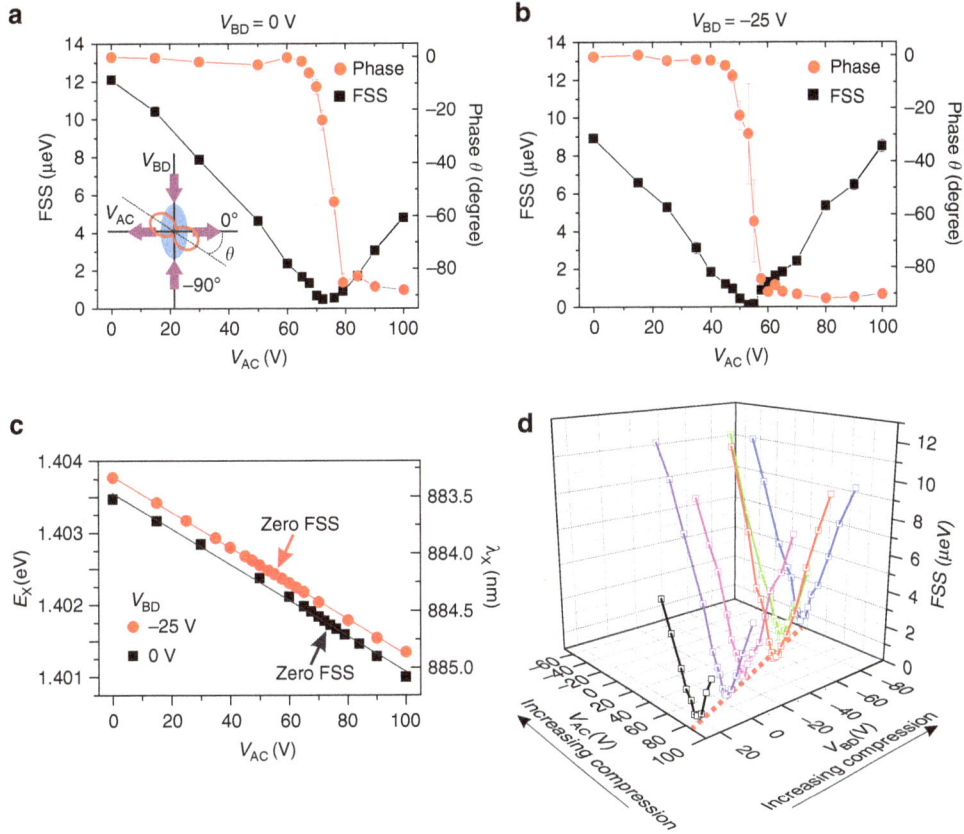

Figure 2 | Anisotropic strain engineering of a QD under orthogonal uniaxial stresses. (a,b) FSS and phase θ are plotted as a function of the voltage V_{AC} at a fixed voltage V_{BD} of 0 and -25 V, respectively. With the increasing voltages on legs A&C, FSS decreases monotonically to around zero and then increases again. And the phase shows an abrupt change from 0 to $-90°$, when FSS reaches the minimum value. The inset in **a** gives the definition of θ. The ellipse indicates an elongated QD with its major axis aligned along a crystallographic direction. The red solid line indicates the exciton polarization. The error bars originate from the fitting process. **(c)** Exciton wavelength is plotted as a function of V_{AC} for the two different V_{BD}. The solid lines are linear fits. The arrows indicate exciton wavelengths at which the FSS are erased. **(d)** The changes in FSS when both V_{BD} and V_{AC} are scanned. The dashed line on the bottom plane indicates a linear shift of the voltage combination (V_{AC}, V_{BD}) at which FSS reaches the minimum values.

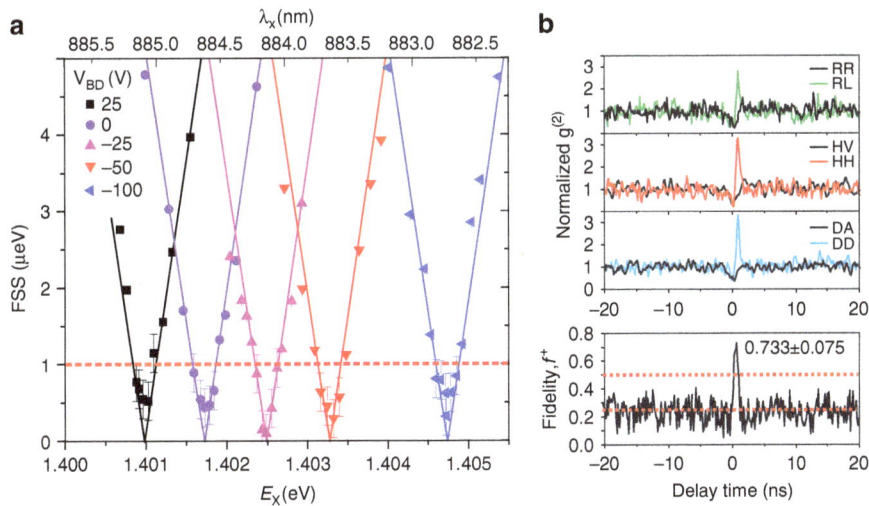

Figure 3 | Independent tunability of exciton wavelength and FSS. (a) FSS is plotted as a function of the exciton wavelength λ_X (energy E_X), at different values of V_{BD}. The solid lines are theoretical fits. In the **k·p** theory we consider the effect of a pair of orthogonal uniaxial stresses applied to an aligned QD. Exciton energy at which FSS ~ 0 is tuned by 3.7 meV. The dashed line is a threshold of 1 μeV for the entangled photons generations. For FSS of below 1 μeV, the error bars of ± 0.25 μeV are indicated (Methods section). **(b)** Polarization correlation spectroscopy, Methods section, is performed on the biexciton and exciton photons, when the QD FSS is tuned to zero. The normalized coincident counts are given for both co-polarized and cross-polarized photons. We have measured a fidelity f^+ of 0.733 ± 0.075 without any background subtraction. The two dashed lines indicate the threshold of 0.5 for the classically correlated light, and the threshold of 0.25 for the uncorrelated light.

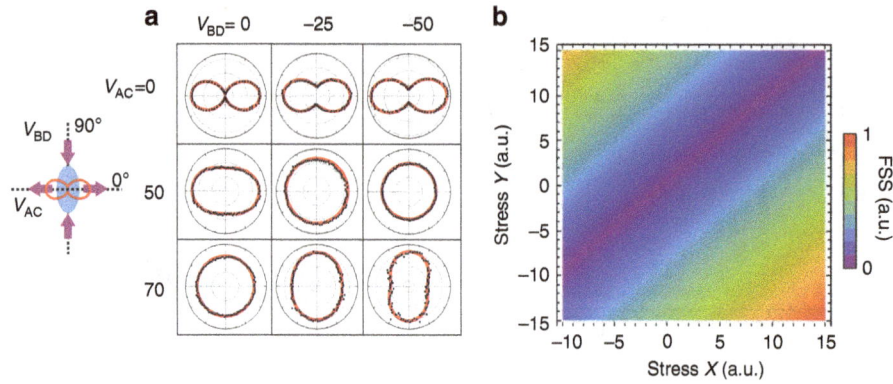

Figure 4 | Semi-quantitative description of the independent tunability of exciton wavelength and FSS. (**a**) Matrix of polarization plots at different voltage combinations. Here the exciton energy is plotted as a function of θ, with the same definition being used in the inset of Fig. 2a. All polar plots have an axis scale of 15 μeV, and the solid red lines represent a fit to the data with a sinusoidal function. The 'circularity' and the direction of the pedals indicate the relative amplitude of FSS and θ, respectively. Circular polarized exciton emission with zero FSS can be observed at three different voltage combinations, that is, at different exciton emission wavelength. No appreciable polarization rotation can be observed at the applied voltages. The schematic shows the stress condition for this type of QDs. (**b**) A density plot of the FSS as a function of the two stress magnitudes X,Y as obtained with the two-level model Hamiltonian. FSS is kept at zero for a range of stress combinations (X,Y), therefore the wavelength at which FSS is zero can be tuned at will.

D, A, R, L denoting the diagonal, anti-diagonal, right-hand circular and left-hand circular polarizations. Clear photon bunching, with a normalized second-order correlation function $g^{(2)}(\tau=0)>3$, can be observed for the co-polarized HH and DD photons, whereas in the circular basis the bunching occurs for the cross-polarized RL photons. The entanglement fidelity f^{+} to the maximally entangled Bell state can be determined from the measurements in Fig. 3b, see Methods section. The peak near the zero time delay yields a fidelity f^{+} of 0.733 ± 0.075 without any background subtraction, which exceeds the threshold of 0.5 for a classically correlated state by >3 s.d. The above results are in line with previous experimental and theoretical works, and verify that highly entangled photons can be generated with our device with large wavelength tunability.

An intuitive understanding of the tuning behaviour can be obtained immediately from a matrix of exciton polarization plots at different voltage combinations, Fig. 4a. The 'circularity' of the polarization pedals indicates the relative magnitude of FSS. By sweeping V_{AC} from 0 to 70 V at V_{BD} of 0 V, an increasing tensile stress is applied along legs A&C and the symmetry is gradually recovered. We observe the 'opening' of the polarization pedal without any appreciable rotation[19,20,31], as predicted by theory[29]. At the voltage combination $(V_{AC}, V_{BD})=(70, 0)$ V, the exciton emissions become circular polarized with a near-zero FSS and polarization-entangled photons are generated. At V_{BD} of -25 V and -50 V, the symmetry is already partially recovered due to the increased compressive stress along the legs B&D, and therefore the symmetry can be fully recovered with less tensile stress, that is, a smaller V_{AC}, along the legs A&C. Due to the symmetry of orthogonal uniaxial stresses, we can observe the same effect when sweeping V_{AC} at fixed voltages of V_{BD}.

Our experimental findings can also be semi-quantitatively understood using the recent theoretical observation, Supplementary Note 1 (ref. 26). The application of two orthogonal stresses of magnitude X and Y can be used as effective knobs to erase the FSS and tune the exciton emission once the stress axis are aligned with the initial exciton polarization. In this case indeed, the FSS scales linearly with the stress anisotropy $\Delta=X-Y$ and eventually vanishes at a critical stress anisotropy $\Delta=\Delta c$. This behaviour is demonstrated in Fig. 4b where we show a density plot of the FSS as a function of the two stress magnitudes X,Y as obtained with the two-level

model Hamiltonian, Supplementary Note 1 (refs 26,29). Furthermore, the remaining independent hydrostatic part $\propto X+Y$ can be used to change at will the exciton wavelength λ_X (energy E_X). This implies a linear interdependence between the exciton emission and the FSS, which is in perfect agreement with the data presented in Fig. 2.

We emphasize that the device presented here holds strong promise for the realization of solid-state scalable QIP platform. First, the wavelength tunability allows the Bell state measurement between two QDs and therefore the swapping of entanglement[24,33]. Also with suitable QDs (see for example Fig. 1d), it is straightforward to couple these sources with atomic vapours[34]. Second, the integrated MEMS on silicon have small footprints and low operation voltages, which solves two of the most challenging problems of the strain engineering technique[15,25]. We can foresee the integration of this device with other photonic structures (for example, with circular ring grating cavities to enhance the photon collection efficiency[35]) and, most intriguingly, with advanced silicon quantum photonic circuits. The envisioned hybrid can exploit the maturity of CMOS technology, and as well as the waveguiding, processing and detection capabilities associated with silicon photonics.

In summary, we have experimentally realized wavelength-tunable entangled photon sources on a III–V/Si chip, which represents an important step towards scalable entangled photon sources based on III–V QDs. The reported device will play an important role not only in building a solid-state quantum network based on entanglement swapping and quantum memories, but also in building advanced quantum photonic circuits for on-chip QIP applications. The MEMS based device features the advantages of sophisticated anisotropic stress control on chip. We envision that it will inspire many other topics in quantum and nano-technologies, and an interesting perspective is to replace the nanomembranes with the emerging two-dimensional materials and to study the strain-dependent photonic and electronic properties.

Methods

Sample growth. The studied sample was grown on a (001) GaAs substrate by solid-source molecular beam epitaxy. Several samples were used in this work, but their general structures are the same. After the deoxidization and GaAs buffer growth, a thin layer of $Al_{0.75}Ga_{0.25}As$ was grown as the sacrificial layer for the nanomembrane release. The InAs QDs were grown by partial capping and annealing, and embedded in the middle of a GaAs layer with a thickness of a few hundred nanometres. The emission range of the QDs is between 880 and 900 nm and we can easily identify low-density regions with $<1\, QD\, \mu m^{-2}$.

Device fabrication. As for the device processing, we start with the QD-embedded nanomembranes. Standard UV photolithography and wet-chemical etching were used to fabricate mesa structures with a size of $80 \times 80\, \mu m^2$. The edge of the nanomembranes was processed along [110] or [1-10] crystal axis of GaAs. The PMN-PT film is $\sim 15\, \mu$ thick. The backside is coated with gold as the bottom/ground contact. And the PMN-PT film is bonded to a silicon substrate via glue (Ibule photonics). We use focused ion beam to define the trenches in the film, and the depth of the trenches is deep enough to penetrate into the silicon substrate. And then the top contact is deposited by E-beam sputtering. After this, the device is undercut by wetting chemical etching to form the suspended PMN-PT thin-film legs.

Then we bond the QD-embedded nanomembranes to the processed PMN-PT/Si substrate using a flip-chip bonder. The edge of nanomembrane is carefully aligned along the strain axis of the PMN-PT actuation legs. To constrain the displacements of the membrane, the same voltage is applied to opposite legs.

Optical measurement. For optical measurements, the device is loaded into a cryostat chamber which is cooled to $\sim 5\, K$. The PL emitted from the QDs is collected by a $\times 100$ microscope objective with numerical aperture of 0.52. The signal is then dispersed by a spectrometer with 1,800 grating and detected by a liquid nitrogen cooled CCD. By inserting a half-wave plate and a linear polarizer directly after the collection lens, polarization-resolved measurements were performed to estimate the FSS. The exciton polarization is determined by aligning the fast optical axis of the polarizer along [1-10] direction of the nano-membrane. With the experimental procedure used in previous works[13,26], we can determine the FSS with an accuracy of sub-μeV. For FSS of below 1 μeV, an error bar of $\pm 0.25\, \mu eV$ is extracted from the 95% prediction band of the sine fit function. As for the polarization correlation spectroscopy, a non-polarizing 50:50 beam splitter is placed directly after the objective to divide the optical paths between two spectrometers, which are then used to detect X and XX separately. After each spectrometer, a Hanbury Brown-Twiss set-up, consisting of a polarizing beam splitter and two high-efficiency single-photon avalanche detectors, is placed. Half- and quarter-wave were used to select the proper polarization basis. The temporal resolution of the system is $\sim 450\, ps$. The entanglement can be quantified by measuring degree of correlation C, which s defined by

$$C_{basis} = \left(g^{(2)}_{XX,X}(\tau) - g^{(2)}_{XX,\bar{X}}(\tau) \right) / \left(g^{(2)}_{XX,X}(\tau) + g^{(2)}_{XX,\bar{X}}(\tau) \right), \qquad (1)$$

where $g^{(2)}_{XX,X}(\tau)$ and $g^{(2)}_{XX,\bar{X}}(\tau)$ are normalized second-order time correlations for co-polarized and cross-polarized XX and X photons, respectively. The fidelity f^+ is calculated by using the formula: $f^+ = (1 + C_{HV} + C_{DA} - C_{RL})/4$, in which C_{HV}, C_{DA} and C_{RL} are degree of correlations in HV, DA and RL bases.

References

1. Pan, J. W. *et al.* Multiphoton entanglement and interferometry. *Rev. Mod. Phys.* **84**, 777–838 (2012).
2. Kwiat, P. G., Waks, E., White, A. G., Appelbaum, I. & Eberhard, P. H. Ultrabright source of polarization-entangled photons. *Phys. Rev. A* **60**, R773–R776 (1999).
3. O'Brien, J. L., Furusawa, A. & Vuckovic, J. Photonic quantum technologies. *Nat. Photon.* **3**, 687–695 (2009).
4. Politi, A., Cryan, M. J., Rarity, J. G., Yu, S. & O'Brien, J. L. Silica-on-silicon waveguide quantum circuits. *Science* **320**, 646–649 (2008).
5. Silverstone, J. W. *et al.* On-chip quantum interference between silicon photon-pair sources. *Nat. Photon.* **8**, 104–108 (2014).
6. Najafi, F. *et al.* On-chip detection of non-classical light by scalable integration of single-photon detectors. *Nat. Commun.* **6**, 5873 (2015).
7. Benson, O., Santori, C., Pelton, M. & Yamamoto, Y. Regulated and entangled photons from a single quantum dot. *Phys. Rev. Lett.* **84**, 2513–2516 (2000).
8. Bayer, M. *et al.* Fine structure of neutral and charged excitons in self-assembled In(Ga)As/(Al)GaAs quantum dots. *Phys. Rev. B* **65**, 195315 (2002).
9. Ellis, D. J. P. *et al.* Control of fine-structure splitting of individual InAs quantum dots by rapid thermal annealing. *Appl. Phys. Lett.* **90**, 011907 (2007).
10. Muller, A., Fang, W., Lawall, J. & Solomon, G. S. Creating polarization-entangled photon pairs from a semiconductor quantum dot using the optical stark effect. *Phys. Rev. Lett.* **103**, 217402 (2009).
11. Stevenson, R. M. *et al.* A semiconductor source of triggered entangled photon pairs. *Nature* **439**, 179–182 (2006).
12. Pooley, M. A. *et al.* Energy-tunable quantum dot with minimal fine structure created by using simultaneous electric and magnetic fields. *Phys. Rev. Appl.* **1**, 024002 (2014).
13. Bennett, A. J. *et al.* Electric-field-induced coherent coupling of the exciton states in a single quantum dot. *Nat. Phys.* **6**, 947–950 (2010).
14. Ghali, M., Ohtani, K., Ohno, Y. & Ohno, H. Generation and control of polarization-entangled photons from GaAs island quantum dots by an electric field. *Nat. Commun.* **3**, 661 (2012).
15. Trotta, R., Wildmann, J. S., Zallo, E., Schmidt, O. G. & Rastelli, A. Highly entangled photons from hybrid piezoelectric-semiconductor quantum dot devices. *Nano Lett.* **14**, 3439–3444 (2014).
16. Zhang, J. *et al.* Strain-tunable entangled-light-emitting diodes with high yield and fast operation speed. *Nat. Commun.* **6**, 10067 (2015).
17. Akopian, N. *et al.* Entangled photon pairs from semiconductor quantum dots. *Phys. Rev. Lett.* **96**, 130501 (2006).
18. Dousse, A. *et al.* Ultrabright source of entangled photon pairs. *Nature* **466**, 217–220 (2010).
19. Muller, M., Bounouar, S., Jons, K. D., Glassl, M. & Michler, P. On-demand generation of indistinguishable polarization-entangled photon pairs. *Nat. Photon.* **8**, 224–228 (2014).
20. Salter, C. L. *et al.* An entangled-light-emitting diode. *Nature* **465**, 594–597 (2010).
21. Lu, C.-Y. & Pan, J.-W. Quantum optics: push-button photon entanglement. *Nat. Photon.* **8**, 174–176 (2014).
22. Harris, N. C. *et al.* Integrated source of spectrally filtered correlated photons for large-scale quantum photonic systems. *Phys. Rev. X* **4**, 041047 (2014).
23. Wang, J., Gong, M., Guo, G. C. & He, L. Towards scalable entangled photon sources with self-assembled InAs/GaAs quantum dots. *Phys. Rev. Lett.* **115**, 067401 (2015).
24. Trotta, R., Martín-Sánchez, J., Daruka, I., Ortix, C. & Rastelli, A. Energy-tunable sources of entangled photons: a viable concept for solid-state-based quantum relays. *Phys. Rev. Lett.* **114**, 150502 (2015).
25. Ding, F. *et al.* Tuning the exciton binding energies in single self-assembled InGaAs/GaAs quantum dots by piezoelectric-induced biaxial stress. *Phys. Rev. Lett.* **104**, 067405 (2010).
26. Trotta, R. *et al.* Universal recovery of the energy-level degeneracy of bright excitons in InGaAs quantum dots without a structure symmetry. *Phys. Rev. Lett.* **109**, 147401 (2012).
27. Plumhof, J. D. *et al.* Strain-induced anticrossing of bright exciton levels in single self-assembled GaAs/Al_{x}Ga_{1-x}As and In_{x}Ga_{1-x}As/GaAs quantum dots. *Phys. Rev. B* **83**, 121302 (2011).
28. Akopian, N., Wang, L., Rastelli, A., Schmidt, O. G. & Zwiller, V. Hybrid semiconductor-atomic interface: slowing down single photons from a quantum dot. *Nat. Photon* **5**, 230–233 (2011).
29. Gong, M., Zhang, W., Guo, G.-C. & He, L. Exciton polarization, fine-structure splitting, and the asymmetry of quantum dots under uniaxial stress. *Phys. Rev. Lett.* **106**, 227401 (2011).
30. Gong, M. *et al.* Statistical properties of exciton fine structure splitting and polarization angles in quantum dot ensembles. *Phys. Rev. B* **89**, 205312 (2014).
31. Hudson, A. J. *et al.* Coherence of an entangled exciton-photon state. *Phys. Rev. Lett.* **99**, 266802 (2007).
32. Fedrizzi, A., Herbst, T., Poppe, A., Jennewein, T. & Zeilinger, A. A wavelength-tunable fiber-coupled source of narrowband entangled photons. *Opt. Express* **15**, 15377–15386 (2007).
33. Troiani, F. Entanglement swapping with energy-polarization-entangled photons from quantum dot cascade decay. *Phys. Rev. B* **90**, 245419 (2014).
34. Duan, L. M., Lukin, M. D., Cirac, J. I. & Zoller, P. Long-distance quantum communication with atomic ensembles and linear optics. *Nature* **414**, 413–418 (2001).
35. Sapienza, L., Davanco, M., Badolato, A. & Srinivasan, K. Nanoscale optical positioning of single quantum dots for bright and pure single-photon emission. *Nat. Commun.* **6**, 7833 (2015).

Acknowledgements

The work was financially supported by the BMBF Q.Com-H (16KIS0106) and the European Union Seventh Framework Programme 209 (FP7/2007-2013) under Grant Agreement No. 601126 210 (HANAS). The authors acknowledge Carmine Ortix for theoretical supports. Part of the data analysis was done by the XRSP3 software developed by Armando Rastelli. We thank Yongheng Huo for providing one of the test QD sample materials. We also thank B. Höfer, C. Jiang, B. Eichler, R. Engelhard, M. Bauer and S. Harazim for discussions and technical supports. F.D. acknowledges Armando Rastelli for the continuous support and his early contributions to this project.

Author contributions

F.D. conceived the experiment and supervised the project together with O.G.S. who directed the research. The devices were designed by Y.C. and F.D. and fabricated by Y.C. with the help from K.J. and Y.Z. Two samples for the project were designed and grown by R.K. Optical experiments were performed by Y.C. and J.Z. The data were analysed by Y.C., F.D. and J.Z. Finite-element modelling simulations were done by

M.Z. and Y.Z. The manuscript was written by Y.C. and F.D. with the inputs from all the authors.

Additional information

Competing financial interests: The authors declare no competing financial interests.

Lasing in silicon–organic hybrid waveguides

Dietmar Korn[1,*], Matthias Lauermann[1,*], Sebastian Koeber[1,2], Patrick Appel[1], Luca Alloatti[1], Robert Palmer[1], Pieter Dumon[3], Wolfgang Freude[1,2], Juerg Leuthold[1,2,†] & Christian Koos[1,2]

Silicon photonics enables large-scale photonic–electronic integration by leveraging highly developed fabrication processes from the microelectronics industry. However, while a rich portfolio of devices has already been demonstrated on the silicon platform, on-chip light sources still remain a key challenge since the indirect bandgap of the material inhibits efficient photon emission and thus impedes lasing. Here we demonstrate a class of infrared lasers that can be fabricated on the silicon-on-insulator (SOI) integration platform. The lasers are based on the silicon–organic hybrid (SOH) integration concept and combine nanophotonic SOI waveguides with dye-doped organic cladding materials that provide optical gain. We demonstrate pulsed room-temperature lasing with on-chip peak output powers of up to 1.1 W at a wavelength of 1,310 nm. The SOH approach enables efficient mass-production of silicon photonic light sources emitting in the near infrared and offers the possibility of tuning the emission wavelength over a wide range by proper choice of dye materials and resonator geometry.

[1] Institute of Photonics and Quantum Electronics (IPQ), Karlsruhe Institute of Technology (KIT), 76131 Karlsruhe, Germany. [2] Institute of Microstructure Technology (IMT), Karlsruhe Institute of Technology (KIT), 76344 Eggenstein-Leopoldshafen, Germany. [3] Department of Information Technology, IMEC, 9000 Gent, Belgium. * These authors contributed equally to this work. † Present address: Laboratory for Electromagnetic Fields and Microwave Electronics (IFH), Swiss Federal Institute of Technology (ETH), Zürich 8092, Switzerland. Correspondence and requests for materials should be addressed to C.K. (email: christian.koos@kit.edu).

Silicon photonics allows fabrication of nanophotonic devices using commercial CMOS facilities and is therefore a highly attractive platform for large-scale photonic integration[1,2]. However, while a wide variety of silicon-based optical and electro-optical devices has been demonstrated over the last years[3], efficient on-chip light sources still represent a challenge[4] due to the indirect bandgap of silicon. Previously reported all-silicon light sources rely on stimulated Raman scattering as a gain mechanism[5–8]. One drawback of these schemes is that they require coupling of external pump lasers to single-mode on-chip waveguides. In early demonstrations, Raman laser cavities were formed by incorporating nanophotonic silicon waveguides into fibre-based off-chip laser cavities to enable synchronization of the cavity round-trip time with a pulsed pump[5,6]. These devices cannot be miniaturized. In contrast to that, silicon photonic Raman lasers with on-chip cavities are compact and enable continuous-wave (CW) lasing[7,8], but require either strong pump lasers in combination with reverse-biased p-i-n-junctions or provide only limited output power in the microwatt range. Hybrid approaches, in which silicon is combined with direct-bandgap III–V compound semiconductors, allow for electrically pumped amplifiers[9] and lasers[10,11], but fabrication requires sophisticated and technologically challenging die-to-wafer bonding processes or advanced technology for the direct growth of III–V quantum dots[12] on silicon. Regarding monolithic integration of light sources on silicon, an electrically pumped CW germanium-on-silicon laser has been demonstrated by using a combination of tensile strain and n-doping of the germanium to enable direct-bandgap transitions in thin germanium layers that are grown on silicon substrates[13]. More recently, lasing has been shown without introducing mechanical strain by using a germanium–tin alloy[14] on silicon. However, fabrication of such devices requires advanced crystal growth techniques and technologically challenging fabrication processes. As an alternative, combinations of erbium-doped active cladding materials and SOI waveguides have been proposed[15,16] and experimentally investigated[17,18]. However, erbium features a rather small emission cross section and hence small gain. As a consequence, lasing in integrated erbium-clad devices has so far only been demonstrated for low-loss silicon nitride waveguides[19], but not for high-index-contrast SOI waveguides. Regarding

peak output power, even the most outstanding on-chip silicon-based lasers are currently limited to ~100 mW or less[12,20,21].

In this work, we demonstrate that lasing can be achieved by combining standard silicon-on-insulator (SOI) waveguides with dye-doped organic cladding materials. This concept of silicon–organic hybrid (SOH) integration is particularly well-suited for flexible and low-cost mass-production of silicon photonic light sources emitting in the near infrared. In a proof-of-principle experiment, we demonstrate pulsed lasing at room temperature with peak output powers of up to 1.1 W at a wavelength of 1,310 nm. Gain is provided by a near-infrared dye that was previously demonstrated to enable lasing in plastic waveguides[22]. More general, exploiting the virtually unlimited variety of organic optical cladding materials, SOH integration allows to complement silicon photonics with novel functionalities while still preserving the strengths of highly standardized CMOS processing[23]. Our proof-of-principle demonstration of SOH light sources complements recent work on SOH integration, comprising high-speed all-optical signal processing[24], broadband electro-optic modulators[25,26] and highly efficient low-power phase shifters[27].

Results

Concept and fabrication of SOH lasers. The basic idea of an SOH laser is illustrated in Fig. 1a. The devices consist of SOI waveguides, which are terminated at both ends with Bragg reflectors[28] and which are covered by a fluorescent organic cladding material suitable for stimulated emission when optically pumped. For efficient light emission, the interaction of the guided optical mode with the active cladding must be maximized. This can be accomplished by using a narrow silicon strip waveguide, for which a large fraction of the guided mode reaches into the cladding, Fig. 1b. Alternatively, a slot waveguide can be used, which consists of two closely spaced silicon rails, Fig. 1c. In both cases, the dominant horizontal electric field component (E_x) of the optical quasi-TE mode experiences strong field discontinuities at the high-index-contrast sidewalls. For the slot waveguide, this leads to an especially pronounced field enhancement within the slot[29], and hence to a strong interaction with the active cladding.

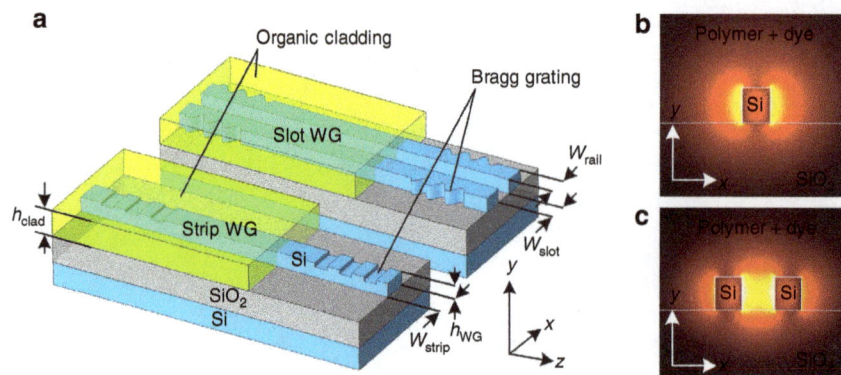

Figure 1 | SOH laser concept. (a) Light is guided by SOI strip or slot waveguides consisting of thin silicon nanowires (width $w_{strip} \approx 150$–500 nm, height $h_{WG} \approx 200$–350 nm) that are optically isolated from the silicon substrate by a thick oxide ($h_{SiO2} \approx 2$ μm). Optical gain is provided by a fluorescent organic cladding material (thickness $h_{clad} \approx 500 \pm 50$ nm), which entirely covers the strip or fills the slot ($w_{rail} \approx 100$–200 nm, $w_{slot} \approx 50$–200 nm). The optical pump is either launched from above or injected into the waveguide at one of the facets. Bragg reflectors can be used to provide wavelength-selective optical feedback. Interaction of the guided light with the active cladding is maximized by the design of the waveguides. **(b)** Dominant electric field component (E_x) of the fundamental quasi-TE mode for a narrow strip waveguide (colour coding: lighter colours for higher magnitude). A large fraction of the guided mode propagates in the cladding. **(c)** Dominant electric field component (E_x) of the fundamental quasi-TE mode for a slot waveguide consisting of two tightly spaced silicon rails. Discontinuities of the dominant horizontal electric field component lead to a strong field enhancement within the slot region and hence to a strong interaction with the active cladding.

We prove the viability of the concept by investigating a simple test structure. To this end, strip and slot waveguides of 4.8-mm length were fabricated using a state-of-the-art SOI CMOS-based process[30]. The waveguides are embedded into a solid active cladding consisting of a poly(methyl methacrylate; PMMA) matrix doped with 1 wt% of the commercially available dye IR26 (refs 22,31) having a maximum fluorescence at 1,150 nm, see Supplementary Fig. 1. The cladding is deposited in a single post-processing step using standard spin-coating techniques. Scanning electron microscope (SEM) images of coated and uncoated samples can be found in Supplementary Fig. 2, showing that the PMMA cladding fills the slot completely without forming any voids. To enable laser operation in a wide wavelength range, we omit the wavelength-selective Bragg reflectors shown in Fig. 1 and exploit spurious back-reflection from cleaved waveguide facets and from on-chip grating coupler (GC) structures, see Fig. 2a,b. For the cleaved facets, power reflection factors between 4 and 8% are estimated. Light emission from the cleaved facets is coupled to lensed standard single-mode fibres (SMF). For coarse alignment of the fibres, we use 1,550-nm light coupled to the SOH waveguide via the GC. The GC is optimized for operation at a wavelength of 1,550 nm and exhibit spurious back-reflection when operated at the laser emission wavelength of 1,310 nm. This reflection amounts to a few per cent and is comparable to that of the cleaved facet. A more detailed description of device fabrication can be found in the section 'Fabrication of SOH lasers' of the Methods.

Lasing could be demonstrated despite the comparatively low quality of the Fabry–Perot laser resonator, underlining the high potential of using dye-doped active claddings as gain media. In the experiment, the devices are pumped from above by a free-space line-focus beam using a pulsed laser with a wavelength of 1,064 nm, a pump pulse duration of 0.9 ns (full width at half maximum, FWHM), and a pulse energy of up to 1.2 mJ at a repetition rate of 13.7 Hz. The duration of the emitted laser pulses amounts to ~ 0.6 ns. Note that this is much longer than the cavity round-trip time of the laser, and pulsed operation is caused solely by the fact that the pump source is switched on only for certain time intervals. The effective lifetime of the excited state amounts to ~ 10 ps and is much shorter than the durations of the pump and the emission pulses. We may hence assume that the lasing process is close to its steady-state. The experimental setup is explained in more detail in the section 'Experimental demonstration of laser emission' of the Methods, which is followed by an estimation of the pump and emission power levels.

Characterization and experimental proof of lasing. We measured the laser output power in the SMF as a function of the pump power for both the strip and the slot waveguide, Fig. 2c,d. In both cases, a clear threshold can be observed at a launched average pump power of approximately 2.3 mW for the strip, and approximately 1.3 mW for the slot waveguide. The absorbed peak power at threshold in the vicinity of the waveguide can be roughly estimated to be 38 W for the strip, and 24 W for the slot waveguide, taking into account specific parameters of the individual waveguides, see the sections 'Experimental demonstration of laser emission' and 'Estimation of emission power levels' of the Methods and Supplementary Note 1 for more details.

The existence of the threshold indicates laser emission. The measured threshold level is in reasonable agreement with theory, see the section 'Consistence of resonator characteristics and threshold pump power' of the Methods for a more detailed discussion. To rule out any laser look-alikes, we investigate

further criteria formulated by Samuel et al.[32] Below threshold, only amplified spontaneous emission is to be seen, which increases exponentially with the pump power, see insets of Fig. 2c,d. Above threshold, the output power increases linearly with the pump power. For very high pump powers, the laser power saturates. The saturation is attributed to absorption bleaching at the pump wavelength and to pump-induced free-carrier absorption (FCA) in the SOI waveguide, see the section 'Optically induced losses and dynamical behaviour' of the Methods for a more detailed discussion.

Moreover, we investigate the emission spectra from the strip and slot waveguides below and above threshold, see Fig. 2e,f. Broadband amplified spontaneous emission can be observed for operation below threshold, see insets of Fig. 2e,f (logarithmic scale). When pumped above threshold, the emission spectrum narrows considerably. In Fig. 2e and f, the observed linewidth appears slightly larger than the resolution bandwidth of the spectrometer (RBW = 5 nm). We attribute this to a multitude of different longitudinal cavity modes which oscillate simultaneously at every pump pulse, see Supplementary Note 2 and Supplementary Fig. 3 for a more detailed description.

Above threshold, the optical output of slot waveguides and of narrow strip waveguides is laterally single-mode, which can be inferred from the observation that there is a single well-defined optimum spot when coupling to a lensed SMF. For strip waveguides, lasing in higher-order lateral modes can be observed for waveguide widths of ~ 300 nm or more as reported in more detail in the next section 'Influence of waveguide geometry'. For the devices shown in Fig. 2, the emitted light is predominantly polarized in the horizontal direction as is expected for lasing of the quasi-TE mode. The polarization extinction ratio (ER) is about 8 dB for both devices in Fig. 2. To confirm that the dye is indeed responsible for lasing, we prepared reference samples without dye in the PMMA cladding. These samples do not show noticeable light emission. Moreover, without the silicon waveguide but with dye in the cladding, only spontaneous emission is observed. These findings exclude any laser look-alikes and confirm the working principle of the SOH laser concept.

Influence of waveguide geometry. Regarding the influence of waveguide geometry on the performance of the SOH lasers, we find that lasing with high output powers can be achieved with a wide range of waveguide dimensions and that the output power is clearly related to the overlap of the guided mode with the active cladding. The geometry-dependent output power levels of different waveguide geometries are shown in Fig. 3. For the strip waveguide, we vary the width, Fig. 3a, whereas for the slot waveguide, the rail width is fixed to 170 nm and the slot width is varied, Fig. 3b. The length of the active section amounts to 4 mm for all devices. As before, the resonator is formed by back-reflection from a cleaved waveguide facet and from a GC operated far from its design wavelength of 1,550 nm. The experimental setup is the same as before and described in the section 'Experimental demonstration of laser emission' of the Methods. The average pump power is fixed to 5 mW. The coloured areas of each bar in Fig. 3 represent the respective contributions of quasi-TE (blue) and quasi-TM polarization (green) to the total output power. For the strip waveguide, the laser power is largest when the strip width is smallest, that is, when the mode fields extend far into the active cladding. The second maximum at $w_{strip} = 375$ nm is due to lasing not only of the fundamental mode, but also of the next higher-order quasi-TE_{10} mode, which also strongly interacts with the cladding. The polarization ER reaches a maximum of $(18 \pm 2$ dB) for the narrowest strip waveguides we investigated.

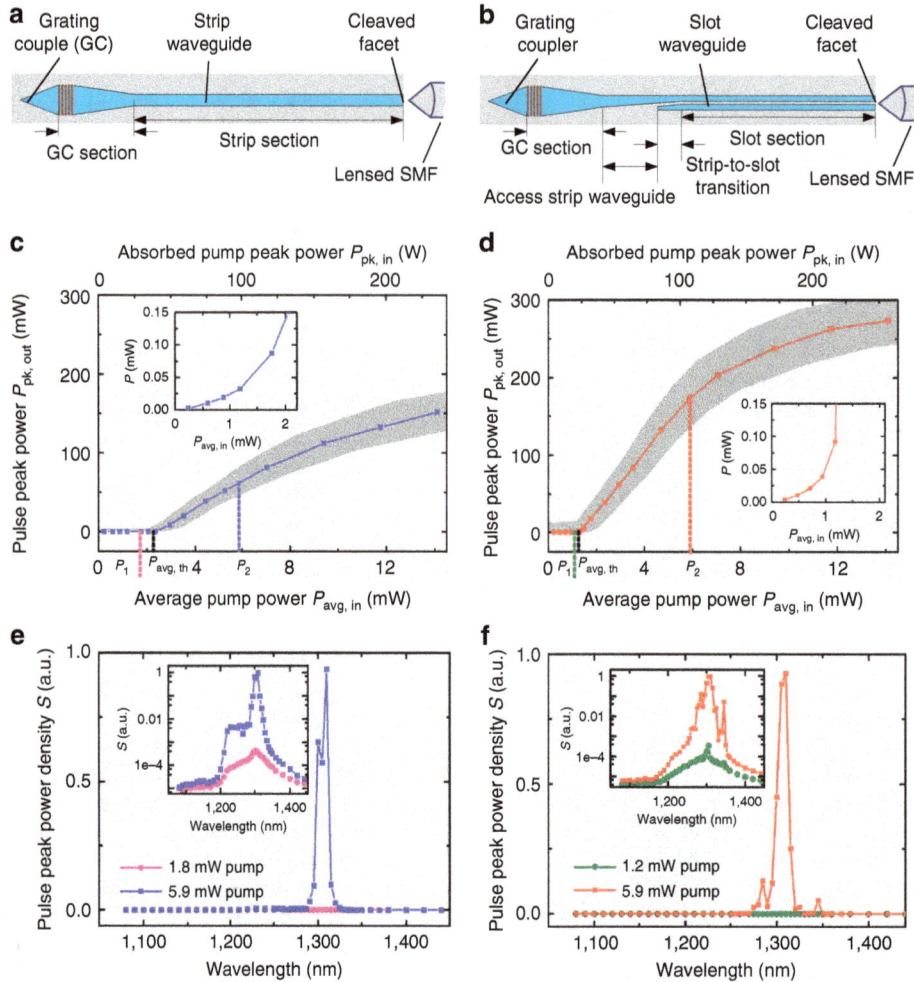

Figure 2 | Experimental proof of lasing in SOH strip and slot waveguides. The cladding consists of the commercially available dye IR26 (ref. 22) dispersed in a PMMA matrix. Cavity mirrors are formed by one cleaved waveguide facet and a GC. The GC is designed for coupling 1,550 nm light from an optical fibre to the strip and exhibits substantial back-reflection at the laser emission wavelength of 1,310 nm. For both the strip and the slot waveguide, the cavities are ~4.8 mm long. The laser output power is measured in a lensed SMF that collects light from the waveguide facet. (**a**) Strip waveguide consisting of a 450 μm long GC section and a 4.3 mm long strip section (waveguide height $h_{WG} \approx 220$ nm, width $w_{strip} \approx 210$ nm). (**b**) Slot waveguide comprising a 450 μm long GC section, a 235 μm long access strip waveguide, a 300 μm long strip-to-slot transition, and a 3.8 mm long slot waveguide section (rail width $w_{rail} \approx 180$ nm, slot width $w_{slot} \approx 215$ nm). (**c**) Peak output power $P_{pk, out}$ (all polarizations) in lensed SMF versus illuminating average pump power $P_{avg, in}$ for the strip-waveguide cavity. A clear pump power threshold of $P_{avg, th} = 2.3$ mW can be observed. The measured incident average pump power (bottom scale) $P_{avg, in}$ is used to calculate the absorbed pulse peak power (top scale) $P_{pk,in}$, taking into account the specific parameters of the waveguide, see the section 'Estimation of pump power levels' of the Methods and Supplementary Note 1. The grey-shaded area indicates an estimate of the accuracy of the measurement. For the uncertainty of the pump power, we use a relative standard error of ±14%, see the section 'Estimation of emission power levels' of the Methods for a more detailed explanation. Regarding the uncertainty of the emitted power, we estimate a relative standard error of ±10% for all pump powers of $P_{p1} = 5$ mW or more. Below pump powers of $P_{p1} = 5$ mW, we assume a constant absolute error which corresponds to the ±10% relative standard error at $P_{p1} = 5$ mW. Note that the grey ranges correspond to a coarse, but conservative estimate of the measurement uncertainties. P_1 and P_2 denote the pump powers for which the spectra in **e** are recorded. (**d**) Peak output power $P_{pk, out}$ in SMF versus incident average pump power $P_{avg, in}$ for the slot-waveguide cavity. The grey-shaded areas indicate again an estimate of the accuracy of the measurements, see the section 'Estimation of emission power levels' of the Methods. A threshold pump power $P_{avg, th} = 1.3$ mW is found, which corresponds to 60% of the threshold for the strip waveguide. P_1 and P_2 denote the pump powers for which the spectra in **f** are recorded. (**e**) Emission spectra below (ASE, magenta) and above threshold (blue) for the strip-waveguide cavity. The spectrum is given in arbitrary units of the spectral pulse peak power density S recorded with a resolution bandwidth of 5 nm (inset with logarithmic scale). Above threshold, the emission spectrum narrows considerably. (**f**) Emission spectra below (ASE, green) and above threshold (red) for the slot-waveguide cavity (inset with logarithmic scale). Also here, the emission spectrum narrows considerably above threshold. For all spectra, the resolution bandwidth amounts to 5 nm to allow for detection of weak ASE and strong laser emission with the same measurement system. High-resolution spectra above lasing threshold have been taken at smaller RBW of 0.2 and 0.05 nm, see Supplementary Note 2 and Supplementary Fig. 3.

In Fig. 3b, we consider slot waveguides and vary the slot width while keeping the rails widths at a constant value of 170 nm. The TE mode dominates laser emission, since interaction with the cladding is enhanced by the electric field discontinuities at the high-index-contrast sidewalls of the slot, as can be seen by comparing the field interaction factors[33] $\Gamma_{clad, TE}$ and $\Gamma_{clad, TM}$ for the two polarizations, see Supplementary Table 1 and Supplementary Note 1. Moreover, the emitted laser power increases with slot width. This is to be expected since larger slot widths lead to both larger field interaction factors of the guided mode with the active cladding and to a larger volume in which dye molecules can interact with the guided mode. For

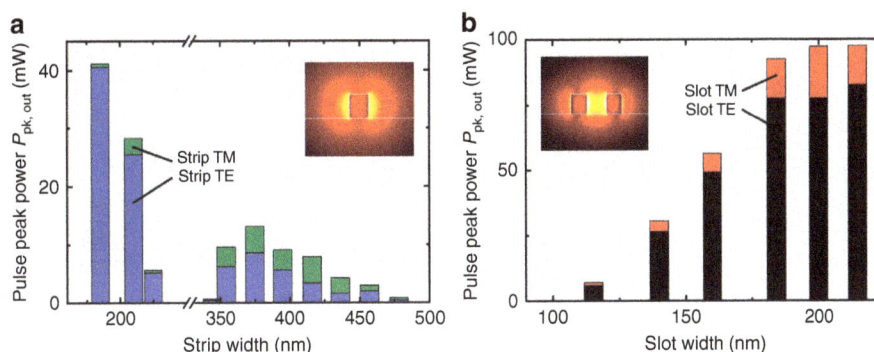

Figure 3 | Geometry-dependent peak output power $P_{pk, out}$ coupled into a lensed SMF for strip and slot waveguides. The resonator relies on back-reflection from one cleaved waveguide facet and from a GC operated far from its design wavelength of 1,550 nm. Quasi-TE and quasi-TM polarizations are measured separately. The average incident pump power is 5 mW for all samples. In the bar diagram, the differently coloured areas represent the contributions of the quasi-TE and the quasi-TM polarization to the total output power; the total bar height corresponds to the total emission. Insets: Dominant electric field magnitudes of the fundamental quasi-TE modes. (**a**) Strip-waveguide cavity. The laser power is largest when the strip width is smallest such that the guided light extends far into the cladding. The secondary maximum at 375 nm is due to lasing of the next higher-order mode (quasi-TE_{10}), which also has a strong overlap with the active cladding, but is not guided for smaller strip widths. (**b**) Slot-waveguide cavity. An increase of the slot width leads to an increase of the field confinement in the cladding and to an expansion of the region in which the active dye interacts with the optical mode. As a consequence, the lasing power increases with slot width. For large slot widths, the fundamental mode is only weakly guided, and the laser power does not increase further. The rail width has only a minor influence (not shown) and is fixed at 170 nm.

very large slot widths, the slot mode is only weakly guided and leaks into the high-index silicon substrate. As a consequence, the output power does not increase further. The polarization ER remains nearly constant and reaches a maximum of 8 ± 2 dB for a slot width of 140 nm. For wider and narrower slots, the ER is slightly smaller.

The optimum choice of the waveguide geometry depends on the desired balance between output power and polarization ER: High power output and a moderate ER when using slot waveguides have to be compared with about half the output power and a high polarization ER obtained from narrow strip waveguides. Using state-of-the-art CMOS fabrication, waveguide dimensions can be reproduced with tolerances of significantly less than 10 nm, which does not influence output power or polarization ER of the SOH lasers to a significant degree. SOH device performance can hence be expected to be resilient against fabrication inaccuracies.

Dynamic emission behaviour. The achievable peak output power of the SOH lasers is remarkable: For an SOH slot waveguide with cleaved facets on both sides, we measured peak output powers of up to 365 mW in the attached SMF, see Fig. 4a,b. The fibre–chip coupling losses are estimated to be (5 ± 1) dB, which leads to peak powers of (30.3 ± 1.0) dBm at the output facet, that is, 1.1 W that could be coupled to an on-chip nanophotonic SOI waveguide. This is the highest peak power emitted from a silicon-based laser with on-chip cavity so far. A more detailed discussion can be found in the section 'Estimation of emission power levels' of the Methods.

The time-dependent emission of the slot waveguide laser is depicted in Fig. 4c for both polarizations, recorded at an average pump power of 5 mW. We observe laser emission into both the quasi-TE and quasi-TM mode, which we attribute to local gain depletion: for large slot widths, the TE and TM modes occupy different cross-sectional domains of the active cladding, see insets in Fig. 4b, and lasing may therefore occur simultaneously in both polarizations. Since the overlap of the quasi-TE slot mode with the active cladding is larger than that of the TM mode, the TE mode experiences higher gain and hence dominates lasing with a polarization ER of 9 dB. The TE and TM emission spectra are

similar—see Supplementary Fig. 3 and Supplementary Note 2 for a more detailed discussion.

Regarding the pulse shapes, we find that the mean FWHM duration of emission amounts to 0.6 ns, which is shorter than the pump pulse FWHM of 0.9 ns. Moreover, the emission pulse features an asymmetric shape and is delayed with respect to the pump pulse. The delay is attributed to the fact that laser emission can only set in once the pump intensity exceeds the threshold level. Note that the relative timing of pump pulse and emission pulses is subject to uncertainties of approximately ± 100 ps due to different propagation delays in the fibre-based measurement setup, see the section 'Optically induced losses and dynamical behaviour' of the Methods for more details. The instantaneous pump power at the onset of laser emission can therefore not be directly associated with the threshold pump power level identified in Fig. 4b. The asymmetric shape of the emission pulse might be caused by nonlinear absorption and subsequent relaxation processes in the active cladding. This aspect requires further investigation.

Discussion

SOH lasers have the potential to cover a broad range of different emission wavelengths between 1.1 and 1.6 μm by using suitable dye materials[34,35]. Due to the high output power, the devices may even be used for exploiting nonlinear optic effects in nanophotonic waveguides. The SOH lasers are remarkably robust: during our experiments, we did not observe significant degradation of the devices, even though they were tested repeatedly over several weeks without taking any specific efforts with respect to encapsulation. This first indication of high stability of the SOH lasers is in good agreement with previous observations, which have shown that photo bleaching of IR26 can be neglected at our pump wavelength of 1,064 nm (ref. 22). A detailed investigation of the stability of SOH lasers is subject of further research.

The devices presented in this paper are first-generation prototypes with considerable room for improvement. In particular, lasing threshold and linewidth of optical emission can be reduced by using optimized Bragg reflectors or ring resonators for optical feedback. Moreover, according to our

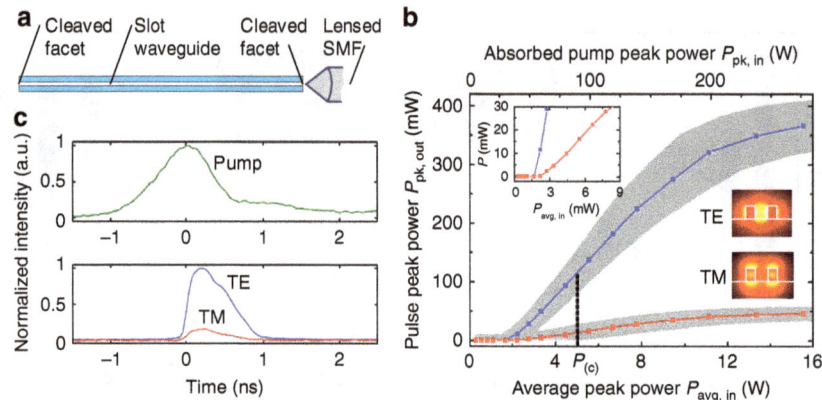

Figure 4 | Lasing in a SOH slot waveguide. In this experiment, cavity mirrors are formed by cleaved waveguide facets on both ends. The cavity length is 3.8 mm, the waveguide height amounts to 220 nm, and for the rail and the slot width, values of $w_{rail} = (160 \pm 15)$ nm and $w_{slot} = (180 \pm 15)$ nm were extracted from scanning electron microscope (SEM) images. (**a**) Schematic top view of the slot waveguide. (**b**) Peak output power in the lensed SMF for quasi-TE and quasi-TM mode versus incident average pump power. The absorbed pump peak power is estimated from the measured incident average pump power, see the section 'Estimation of pump power levels' of the Methods. The grey-shaded areas indicate an estimate of the accuracy of the measurement. For the uncertainty of the pump power, we assume a ±14% relative standard error. Regarding the uncertainty of the emitted power, we estimate a relative standard error of ±10% for all pump powers of $P_{p1} = 5$ mW or more. Below pump powers of $P_{p1} = 5$ mW, we assume a constant absolute error which corresponds to the ±10% relative standard error at $P_{p1} = 5$ mW. More details can be found in the section 'Estimation of emission power levels' of the Methods. Note that the grey ranges correspond to a coarse, but conservative estimate of the measurement uncertainties. Inset: zoom-in of pulse peak power at low pump powers, demonstrating sharp thresholds for both TE and TM mode. (**c**) Temporal shape of the pump pulse at an average power of 5 mW (green) and of the corresponding emission pulses (TE, blue; TM, red). The shape of the pump pulse was measured by averaging over 16 pulses and normalizing to a peak value of 1. Likewise, the emission pulses were measured in the SMF and averaged over 16 pulses. In the plot, the peak of the TE emission has been normalized to 1, and the TM emission is plotted at the same scale. The exact delay between pump and emission cannot be exactly determined due to modal and chromatic dispersion in the standard SMF. The peak pump power was determined with a relative standard error of ±14%; for the peak power of the emitted pulse the relative standard error is ±10%, see the sections 'Estimation of emission power levels' and 'Estimation of pump power levels' of the Methods for a more detailed discussion.

study of the laser dynamics, we expect that better efficiency and lower threshold can be achieved by avoiding FCA as an important loss mechanism of the cavity. To this end, one might consider dyes that allow for pump wavelengths above the absorption edge of silicon[35]. Moreover, the pump efficiency can be improved considerably by guiding the pump light along the SOI waveguide to concentrate it in the active zone. This could be achieved by using an additional polymer waveguide around the SOI waveguide. High duty cycles or CW emission are in general difficult to achieve in dye lasers due to triplet-state excitation and subsequent photo-induced degeneration. This deficiency could be overcome by doping the matrix material with triplet-state quenching or triplet-trapping species of molecules[36], by using optofluidic concepts[37] or by choosing other gain materials such as lanthanide ions or colloidal quantum dots[38,39] that might even be suited for direct electrical pumping[40].

Nevertheless, even without CW operation, SOH lasers enable greatly simplified one-step fabrication processes for realizing thousands of light sources directly integrated into silicon photonic circuitry. Such light sources lend themselves to a wide range of applications such as biosensing[41], where pulsed operation with low-duty cycles is sufficient, where cost-efficient mass fabrication is essential to enable disposable chips for one-time use, and where pump efficiency is secondary. Moreover, the high peak power of the SOH lasers might open interesting opportunities in nonlinear infrared spectroscopy. Further investigation of the dynamics, optimization of the active cladding, and the use of better resonators should help enlarging the application range. We therefore believe that the present approach will be the basis for a novel class of silicon photonic on-chip sources that stand out due to their high peak output power and ease of fabrication.

Methods

Fabrication of SOH lasers. Waveguides were fabricated on SOI wafers from SOITEC using a CMOS pilot line based on 193-nm deep-ultra-violet lithography[30]. All waveguides have a height of $h_{WG} = 220$ nm and are optically isolated from the silicon substrate by a buried oxide (SiO_2) layer of thickness $h_{SiO2} = 2$ µm.

The gain medium is deposited on the silicon waveguides in a single post-processing step by spin-coating. The active organic cladding consists of a PMMA matrix which is doped with 1 wt% of the commercially available dye IR26 (ref. 22). The final thickness of the cladding amounts to $h_{clad} \approx (500 \pm 50)$ nm. The measured absorption and fluorescence spectra of a liquid dye solution are depicted[31] in Supplementary Fig. 1, exhibiting a fluorescence emission peak at 1,130 nm. When using the dyes in an extended waveguide structure, the emission peak of IR26 shifts to ~1,300 nm due to self-absorption along the waveguide in the overlap region of the emission and the absorption spectra[42]. This is in good agreement with the laser emission wavelength observed in ref. 22.

Experimental demonstration of laser emission. The experimental setup is depicted in Fig. 5. The SOH devices are pumped from top by a pulsed laser at a wavelength of 1,064 nm with a duty cycle of approximately $p_t = 1.23 \times 10^{-8}$. The FWHM of the pump pulse amounts to 0.9 ns, the repetition frequency is 13.7 Hz. The incident pump power is controlled by adjusting the angle of a half-wave plate in front of a polarizing beam splitter. The pump light is polarized in a direction perpendicular to the waveguide axis and focused on the waveguide under test using a cylindrical lens, see Fig. 5a.

To measure emission from the SOH device, a lensed SMF is placed near the facet, denoted as 'Fibre 2' in Fig. 5a. The fibre collects the emitted light with an estimated coupling loss of ~5 ± 1 dB. By coupling an auxiliary light beam at 1,550 nm through the on-chip GC to the SOH waveguide (Fibre 1), we can facilitate the alignment of the lensed Fibre 2 with respect to the waveguide facet. Polarization-maintaining fibres are used throughout the setup, and Fibre 2 is aligned such that the quasi-TE and quasi-TM emission of the SOH laser is coupled to the slow and the fast axis of the PM fibre, respectively. To characterize the laser emission, we use two different detection paths in our setup: A 'high-sensitivity detection' path, corresponding to the upper part in Fig. 5b, and a 'fast-detection' path, represented by the lower part in Fig. 5b.

The high-sensitivity path allows to measure input–output power characteristics and spectral properties of the laser emission. To this end, we use a monochromator and a highly sensitive photodetector with a large dynamic range, followed by an electrical low-pass filter for noise reduction and a standard oscilloscope, see Fig. 5b.

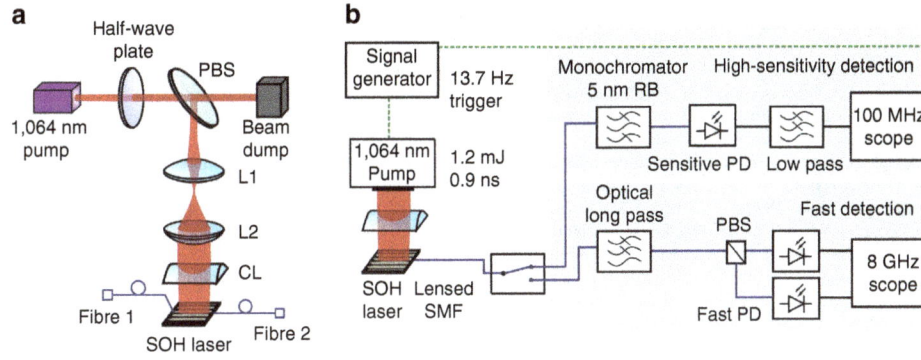

Figure 5 | Measurement setup. (a) Pump light at 1,064 nm is focused on the SOH waveguide using a cylindrical lens. Pump power is adjusted by sending the linearly polarized light from the pump laser through a half-wave plate and a polarizing beam splitter. Fibre 1 (cleaved SMF illuminating a GC) is used only to facilitate coarse alignment of Fibre 2 (lensed SMF) by using 1,550 nm light. **(b)** Emission from the SOH laser is collected by the lensed fibre (Fibre 2), which is connected to different detector setups by an optical switch. The upper path is used for high-sensitivity detection. It contains a monochromator and a slow but highly sensitive photodetector to record weak fluorescence. The sensitive PD has a low bandwidth, and a consecutive electrical low-pass filter is used to further suppress noise. The lower 'fast-detection' path is used for time- and polarization-resolved measurements. It is equipped with fast PDs. Residual pump power is blocked by an optical long-pass filter.

The oscilloscope is triggered by the emission of the pump laser and averages over 16 subsequent pulses. Due to the electrical low-pass filter and the bandwidth limitations of both the photodetector and the oscilloscope, the recorded electrical pulse is strongly widened compared with its optical counterpart. However, the peak of the recorded electrical pulse still remains proportional to the received optical power. This setup allows measuring the wavelength-resolved emission spectrum. For high output powers, an attenuator (not shown) was inserted in front of the photodiode.

Time-resolved measurements are made with the fast-detection path. An optical long-pass filter blocks spurious pump light that might be scattered into the lensed fibre, and a polarization beam splitter is used to separate the two polarization states for individual detection. Light pulses with a duration in the (sub-)ns-range are detected with fast photodiodes (NewFocus 25-GHz model 1434, NewFocus 45 GHz model 1014). A high-speed oscilloscope (Tektronix DPO 70804B, 8 GHz bandwidth, 25 GSa s^{-1}) is used to record time-resolved traces. The traces displayed in Fig. 4c have been obtained by averaging over 16 subsequent pulses. We find an average pump pulse duration of 0.9 ns FWHM with a standard error of 0.13 ns (15%). The durations of the emitted SOH laser pulses are shorter than that of the pump pulse. For quasi-TE polarization, the mean FWHM duration amounts to 0.6 ns with a standard error of ± 0.06 ns (10%).

Estimation of emission power levels. For high output powers above the lasing threshold, the peak power levels in the output fibre were measured using the fast-detection path of the setup depicted in Fig. 5, taking into account the responsivity of the fast photodiode and the optical and electrical losses of the various components. To obtain a lower boundary for the on-chip power levels, we assume that the total fibre–chip coupling losses are as low as 5 dB (factor 3.2). This value was estimated from reference measurements at 1,550 nm; the actual losses at 1,310 nm may be slightly higher. The coupling factor also includes losses of 6% due to reflection from the waveguide facet. A measured SOH laser peak power of 365 mW in the SMF hence corresponds to a laser peak power of at least 365 mW × 3.2 × 0.94 = 1.1 W which is coupled out from the waveguide facet and which could be used in an on-chip device that is connected to the SOH laser. To estimate the stochastic variations of the measured emission power, the high-speed detection path depicted in Fig. 5 is used. We record subsequent emission pulses from an SOH slot waveguide similar to the one depicted in Fig. 4a, pumped at powers of $P_{p1} = 5$ mW and $P_{p2} = 15$ mW, both of which are well above the threshold pump power of $P_{p,th} = 2$ mW. At $P_{p1} = 5$ mW, we find relative standard errors of approximately ± 5%, and at $P_{p2} = 15$ mW, the relative standard error amounts to ± 10%. For a conservative estimate, we assume that the relative standard error of the emitted power is ± 10% for all pump powers of $P_{p1} = 5$ mW or more. Below pump powers of $P_{p1} = 5$ mW, we further assume a constant absolute error which corresponds to the ± 10% relative standard error at $P_{p1} = 5$ mW. The range of negative powers is discarded, leading to the grey-shaded areas in Fig. 4b The grey-shaded areas in Fig. 2c,d were constructed in a similar way: For pump powers above $P_{p1} = 5$ mW, we assume a ± 10% relative error, whereas for pump powers below 5 mW, we use a constant absolute error which corresponds to the ± 10% relative standard error at $P_{p1} = 5$ mW. These uncertainty ranges are a coarse, but conservative estimate, which can only give a rough impression of the uncertainties of the measurement data.

For spectrally resolved measurements or for small power levels below the laser threshold, we use the high-sensitivity detection path of our setup. The peak power levels of the deformed pulses in the high-sensitivity path are calibrated by comparison with the corresponding peaks of the true pulse shapes in the

fast-detection path using medium power levels that can reliably be detected in both paths.

Estimation of pump power levels. While the total average pump power is directly accessible by measurement, the absorbed peak pump power needs to be estimated based on further assumptions. The elliptical Gaussian pump spot features a major axis of 8 mm and a minor axis of 0.3 mm, both defined by the FWHM of the intensity on the chip surface. This is much larger than the active area of the SOH waveguide, defined by the region in which pumped dye molecules interact with the lasing waveguide mode. Considering the example of the device depicted in Fig. 4, the length $l_{act, region} = 3.8$ mm of the active region is defined by the length of the slot waveguide section, and the width is estimated to the TE mode field diameter $MFD_x = 0.77$ µm in the lateral direction. The fraction of light that overlaps with the active zone is estimated by integrating the two-dimensional Gaussian distribution over the rectangle of MFD_x and waveguide length in the (x, z)-plane. This integral amounts to $p_{xz} = 0.0027$. To estimate the fraction p_y of pump light absorbed in the active cladding, we need to determine the corresponding absorption coefficient. From a direct transmission measurement using a 1.1-µm-thick IR26 dye-doped polymer layer on glass with the same dye concentration as the cladding material, the absorption cross section of the dispersed dye molecules is found to be $\sigma_p = 1.7 \times 10^{-16}$ cm^2. This is in fair agreement with the value $\sigma_p = 5 \times 10^{-16}$ cm^2 measured in a solution of the dye in 1, 2-dichloroethane[43]. The thickness of the cladding $h_{clad} = (500 \pm 50)$ nm has been measured using a profilometer. Using $\sigma_p = 1.7 \times 10^{-16}$ cm^2 and a dye molecule concentration of $n = 10^{19}$ cm^3, a value of $p_y = 1 - \exp(-\sigma_p N h_{clad}) = 0.08$ is found. The dye molecule number density N is derived from the measured mass ratio before mixing the PMMA matrix with the IR26 dyes. The total percentage of pump light absorbed in the active region is therefore $p_{xyz} = p_{xz} \times p_y = 0.022\%$. Using the measured pump pulse shape and the duty cycle, we find a ratio of average pump power to peak pump power of $p_{avg/peak} = 1.23 \times 10^{-8}$, which leads to a ratio of average incident pump power to 'absorbed' peak pump power of $p = p_{avg/peak}/p_{xyz} = 5.6 \times 10^{-5}$. This ratio is used to relate the top and the bottom power scales in Fig. 4b. Consequently, the average incident threshold pump power of 1.8 mW leads to an estimate of the absorbed peak pump power of 32 W. The same method was used to relate the top and bottom power scales in Fig. 2c,d; the corresponding ratios of average incident power to absorbed peak power are listed in Supplementary Table 1. To estimate the variation of the measured pump power, a fraction of the pump pulse is coupled to a fibre and fed to a high-speed photodiode. From the measurements we find that the standard error of the peak pump power is ∼14%.

Consistence of resonator characteristics and threshold pump power. The measured threshold pump powers of the SOH lasers are in reasonable agreement with the losses of the cavities. This is demonstrated by analysing the round-trip losses of a Fabry–Perot resonator with two cleaved facets as used in Fig. 4, and by relating them to the material gain of the active cladding.

The resonator round-trip losses are estimated by measuring the Fabry–Perot fringes in the transmission spectrum of the resonator and by evaluating the fringe contrast, see Supplementary Note 3 for a more detailed discussion. For TE polarization, we find a contrast ratio C of ∼0.5 dB between the transmission maxima and the adjacent minima, see Supplementary Fig. 4. According to Supplementary Note 3, this corresponds to a total round-trip loss of $10 \log_{10}(a^2 R^2) = 30.8$ dB, where R denotes the power reflection factor at each facet and where a is the single-pass power transmission factor in the 3.8-mm long waveguide. This result is in good agreement with a bottom-up consideration: we use a finite-element solver[44] to calculate the back-reflection R from the cleaved facet of an SOH waveguide, leading to a value of 6% (-12.2 dB), see Supplementary Table 1. Given the resonator length of $l = 3.8$ mm and the total round-trip loss of 30.8 dB, we hence estimate a propagation loss of ~ 0.9 dB mm^{-1} for the slot waveguide. This is in accordance with typically measured propagation losses of slot waveguides[45] which are of the order of 1 dB mm^{-1}.

At threshold, the round-trip losses of the resonator must be compensated by the round-trip amplification. For TE polarization, this requires a waveguide gain $\Gamma_{clad,TE}\, g = -\log(aR)/l$ corresponding to 4.1 dB mm^{-1}, where $\Gamma_{clad,TE} = 0.78$ denotes the field interaction factor of the guided mode with the active cladding, see Supplementary Note 1 for more details. Laser emission in the dye cladding is governed by a transition that has a radiative lifetime[43] of the order of 14 ns and a fluorescence quantum efficiency ϕ ranging from 0.02 to 0.1%, see (refs 43,46). The effective lifetime of the excited state hence amounts to $\phi\tau \approx 3$–14 ps—much shorter than the durations of the pump and the emission pulses. For estimating the pump intensity I_{thresh} at threshold, we may hence use steady-state approximations of the rate equations as described in detail in ref. 47 and Supplementary Note 3. This results in the relation

$$I_{thres} = \frac{hc}{\lambda_p \sigma_p \tau \phi}\left(\frac{\Gamma_{clad,TE}g}{\Gamma_{clad,TE}N\sigma_e - \Gamma_{clad,TE}g}\right), \qquad (1)$$

where $\lambda_p = 1{,}064$ nm is the pump wavelength, $\sigma_P = 1.7 \times 10^{-16}$ cm^2 denotes the measured absorption cross section at this wavelength, N denotes the volume density of dye molecules, $\tau = 14.4$ ns is the radiative lifetime, $\sigma_e = 0.5 \times 10^{-16}$ cm^2 is the emission cross-section[43], and ϕ is the fluorescence quantum efficiency with typical values ranging from 0.02 to 0.1%, see (refs 43,46), as specified for a liquid solution of the dye molecules.

When applied to the TE emission of the device depicted in Fig. 4, equation (1) leads to theoretically estimated threshold peak pump intensities ranging from 1.9 to 9.5 mW cm^{-2}. This is in reasonable with agreement our experimental estimation of the threshold peak pump intensity of 13.7 mW cm^{-2}. This estimation is based on the launched average threshold pump power of ~ 1.8 mW, the overlap $p_{xz} = 0.0027$ of the active area with the Gaussian pump spot in the x, z-plane, the pump duty cycle of approximately $p_t = 1.23 \times 10^{-8}$, and the area of the active zone having a length of $l = 3.8$ mm and a width of MFD$_x = 0.77$ μm.

The deviations between the measured and the predicted the peak pump intensity is attributed to large uncertainties of the quantum efficiency ϕ. Previously published figures range from 0.02 to 0.1% and were measured in liquid dye solutions, see refs 43,46, whereas we use the dyes in a solid polymer matrix. The measured value of 13.7 mW cm^{-2} for the peak pump intensity can be reproduced by equation (1) when assuming a quantum efficiency of $\phi = 0.014$%—which is comparable to the values obtained for liquid dye solutions. In addition, it turns out that FCA may additionally increase the cavity losses, see the section 'Optically induced losses and dynamical behaviour' of the Methods for more details. This would explain the fact that the experimentally measured threshold is slightly larger than the theoretically predicted value and lead to quantum efficiencies that are even closer to previously published values.

For TM polarization, the measured contrast of the Fabry–Perot fringes is comparable to that for TE polarization. Both polarizations hence experience similar cavity losses. Figure 4b shows a slightly increased threshold pump power of the TM compared with the TE mode—this is attributed to a reduced field interaction factor of $\Gamma_{clad,TM} = 0.42$ in the cladding compared with $\Gamma_{clad,TE} = 0.78$. Moreover, the TM mode experiences higher FCA than the TE mode due to a stronger field interaction with the silicon waveguide core, see the section 'Optically induced losses and dynamical behaviour' below.

Optically induced losses and dynamical behaviour. The dynamical behaviour of the laser emission is depicted in Fig. 4c. In this figure, the relative timing of the pump pulse and the emission pulses is subject to uncertainties: The various traces for the pump pulse, the TE emission and the TM emission were measured by an oscilloscope and a photodetector connected to the chip by standard SMF (G.652). For measuring the TE and the TM emission pulse, light was collected from the same fibre facet, and we may assume that both pulse trains experience the same propagation delay in the fibre. This is different for the pump—for measuring the pump pulse trace, we first had to remove the long-pass filter that was used to suppress residual pump light before it reaches the detector. We then moved the lensed fibre (Fibre 2 in Fig. 5) laterally to collect a small portion of 1,064 nm pump light scattered from the surface of the chip. The group delay of the pump pulses from the fibre tip to the detector is slightly different than that of the emission pulses since the optical setup had to be changed slightly and since the optical fibre is operated below its single-mode cutoff wavelength of 1,260 nm. This leads to higher-order mode propagation and hence to further uncertainties of the group delay. The overall uncertainty in relative timing between the pump and the emission pulses is estimated to be ± 100 ps.

We also investigated the dynamics of intra-cavity losses at the emission wavelength of 1,310 nm. The influence of two-photon absorption (TPA) of the emitted light and TPA-induced FCA can be neglected, see Supplementary Note 4. As the only relevant loss mechanism, we identify FCA induced by direct absorption of 1,064 nm pump light in the silicon waveguide core: during the pump pulse, free

carriers accumulate within the core of the silicon waveguide, thereby leading to absorption and considerably increasing the optical losses of the resonator also at the emission wavelength. For a rough quantitative estimate, we assume a linear absorption coefficient of 10 cm^{-1} for the 1,064 nm pump light in the silicon waveguide core[48]. During pumping, photons absorbed in the waveguide create pairs of free carriers with an effective lifetime[45] of the order of 1 ns. Similarly to the considerations made for the active region of the SOH laser, the fraction of pump light that overlaps with the silicon waveguide is estimated to be $p_{xz, Si} = 0.0011$, and the fraction of pump light absorbed in the 220 nm high silicon waveguide core is estimated to $p_{y, Si} = 0.00022$. Using these values, the free-carrier density would reach 6.6×10^{17} cm^{-3} for an average pump power of 1.8 mW, corresponding to the threshold of the laser depicted in Fig. 4. For this carrier density, an empirical model[48] allows us to roughly estimate an upper limit of the FCA-related propagation loss of ~ 5 dB mm^{-1} in the silicon core at the end of the pump pulse. Additional losses of this magnitude may significantly reduce the quality of the optical resonator during pumping and lead to an increased threshold. This is consistent with the observation that the experimentally measured threshold is slightly larger than the theoretically predicted value. We expect that in future devices, FCA can be mitigated by pumping at infrared wavelengths, which are not absorbed in the SOI waveguide core, or by using reverse-biased p-i-n structures that remove free carriers from the silicon core of the waveguides[7]. That would allow to considerably reduce threshold pump powers and to increase the slope efficiencies of the devices.

Summary of resonator and laser emission characteristics. For the quantitative estimations in this paper, various waveguide and resonator parameters are used. These parameters are summarized in Supplementary Table 1 along with threshold and emission power levels of the respective devices. The values are obtained either from experiments or from numerical simulations, for example, for the case of the field interaction factor, effective area[49] and mode field diameter. The underlying mathematical relations are given in Supplementary Note 1.

References

1. Jalali, B. & Fathpour, S. Silicon photonics. *J. Lightw. Technol.* **24**, 4600–4615 (2006).
2. Hochberg, M. & Baehr-Jones, T. Towards fabless silicon photonics. *Nat. Photon.* **4**, 492–494 (2010).
3. Liang, D. & Bowers, J. E. Recent progress in lasers on silicon. *Nat. Photon.* **4**, 511–517 (2010).
4. Bowers, J. E. *et al.* Hybrid silicon lasers: the final frontier to integrated computing. *Opt. Photon. News* **21**, 28–33 (2010).
5. Boyraz, O. & Jalali, B. Demonstration of a silicon Raman laser. *Opt. Express* **12**, 5269 (2004).
6. Boyraz, O. & Jalali, B. Demonstration of directly modulated silicon Raman laser. *Opt. Express* **13**, 796 (2005).
7. Rong, H. *et al.* Low-threshold continuous-wave Raman silicon laser. *Nat. Photon.* **1**, 232–237 (2007).
8. Takahashi, Y. *et al.* A micrometre-scale Raman silicon laser with a microwatt threshold. *Nature* **498**, 470–474 (2013).
9. Park, H. *et al.* A hybrid AlGaInAs-silicon evanescent amplifier. *IEEE Photon. Technol. Lett.* **19**, 230–232 (2007).
10. Fang, A. W. *et al.* Electrically pumped hybrid AlGaInAs-silicon evanescent laser. *Opt. Express* **14**, 9203–9210 (2006).
11. Liang, D. *et al.* Hybrid silicon evanescent approach to optical interconnects. *Appl. Phys. A Mater. Sci. Process* **95**, 1045–1057 (2009).
12. Chen, S. M. *et al.* 1.3 um InAs/GaAs quantum-dot laser monolithically grown on Si substrates operating over 100°C. *Electron. Lett.* **50**, 1467–1468 (2014).
13. Camacho-Aguilera, R. E. *et al.* An electrically pumped germanium laser. *Opt. Express* **20**, 11316–11320 (2012).
14. Wirths, S. *et al.* Lasing in direct-bandgap GeSn alloy grown on Si. *Nat. Photon.* **9**, 88–92 (2015).
15. Barrios, C. A. & Lipson, M. Electrically driven silicon resonant light emitting device based on slot-waveguide. *Opt. Express* **13**, 10092–10101 (2005).
16. Pintus, P., Faralli, S. & Pasquale, F. D. Low-threshold pump power and high integration in Al2O3:Er slot waveguide lasers on SOI. *IEEE Photon. Technol. Lett.* **22**, 1428–1430 (2010).
17. Tengattini, A. *et al.* Toward a 1.54 um electrically driven Erbium-doped silicon slot waveguide and optical amplifier. *J. Lightw. Technol.* **31**, 391–397 (2013).
18. Isshiki, H., Jing, F., Sato, T., Nakajima, T. & Kimura, T. Rare earth silicates as gain media for silicon photonics. *Photonics Research* **2**, A45 (2014).
19. Hosseini, E. S. *et al.* CMOS-compatible 75 mW erbium-doped distributed feedback laser. *Opt. Lett.* **39**, 3106 (2014).
20. Sun, X. *et al.* Electrically pumped hybrid evanescentSi/InGaAsP lasers. *Opt. Lett.* **34**, 1345 (2009).
21. Tanaka, S. *et al.* High-output-power, single-wavelength silicon hybrid laser using precise flip-chip bonding technology. *Opt. Express* **20**, 28057 (2012).

22. Morishita, T., Yamashita, K., Yanagi, H. & Oe, K. 1.3μm solid-state plastic laser in dye-doped fluorinated-polyimide waveguide. *Appl. Phys. Express* **3**, 092202 (2010).

23. Koos, C. *et al.* Silicon-organic hybrid (SOH) and plasmonic-organic hybrid (POH) integration. *J. Lightw. Technol.* doi:10.1109/JLT.2015.2499763 (2015).

24. Koos, C. *et al.* All-optical high-speed signal processing with silicon–organic hybrid slot waveguides. *Nat. Photon.* **3**, 216–219 (2009).

25. Lauermann, M. *et al.* Low-power silicon-organic hybrid (SOH) modulators for advanced modulation formats. *Opt. Express* **22**, 29927 (2014).

26. Koeber, S. *et al.* Femtojoule electro-optic modulation using a silicon–organic hybrid device. *Light Sci. Appl.* **4**, e255 (2015).

27. Pfeifle, J., Alloatti, L., Freude, W., Leuthold, J. & Koos, C. Silicon-organic hybrid phase shifter based on a slot waveguide with a liquid-crystal cladding. *Opt. Express* **20**, 15359–15376 (2012).

28. Wang, X., Grist, S., Flueckiger, J., Jaeger, N. A. F. & Chrostowski, L. Silicon photonic slot waveguide Bragg gratings and resonators. *Opt. Express* **21**, 19029 (2013).

29. Xu, Q., Almeida, V. R., Panepucci, R. R. & Lipson, M. Experimental demonstration of guiding and confining light in nanometer-sizelow-refractive-index material. *Opt. Lett.* **29**, 1626–1628 (2004).

30. Selvaraja, S. K., Bogaerts, W., Dumon, P., Van Thourhout, D. & Baets, R. Subnanometer linewidth uniformity in silicon nanophotonic waveguide devices using CMOS fabrication technology. *IEEE J. Sel. Topics Quantum Electron* **16**, 316–324 (2010).

31. Kranitzky, W., Kopainsky, B., Kaiser, W., Drexhage, K. H. & Reynolds, G. A. A new infrared laser dye of superior photostability tunable to 1.24 μm with picosecond excitation. *Opt. Commun.* **36**, 149–152 (1981).

32. Samuel, I. D. W., Namdas, E. B. & Turnbull, G. A. How to recognize lasing. *Nat. Photon.* **3**, 546–549 (2009).

33. Brosi, J.-M. *et al.* High-speed low-voltage electro-optic modulator with a polymer-infiltrated silicon photonic crystal waveguide. *Opt. Express* **16**, 4177–4191 (2008).

34. Elsaesser, T. & Kaiser, W. in *Dye Lasers: 25 Years.* (ed. Stuke, M.) **70**, 95–109 (Springer, 1992).

35. Zhang, J. & Zhu, Z. Novel heptamethine thiapyrylium infrared laser dyes of superior photostability tunable from 1.35 to 1.65 μm. *Opt. Commun.* **113**, 61–64 (1994).

36. Zhang, Y. & Forrest, S. R. Existence of continuous-wave threshold for organic semiconductor lasers. *Phys. Rev. B* **84**, 241301 (2011).

37. Schmidt, H. & Hawkins, A. R. The photonic integration of non-solid media using optofluidics. *Nat. Photon.* **5**, 598–604 (2011).

38. Schaller, R. D., Petruska, M. A. & Klimov, V. I. Tunable near-infrared optical gain and amplified spontaneous emission using PbSe nanocrystals. *J. Phys. Chem. B* **107**, 13765–13768 (2003).

39. Rogach, A. L., Eychmüller, A., Hickey, S. G. & Kershaw, S. V. Infrared-emitting colloidal nanocrystals: synthesis, assembly, spectroscopy, and applications. *Small* **3**, 536–557 (2007).

40. Heo, J., Jiang, Z., Xu, J. & Bhattacharya, P. Coherent and directional emission at 1.55 μm from PbSe colloidal quantum dot electroluminescent device on silicon. *Opt. Express* **19**, 26394 (2011).

41. Iqbal, M. *et al.* Label-free biosensor arrays based on silicon ring resonators and high-speed optical scanning instrumentation. *IEEE J. Sel. Top. Quantum Electron.* **16**, 654–661 (2010).

42. Casalboni, M. *et al.* 1.3 μm light amplification in dye-doped hybrid sol-gel channel waveguides. *Appl. Phys. Lett.* **83**, 416 (2003).

43. Benfey, D. P., Brown, D. C., Davis, S. J., Piper, L. G. & Foutter, R. F. Diode-pumped dye laser analysis and design. *Appl. Opt.* **31**, 7034–7041 (1992).

44. CST - Computer Simulation Technology. 3d electromagnetic simulation software. Available at https://www.cst.com/.

45. Vallaitis, T. *et al.* Optical properties of highly nonlinear silicon-organic hybrid (SOH) waveguide geometries. *Opt. Express* **17**, 17357–17368 (2009).

46. Semonin, O. E. *et al.* Absolute photoluminescence quantum yields of IR-26 Dye, PbS, and PbSe quantum dots. *J. Phys. Chem. Lett.* **1**, 2445–2450 (2010).

47. Shank, C. V. Physics of dye lasers. *Rev. Mod. Phys.* **47**, 649–657 (1975).

48. Vardanyan, R. R., Dallakyan, V. K., Kerst, U. & Boit, C. Modeling free carrier absorption in silicon. *J. Contemp. Phys.* **47**, 73–79 (2012).

49. Koos, C., Jacome, L., Poulton, C., Leuthold, J. & Freude, W. Nonlinear silicon-on-insulator waveguides for all-optical signal processing. *Opt. Express* **15**, 5976–5990 (2007).

Acknowledgements

This work was supported by the European Research Council (ERC Starting Grant 'EnTeraPIC', number 280145), the Alfried Krupp von Bohlen und Halbach Foundation, the EU-FP7 projects SOFI (grant 248609) and PhoxTrot, the Center for Functional Nanostructures (CFN) of the Deutsche Forschungsgemeinschaft (DFG), the Karlsruhe Nano-Micro Facility (KNMF), the Karlsruhe School of Optics and Photonics (KSOP), the Initiative and Networking Fund of the Helmholtz Association, and the Helmholtz International Research School for Teratronics (HIRST). We acknowledge support by Deutsche Forschungsgemeinschaft and Open Access Publishing Fund of Karlsruhe Institute of Technology. We are grateful for technological support by the Light Technology Institute (LTI) at Karlsruhe Institute of Technology, and by ePIXfab (silicon photonics platform).

Author contributions

D.K. and M.L. designed and performed the experiments, fabricated SOH devices, analysed the data and wrote the manuscript. S.K. and P.A. supported experiments and device fabrication and analysed the data. L.A. designed the SOI waveguides. R.P. and D.K. numerically determined reflection factors. P.D. coordinated chip tapeout and fabrication. J.L. supported analysis of the data. C.K. and W.F. designed the experiments, supported analysis of the data and wrote the manuscript. The manuscript was reviewed by all authors.

Additional information

Enhanced nonlinear interactions in quantum optomechanics via mechanical amplification

Marc-Antoine Lemonde[1], Nicolas Didier[1,2] & Aashish A. Clerk[1]

The quantum nonlinear regime of optomechanics is reached when nonlinear effects of the radiation pressure interaction are observed at the single-photon level. This requires couplings larger than the mechanical frequency and cavity-damping rate, and is difficult to achieve experimentally. Here we show how to exponentially enhance the single-photon optomechanical coupling strength using only additional linear resources. Our method is based on using a large-amplitude, strongly detuned mechanical parametric drive to amplify mechanical zero-point fluctuations and hence enhance the radiation pressure interaction. It has the further benefit of allowing time-dependent control, enabling pulsed schemes. For a two-cavity optomechanical set-up, we show that our scheme generates photon blockade for experimentally accessible parameters, and even makes the production of photonic states with negative Wigner functions possible. We discuss how our method is an example of a more general strategy for enhancing boson-mediated two-particle interactions and nonlinearities.

[1] Department of Physics, McGill University, 3600 rue University, Montreal, Quebec, Canada H3A 2T8. [2] Départment de Physique, Université de Sherbrooke, 2500 Boulevard de l'Université, Sherbrooke, Québec, Canada J1K 2R1. Correspondence and requests for materials should be addressed to A.A.C. (email: clerk@physics.mcgill.ca).

The field of quantum cavity optomechanics aims at synthesizing quantum states of light and motion using radiation pressure, the fundamental nonlinear interaction between photons and phonons. Considerable effort is currently devoted to reaching the true quantum regime, where nonlinear signatures are observed at the single-photon level[1,2]. In the canonical system of a cavity comprising a movable mirror, the quantum nonlinear regime requires the single-photon coupling constant g to be comparable to both the mechanical resonator frequency ω_M, as well as the cavity-damping rate κ (refs 3–6). Current experiments are still far from this regime.

The simplest strategy to enhance the optomechanical interaction is to coherently drive the cavity. This approach has facilitated a wide variety of interesting phenomena, ranging from ground-state cooling of the mechanical resonator[7,8] to mechanically mediated state transfer[9], and the generation of squeezed light[10–12]. The optomechanical interaction is however effectively linearized in this strong driving regime, and hence there is generally no enhancement of quantum nonlinear effects. For enhanced nonlinearity, one can tune the strong drive so that the weak residual optomechanical nonlinearity becomes resonant[13,14]. The quantum regime is then reached for $g \sim \kappa$, where the damping rate of the cavity κ can be much smaller than ω_M. A similar enhancement of quantum nonlinear effects is found in undriven two-cavity set-ups[15], where the energy difference between the optical modes is set to render the nonlinear optomechanical interaction resonant[16] or nearly resonant[17,18]. Enhancement of the nonlinearity has also been proposed in a transient scheme[19] and in optomechanical arrays[20]. Experimentally, these approaches are still not sufficient: for systems in the optimal good cavity regime ($\omega_M > \kappa$), the largest achieved couplings g are at most a percent of κ (refs 1,2,21).

In this paper, we present a new method for enhancing the single-photon optomechanical interaction for systems deep in the well-resolved sideband regime. It enables true quantum non-linearity even when the single-photon coupling g is much smaller than the cavity-damping rate κ. Crucially, our scheme results in a tunable nonlinearity, and only requires additional linear resources: it does not require a coupling to an auxiliary quantum nonlinear system (like a qubit[22–24]). The key idea is to use detuned parametric driving of the mechanics to increase the effective scale of mechanical zero-point position fluctuations x_{zpf}. This amplification directly enhances the coupling strength (as $g \propto x_{zpf}$), while the large detuning allows the mechanics to still effectively mediate a photon–photon interaction. So far, parametric mechanical driving has been studied only in the linearized regime of optomechanics[25–27].

Combined with the resonant enhancement possible in two-cavity set-ups[16–18], our novel approach lets one reach the quantum regime in current state-of-the-art experiments ($g \sim 10^{-2}\kappa$). In addition, by controlling the parametric-drive amplitude, the nonlinear interaction can be rapidly turned on and off in time, greatly extending its utility. We stress that due to the fundamental asymmetry between photons and phonons in the optomechanical interaction, parametrically driving the cavity[28] does not enhance single-photon quantum effects. While such photonic parametric driving generates an enhanced nonlinearity, this nonlinearity necessarily involves states with large photon numbers (that is, squeezed Fock states), reducing its utility (Supplementary Fig. 1). As we discuss in detail, parametrically driving the mechanics results in very different physics and a true enhancement of single-photon nonlinearity.

The approach outlined in our work is a particular example of a general strategy for enhancing two-particle interactions using only linear resources. It could thus have applications to continuous variable quantum information processing, where

strong nonlinearities are crucial for universal control, but often difficult to achieve[29]. In our optomechanical system, the mechanical resonator mediates an effective retarded interaction between photons[3,4,17,30]. Our scheme enhances this interaction by using a parametric drive to manipulate the mechanical dynamics. Similar improvements can be obtained in any system where bosonic modes mediate a two-particle interaction: by parametrically driving the intermediate modes, interactions can be greatly enhanced (see, for example, phonon-mediated electron–electron interactions in superconductivity[31–33]). An intuitive picture of the physics is provided by the effective Keldysh action describing the cavity photons in our system. This approach explicitly connects the nonlinear interaction to the mechanical Green's functions, and shows how a large detuning of the parametric drive is important to get a time-local interaction.

Results

System. We consider an optomechanical (OM) system consisting of two optical modes coupled to a single mechanical resonator (MR) via radiation pressure (cf. Fig. 1), where the interaction is of the form $g(\hat{a}_2^\dagger \hat{a}_1 + \hat{a}_1^\dagger \hat{a}_2)(\hat{b} + \hat{b}^\dagger)$. Here $\hat{a}_{1,2}$ and $\hat{b}$ are the annihilation operators of the optical modes 1, 2 and the MR, respectively. Such three-mode OM systems have been discussed extensively in the literature[16–18] and have been realized experimentally[34–36]. As already discussed, if one tunes the mode splitting $\omega_{21} \equiv \omega_2 - \omega_1$ to make the optomechanical interaction resonant, quantum nonlinear effects can be observed when the OM coupling g is comparable to the damping rate κ of the cavities[16–18].

We wish to enhance this generic system so that single-photon quantum effects are possible even when $g \ll \kappa$. To that end, we introduce a strongly detuned parametric drive to the mechanics. The generic system Hamiltonian then reads

$$\hat{H} = \Delta \hat{b}^\dagger \hat{b} - \frac{1}{2}\left(\lambda \hat{b}^2 + \lambda^* \hat{b}^{\dagger 2}\right) + g\left[\hat{a}_2^\dagger \hat{a}_1 \hat{b} e^{-i\delta t} + \text{H.c.}\right] \quad (1)$$

Here we work in an interaction picture with respect to the free cavity Hamiltonians and, for the mechanics, with respect to the pump frequency ω_p. The parameter λ is the parametric-drive strength, $\Delta \equiv \omega_M - \omega_p$ and $\delta = \omega_p - \omega_{21}$. We have assumed $\omega_p + \omega_{21}$ large enough to neglect highly non-resonant interaction terms; this approximation is always valid for the parameters considered in this work (Supplementary Note 2). In what follows, we always stay in the regime where the MR is stable even without dissipation, that is, $\lambda < \Delta$. The quadratic part of $\hat{H}$ is then diagonal when expressed in terms of the Bogoliubov mode $\hat{\beta}$, defined as $\hat{\beta} = \hat{b}\cosh r - \hat{b}^\dagger \sinh r$, with energy $E_\beta = \Delta/\cosh 2r$. The

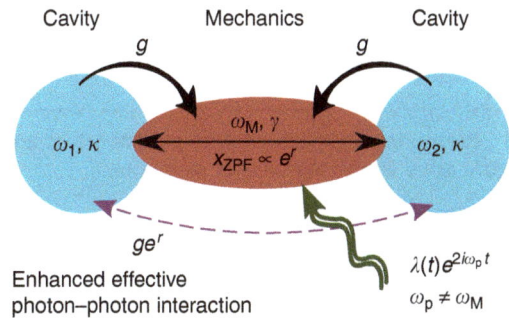

Figure 1 | Sketch of the system. Tunnelling between two optical cavities (blue circles) is mediated by a mechanical mode (red ellipse) on which a large-amplitude, strongly detuned parametric drive is applied to amplify its $\hat{X}$ quadrature. This scheme results in an exponential enhancement of the single-photon coupling constant g, thereby amplifying the resulting effective photon–photon interaction.

parameter r is set by the parametric-drive strength, $\tanh 2r = \lambda/\Delta$. Experimentally, detuned parametric drives have already been employed in optomechanics set-ups[37,38], and are particularly compatible with recent state-of-the-art electromechanical set-ups[39]. Large amounts of mechanical parametric amplification has also recently been obtained in optomechanics ($> 30\,\mathrm{dB}$), albeit in a non-stationary regime[40]. In general, the maximum value of Δ will be constrained by ω_M and the desired amount of amplification, though the form of this constraint depends on the particular implementation of the mechanical parametric drive (for example, spring constant modulation or the auxiliary cavity method, Supplementary Note 3).

Enhanced, tunable nonlinear interactions. The detuning δ can be chosen to select the nature of the nonlinear interaction that is effectively amplified. Taking $\delta = 0$ gives rise to the interaction

$$\hat{H}_\mathrm{SRP} = E_\beta \hat{\beta}^\dagger \hat{\beta} + \tilde{g}\left(\hat{a}_2^\dagger \hat{a}_1 + \hat{a}_1^\dagger \hat{a}_2\right)\left(\hat{\beta} + \hat{\beta}^\dagger\right) + \hat{H}'_\mathrm{SRP} \quad (2)$$

For large amplification (that is, for $e^{2r} \gg 1$) and a state where the Bogoliubov mode is not strongly squeezed, the term $\hat{H}'_\mathrm{SRP} = \frac{1}{2}ge^{-r}(\hat{\beta} - \hat{\beta}^\dagger)(\hat{a}_2^\dagger \hat{a}_1 - \hat{a}_1^\dagger \hat{a}_2)$ can be dropped and $\hat{H}_\mathrm{SRP}$ becomes similar to the standard radiation pressure interaction in the two-cavity OM system[16–18]. While the effects of $\hat{H}'_\mathrm{SRP}$ are negligible for the parameters considered in this work, we keep their contributions in all subsequent numerical results. We now however have an exponentially enhanced effective single-photon coupling constant

$$\tilde{g} = \frac{1}{2}ge^r \gg g \quad (3)$$

This enhancement is a direct consequence of the parametric drive: it amplifies the vacuum fluctuations of the mechanical $\hat{X} \equiv \hat{b} + \hat{b}^\dagger$ quadrature, and thus enhances the coupling of the cavities to this quadrature. The effective photon–photon interaction induced by equation (2) is further enhanced compared with a standard single-cavity OM set-up, as the Bogoliubov mode energy E_β is also tunable and can be made much smaller than ω_M. However, one also needs this mechanically mediated interaction to be sufficiently time-local; as shown below, this further constrains $E_\beta > \tilde{g}, \kappa$. The induced photon–photon interaction thus scales as $\tilde{g}^2/E_\beta$, as opposed to $\sim g^2/\omega_\mathrm{M}$ in a standard OM cavity[3,4]. We stress that only the amplification effect of the parametric drive is crucial here. This means that the mechanics does not have to be in a vacuum-squeezed state (that is, the Bogoliubov mode can have a thermal population).

If one instead tunes frequencies so that $\delta \approx E_\beta > \tilde{g}, \kappa$, one can make an additional rotating wave approximation, yielding the interaction ($e^{2r} \gg 1$)

$$\hat{H}_\mathrm{PAT} = \left(E_\beta - \delta\right)\hat{\beta}^\dagger \hat{\beta} + \tilde{g}\left(\hat{a}_2^\dagger \hat{a}_1 \hat{\beta} + \hat{a}_1^\dagger \hat{a}_2 \hat{\beta}^\dagger\right) \quad (4)$$

This is a phonon-assisted photon-tunnelling interaction, with an enhanced interaction strength $\tilde{g}$ again given by equation (3). This form of interaction (without any parametric enhancement) has been studied in the resonant regime ($E_\beta = \delta$)[16], as well as in the detuned regime ($\tilde{g}, \kappa < |E_\beta - \delta| \ll |E_\beta + \delta|$)[17]. While tuning the parametric-drive frequency lets us pick the form of the effective nonlinear interaction, tuning its amplitude lets us control the interaction strength. As discussed below, the possibility to modulate the interaction strength in time is extremely useful to prepare the β mode in the desired state while preventing mechanical heating.

We stress that our scheme allows in principle arbitrarily large nonlinearity enhancements in optomechanics using only additional linear resources (that is, a large parametric drive that is strongly detuned). In practice, the achievable enhancement will be limited by the maximum detuning Δ possible (needed to ensure $E_\beta > \tilde{g}, \kappa$) and by the stability of the parametric drive (one should not cross the instability threshold). Thus, if one wants to use large r values to enhance the interaction, the requirement that $E_\beta > \kappa, \tilde{g}$ implies that the system must be deep in the well-resolved sideband regime. The requirements on the parametric drive and the achievable amplification r are summarized in Table 1. For a standard realization of mechanical parametric driving via the modulation of the spring constant, Δ is furthered constrained by ω_M and the amount of amplification. A much weaker constraint applies if the parametric drive is realized using an auxiliary cavity (Supplementary Note 3). Despite these caveats, our approach represents a practically attractive route towards single-photon strong coupling, given the difficulty of engineering systems with an intrinsically large value of g.

Dissipation and mechanical state preparation. In addition to the coherent dynamics described by equation (1), we take into account the coupling of the MR and both optical modes to Markovian baths; these cause the cavities to be damped at a rate κ, and the mechanics at a rate γ. In the presence of a parametric drive, the noise coming from the MR bath is also amplified. In the weak mechanical dissipation limit ($\gamma \ll E_\beta$), a MR bath of thermal occupancy $\bar{n}_\mathrm{M}^\mathrm{th}$ corresponds to a bath for the β mode of effective temperature $\bar{n}_\beta^\mathrm{th} = \bar{n}_\mathrm{M}^\mathrm{th}\cosh 2r + \sinh^2 r$. For mechanical excitations off-resonant with the optical modes, that is, $E_\beta - \delta > \kappa, \tilde{g}$ and $e^{2r} \gg 1$, the cavities are heated through the OM interaction at a rate $\Gamma \propto \gamma[\tilde{g}e^r/(E_\beta - \delta)]^2(2\bar{n}_\mathrm{M}^\mathrm{th} + 1)$ (Supplementary Note 4); left unchecked, this heating could corrupt any nonclassical behaviour induced by the enhanced single-photon OM interaction. To circumvent amplified noise from the mechanical bath, a possible strategy is to add an optical mode to the system, and use it to keep the Bogoliubov mode in its ground state via dissipative squeezing[41–47]. This steady-state technique has recently been implemented experimentally[48–50].

Table 1 | Parameter regime needed to get important amplification of the single-photon coupling constant $\tilde{g} = ge^r/2$ using a parametric drive on the mechanical resonator.

Necessary parameter regime	
Good cavity limit	$\kappa \ll \omega_\mathrm{M}$
Large parametric-drive detuning	$\kappa \ll \Delta$
Strong parametric drive	$\lambda \to \Delta$

Optimal enhanced interaction	Regime
Requirement for local-in-time interaction	
$\tilde{g}_\mathrm{opt} \approx (g/2)^{2/3}\Delta^{1/3}$	$(g/2)^2\Delta > \kappa^3$
$\tilde{g}_\mathrm{opt} \approx g\sqrt{\Delta/2\kappa}$	$(g/2)^2\Delta < \kappa^3$

The enhanced single-photon coupling that leads to maximal photon blockade, which corresponds to having an effectively local-in-time photon–photon interaction, is denoted as $\tilde{g}_\mathrm{opt}$.

As an alternative to using an additional optical mode, one can instead take advantage of the tunability of the parametric drive. Indeed, one can first turn on the parametric drive on a timescale τ_{on} short enough to avoid significant perturbation of the initial photon state, $\tau_{\text{on}} < 1/\kappa, 1/\tilde{g}$. Then, one can let the system interact for a time τ sufficient to observe nonclassical signatures, $\tau > 1/\tilde{g}$. This protocol has to be performed in a total time short enough to avoid unwanted cavity heating, $\tau_{\text{on}} + \tau < 1/\Gamma$. This is possible given that Γ remains $\ll \tilde{g}$ even for large enhancement factors $e^{2r} \gg 1$, as the intrinsic mechanical damping γ is extremely low in state-of-the-art experiments.

In such pulsed schemes, it is crucial that the initial ramp of the parametric-drive amplitude prepares the β mode reasonably close to its ground state; this is needed to obtain the radiation pressure interaction in equation (2) (i.e. the effects of $\hat{H}'_{\text{SRP}}$ remain negligible). If the mechanics $\hat{b}$ remains in its ground state, as it would occur for an abrupt turn on of the parametric drive, then the β mode is in a highly squeezed state and this squeezing completely negates the exponential enhancement of the interaction in equation (2). While an adiabatic protocol would prevent β-mode squeezing, it would be too slow to prevent important perturbation of initial cavity states. Indeed, for $E_\beta \sim \kappa, \tilde{g}$, adiabaticity is ensured for turn-on times much longer than $1/E_\beta \sim 1/\kappa, 1/\tilde{g}$. An appropriate solution is to use the so-called 'counterdiabatic' or 'transitionless' driving (TD) protocols[51–53]. These require one to control the amplitude and the phase of the parametric drive, $\lambda(t) = \lambda_0(t) + i\lambda_1(t)$ (cf. Fig. 2a). The term $\lambda_0(t)$ defines the instantaneous Bogoliubov mode of interest through $\tanh 2r(t) = \lambda_0(t)/\Delta$, with $\lambda(0) = 0$ and $\lambda(\tau_{\text{on}}) = \lambda$. The correction $\lambda_1(t) = -\dot{r}(t)$ ensures that the MR stays in the ground state of the instantaneous Bogoliubov mode despite non-adiabatic effects. In Fig. 2b, we show the evolution of the β mode without nonlinear interaction ($g = 0$) and for a MR initially in its ground state. Using TD, the final β mode is prepared in its ground state for

$\tau_{\text{on}} \ll 1/\kappa$. Such an ideal preparation is not possible if one just suddenly turns on $\lambda(t)$.

Standard radiation pressure interaction. We focus in the remainder of the paper on the case where the relative detuning $\delta = 0$, such that the OM interaction is described by $\hat{H}_{\text{SRP}}$, equation (2); we further take parameters such that $E_\beta > \kappa, \tilde{g}$ to ensure a sufficiently wide-bandwidth mechanically mediated photon–photon interaction. This two-photon interaction can be understood as an effective 'feedback' process: the photonic system first displaces the MR and then this displacement results in an effective forcing of the photonic system[3,4,17,30]. The conventional approach to describing such an interaction uses a polaron transformation $\hat{U} = \exp\{(\tilde{g}/E_\beta)(\hat{a}_2^\dagger \hat{a}_1 + \hat{a}_1^\dagger \hat{a}_2)(\hat{\beta}^\dagger - \hat{\beta})\}$ on $\hat{H}_{\text{SRP}}$, leading to the polaron Hamiltonian $\hat{H}_{\text{SRP}}^{\text{pol}} = \hat{U}\hat{H}_{\text{SRP}}\hat{U}^\dagger$

$$\hat{H}_{\text{SRP}}^{\text{pol}} = E_\beta \hat{\beta}^\dagger \hat{\beta} - \Lambda\left(\hat{a}_2^\dagger \hat{a}_1 + \hat{a}_1^\dagger \hat{a}_2\right)^2 \qquad (5)$$

$$= E_\beta \hat{\beta}^\dagger \hat{\beta} - \Lambda\left(\hat{a}_s^\dagger \hat{a}_s - \hat{a}_a^\dagger \hat{a}_a\right)^2 \qquad (6)$$

with $\Lambda = \tilde{g}^2/E_\beta$. Equation (6) is written in the symmetric/antisymmetric photonic basis, defined by the modes $\hat{a}_{s,a} = (\hat{a}_1 \pm \hat{a}_2)/\sqrt{2}$. When only one of the photonic modes is driven (symmetric or antisymmetric)[36], the nonlinearity is a Kerr interaction and the physics of the radiation pressure interaction in a single-cavity OM system is recovered. As described in refs 3,4, the polaron transformation only diagonalizes the Hamiltonian of the closed system. When including dissipation or a drive, the finite displacement of the β mode caused by the photons has to be accounted for ($\langle\hat{\beta}\rangle = \tilde{g}/E_\beta\langle\hat{a}_s^\dagger \hat{a}_s - \hat{a}_a^\dagger \hat{a}_a\rangle$). As a result, when a photon enters or leaves a cavity, it generates phonon sideband excitations (that is, excitations of the β mode); this is analogous to standard Franck Condon physics.

Photon blockade. The photon–photon interaction in equation (6) can lead to photon blockade, a quantum phenomenon characterized by a strong suppression of the probability of having more than one photon in the cavity together with antibunched photon statistics. It has been thoroughly studied in the single-cavity set-up[3]; here we highlight the advantage of parametrically driving the MR. Photon blockade is typically quantified by the equal-time intensity correlation function $g_a^{(2)}(0) = \langle\hat{a}^\dagger \hat{a}^\dagger \hat{a}\hat{a}\rangle/\langle\hat{a}^\dagger \hat{a}\rangle^2$ that drops below the classical bound, $g_a^{(2)}(0) < 1$. Note that, although $g_a^{(2)}(0) < 1$ can be obtained with Gaussian states obeying a linear dynamics[54], here the $g_a^{(2)}(0)$ suppression cannot be reproduced if the interaction $\hat{H}_{\text{SRP}}$ is linearized (Supplementary Fig. 2).

The intensity correlation of the symmetric mode, $g_{a_s}^{(2)}(0)$, is calculated in presence of a weak probe drive on $\hat{a}_s$. We use a standard quantum master equation to describe the coherent dynamics governed by $\hat{H}_{\text{SRP}}$ and the dissipation to zero-temperature baths of the $\hat{\beta}$ and $\hat{a}_{1,2}$ modes. We thus assume that the MR is either cooled using dissipative squeezing, or has been prepared in its ground state via the TD protocol. The resulting $g_{a_s}^{(2)}(0)$, with and without mechanical parametric drive, are compared in Fig. 3a. The parametric drive markedly reduces $g_{a_s}^{(2)}(0)$, especially in the experimentally accessible regime $g < 0.1\kappa$, for example, for $g = 0.1\kappa$, $g_{a_s}^{(2)}(0) \approx 0.8$ ($g_{a_s}^{(2)}(0) \approx 0.3$) for 20 dB (30 dB) of amplification while $g_{a_s}^{(2)}(0) \approx 0.999$ without parametric drive. In the limit $\tilde{g} > \kappa$, $g_{a_s}^{(2)}(0)$ is minimized for $E_\beta = 2\tilde{g}$, that is, for a parametric detuning $\Delta = \frac{1}{2}ge^{3r}$ (cf. insets of Fig. 3). For 20 dB of amplification and $g \sim 0.1\kappa$, this implies $\Delta \sim 100\kappa$. The optimal E_β corresponds to the situation where, in the polaron

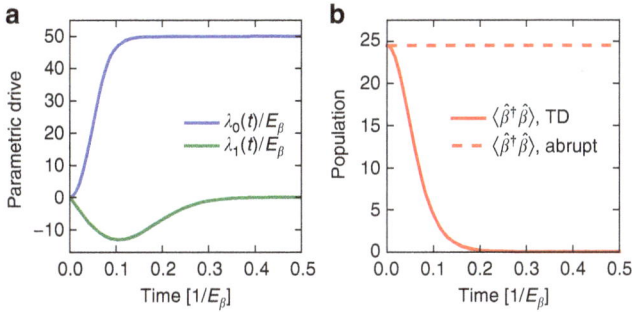

Figure 2 | Initialization of the mechanical state. Fast turn on of mechanical parametric driving using the TD scheme (see main text). (**a**) Time dependence of the parametric driving strength $\lambda(t) = \lambda_0(t) + i\lambda_1(t)$, corresponding to a Gaussian profile for the instantaneous amplification factor $r(t)$ [$\tanh 2r(t) = \lambda_0(t)/\Delta$]. The final value of $\lambda(t)$ corresponds to $e^{2r} = 20$ dB. The pulse is chosen to ramp up the parametric drive in a time much shorter than the inverse Bogoliubov-mode energy E_β. (**b**) Evolution of the mechanical state, as characterized by the population of the Bogoliubov mode $\hat{\beta}$. The solid red line is for the TD approach, showing preparation of a pure squeezed state (characterized by no β-mode excitations) in a time $\sim 0.1/E_\beta$. In contrast, a sudden (step-function) turn on of the parametric drive results in β mode being far from its ground state (red dashed line). The TD protocol plays a crucial role in our scheme, as it allows a rapid turn on of the mechanically mediated photon–photon interaction, without any spurious effects resulting from a large initial β-mode population. Neither a purely adiabatic protocol nor a sudden diabatic approach would be sufficient. Here the mechanical dissipation is $\gamma = 10^{-4}E_\beta$ and $g = 0$, but the results are unchanged for $g \neq 0$ and a sufficiently small γ.

a

b

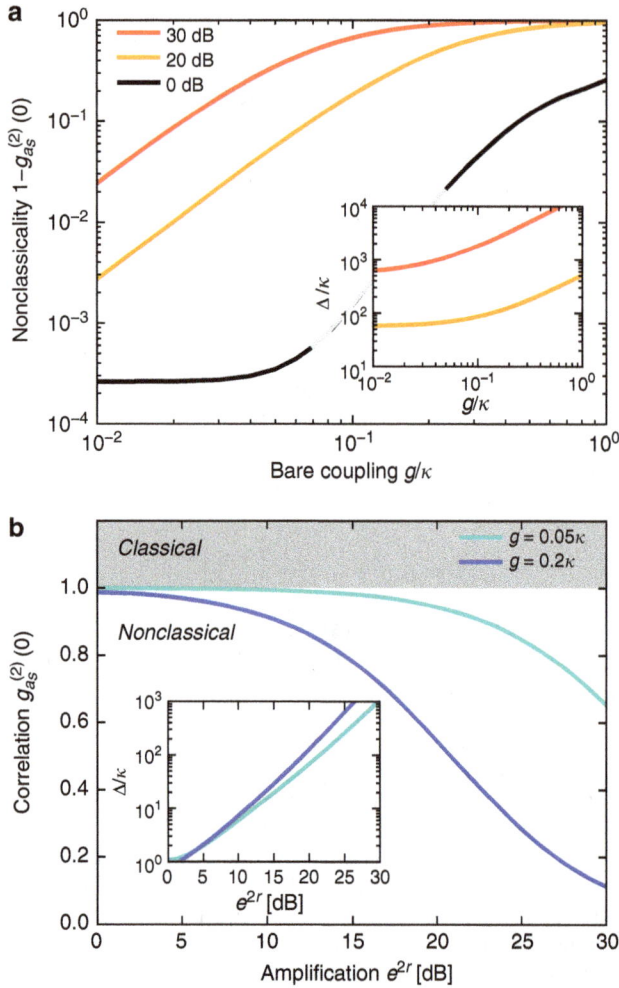

Figure 3 | Nonclassical photon intensity correlations. Intensity correlation function $g_{a_s}^{(2)}(0) = \langle \hat{a}_s^\dagger \hat{a}_s^\dagger \hat{a}_s \hat{a}_s \rangle / \langle \hat{a}_s^\dagger \hat{a}_s \rangle^2$, for a mechanical parametric drive yielding the optomechanical interaction $\hat{H}_{SRP}$ (cf. equation (2)), and for a weak coherent probe tone applied to the cavities. In **a**, $g_{a_s}^{(2)}(0)$ is plotted as a function of g for different values of the amplification factor e^{2r}; in **b** it is plotted as a function of e^{2r} for a fixed value of g. For each value of g and r, the probe frequency and the parametric-drive detuning Δ are optimized to minimize $g_{a_s}^{(2)}(0)$ (Δ is plotted in insets). The amplitude of the weak probe is kept fixed. (**a**) The violation of the classical bound $g_{a_s}^{(2)}(0) \geq 1$ is enhanced in our scheme: the presence of the mechanical parametric drive leads to significant suppression of $g_{a_s}^{(2)}(0)$ for experimentally accessible couplings $g \sim 0.01\kappa$. (**b**) Mechanical parametric driving brings the optical field deep into the nonclassical region even for $g = 0.05\kappa$, leading to non-Gaussian optical fields (Supplementary Fig. 2). For these results, dissipative squeezing is used, with a damping rate $\gamma_\beta = 0.001\kappa$.

picture (cf. equation (6)), the state with two symmetric photons ($|2, 0, 0\rangle$) and the corresponding first phonon sideband ($|2, 0, 1\rangle = \hat{\beta}^\dagger |2, 0, 0\rangle$) are equally detuned from the one-photon state ($|1, 0, 0\rangle$). The intensity correlations are described to a good approximation by $g_{a_s}^{(2)}(0) \approx 1/[1 + 0.8(\tilde{g}/\kappa)^2]$. Increasing the parametric drive is thus, in principle, always beneficial since, for any coupling $g < \kappa$, there is always an amplification strength r that leads to a desired $g_{a_s}^{(2)}(0) < 1$. For instance, $g_{a_s}^{(2)}(0) = 0.5$ is obtained for $e^r \approx 2.2\kappa/g$.

Negative Wigner functions. The possibility for time-dependent control of the photon–photon interaction in our system opens the

door to a wealth of interesting functionalities. Perhaps the most demanding challenge is the production of photon states exhibiting strongly negative Wigner functions. We show here how this can be accomplished in our set-up, in a manner that produces negativity both in the states of intracavity and propagating photons. Crucially, this can be done using a bare coupling g that is still smaller than the cavity-damping rate ($g/\kappa \sim 0.3$). We stress that this kind of negative Wigner function generation would be essentially impossible without mechanical parametric driving: not only would one require a g that is at least an order of magnitude larger, one would need some alternate means for controlling it in time. Our scheme thus significantly lowers the level of experimental improvement needed for generating negative photonic Wigner functions.

One first prepares cavity 1 in a low-amplitude coherent state using a classical laser drive while cavity 2 remains in vacuum (Fig. 4a). The mechanical parametric drive is off during this step, so that there is essentially no photonic nonlinearity. Once this initial cavity state is prepared, the cavity drive is turned off, and the photonic interaction is amplified by ramping up the mechanical parametric drive. The TD scheme described earlier allows this turn-on step to be completed in a time $\tau_{on} \ll 1/\kappa, 1/\tilde{g}$, that is, fast enough to be effectively instantaneous to the photons. At the same time, this scheme ensures that the β mode is prepared in its ground state.

We pick a frequency detuning of the parametric drive $\delta = 0$ to realize the two-photon tunnelling interaction $\hat{H}_{SRP}$ (c.f. equation (2)). The effective Hamiltonian $\hat{H}_{SRP}^{pol}$ leads (in the absence of dissipation) to a periodic evolution with the characteristic time $\tau_{int} = 2\pi/\Lambda$, with $\Lambda = \tilde{g}^2/E_\beta$. If $\Lambda \gg \kappa$ (possible with large enough parametric driving), one finds that the cavity-1 state is strongly nonclassical at $\tau \sim \tau_{int}/4$, characterized by a Wigner function exhibiting large amounts of negativity, while the MR practically stays in the a pure squeezed state (Fig. 4c). This can be easily understood by considering the effects of the two-photon tunnelling interaction that is mediated by the mechanics, $\hat{a}_2^\dagger \hat{a}_2^\dagger \hat{a}_1 \hat{a}_1 + \text{H.c.}$ As cavity 2 starts in vacuum and cavity 1 has negligible probability for having more than two photons, this term initially transfers two photons from the first to the second cavity in a time $\tau_{int}/8$. The two-photon Fock state of cavity 2 then gets weakly populated and its Wigner function is reminiscent of a low-amplitude squeezed state (Fig. 4e). After an additional evolution for a time $\tau_{int}/8$, these two photons return to cavity 1, with an overall π-phase shift. This phase shift of the two-photon component of the cavity-1 state (with respect to the one photon component) leads to negativity in the Wigner function (Fig. 4c).

Next, at the special time $\tau_{int}/4$ where the cavity-1 state is maximally nonclassical, the parametric drive is rapidly turned off. By using the reverse of our TD protocol (cf. Fig. 4g), this can be done in such a way that the MR returns to its ground state. At this stage, the nonlinear optomechanical interaction is almost completely suppressed: not only is its magnitude greatly diminished, but it is now no longer resonant, such that any residual effects will scale as $g^2/\omega_M \ll \kappa$ (Supplementary Fig. 3). Finally, in the ideal case where internal cavity losses are weak, the nonclassical cavity-1 state is converted perfectly to a propagating mode in the cavity-1 input–output waveguide with an exponential profile. We thus have generated a nonclassical, propagating photonic state, using an underlying weak single-photon optomechanical coupling and the additional linear resource of a parametric drive. We stress that the ability to rapidly turn the mechanically mediated nonlinear interaction on and off is crucial to being able to do this experiment.

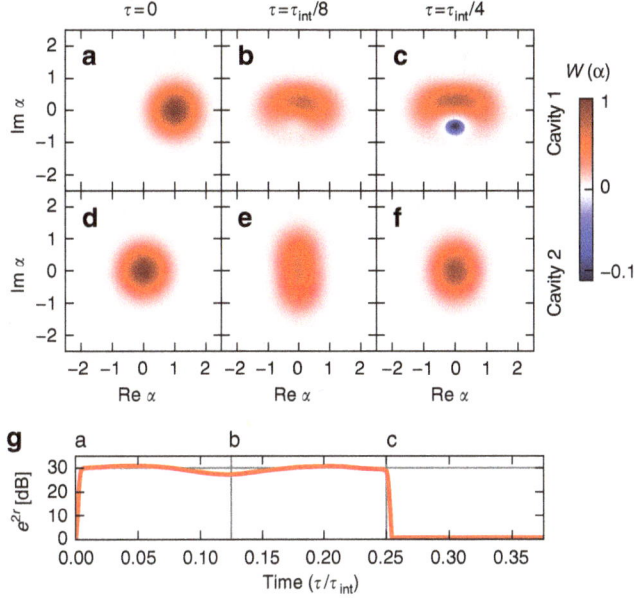

Figure 4 | Emergence of negative photonic Wigner functions from enhanced optomechanical coupling. Results illustrating the pulsed protocol described in the main text, for a mechanical parametric drive yielding the optomechanical interaction $\hat{H}_{SRP}$ (c.f. equation (2)). The effective photon-photon interaction strength $\Lambda = \tilde{g}^2/E_\beta$ sets the characteristic time $\tau_{int} = 2\pi/\Lambda$. The Wigner function of cavity 1 (2) is plotted in **a-c** (**d-f**) at three characteristic times. Negative (positive) values of the Wigner functions are plotted in blue (red). (**a,d**) Initial state, where cavity 1 is initially displaced by $\alpha_1 = 1$, cavity 2 is in vacuum. The parametric drive is then switched on using the transitionless driving scheme with a short turn-on time $\tau_{on} = \tau_{int}/400$ and a Gaussian profile for $r(t)$. The corresponding mechanical amplification strength is plotted in **g**, where a, b and c refer respectively to $\tau = 0$, $\tau = \tau_{int}/8$ and $\tau = \tau_{int}/4$. (**c**) Negativity in the cavity-1 Wigner function is maximal at $\tau = \tau_{int}/4$. As discussed in the main text (and shown in **g**), the parametric drive can then be turned off with the TD scheme, and the cavity-1 state will be emitted into the cavity-1 input-output waveguide, resulting in a propagating photonic state with a negative Wigner function. (**g**) The MR stays in a squeezed state when the parametric drive is on ($\tau_{on} < \tau < \tau_{int}/4$). Parameters here are $g = 0.3\kappa$, mechanical damping $\gamma = 10^{-4}\kappa$ and mechanical bath occupancy $\bar{n}_M^{th} = 0.5$; the parametric-drive strength and detuning are chosen to yield an amplification factor $e^{2r} = 30$ dB and $E_\beta = 2\tilde{g}$. The resulting Kerr-interaction strength is then $\Lambda \approx 2.4\kappa$ and the rate at which the mechanical noise heats the cavities is $\Gamma \sim \kappa/40$. If one reduces the amplification factor to 25 dB, the negativity is lost; this highlights the crucial role of the parametric driving.

Engineered MR response function. While our treatment so far is rigorous, the origin of the enhanced OM interaction may still seem somewhat mysterious. An alternate approach that provides a more intuitive picture, and that is more easily generalized to more complex systems, is based on deriving an effective Keldysh action for the cavity photons. In this approach, one clearly sees that the mechanical resonator mediates a time-non-local effective photon–photon interaction that depends crucially on the retarded Green's functions of the mechanics.

Indeed, by integrating out the mechanical degree of freedom in the Keldysh action obtained for the interaction $\hat{H}_{SRP}$, one gets an exact effective action describing two distinct time-non-local photon–photon interactions. These interactions can equivalently be captured by writing the equations of motion for the cavity fields; for cavity 1, the interaction term in the

equation of motion is

$$\dot{\hat{a}}_1(t) = \int_{-\infty}^{+\infty} dt' 2i\left\{\Lambda(t-t')\hat{a}_2^\dagger(t')\hat{a}_1(t')\hat{a}_2(t)\right.$$
$$\left. + \tilde{\Lambda}(t-t')\hat{a}_1^\dagger(t')\hat{a}_2(t')\hat{a}_2(t)\right\} + \dots \quad (7)$$

with $\Lambda(t) = \frac{1}{2}g^2 G_b^R(t)$ and $\tilde{\Lambda}(t) = \frac{1}{2}g^2 \tilde{G}_b^R(t)$. Here $G_b^R(t) = -i\theta(t)\langle[\hat{b}(t), \hat{b}^\dagger(0)]\rangle$ and $\tilde{G}_b^R(t) = -i\theta(t)\langle[\hat{b}(t), \hat{b}(0)]\rangle$ are, respectively, the non-interacting diagonal and off-diagonal retarded mechanical Green's functions. The role of the parametric drive is to render the off-diagonal element non-zero and amplify $G_b^R(t)$ and $\tilde{G}_b^R(t)$. For large amplification, $G_b^R(t) = \tilde{G}_b^R(t)$ with, in the Fourier domain

$$\tilde{G}_b^R[\omega] = \frac{e^{2r}}{4}\left[\frac{1}{\omega - E_\beta + i\gamma/2} - \frac{1}{\omega + E_\beta + i\gamma/2}\right] \quad (8)$$

In the limit where $E_\beta \gg \kappa, \tilde{g}$, the frequency dependence of the interaction is not important on the relevant energy scales of the system and can be neglected. In this case, the situation is similar to an instantaneous interaction and one recovers the polaron picture of equation (6). This description clearly shows the general idea: to amplify the effective photon–photon interaction, one has to engineer the dynamics, that is, the response function, of the MR.

Discussion

We have studied a two-cavity OM system, showing that parametrically driving the MR exponentially enhances the nonlinear OM interaction. One can thus reach the much-coveted single-photon strong coupling regime starting from an extremely weak bare interaction g. This allows photon blockade and non-Gaussian state generation even when $g \ll \kappa$. This new scheme further benefits from its controllability: one can choose the nature of the nonlinear interaction to amplify as well as modulate in time its strength. Our work suggests more general approaches for enhancing bosonic-mediated interactions and nonlinearities through simple parametric driving.

Methods

Transitionless driving. We give here more details about the 'counterdiabatic' or 'transitionless' driving protocols[51-53]. These imply controlling the amplitude and the phase of the parametric drive, such that $\lambda(t) = \lambda_0(t) + i\lambda_1(t)$, with $\lambda(0) = 0$ and $\lambda(\tau_{on}) = \lambda$. Defining the instantaneous unitary transformation $\hat{U}(t) = \exp[\frac{1}{2}r(t)(\hat{b}^2 - \hat{b}^{\dagger 2})]$ with $\tanh 2r(t) = \lambda_0(t)/\Delta$ and considering only the parametrically driven MR, that is, $\hat{H}$ given by equation (1) with $g = 0$, the transformed Hamiltonian is

$$\hat{\tilde{H}}(t) = \hat{U}(t)\hat{H}\hat{U}^\dagger(t) + i\dot{\hat{U}}(t)\hat{U}^\dagger(t) \quad (9)$$

$$= \frac{\Delta}{\cosh 2r(t)}\hat{b}^\dagger\hat{b} + i\frac{1}{2}[\lambda_1(t) + \dot{r}(t)]\left(\hat{b}^{\dagger 2} - \hat{b}^2\right) \quad (10)$$

Consequently, starting from the mechanical ground state, a parametric drive modulated with $\lambda_1(t) = -\dot{r}(t)$ ensures that (without dissipation) the instantaneous Bogoliubov mode $\hat{\beta}(t) = \cosh r(t)\hat{b} - \sinh r(t)\hat{b}^\dagger$ stays in its ground state. At the end of the protocol, the desired β mode is thus in a vacuum state. Considering dissipation, we show in the main text that it is still possible to prepare the final Bogoliubov mode $\hat{\beta}$ in the same state as the initial MR state in a time τ_{on} much faster than any other timescales of the system (Fig. 2).

Quantum master equation. To obtain the $g_a^{(2)}$ correlation function and the Wigner function of the cavities state, we use a standard quantum master equation approach[55]. The coherent dynamics is governed by $\hat{H}_{SRP}$ (cf. equation (2)) and the coupling of the cavities to zero-temperature baths is described with the Lindbladians $\hat{\mathcal{L}} = \kappa D[\hat{a}_1] + \kappa D[\hat{a}_2]$, where $D[\hat{a}] \cdot = \hat{a} \cdot \hat{a}^\dagger - \frac{1}{2}\{\hat{a}^\dagger\hat{a}, \cdot\}$. Concerning the mechanical Lindbladian, we consider two configurations: either the Bogoliubov mode is cooled down to its ground state with dissipative squeezing or a TD scheme is used with a MR initially in a thermal state (population $\bar{n}_M^{th}$). The dissipative squeezing protocol is modelled with the Lindbladian $\gamma_\beta D[\hat{\beta}]$, where γ_β is the coupling rate to the engineered reservoir that squeezes the mechanics, and is used to obtain the results presented in Fig. 3. For these results, we consider a drive on the cavities, $\hat{H}_{drive} = \epsilon(\hat{a}_1 e^{i\omega_{d1}t} + \hat{a}_2 e^{i\omega_{d2}t}) + $ H.c.; the drive is used to probe the intensity correlations. Meanwhile, the transitionless driving scheme is used in the protocol that leads to negative Wigner functions of the optical mode (cf. Fig. 4) and corresponds to a Lindbladian $\gamma\left(1+\bar{n}_{\mathrm{M}}^{\mathrm{th}}\right)D[\hat{b}]+\gamma\bar{n}_{\mathrm{M}}^{\mathrm{th}}D[\hat{b}^{\dagger}]$. The parametric-drive strength is turned on continuously, with $\lambda(t)$ derived above, and we consider an initial coherent state in cavity 1. The density matrix $\hat{\rho}$ in these two situations is then obtained from the quantum master equation

$$\dot{\hat{\rho}} = -i[\hat{H},\hat{\rho}] + \hat{L}\hat{\rho} \tag{11}$$

The intensity correlations $g_a^{(2)}(0)$ are calculated from the steady-state value of $\hat{\rho}$. For given values of the coupling g and amplification strength e^{2r}, the detuning Δ and the drive frequency are optimized to minimize the intensity correlations. The results are plotted in Fig. 3.

Effective Keldysh action. As explained in the main text, the non-zero response time of the MR results in a time-non-local photon–photon interaction. To describe this physics, we calculate the action of the system in the same interaction picture used for equations (2)–(4), that is, in a frame where the Hamiltonian is not explicitly time-dependent. Since the OM system is driven and subject to dissipation, the Keldysh formalism is well adapted to study this out-of-equilibrium system[56]; a detailed example of the Keldysh formalism in OM systems is presented in ref. 14. In this approach, each annihilation operator used in the Hamiltonian-based description is mapped onto two time-dependent fields: a classical (cl) and a quantum (q) field.

For the cavities ($\sigma = 1, 2$) and MR fields

$$\mathbf{a}_\sigma^\dagger \equiv \left(a_{\sigma,\mathrm{cl}}^*(t), a_{\sigma,\mathrm{cl}}(t), a_{\sigma,\mathrm{q}}^*(t), a_{\sigma,\mathrm{q}}(t)\right) \tag{12}$$

$$\mathbf{b}^\dagger \equiv \left(b_{\mathrm{cl}}^*(t), b_{\mathrm{cl}}(t), b_{\mathrm{q}}^*(t), b_{\mathrm{q}}(t)\right) \tag{13}$$

the Keldysh action that describes the full OM system studied in the main text (cf. equation (1)) has the general following form

$$S_{\mathrm{tot}} = \sum_{\sigma=1,2} \int_{-\infty}^{\infty}\int_{-\infty}^{\infty} dt\,dt' \left[\mathbf{a}_\sigma^\dagger(t)\check{\mathbf{G}}_{a_\sigma}^{-1}(t-t')\mathbf{a}_\sigma(t') + \mathbf{b}^\dagger(t)\check{\mathbf{G}}_b^{-1}(t-t')\mathbf{b}(t')\right]$$
$$+ \frac{g}{\sqrt{2}}\sum_{ijk=\mathrm{cl,q}}\zeta_{ijk}\int_{-\infty}^{\infty}dt\left[a_{2,i}^*(t)a_{1,j}(t)b_k(t)+\mathrm{c.c.}\right]$$
$$\tag{14}$$

Here $\zeta_{ijk}=1$ if there is an odd number of quantum fields and 0 otherwise.

In equation (14), the two first terms represent the Gaussian action that governs the non-interacting dynamics (that is, $g=0$). It involves the non-interacting cavities (MR) Green's functions $\check{\mathbf{G}}_{a_\sigma}(t-t')$ $[\check{\mathbf{G}}_b(t-t')]$. Here the most general Green's functions are 4×4 matrices of the form

$$\check{\mathbf{G}}_b(t) = \begin{pmatrix} \mathbf{G}_b^{\mathrm{K}}(t) & \mathbf{G}_b^{\mathrm{R}}(t) \\ \mathbf{G}_b^{\mathrm{A}}(t) & 0 \end{pmatrix} \tag{15}$$

$$\mathbf{G}_b^{\mathrm{R}}(t) = \left[\mathbf{G}_b^{\mathrm{A}}(-t)\right]^{\mathrm{T}} = \begin{pmatrix} G_b^{\mathrm{R}}(t) & \tilde{G}_b^{\mathrm{R}}(t) \\ \left[\tilde{G}_b^{\mathrm{R}}(t)\right]^* & \left[G_b^{\mathrm{R}}(t)\right]^* \end{pmatrix} \tag{16}$$

$$\mathbf{G}_b^{\mathrm{K}}(t) = \begin{pmatrix} G_b^{\mathrm{K}}(t) & \tilde{G}_b^{\mathrm{K}}(t) \\ -\left[\tilde{G}_b^{\mathrm{K}}(-t)\right]^* & G_b^{\mathrm{K}}(-t) \end{pmatrix} \tag{17}$$

The retarded Green's functions encode information on the single-particle density of states, and also describe linear response of the system to external perturbations:

$$G_b^{\mathrm{R}}(t) = -i\theta(t)\left\langle\left[\hat{b}(t),\hat{b}^\dagger(0)\right]\right\rangle \tag{18}$$

$$\tilde{G}_b^{\mathrm{R}}(t) = -i\theta(t)\left\langle\left[\hat{b}(t),\hat{b}(0)\right]\right\rangle \tag{19}$$

The Keldysh Green functions encode information on the distribution functions:

$$G_b^{\mathrm{K}}(t) = -i\left\langle\left\{\hat{b}(t),\hat{b}^\dagger(0)\right\}\right\rangle \tag{20}$$

$$\tilde{G}_b^{\mathrm{K}}(t) = -i\left\langle\left\{\hat{b}(t),\hat{b}(0)\right\}\right\rangle \tag{21}$$

As the action of equation (14) only has linear and quadratic terms in the mechanical fields, the MR can exactly be integrated out[56]. The resulting action that describes only photonic degrees of freedom is

$$S_{\mathrm{eff}} = S_{a_1}^0 + S_{a_2}^0 + \frac{g^2}{8}\int\int_{\mathbb{R}^2} dt\,dt'\left[s^{\mathrm{R}}(t,t')+s^{\mathrm{K}}(t,t')\right] \tag{22}$$

$$s^{\mathrm{R,K}}(t,t') = \sum_{ijkl=\mathrm{cl,q}}\zeta_{ijkl}^{\mathrm{R,K}}\left[s_{ijkl}^{\mathrm{R,K}}(t,t')+\mathrm{c.c.}\right], \tag{23}$$

$$s_{ijkl}^{\mathrm{R,K}}(t,t') = a_{2,i}^*(t)a_{1,j}(t)G_b^{\mathrm{R,K}}(t-t')a_{2,k}(t')a_{1,l}^*(t')$$
$$+ a_{2,i}^*(t)a_{1,j}(t)\tilde{G}_b^{\mathrm{R,K}}(t-t')a_{2,k}^*(t')a_{1,k}(t') \tag{24}$$

Here $\zeta_{ijkl}^{\mathrm{R}}=1$ if the interaction term has an odd number of quantum fields and $\zeta_{ijkl}^{\mathrm{R}}=0$ otherwise, while $\zeta_{ijkl}^{\mathrm{K}}=1$ if there is both one quantum field between the i, j components and one quantum field between the k, l component, that is, a total of two quantum fields, and $\zeta_{ijkl}^{\mathrm{K}}=0$ otherwise. The first two terms of equation (22) represent the non-interacting cavities, the third term describes the coherent time-non-local photon–photon interaction while the fourth term describes the extra noise that perturbs the cavities due to their interaction with the MR. As one can see, the diagonal (off-diagonal) MR Green function $G_b^{\mathrm{R}}(t)$ $\left[\tilde{G}_b^{\mathrm{R}}(t)\right]$ mediates a cross Kerr type interaction (two-photon tunnelling) between the cavities. From this effective action, it is clear that modifying the MR Green's functions leads to a modification of the effective photon–photon interaction.

Finally, following ref. 56, one can show that the interaction term s^{R} in the action of equation (22) is equivalent, in the cavity effective equation of motion, to the contribution highlighted in equation (7). A less-elegant alternative approach to obtain this effective equation of motion is to first solve the Heisenberg–Langevin equation for $\hat{\beta}$. This solution is used to eliminate the $\hat{\beta}$ from the cavities' Heisenberg–Langevin equations. The effective photon–photon interaction, as well as the additional nonlinear noise term then explicitly appear. Another method is to derive an effective Markovian quantum master equation for the optical modes by adiabatically eliminating the MR degrees of freedom[55]. The validity of the adiabatic elimination relies on having a strongly damped MR or a weak ratio $\tilde{g}/E_\beta$. In contrast, the effective Keldysh action derived here is exact and can thus capture non-Markovian effects.

References

1. Aspelmeyer, M., Kippenberg, T. J. & Marquardt, F. *Cavity Optomechanics* (Springer, 2014).
2. Aspelmeyer, M., Kippenberg, T. J. & Marquardt, F. Cavity optomechanics. *Rev. Mod. Phys.* **86,** 1391–1452 (2014).
3. Rabl, P. Photon blockade effect in optomechanical systems. *Phys. Rev. Lett.* **107,** 063601 (2011).
4. Nunnenkamp, A., Børkje, K. & Girvin, S. Single-photon optomechanics. *Phy. Rev. Lett.* **107,** 063602 (2011).
5. Kronwald, A., Ludwig, M. & Marquardt, F. Full photon statistics of a light beam transmitted through an optomechanical system. *Phys. Rev. A* **87,** 013847 (2013).
6. Kronwald, A. & Marquardt, F. Optomechanically induced transparency in the nonlinear quantum regime. *Phys. Rev. Lett.* **111,** 133601 (2013).
7. Teufel, J. *et al.* Sideband cooling of micromechanical motion to the quantum ground state. *Nature* **475,** 359–363 (2011).
8. Chan, J. *et al.* Laser cooling of a nanomechanical oscillator into its quantum ground state. *Nature* **478,** 89–92 (2011).
9. Palomaki, T. A., Harlow, J. W., Teufel, J. D., Simmonds, R. W. & Lehnert, K. W. Coherent state transfer between itinerant microwave fields and a mechanical oscillator. *Nature* **495,** 210–214 (2013).
10. Brooks, D. W. *et al.* Non-classical light generated by quantum-noise-driven cavity optomechanics. *Nature* **488,** 476–480 (2012).
11. Safavi-Naeini, A. H. *et al.* Squeezed light from a silicon micromechanical resonator. *Nature* **500,** 185–189 (2013).
12. Purdy, T. P., Yu, P. L., Peterson, R. W., Kampel, N. S. & Regal, C. A. Strong optomechanical squeezing of light. *Phys. Rev. X* **3,** 031012 (2013).
13. Lemonde, M.-A., Didier, N. & Clerk, A. A. Nonlinear interaction effects in a strongly driven optomechanical cavity. *Phys. Rev. Lett.* **111,** 053602 (2013).
14. Lemonde, M.-A. & Clerk, A. A. Real photons from vacuum fluctuations in optomechanics: the role of polariton interactions. *Phys. Rev. A* **91,** 033836 (2015).
15. Bhattacharya, M., Uys, H. & Meystre, P. Optomechanical trapping and cooling of partially reflective mirrors. *Phys. Rev. A* **77,** 033819 (2008).
16. Komar, P. *et al.* Single-photon nonlinearities in two-mode optomechanics. *Phys. Rev. A* **87,** 013839 (2013).
17. Ludwig, M., Safavi-Naeini, A. H., Painter, O. & Marquardt, F. Enhanced quantum nonlinearities in a two-mode optomechanical system. *Phys. Rev. Lett.* **109,** 063601 (2012).
18. Liao, J.-Q., Law, C. K., Kuang, L.-M. & Nori, F. Enhancement of mechanical effects of single photons in modulated two-mode optomechanics. *Phys. Rev. A* **92,** 013822 (2015).
19. Xu, X., Gullans, M. & Taylor, J. M. Quantum nonlinear optics near optomechanical instabilities. *Phys. Rev. A* **91,** 013818 (2015).
20. Xuereb, A., Genes, C. & Dantan, A. Strong coupling and long-range collective interactions in optomechanical arrays. *Phys. Rev. Lett.* **109,** 223601 (2012).
21. Chan, J., Safavi-Naeini, A. H., Hill, J. T., Meenehan, S. & Painter, O. Optimized optomechanical crystal cavity with acoustic radiation shield. *Appl. Phys. Lett.* **101,** 081115 (2012).
22. Armour, A. D., Blencowe, M. P. & Schwab, K. C. Entanglement and decoherence of a micromechanical resonator via coupling to a cooper-pair box. *Phys. Rev. Lett.* **88,** 148301 (2002).
23. Pirkkalainen, J. M. *et al.* Cavity optomechanics mediated by a quantum two-level system. *Nat. Commun.* **6,** 6981 (2015).

24. Didier, N., Pugnetti, S., Blanter, Y. M. & Fazio, R. Detecting phonon blockade with photons. *Phys. Rev. B* **84**, 054503 (2011).
25. Szorkovszky, A., Doherty, A. C., Harris, G. I. & Bowen, W. P. Mechanical squeezing via parametric amplification and weak measurement. *Phys. Rev. Lett.* **107**, 213603 (2011).
26. Szorkovszky, A., Clerk, A. A., Doherty, A. C. & Bowen, W. P. Detuned mechanical parametric amplification as a quantum non-demolition measurement. *New J. Phys.* **16**, 043023 (2014).
27. Farace, A. & Giovannetti, V. Enhancing quantum effects via periodic modulations in optomechanical systems. *Phys. Rev. A* **86**, 013820 (2012).
28. Lü, X.-Y. *et al.* Squeezed optomechanics with phase-matched amplification and dissipation. *Phys. Rev. Lett.* **114**, 093602 (2015).
29. Braunstein, S. L. & van Loock, P. Quantum information with continuous variables. *Rev. Mod. Phys.* **77**, 513–577 (2005).
30. Bose, S., Jacobs, K. & Knight, P. L. Preparation of nonclassical states in cavities with a moving mirror. *Phys. Rev. A* **56**, 4175–4186 (1997).
31. Hakioğlu, T. & Türeci, H. Correlated phonons and the T_c-dependent dynamical phonon anomalies. *Phys. Rev. B* **56**, 11174–11183 (1997).
32. Misochko, O. V., Hu, J. & Nakamura, K. G. Controlling phonon squeezing and correlation via one- and two-phonon interference. *Phys. Lett. A* **375**, 4141–4146 (2011).
33. Misochko, O. V. Nonclassical states of lattice excitations: squeezed and entangled phonons. *Phys. Usp.* **56**, 868 (2013).
34. Thompson, J. D. *et al.* Strong dispersive coupling of a high-finesse cavity to a micromechanical membrane. *Nature* **452**, 72–75 (2008).
35. Grudinin, I. S., Lee, H., Painter, O. & Vahala, K. J. Phonon laser action in a tunable two-level system. *Phys. Rev. Lett.* **104**, 083901 (2010).
36. Safavi-Naeini, A. H. & Painter, O. Proposal for an optomechanical traveling wave phonon–photon translator. *New J. Phys.* **13**, 013017 (2011).
37. Szorkovszky, A., Brawley, G. A., Doherty, A. C. & Bowen, W. P. Strong thermomechanical squeezing via weak measurement. *Phys. Rev. Lett.* **110**, 184301 (2013).
38. Mari, A. & Eisert, J. Gently modulating optomechanical systems. *Phys. Rev. Lett.* **103**, 213603 (2009).
39. Andrews, R. W., Reed, A. P., Cicak, K., Teufel, J. D. & Lehnert, K. W. Quantum-enabled temporal and spectral mode conversion of microwave signals. *Nat. Commun.* **6**, 10021 (2015).
40. Patil, Y. S., Chakram, S., Chang, L. & Vengalattore, M. Thermomechanical two-mode squeezing in an ultrahigh-q membrane resonator. *Phys. Rev. Lett.* **115**, 017202 (2015).
41. Cirac, J. I., Parkins, A. S., Blatt, R. & Zoller, P. 'dark' squeezed states of the motion of a trapped ion. *Phys. Rev. Lett.* **70**, 556–559 (1993).
42. Rabl, P., Shnirman, A. & Zoller, P. Generation of squeezed states of nano-mechanical resonators by reservoir engineering. *Phys. Rev. B* **70**, 205304 (2004).
43. Parkins, A. S., Solano, E. & Cirac, J. I. Unconditional two-mode squeezing of separated atomic ensembles. *Phys. Rev. Lett.* **96**, 053602 (2006).
44. Dalla Torre, E. G., Otterbach, J., Demler, E., Vuletic, V. & Lukin, M. D. Dissipative preparation of spin squeezed atomic ensembles in a steady state. *Phys. Rev. Lett.* **110**, 120402 (2013).
45. Tan, H., Li, G. & Meystre, P. Dissipation-driven two-mode mechanical squeezed states in optomechanical systems. *Phys. Rev. A* **87**, 033829 (2013).
46. Didier, N., Qassemi, F. & Blais, A. Perfect squeezing by damping modulation in circuit quantum electrodynamics. *Phys. Rev. A* **89**, 013820 (2014).
47. Kronwald, A., Marquardt, F. & Clerk, A. A. Arbitrarily large steady-state bosonic squeezing via dissipation. *Phys. Rev. A* **88**, 063833 (2013).
48. Wollman, E. E. *et al.* Quantum squeezing of motion in a mechanical resonator. *Science* **349**, 952–955 (2015).
49. Pirkkalainen, J.-M., Damskägg, E., Brandt, M., Massel, F. & Sillanpää, M. A. Squeezing of quantum noise of motion in a micromechanical resonator. *Phys. Rev. Lett.* **115**, 243601 (2015).
50. Lecocq, F., Clark, J. B., Simmonds, R. W., Aumentado, J. & Teufel, J. D. Quantum nondemolition measurement of a nonclassical state of a massive object. *Phys. Rev. X* **5**, 041037 (2015).
51. Demirplak, M. & Rice, S. A. Adiabatic Population Transfer with Control Fields. *J. Phys. Chem. A* **107**, 9937–9945 (2003).
52. Demirplak, M. & Rice, S. A. On the consistency, extremal, and global properties of counterdiabatic fields. *J. Chem. Phys.* **129**, 154111 (2008).
53. Berry, M. V. Transitionless quantum driving. *J. Phys. A* **42**, 365303 (2009).
54. Lemonde, M.-A., Didier, N. & Clerk, A. A. Antibunching and unconventional photon blockade with gaussian squeezed states. *Phys. Rev. A* **90**, 063824 (2014).
55. Gardiner, C. & Zoller, P. *Quantum Noise: A Handbook of Markovian and Non-Markovian Quantum Stochastic Methods with Applications to Quantum Optics.* (Springer Series in Synergetics, Springer, 2004).
56. Kamenev, A. *Field Theory of Non-Equilibrium Systems* (Cambridge Univ. Press, 2011).

Acknowledgements

This work was supported by NSERC.

Authors contributions

All authors participated in the conception and planning of the project and were involved in the analysis and interpretation of the results. In particular, M.-A.L. and N.D. led the derivation of theoretical results (with assistance from A.A.C.), and performed all numerical simulations. M.-A.L., N.D. and A.A.C. wrote the manuscript; A.A.C. supervised the project.

Additional information

Frequency and bandwidth conversion of single photons in a room-temperature diamond quantum memory

Kent A.G. Fisher[1], Duncan G. England[2], Jean-Philippe W. MacLean[1], Philip J. Bustard[2], Kevin J. Resch[1] & Benjamin J. Sussman[2,3]

The spectral manipulation of photons is essential for linking components in a quantum network. Large frequency shifts are needed for conversion between optical and telecommunication frequencies, while smaller shifts are useful for frequency-multiplexing quantum systems, in the same way that wavelength division multiplexing is used in classical communications. Here we demonstrate frequency and bandwidth conversion of single photons in a room-temperature diamond quantum memory. Heralded 723.5 nm photons, with 4.1 nm bandwidth, are stored as optical phonons in the diamond via a Raman transition. Upon retrieval from the diamond memory, the spectral shape of the photons is determined by a tunable read pulse through the reverse Raman transition. We report central frequency tunability over 4.2 times the input bandwidth, and bandwidth modulation between 0.5 and 1.9 times the input bandwidth. Our results demonstrate the potential for diamond, and Raman memories in general, as an integrated platform for photon storage and spectral conversion.

[1] Institute for Quantum Computing and Department of Physics and Astronomy, University of Waterloo, 200 University Avenue West, Waterloo, Ontario, Canada N2L 3G1. [2] National Research Council of Canada, 100 Sussex Drive, Ottawa, Ontario, Canada K1A 0R6. [3] Department of Physics, University of Ottawa, 150 Louis Pasteur, Ottawa, Ontario, Canada K1N 6N5. Correspondence and requests for materials should be addressed to K.J.R. (email: kresch@uwaterloo.ca) or to B.J.S. (email: ben.sussman@nrc.ca).

The fragility of the quantum state is a challenge facing all quantum technologies. Great efforts have been undertaken to mitigate the deleterious effects of decoherence by isolating quantum systems, for example, by cryogenically cooling and isolating in vacuum. State-of-the-art decoherence times are now measured in hours[1]. An alternative approach is to build quantum technologies that execute on ultrafast timescales—as short as femtoseconds—such that operations can be completed before decoherence overwhelms unitarity. A shining example is the Raman quantum memory[2-5], which can absorb single photons of femtosecond duration and release them on demand several picoseconds later[6]. While picosecond storage times are not appropriate for conventional quantum memory applications such as long-distance communication, it has been suggested[2] that Raman quantum memories can find additional uses such as frequency and bandwidth conversion.

Controlling the spectral properties of single photons is essential for a wide array of emerging optical quantum technologies spanning quantum sensing[7], quantum computing[8] and quantum communications[9]. Essential components for these technologies include single-photon sources[10], quantum memories[11], waveguides[12] and detectors[13]. The ideal spectral operating parameters (wavelength and bandwidth) of these components are rarely similar; thus, frequency conversion and spectral control are key enabling steps for component hybridization[14]. Beyond hybridization, frequency conversion is an area of emerging interest in quantum optical processing. The frequency degree of freedom can be used along side conventional encoding in, for example, polarization, or time-bin, to build quantum states of higher dimensionality[15].

Spectral control is a mature field in ultrafast optics where phase- and amplitude-shaping of a THz-bandwidth pulse can be achieved using passive pulse-shaping elements in the Fourier plane[16]. Meanwhile, a range of nonlinear optical techniques[17] such as second harmonic generation (SHG), sum- and difference-frequency generation, four-wave mixing and Raman scattering are routinely employed to shift the frequency of laser pulses.

Extending these frequency conversion techniques into the quantum regime is a critical task for many quantum technologies but is made difficult by the low intensity of single photons, and the sensitivity of quantum states to loss and noise. Despite these challenges, quantum frequency conversion[18] has been demonstrated in a number of systems including waveguides in nonlinear crystals[19-24], photonic crystal fibres[25] and atomic vapour[26]. Similarly, photon bandwidth compression has been shown using chirped-pulse upconversion[27]. Full control over the spectral properties of single photons[28] has been proposed using second-[14,29] and third-order[30] optical nonlinearities.

Large frequency shifts, such as those achieved using sum- and difference-frequency generation, are desirable to convert photons to, and from, the telecommunication band. Meanwhile, smaller shifts can be useful for frequency-multiplexing in several closely spaced bins. This concept is widely used in classical fibre optics, where wavelength division multiplexing is employed to achieve data rates far beyond that which could be achieved with monochromatic light[31]. However, in quantum optics the utility of frequency multiplexing has only recently been explored[15,32-34]. In a frequency-multiplexed quantum architecture, it will be critical to build components that can add, drop, and manipulate different frequency bins: it has been proposed that frequency-selective quantum memories can perform this task[15,33]. As with classical wavelength division multiplexing, the frequency bins will likely be closely spaced so that small shifts around a central frequency will be required.

In this article, we demonstrate the use of a Raman quantum memory to perform quantum frequency conversion; we manipulate the spectral properties of THz-bandwidth photons using a memory in the optical phonon modes of diamond[6]. Crucially, the quantum properties of the photon must be maintained even while the carrier frequency and bandwidth are modified. In our demonstration, a signal photon is mapped into an optical phonon by the write pulse, and then retrieved with its spectral properties modified according to the properties of the read pulse.

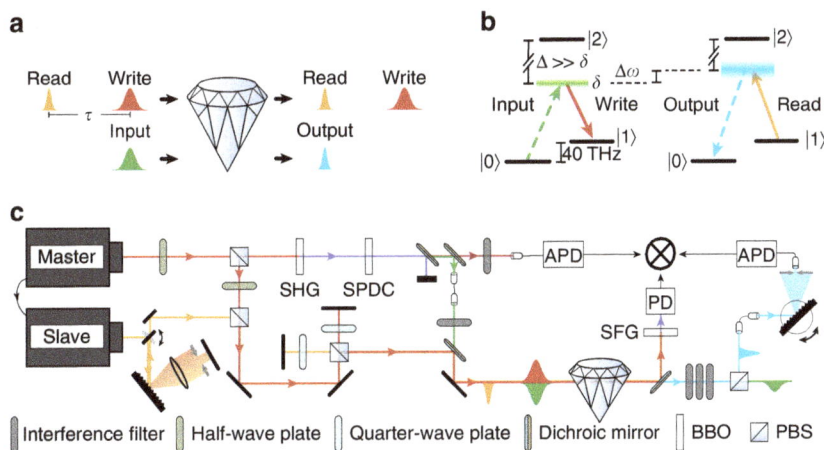

Figure 1 | Concept and experiment. (**a**) Input signal photons stored in the diamond by the strong write pulse can be retrieved with modified spectral properties upon output. The output spectrum is controlled by the spectrum of the read pulse. (**b**) Photons are Raman-absorbed to create optical phonons ($|1\rangle$), 40 THz above the ground state ($|0\rangle$). A read pulse of tunable wavelength and bandwidth retrieves the photon, determining its spectrum. Here, Δ is the detuning from the conduction band ($|2\rangle$), δ is the input photon bandwidth, $\Delta\omega$ is the detuning between input and output frequencies. (**c**) The master laser (red) is split between the write field and photon source. In the photon source, frequency-doubled laser light pumps SPDC and heralded input signal photons (green) are generated. Signal photons are Raman-absorbed into the optical phonon modes in the diamond via the write field. The slave laser (orange) emits the read field which retrieves the photon from the diamond after time τ. The output signal photon (blue) spectrum is measured on a monochromator. The SFG of read and write pulses triggers the experiment. Coincident detections of output, herald photons and SFG events are measured by a coincidence logic unit. SFG, second-harmonic generation; SPDC, spontaneous parametric downconversion; APD, avalanche photodiode; PD, photodiode; SFG, sum-frequency generation; BBO, β-barium borate; PBS, polarizing beamsplitter.

Results

Experiment. The frequency converter is based on a quantum memory[6] modelled by the Λ-level system shown in Fig. 1b, where an input signal photon (723.5 nm centre wavelength and bandwidth $\delta = 4.1$ nm full-width at half-maximum (FWHM)) and a strong write pulse (800, 5 nm FWHM) are in Raman resonance with the optical phonon band (frequency 40 THz). The large detuning of both fields from the conduction band (detuning $\Delta \approx 950$ THz) allows for the storage of high-bandwidth photons, while the memory exhibits a quantum-level noise floor even at room temperature[6]. The input signal photon is stored in the memory by Raman absorption with the write pulse, creating an optical phonon. After a delay τ, the read pulse annihilates the phonon and creates a modified output photon. By tuning the wavelength and bandwidth of the read pulse, we convert the wavelength of the input signal photon over a range of 17 nm as well as performing bandwidth compression to 2.2 nm and expansion to 7.6 nm (FWHM). The diamond memory is ideally suited to this task, offering low-noise frequency manipulation of THz-bandwidth quantum signals at a range of visible and near-infrared wavelengths in a robust room-temperature device[35].

The experimental setup is shown in Fig. 1c. The master laser for the experiment is a Ti:sapphire oscillator producing 44 nJ pulses at a repetition rate of 80 MHz and a central wavelength of 800 nm. This beam is split in two parts: the photon-source pump and the write field. In the photon source, the SHG of the laser light pumps collinear type-I spontaneous parametric down-conversion (SPDC) in a β-barium borate (BBO) crystal, generating photons in pairs, with one at 723.5 nm (input signal) and the other at 894.6 nm (herald). The herald photon is detected on an avalanche photodiode (APD), while the input is spatially and spectrally filtered and overlapped with the orthogonally polarized write pulse on a dichroic mirror. The input signal photon and write field are incident on the $\langle 100 \rangle$ face of the diamond and the input is Raman-absorbed.

The photon is retrieved from the diamond using a read pulse produced by a second Ti:sapphire laser (slave), whose repetition rate is locked to the master, but whose frequency and bandwidth can be independently modified. In this experiment we vary the

read field wavelength between 784 and 814 nm, and its bandwidth between 2.1 and 12.1 nm FWHM. To narrow the bandwidth of the read pulse, a folded-grating 4f-system[16] with a narrow slit is used, while in all other configurations the 4f line is removed. The read pulse is then overlapped with the write on a polarizing beamsplitter, arriving at the diamond a time τ after storage. The horizontally polarized read pulse retrieves a vertically polarized photon (output) from the diamond with spectral shape close to that of the read pulse, blue-shifted by the phonon frequency (40 THz).

The read and write pulses are separated from the signal photons after the diamond by a dichroic mirror; sum-frequency generation (SFG) of the pulses is detected on a fast-photodiode (PD) and used to confirm their successful overlap. Frequency-converted output photons are separated from any unstored input photons by a polarizing beamsplitter, coupled into a single-mode fibre and directed to a monochromator. The spectrally filtered output from the monochromator is coupled into a multi-mode fibre and detected on an APD. Coincident detections between output, herald and read–write SFG events are measured; the experiment is triggered by the joint detection of a herald photon and an SFG signal. (see Methods for further details.)

Frequency shifts. Frequency conversion of the signal photon is observed by tuning the slave laser wavelength. We vary this from 784 to 812 nm and measure the output photon wavelength using the monochromator, recording threefold coincidence events. The resulting output spectra, with the read pulse centred at 792 and 808 nm, are shown in Fig. 2a (hollow circles). The spectrum of each read pulse (solid lines), blue-shifted by the phonon frequency, is plotted alongside the relevant output photon spectrum to show how the photon spectrum is determined by the read pulse. We find the peaks of the output spectra to be 716 and 728 nm with bandwidths 3.3 and 3.5 nm, FWHM respectively, making the output spectrally distinguishable from the input (green).

Following retrieval, the time-correlations characteristic of SPDC photon-pairs are preserved. This is measured by scanning the electronic delay between the signal and herald detection

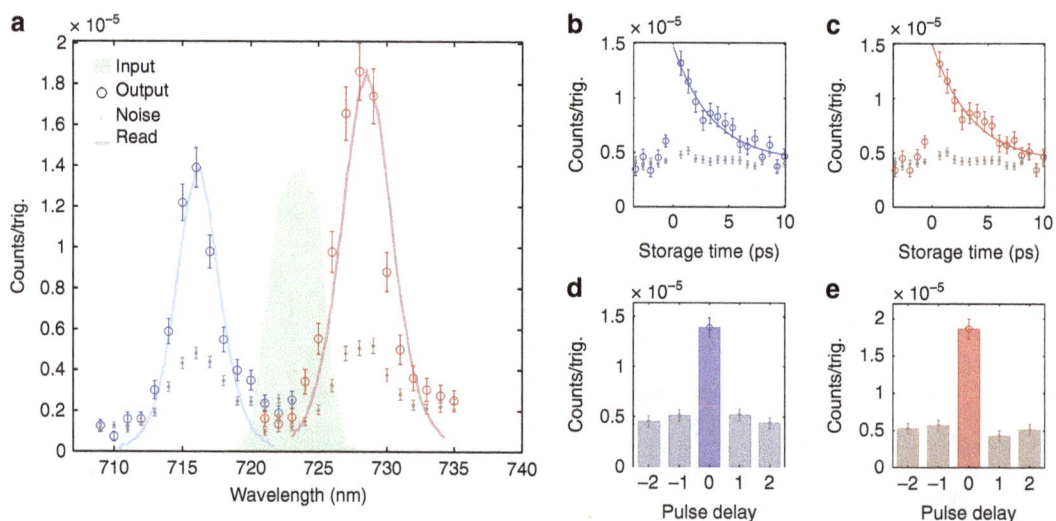

Figure 2 | Frequency conversion. (**a**) The measured blue- and red-shifted output photon spectra (hollow circles), and noise (dots), when the read beam is tuned to 792 and 808 nm, respectively. Corresponding read beam spectra, blue-shifted by the phonon frequency, are shown (solid lines) for reference along with the input photon spectrum (green). (**b**) Retrieved blue- and (**c**) Red-shifted signal (hollow circles), and noise (dots), as read–write delay is scanned. An exponential fit gives a phonon lifetime of 3.5 ps. (**d**) Coincidence detection events between blue- and (**e**) red-shifted output and herald photons while scanning the electronic delay between them, as measured at the peak of the spectrum. (**a**–**e**) Error bars show 1 standard deviation calculated assuming Poissonian noise.

events in steps of 12.5 ns (the time between adjacent oscillator pulses) and counting coincident detections. Results are shown in Fig. 2d,e for blue- and red-shifted cases, respectively. We quantify this using the two-mode intensity cross-correlation function between output signal and herald fields given by $g^{(2)}_{s,h} = P_{s,h}/(P_s P_h)$. Here, $P_s(P_h)$ is the probability of detecting a photon in the signal (herald) mode, and $P_{s,h}$ is the probability of measuring a joint detection. A measurement of $g^{(2)}_{s,h} > 2$ indicates non-classical correlation[36,37] (see Methods), whereas uncorrelated photon detections, for example, from noise, give $g^{(2)}_{s,h} = 1$. We calculate the values of $g^{(2)}_{s,h}$ at the peak of the blue- and red-shifted spectra to be 2.7 ± 0.2 and 3.4 ± 0.3, respectively.

Figure 2b,c shows the blue-(red-)shifted photon retrieval rate as a function of the optical delay τ between read and write pulses. An exponential function is fit to the data and we find a memory lifetime of 3.5 ps as expected from the lifetime of the optical phonon[35,38]. This exceeds 12 times the duration of the input photon (see Methods). Also plotted in Fig. 2a–c are the measured coincidences due to noise (dots), which are measured by taking the average of the ± 12.5 and ± 25 ns time-bins as shown in Fig. 2d,e. Noise comes from two processes: four-wave mixing[35]; and read pulses scattering from thermally populated phonons producing anti-Stokes light[6,35]. The latter can be reduced by an order of magnitude by cooling the diamond to $-60\,°\mathrm{C}$.

Figure 3 shows $g^{(2)}_{s,h}$ measured at the peak of each output signal spectrum as the read wavelength is tuned over a 30 nm range. We find that blue- and red-shifted single photons maintain non-classical correlations over a 17 nm range. Since noise is uncorrelated with herald photons, we expect the noise to have cross-correlation $g^{(2)} = 1$. The $g^{(2)}_{s,h}$ will increase from 1 in proportion to the signal-to-noise ratio (see Methods for further details),

$$g^{(2)}_{s,h} \approx 1 + \frac{\eta_h \eta_{fc}(\Delta\omega)}{P_n}. \tag{1}$$

Here $\eta_h = P_{s,h}/P_h = 0.13\%$ is the photon heralding efficiency in the signal arm including the monochromator, $P_n = 3.8 \times 10^{-6}$ is the probability of detecting a noise photon, and $\eta_{fc}(\Delta\omega)$ is the conversion efficiency as a function of frequency detuning, $\Delta\omega$, between input and output photons. The conversion efficiency

$\eta_{fc}(\Delta\omega) = \eta_{fc}(0) \times \mathrm{sinc}^2(\Delta kL/2)$, where $L = 2.3$ mm is the length of the diamond along the propagation axis, and $\Delta\mathbf{k} = \mathbf{k}_i - \mathbf{k}_o + \mathbf{k}_r - \mathbf{k}_w$ is the phase mismatch between the input signal (i), output signal (o), read (r) and write (w) fields because of material dispersion in diamond[35]. The conversion efficiency was measured to be $\eta_{fc}(0) = 1.1\%$, inside the diamond at zero detuning. As the diamond is not anti-reflection coated, a 17% reflection loss occurs at each face.

Inserting experimental parameters into equation 1 returns $g^{(2)}_{s,h} \approx 1 + 3.7 \times \mathrm{sinc}^2(\Delta kL/2)$, which is plotted along side data in Fig. 3 (solid line). The close agreement with experiment suggests that the limitation on frequency conversion comes primarily from phase-matching conditions. We then expect that the range of frequency conversion in diamond can be extended by modifying the phase-matching conditions. This could be achieved by shortening the diamond crystal, or by employing non-collinear beam geometries[39]. In the current configuration, the maximum conversion efficiency is limited to $\sim 1\%$ due to the efficiency of the quantum memory[6]. However, we note that this could be improved with increased intensity in read and write pulses, or by increasing the Raman coupling, for example, by the use of a waveguide.

Bandwidth manipulation. Bandwidth conversion is observed by tuning the slave laser bandwidth. With the read pulse wavelength centred at 801 nm, its bandwidth could be tuned from 12.1 to 2.1 nm FWHM using a slit in a grating $4f$ line. Figure 4a,b shows the resulting narrowed (expanded) output photon spectrum with the corresponding read pulse spectrum, blue-shifted by the phonon frequency, and the input signal photon spectrum for reference. The resulting narrowed and expanded photon bandwidths are 2.2 and 7.6 nm, FWHM, respectively. Figure 4c,d shows the conservation of timing correlations between bandwidth-narrowed (-expanded) photons and herald photons, respectively. We measure $g^{(2)}_{s,h} = 2.6 \pm 0.2$ in the narrowed bandwidth case, $g^{(2)}_{s,h} = 2.9 \pm 0.2$ in the expanded bandwidth case, showing that bandwidth-converted output light from the diamond maintains non-classical correlations with herald photons.

Discussion

We have demonstrated ultrafast quantum frequency manipulation by adjusting the central wavelength and spectrum, of THz-bandwidth heralded single photons. We achieve this spectral control using a modified Raman quantum memory in diamond. The single photons are written to the memory from one spectral mode, and recalled to another. Critically—and unlike frequency conversion based on, for example, amplification—the non-classical photon statistics in our demonstration were retained after spectral manipulation. Diamond therefore offers low-noise THz-bandwidth storage and frequency control of single photons on a single, robust, room-temperature platform. We have demonstrated frequency conversion of a single polarization; two memories could be used in parallel to convert a polarization qubit. Quantum memories for long-distance quantum communication typically demand long storage times at the expense of high bandwidth; this application leverages a high-bandwidth memory where long storage time is not relevant. We believe that this system could find use in a number of applications, including entangling two photon-pair sources of different colour; reducing the bandwidth of the ubiquitous ultrafast SPDC photon source, without losing photons (as occurs in passive filtering); broadening the bandwidth of an SPDC photon source to match a chosen material system (not possible with a passive filtering system); and increasing the dimensionality of quantum encoded information[15]. Ultimately, we believe that arbitrary optical

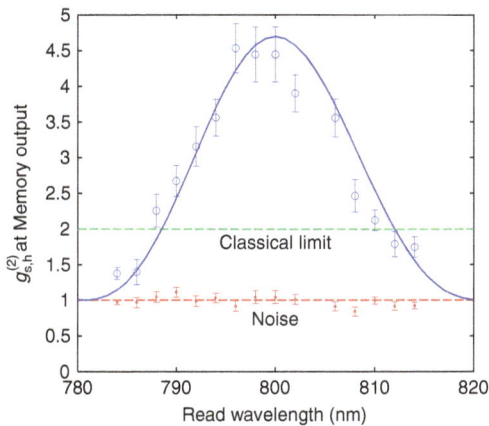

Figure 3 | Range of frequency conversion. Measured $g^{(2)}_{s,h}$ of frequency-shifted photons. Frequency conversion, tuning the read beam over a 30 nm range, is observed. Non-classical statistics, that is, $g^{(2)}_{s,h} > 2$, are maintained over a 18 nm range. The estimated $g^{(2)}_{s,h}$ (solid line) which depends on $\mathrm{sinc}^2(\Delta kL/2)$ agrees well with experimental data suggesting that the range of frequency conversion is determined by phase-matching conditions. Error bars show 1 standard deviation calculated assuming Poissonian noise.

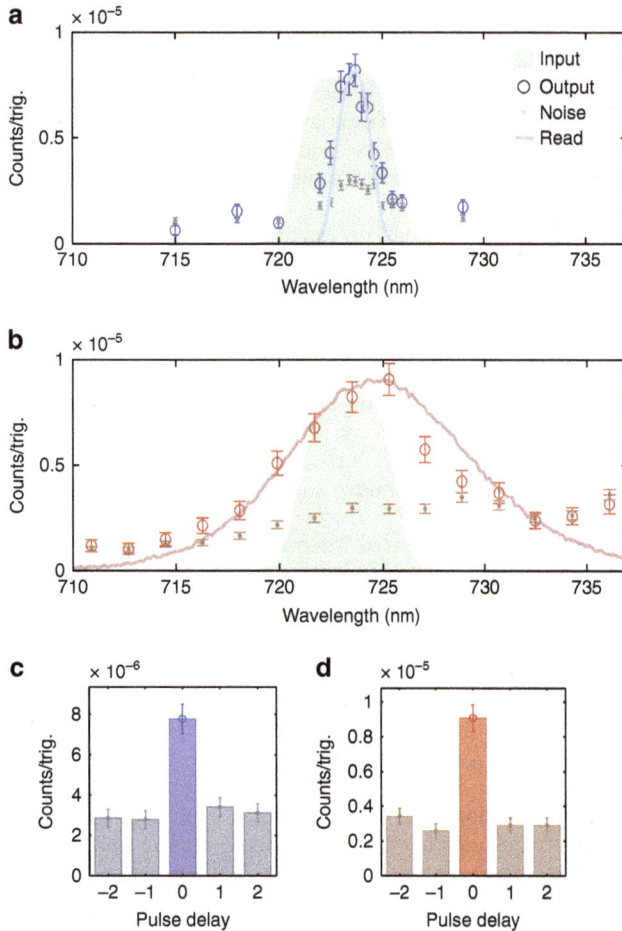

Figure 4 | Bandwidth conversion. (**a**) Narrowed output spectrum (hollow circles) and noise (dots) with read beam FWHM at 2.1 nm. (**b**) Expanded output photon spectrum (hollow circles) and noise (dots) with read beam FWHM at 12.1 nm. (**a,b**) Corresponding read beam spectra, blue-shifted by the phonon frequency, are shown (solid lines) for reference along with the input photon spectrum (green). (**c**) Coincidence detection events between bandwidth narrowed output photons, and (**d**) expanded output photons and heralds while scanning the electronic delay between them, as measured at the peak of the spectrum. (**a–d**) Error bars show 1 standard deviation calculated assuming Poissonian noise.

function generation[40]—with both classical and quantum light—and associated signal processing capabilities will be a platform for future technology development. We expect that the large-bandwidth nonlinear optical conversion of Raman-based quantum memories will find use in implementations of these generators.

Methods

Photon source. Laser light is frequency-doubled by type-I SHG in a 1 mm BBO crystal before pumping collinear type-I SPDC in a second 1 mm BBO crystal. Horizontally polarized photon pairs are generated at 894.6 and 723.5 nm. Remaining pump light is filtered out and photon pairs are separated by a 801 nm long-pass dichroic mirror. The 894.6 nm photon passes through a 5 nm interference filter, is coupled into a single-mode fibre, and detected on an APD. A detection heralds the presence of the 723.5 nm photon, which is spatially filtered in a single-mode fibre and spectrally filtered by an interference filter with bandwidth of 5 nm (FWHM). The input and write pulses are overlapped using a 750 nm shortpass dichroic mirror. The input and write pulses are focused into the diamond by an achromatic lens of focal length 6 cm.

Diamond. The diamond is a high-purity, low birefringence crystal grown by chemical vapour deposition by Element Six Ltd. The crystal is 2.3 mm long, cut along the ⟨100⟩ lattice direction, and polished on two sides.

Storage time. Absorption of the input photon by the diamond lattice is observed by an 18% dip in input-herald coincidences when the input photon and write field arrive at the diamond simultaneously. The duration of the input photon can be deconvolved from the width of the absorption dip, 346 fs. With write pulses 190 fs in duration, the input photon pulse duration time is $\sqrt{346^2 - 190^2}$=289 fs, assuming transform-limited Gaussian pulses. The characteristic storage time of the diamond memory is 3.5 ps, found from an exponential fit to storage data, over 12 times the duration of the input pulse.

Laser locking. The repetition rate of the slave laser is locked to that of the master using a Spectra Physics Lok-to-Clock device. We send read and write beams through a cross-correlator (type-II SFG in a 1 mm BBO crystal) and detect the resulting signal on a PD confirming that the time difference between the two pulses is $\leq$ 200 fs. We measure a typical SFG signal rate of 2.5 MHz; we use this signal to trigger the experiment.

Monochromator. The monochromator (Acton SP2300) is comprised of a 1,200 g mm^{-1} grating between two 30 cm focal length spherical mirrors. The output is coupled to a multi-mode fibre (105 μm core). The apparatus has a spectral resolution of 1.1 nm and an overall efficiency of 10% at 723 nm.

Cross-correlation function. The cross-correlation function between the herald and frequency-converted light is given by $g_{s,h}^{(2)} = P_{s,h}/(P_s P_h)$. Classically, $g_{s,h}^{(2)}$ is upper-bounded by a Cauchy–Schwarz inequality[36,37] $g_{s,h}^{(2)} \leq \sqrt{g_{s,s}^{(2)} g_{h,h}^{(2)}}$. Here, the terms on the right-hand side are the intensity auto-correlation functions for the output signal and herald fields, which we assume, being produced by SPDC, follow thermal statistics and have $g_{s,s}^{(2)} = g_{h,h}^{(2)} = 2$. Adding any uncorrelated noise would strictly lower terms on the right-hand side towards 1. To model the effect of noise on this measurement, we assume that the signal is made up of a mixture of noise photons (detected with probability $P_n = 3.8 \times 10^{-6}$) and frequency-converted photons (probability P_γ), such that

$$P_s = P_\gamma + P_n \approx P_h \eta_h \eta_{fc}(\Delta\omega) + P_n, \tag{2}$$

$$P_{s,h} = P_{\gamma,h} + P_{n,h} \approx P_h \eta_h \eta_{fc}(\Delta\omega) + P_n P_h, \tag{3}$$

where $\eta_h = 1.3 \times 10^{-3}$ is the heralding efficiency which equates to the collection efficiency of the entire signal arm, including the monochromator, and $\eta_{fc}(\Delta\omega)$ is the efficiency of the quantum frequency conversion. This returns

$$g_{s,h}^{(2)} = \frac{\eta_h \eta_{fc}(\Delta\omega) + P_n}{P_h \eta_h \eta_{fc}(\Delta\omega) + P_n}, \tag{4}$$

from which equation 1 follows, given that $P_h \eta_h \eta_{fc}(\Delta\omega) \ll P_n$.

Background subtraction. When measured at the photon source the input and herald photon cross-correlation is $g_{in,h}^{(2)}$=164. Because of imperfect polarization extinction the input photon can, with low probability, traverse the monochromator and be detected, thereby artificially inflating the measured $g_{s,h}^{(2)}$ of the converted output. For this reason we make a measurement with no read/write pulses present and subtract these counts from the output signal when read/write pulses are present to portray an accurate value of $g_{s,h}^{(2)}$. As an example, in Fig. 3, the peak count rate is 19×10^{-6} photons per pulse compared with a background of 3×10^{-6} photons per pulse.

References

1. Zhong, M. *et al.* Optically addressable nuclear spins in a solid with a six-hour coherence time. *Nature* **517,** 177–180 (2015).
2. Nunn, J. *et al.* Mapping broadband single-photon wave packets into an atomic memory. *Phys. Rev. A* **75,** 011401(R) (2007).
3. Reim, K. *et al.* Towards high-speed optical quantum memories. *Nat. Photon.* **4,** 218–221 (2010).
4. Bustard, P. J., Lausten, R., England, D. G. & Sussman, B. J. Toward quantum processing in molecules: A THz-bandwidth coherent memory for light. *Phys. Rev. Lett.* **111,** 083901 (2013).
5. Michelberger, P. S. *et al.* Interfacing GHz-bandwidth heralded single photons with a warm vapour Raman memory. *New J. Phys.* **17,** 043006 (2015).
6. England, D. G. *et al.* Storage and retrieval of THz-bandwidth single photons using a room-temperature diamond quantum memory. *Phys. Rev. Lett.* **114,** 053602 (2015).
7. Giovannetti, V., Lloyd, S. & Maccone, L. Quantum-enhanced measurements: beating the standard quantum limit. *Science* **306,** 1330–1336 (2004).
8. Knill, E., Laflamme, R. & Milburn, G. J. A scheme for efficient quantum computation with linear optics. *Nature* **409,** 46–52 (2001).
9. Duan, L.-M., Lukin, M. D., Cirac, J. I. & Zoller, P. Long-distance quantum communication with atomic ensembles and linear optics. *Nature* **414,** 413–418 (2001).

10. Kurtsiefer, C., Mayer, S., Zarda, P. & Weinfurter, H. Stable solid-state source of single photons. *Phys. Rev. Lett.* **85**, 290–293 (2000).

11. Kozhekin, A. E., Mølmer, K. & Polzik, E. Quantum memory for light. *Phys. Rev. A.* **62**, 033809 (2000).

12. Politi, A., Cryan, M. J., Rarity, J. G., Yu, S. & O'Brien, J. L. Silica-on-silicon waveguide quantum circuits. *Science* **320**, 646–649 (2008).

13. Lita, A. E., Miller, A. J. & Nam, S. W. Counting near-infrared single-photons with 95% efficiency. *Opt. Express* **16**, 3032–3040 (2008).

14. Kielpinski, D., Corney, J. F. & Wiseman, H. M. Quantum optical waveform conversion. *Phys. Rev. Lett.* **106**, 130501 (2011).

15. Humphreys, P. C. *et al.* Continuous-variable quantum computing in optical time-frequency modes using quantum memories. *Phys. Rev. Lett.* **113**, 130502 (2014).

16. Weiner, A. M. Femtosecond pulse shaping using spatial light modulators. *Rev. Sci. Instrum.* **71**, 1929–1960 (2000).

17. Boyd, R. W. *Nonlinear Optics* (Academic Press, 2008).

18. Kumar, P. Quantum frequency conversion. *Opt. Lett.* **15**, 1476 (1990).

19. Rakher, M. T., Ma, L., Slattery, O., Tang, X. & Srinivasan, K. Quantum transduction of telecommunications-band single photons from a quantum dot by frequency upconversion. *Nat. Photon.* **4**, 786–791 (2010).

20. Rakher, M. T. *et al.* Simultaneous wavelength translation and amplitude modulation of single photons from a quantum dot. *Phys. Rev. Lett.* **107**, 083602 (2011).

21. Tanzilli, S. *et al.* A photonic quantum information interface. *Nature* **437**, 116–120 (2005).

22. Ikuta, R. *et al.* Wide-band quantum interface for visible-to-telecommunication wavelength conversion. *Nat. Commun.* **2**, 537 (2011).

23. De Greve, K. *et al.* Quantum-dot spin-photon entanglement via frequency downconversion to telecom wavelength. *Nature* **491**, 421–425 (2012).

24. Guerreiro, T. *et al.* Nonlinear interaction between single photons. *Phys. Rev. Lett.* **113**, 173601 (2014).

25. McGuinness, H. J., Raymer, M. G., McKinstrie, C. J. & Radic, S. Quantum frequency translation of single-photon states in a photonic crystal fiber. *Phys. Rev. Lett.* **105**, 093604 (2010).

26. Dudin, Y. O. *et al.* Entanglement of light-shift compensated atomic spin waves with telecom light. *Phys. Rev. Lett.* **105**, 260502 (2010).

27. Lavoie, J., Donohue, J. M., Wright, L. G., Fedrizzi, A. & Resch, K. J. Spectral compression of single photons. *Nat. Photon.* **7**, 363–366 (2013).

28. Raymer, M. G. & Srinivasan, K. Manipulating the color and shape of single photons. *Phys. Today* **65**, 32–37 (2012).

29. Brecht, B., Eckstein, A., Christ, A., Suche, H. & Silberhorn, C. From quantum pulse gate to quantum pulse shaper—engineered frequency conversion in nonlinear optical waveguides. *New J. Phys.* **13**, 065029 (2011).

30. McKinstrie, C. J., Mejling, L., Raymer, M. G. & Rottwitt, K. Quantum-state-preserving optical frequency conversion and pulse reshaping by four-wave mixing. *Phys. Rev. A* **85**, 053829 (2012).

31. Brackett, C. A. Dense wavelength division multiplexing networks: principles and applications. *IEEE J. Sel. Areas Commun.* **8**, 948–964 (1990).

32. Sinclair, N. *et al.* Spectral multiplexing for scalable quantum photonics using an atomic frequency comb quantum memory and feed-forward control. *Phys. Rev. Lett.* **113**, 053603 (2014).

33. Campbell, G. T. *et al.* Configurable unitary transformations and linear logic gates using quantum memories. *Phys. Rev. Lett.* **113**, 063601 (2014).

34. Donohue, J. M., Lavoie, J. & Resch, K. J. Ultrafast time-division demultiplexing of polarization-entangled photons. *Phys. Rev. Lett.* **113**, 163602 (2014).

35. England, D., Bustard, P., Nunn, J., Lausten, R. & Sussman, B. J. From photons to phonons and back: A THz optical memory in diamond. *Phys. Rev. Lett.* **111**, 243601 (2013).

36. Loudon, R. *The Quantum Theory of Light* (Oxford University Press, 2004).

37. Clauser, J. F. Experimental distinction between the quantum and classical field-theoretic predictions for the photoelectric effect. *Phys. Rev. D* **9**, 853–860 (1974).

38. Lee, K. C. *et al.* Macroscopic non-classical states and terahertz quantum processing in room-temperature diamond. *Nat. Photon.* **6**, 41–44 (2012).

39. Eckbreth, A. C. BOXCARS: Crossed-beam phase-matched CARS generation in gases. *Appl. Phys. Lett.* **32**, 421–423 (1978).

40. Kowligy, A. S. *et al.* Quantum optical arbitrary waveform manipulation and measurement in real time. *Opt. Express* **22**, 27942–27957 (2014).

Acknowledgements

We thank Matthew Markham and Alastair Stacey of Element Six Ltd. for the diamond sample. We also thank Rune Lausten, Paul Hockett, John Donohue, Michael Mazurek and Khabat Heshami for fruitful discussions. Doug Moffatt and Denis Guay provided important technical assistance. This work was supported by the Natural Sciences and Engineering Research Council of Canada, Canada Research Chairs, Ontario Centres of Excellence, and the Ontario Ministry of Research and Innovation Early Researcher Award.

Author contributions

K.A.G.F., D.G.E. and J.-P.W.M. performed the experiment and analysed the data. All authors contributed to the final manuscript.

Additional information

Competing financial interests: The authors declare no competing financial interests.

Diffractive imaging of a rotational wavepacket in nitrogen molecules with femtosecond megaelectronvolt electron pulses

Jie Yang[1], Markus Guehr[2,3], Theodore Vecchione[4], Matthew S. Robinson[1], Renkai Li[4], Nick Hartmann[4], Xiaozhe Shen[4], Ryan Coffee[4], Jeff Corbett[4], Alan Fry[4], Kelly Gaffney[4], Tais Gorkhover[4], Carsten Hast[4], Keith Jobe[4], Igor Makasyuk[4], Alexander Reid[4], Joseph Robinson[4], Sharon Vetter[4], Fenglin Wang[4], Stephen Weathersby[4], Charles Yoneda[4], Martin Centurion[1] & Xijie Wang[4]

Imaging changes in molecular geometries on their natural femtosecond timescale with sub-Angström spatial precision is one of the critical challenges in the chemical sciences, as the nuclear geometry changes determine the molecular reactivity. For photoexcited molecules, the nuclear dynamics determine the photoenergy conversion path and efficiency. Here we report a gas-phase electron diffraction experiment using megaelectronvolt (MeV) electrons, where we captured the rotational wavepacket dynamics of nonadiabatically laser-aligned nitrogen molecules. We achieved a combination of 100 fs root-mean-squared temporal resolution and sub-Angstrom (0.76 Å) spatial resolution that makes it possible to resolve the position of the nuclei within the molecule. In addition, the diffraction patterns reveal the angular distribution of the molecules, which changes from prolate (aligned) to oblate (anti-aligned) in 300 fs. Our results demonstrate a significant and promising step towards making atomically resolved movies of molecular reactions.

[1] Department of Physics and Astronomy, University of Nebraska-Lincoln, 855 N 16th Street, Lincoln, Nebraska 68588, USA. [2] PULSE Institute, SLAC National Accelerator Laboratory, Menlo Park, California 94025, USA. [3] Institute of Physics and Astronomy, Potsdam University, Potsdam 14476, Germany. [4] SLAC National Accelerator Laboratory, Menlo Park, California 94025, USA. Correspondence and requests for materials should be addressed to M.G. (email: mguehr@uni-potsdam.de) or to M.C. (email: martin.centurion@unl.edu) or to X.W. (email: wangxj@slac.stanford.edu).

The dynamical behaviour of molecules is governed by the complex interplay of a correlated system of nuclei and electrons which interact via Coulomb and exchange forces. Determining how the individual nuclei within a molecule move relative to one another during a molecular transformation represents a key step to understanding chemical reactivity. Developments in the field of femtochemistry enabled capturing the motion of nuclear wavepackets using purely spectroscopic measurements with femtosecond laser pulses[1–8]. For small molecules, structural information on the nuclear geometry can be indirectly inferred based on exact knowledge of spectral transitions involved in the probe process. For larger molecules, this inference becomes unfeasible. Diffraction techniques, which provide direct access to the position of each atom within a molecule, are a more effective approach. Over the last few decades, great strides have been taken to observe ultrafast dynamics with ultrashort X-ray pulses from synchrotrons and free electron lasers (FELs)[9–12] and via diffraction of short electron pulses[13–16]. With the advent of FELs, X-ray diffraction experiments have now reached a sub-100 fs temporal resolution. The spatial resolution is sufficient to observe larger-scale molecular processes like light-induced opening of a six-membered ring, but not to resolve individual atoms because of the limited wavelength available in FELs[12]. Laser-induced electron diffraction[17,18] and photoelectron holography[19] can also provide spatial and temporal resolution simultaneously, but require significant theoretical input to retrieve the interatomic distances. Ultrafast electron diffraction (UED) from gas-phase molecules has achieved sub-Angstrom spatial resolution but the temporal resolution has not been sufficient to observe molecular geometry changes on the femtosecond timescale[20,21]. Here we report a crucial advance in gas-phase UED using multi-MeV relativistic electron pulses to achieve a temporal resolution of 100 fs root-mean-squared (RMS), or 230 fs full-width at half-maximum (FWHM) that opens the door to observing the motion of individual nuclei that result from structural changes during photochemical reactions of isolated molecules.

The majority of UED experiments have been performed on solid-state samples[22,23], which are ideal to understand collective effects in condensed media such as superconductivity, heat transport and magnetism. Gaseous molecules are ideal to study prototypical processes in chemistry. They also provide a direct link between experiments and quantum chemical calculations, which can be performed on the highest level for isolated molecules[23–25]. Early gas phase UED studies employed stroboscopic electron diffraction[26]. A breakthrough was achieved when picosecond electron beams became available in late 1980s (ref. 27). Zewail and co-workers were able to resolve non-equilibrium molecular structures[28], transient molecular structures[15] and radiationless dark structures[16] using picosecond UED. However, in order to resolve molecular geometry changes in real time, a better temporal resolution is required. In the context of the crucial excited state photoisomerization reactions[24], a 200-fs resolution is suitable for exploring the nuclear dynamics of the isolated azobenzene isomerization[29]. In addition, this temporal resolution is sufficient to explore photoprotection processes of nucleobases[30–33].

To improve UED temporal resolution, two major effects must be overcome: space-charge repulsion between electrons[34] and velocity mismatch[35] resulting from the electron pulse lagging behind light pulses used for the molecular excitation. Both of these limitations can be minimized using relativistic MeV electrons[36–45]. The longitudinal space-charge pulse elongation is proportional to $1/\beta^2\gamma^5$, where $\beta = v/c$, $\gamma = (1-\beta^2)^{-1/2}$ is the Lorentz factor, and v and c are the speed of electrons and the speed of light in vacuum, respectively[46]. Concerning the

velocity mismatch, electrons with 3.7 MeV kinetic energy travels at $v = 0.993c$. This results in only 5 fs delay with respect to an optical pulse for a typical 200 μm interaction length.

Here we present the ultrafast laser-induced rotational dynamics of N_2 molecules in the gas phase. We impulsively excited a rotational wavepacket in a N_2 gas sample using a 35-fs FWHM laser pulse, spectrally centred at 800 nm, which interacts non-resonantly with the anisotropic molecular polarizability tensor. In the impulsive alignment regime, the alignment laser pulse duration is much shorter than the rotational period of the molecule, and the ensemble reaches its maximum degree of alignment after the interaction with the laser pulse. The phase evolution of the rotational wavepacket results in rotational revivals, with molecules alternating between aligned and anti-aligned directions[47,48]. Previous efforts to investigate rotational wavepackets with UED have captured dynamics on picosecond timescales[49,50]. The rotational dynamics of laser-aligned N_2 has been previously observed with optical birefringence[51], strong field ionization[52], high harmonic generation[53,54] and Auger electron spectroscopy[55]. We have observed the temporal evolution of the full wavepacket revival with an 8.35 ps period by quantifying the anisotropy in the diffraction patterns. The temporal resolution was determined by a fitting routine using the measured dynamics. Furthermore, we have retrieved molecular images of the aligned and anti-aligned molecular ensemble with atomic resolution.

Results

Experimental layout and static diffraction. The experimental layout is shown in Fig. 1. The electron pulse (blue) is diffracted from the nitrogen gas jet (grey), which is introduced into the vacuum chamber using a pulsed nozzle (black). The diffraction pattern is captured by a phosphor screen and a detector. The 800-nm alignment laser pulse (red) is directed to the target and removed from the vacuum chamber by two holey mirrors at a 45° angle to the electron beam. The full setup is discussed in detail elsewhere[56]. The electron and the laser beam have a small angle of ∼5°. This makes the velocity of the laser 0.996c along the electron beam direction, very close to 0.993c, the velocity of

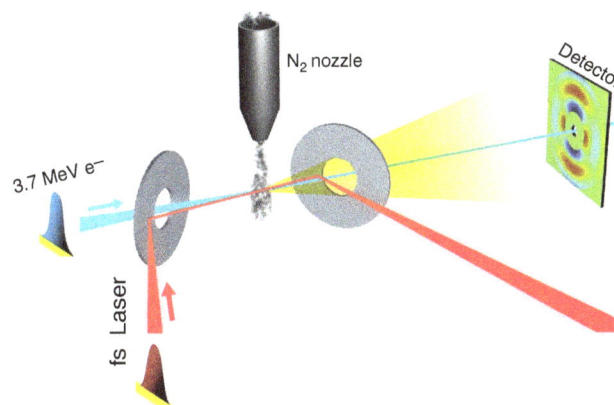

Figure 1 | Experimental layout. A sketch of the experimental setup. A 3.7-MeV pulsed electron beam (blue) is directed towards a nitrogen gas jet (grey). The gas jet is introduced into the vacuum chamber using a pulsed nozzle (black). The pump laser pulse (red) is deflected by two ring-shaped mirrors. The laser propagates at a small angle (∼5°) with respect to the electron beam as it traverses the target, and is then deflected away from the detector by the second mirror. The electron diffraction pattern is recorded with a phosphor screen located 3.1 m downstream from the interaction region. The unscattered electron beam is transmitted through a hole in the phosphor screen.

the electron beam. Therefore, the velocity mismatch is nearly eliminated. More details of the experiment can be found in the Experimental Setup section in the Methods.

The diffraction pattern is commonly expressed as a function of the momentum transfer

$$s = \frac{4\pi}{\lambda}\sin(\theta/2) \qquad (1)$$

where λ is the wavelength of the electron beam and θ is the angle between the scattered and transmitted electrons. For a 3.7-MeV electron beam, $\lambda = 0.30$ pm. The total scattering intensity I_{tot} is the sum of the atomic scattering intensity I_{at} and the molecular scattering intensity I_{mol}. I_{at} is defined as

$$I_{at} = \sum_{i=1}^{N} |f_i(s)|^2 \qquad (2)$$

where N is the number of atoms in the molecule and f_i is the elastic scattering amplitude for the ith atom. For MeV electrons, f_i can be calculated using the ELSEPA program[57].

The structural information of the molecule is encoded in the molecular scattering intensity, I_{mol}, given by

$$I_{mol} = \sum_{i=1}^{N} \sum_{j \neq i}^{N} |f_i(s)||f_j(s)|\cos(\eta_i - \eta_j)\frac{\sin(sr_{ij})}{sr_{ij}} \qquad (3)$$

where η_i is the scattering phase of the ith atom and r_{ij} is the distance between the ith and the jth atoms[20].

The so-called modified diffraction intensity is defined as

$$sM(s) = s\frac{I_{mol}(s)}{I_{at}(s)} \qquad (4)$$

The spatial resolution, δ, is determined by the maximum measured s value s_{max} using the formula

$$\delta = 2\pi/s_{max} \qquad (5)$$

In Fig. 2a, the red curve shows an azimuthally averaged raw pattern after subtraction with a dark background pattern that is taken with the electron beam turned off. The azimuthal average of the subtracted background pattern is shown in the green curve in Fig. 2a, and the black dashed curve shows the experimental scattering background, determined using a standard fitting procedure[20]. This fitted background includes the atomic scattering I_{at}, scattering from background gas and other types of background scatterings. The details of the bond length

determination are explained in the Bond Length Determination from Static Pattern section in the Methods. Figure 2b shows the azimuthally averaged experimental and theoretical modified diffraction intensity for static diffraction pattern. The bond length was determined to be 1.073 ± 0.027 Å, in agreement with the previously measured N_2 bond length of 1.098 Å (ref. 58). The 2.5% measurement uncertainty is due to the uncertainty in the calibration of the sample-to-detector distance and electron energy, as explained in the Experimental Setup section in the Methods. It should be noted that for time-resolved experiments usually the ground-state structure is known, in which case the static diffraction patterns can be used as a calibration. We show later that the interatomic distance can be determined very accurately from diffraction of transiently aligned molecules if the static diffraction is used as a calibration. In this experiment, scattering signal is available in the region between 3.5 and 12 Å^{-1}. The fitting procedure relies on the zeros of $sM(s)$ and this reduces the available data in Fig. 2 to $s > 4.5$ Å^{-1}. The measurement agrees well with the simulation up to $s \sim 12$ Å^{-1}.

Temporal evolution of N_2 alignment. For impulsive alignment, the full rotational revival is expected at $t = 1/2cB$, where B is the rotational constant and c is the speed of light in vacuum. For N_2 molecules, the full revival is at 8.35 ps ($B = 1.998$ cm^{-1}). The rapid evolution of the angular distribution can be used to determine the temporal resolution of the measurement technique.

Diffraction patterns from a molecular ensemble aligned with a polarization in the detector plane are not circularly symmetric, contrary to the static case in equation (3). The anisotropy, $a(t)$, in a diffraction pattern can be used to trace the temporal evolution of alignment[49]. In addition, $a(t)$ is a self-normalized parameter that is extracted directly from the diffraction patterns (see Data Processing of Diffraction Patterns and Anisotropy in the Methods). Figure 3 shows the temporal evolution of the simulated and experimentally measured $a(t)$ values. The experimental data are recorded with 100 fs steps in the delay between laser and electron pulses, and at each point data are collected for 2 min (14,400 shots at a repetition rate of 120 Hz). The simulation is composed of two parts: an impulsive alignment simulation that calculates the angular distribution at different delay times[59], followed by a diffraction pattern simulation based on the modelled angular distribution[60]. The anisotropy of the simulated patterns is calculated using the same method as for the experimental patterns. The details of the

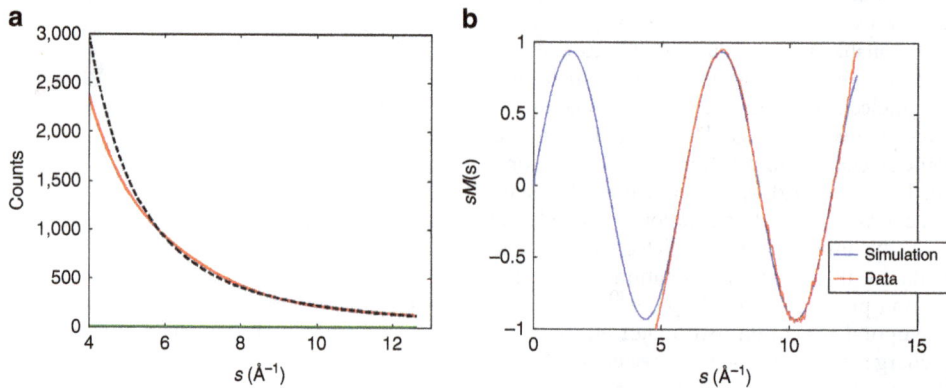

Figure 2 | Static N_2 diffraction. (a) The red curve shows an azimuthally averaged raw experimental diffraction pattern after subtraction with a dark background pattern that is taken without the electron beam. The green curve shows the azimuthal average of the subtracted dark background pattern. The dashed black curve shows the fitted background, including the atomic scattering I_{at} and other background scattering. The vertical axis is average detector counts per pixel per minute of exposure time, averaged over ~100 min. The green curve varies between 0.8 and 2.5 counts. (b) The theoretical (blue) and experimental (red) modified diffraction intensity sM from N_2 gas, which shows the enhanced diffraction rings. The experimental sM is calculated from the diffraction pattern in part (a).

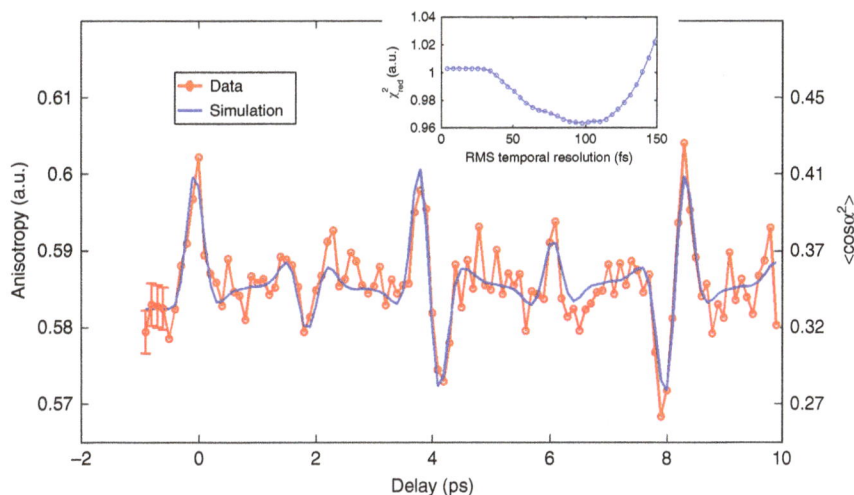

Figure 3 | Temporal evolution of the N_2 rotational wavepacket. Anisotropy in the diffraction patterns from experimental data (red) and simulation (blue) versus time. Statistical error bars for the first few points (before alignment) are shown to illustrate the uncertainty of this measurement. The right-hand side axis gives the degree of alignment $<\cos^2\alpha>$ for the simulated curve. The simulation parameters of initial rotational temperature, alignment laser fluence, temporal resolution and rescaling factor are obtained from a fitting routine. Each data point is accumulated over 2 min. Reduced χ^2 error versus RMS temporal resolution in the four-parameter fit is shown in the inset.

simulation are explained in the Alignment Simulation and Diffraction Pattern Simulation section in the Methods. For diatomic molecules, the rotational states can be described by $|J, M>$, where J and M are the quantum numbers of the total and the z-component of the angular momentum, respectively. The angular distribution of $|J, M>$ is given by Laplace's spherical harmonics Y_J^m. Before the excitation, the ensemble is described by a Boltzmann distribution of rotational states taking into account the exact nuclear statistics. Initially, the angular distribution is isotropic. For the initial rotational temperature of our sample of 54 K, all values of J up to 10 are considerably populated before excitation. The laser pulse interacts with the polarizability tensor of the molecules, which is highest along the internuclear axis. Classically, it induces a dipole that minimizes its energy by aligning the molecular axis parallel to the laser field. Quantum mechanically, the laser induces transitions of $\Delta J = 0, \pm 2$ and $\Delta M = 0$. After excitation, the maximum value of J is about 14. The phase of each rotational state evolves with time. When the rotational states are in phase, the angular distribution becomes aligned or anti-aligned with respect to the laser polarization axis.

The temporal evolution of the alignment has a rich structure that varies on a fast timescale. For example, at the half-revival (delay of 4 ps), the distribution changes from aligned to anti-aligned in 300 fs. The alignment peak corresponds to a prolate angular distribution, with the long axis along the direction of the laser polarization. The angular distribution during anti-alignment is oblate, with the molecules preferentially lying in a plane perpendicular to the laser polarization. The full revival at around 8 ps shows a similarly fast transition from oblate to prolate distribution. In between the revivals, the anisotropy of the diffraction pattern captures additional dynamics.

We have used the fast-changing distribution to characterize the temporal resolution of the measurement. The shape of the aligned molecular ensemble is determined by the initial temperature of the molecules and the fluence of the alignment laser pulse[61]. In this experiment, the laser fluence was measured to be 2.0 J cm^{-2} and the initial temperature was estimated to be 65 K, using a supersonic expansion model[62]. At these parameters, the wavepacket revivals are relatively sharp compared with the temporal resolution. The limited temporal resolution effectively blurs out the rotational dynamics and has a significant effect on

the observed structure. We performed first a two-parameter χ^2 fitting by fixing the laser fluence and initial rotational temperature to the measured and calculated values, and then a full four-parameter fit where all parameters were allowed to vary.

The two-parameter fit, varying only the temporal resolution and a re-scaling factor that accounts for the spatial overlap between laser and electron pulse, returned a temporal resolution of 85 fs RMS (200 fs FWHM). The four-parameter fit, which varies temporal resolution in addition to initial rotational temperature, laser fluence and the re-scaling factor, accounts for uncertainties in the laser fluence, initial temperature and the fraction of molecules excited by the laser. In the simulation, we assume that all excited molecules are exposed to the same laser intensity. This method achieved a best fit resolution of 100 fs RMS (230 fs FWHM), with an initial temperature of 54 K and a laser fluence of 1.8 J cm^{-2}, which are comparable to our initial estimates for the two-parameter fit. The re-scaling factor was 0.42, meaning that the best match to the experimental data is when the simulation assumes that 42% of the molecules in the diffraction volume are excited by the laser. The simulation shown in Fig. 3 shows the best fit for all four parameters. The left axis in the figure shows the anisotropy in the diffraction pattern, and the axis on the right shows the degree of alignment quantified by the $<\cos^2\alpha>$ value from simulation, where α is the angle between the molecular axis and the laser polarization direction. The reduced χ^2 error in the four-parameter fit versus the RMS temporal resolution is shown as an inset in Fig. 3. We obtained a 0.96 reduced χ^2, indicating a good fit. More details of the fitting are explained in the Temporal Evolution Fitting section in the Methods.

The 100-fs (RMS) overall temporal resolution of this experiment is consistent with our expectations, based on the performance study of this machine[56]. A simulation showed that the electron bunch length was 70 fs RMS at the interaction region. Measurements of the phase and amplitude stability of the radio-frequency (RF) gun lead to an expected time of arrival jitter of 50 fs RMS[56]. The calculated overall temporal resolution was then 87 fs RMS, or 205 fs FWHM, which is close to the measured value.

Molecular images with different angular distribution. High-resolution molecular diffraction images were retrieved for prolate and oblate ensembles at the half revival. Diffraction patterns with adequate signal-to-noise ratio were recorded with 60 and 90 min of integration time for oblate and prolate distributions, respectively. We use diffraction-difference patterns to remove the experimental background and the diffraction signal from unexcited molecules. The diffraction intensity difference is given by $\Delta I(t) = I(t) - I(t = -5\,\mathrm{ps})$, where $t = 0$ corresponds to the maximum of the first alignment peak after laser. Before $t = -0.4\,\mathrm{ps}$ when the pump laser arrives, the angular distribution is isotropic. In the two-dimensional (2D) diffraction patterns, the

$s_{max} = 8.3\,\text{Å}^{-1}$, corresponding to a spatial resolution of 0.76 Å. This makes it possible to observe molecular structures with resolution better than the shortest possible bond lengths.

Figure 4 shows the experimental (left panels) and simulated (right panels) 2D diffraction patterns and their corresponding Fourier transforms for prolate and oblate distribution at the half revival. Figure 4a,b shows the experimental and simulated diffraction-difference patterns ΔI for the prolate distribution. The diffraction pattern was captured at a time delay of 3.8 ps after the first alignment peak (Fig. 3a). The diffraction-difference

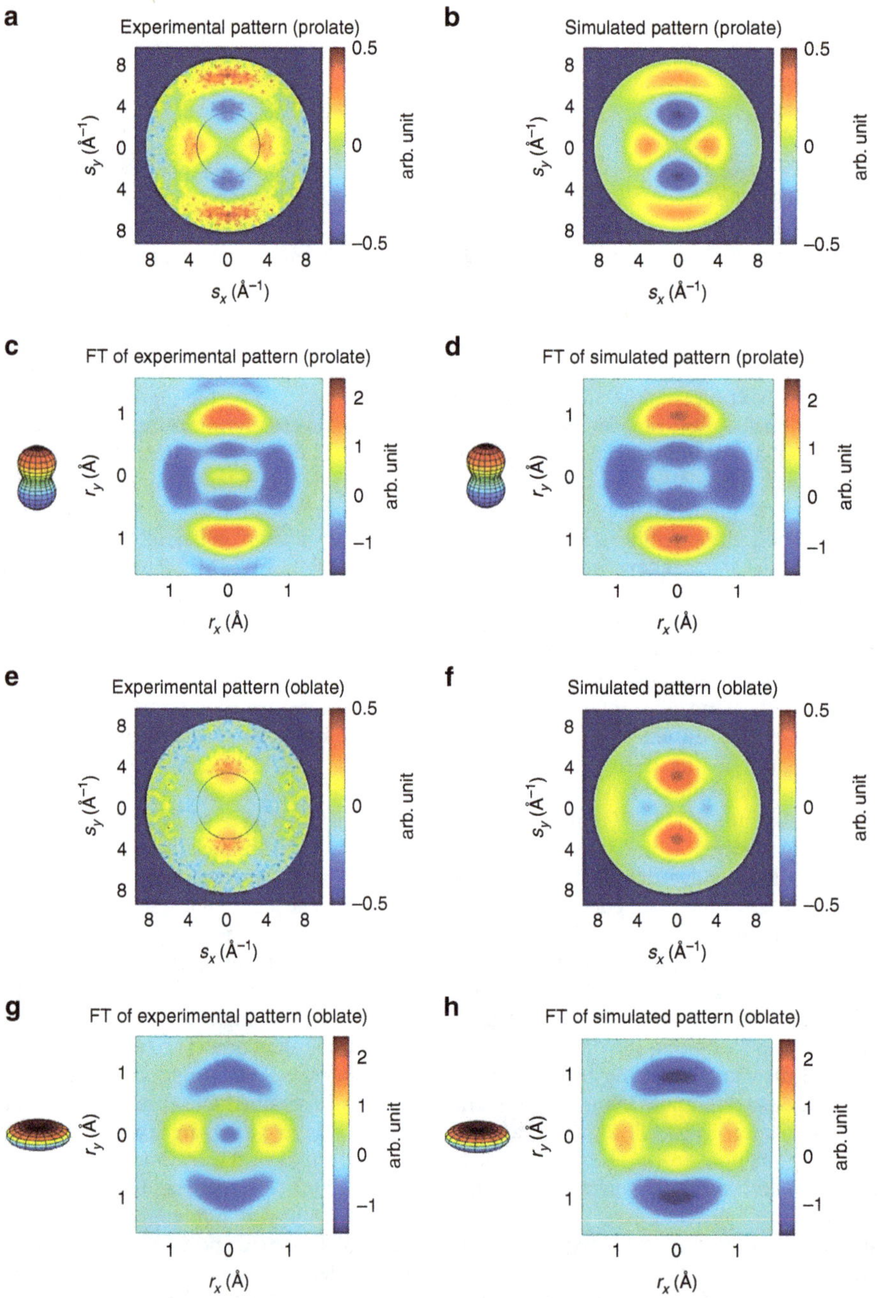

Figure 4 | 2D N₂ diffraction patterns at half revival. (a) Experimentally measured and **(b)** simulated diffraction-difference patterns of the prolate distribution. Images shown in **c,d** are Fourier transforms of **a,b**, respectively. The Fourier transform of the diffraction-difference patterns show the changes in the angular distribution of the molecules. The positive regions (red colour) indicate where the population has increased and the negative regions (blue colour) indicate where the population has decreased. **(e)** Experimentally measured and **(f)** simulated diffraction-difference pattern of the oblate distribution. Images shown in **g,h** are Fourier transforms of **e,f**, respectively. In patterns **(a,e)**, the data inside the black circles are missing due to the beam stop. They are obtained by extrapolating the pattern and letting the counts smoothly go to zero towards the centre. For illustrative purpose, angular distributions are shown on the side of panel **(c,d,g,h)** for visual guidance. In these angular distributions, the colour code indicates polar angle.

pattern is anisotropic as a result of molecular alignment. The experimental pattern shows excellent agreement with the simulation for a range of $s = 3.5\text{--}8.3\,\text{Å}^{-1}$.

Figure 4c,d shows the Fourier transforms of the difference signals in Fig. 4a,b, respectively. The Fourier transform of the diffraction pattern displays the autocorrelation of the molecular structure, convolved with the angular distribution and projected onto the detector plane. The centre of the Fourier transforms goes to zero because they are generated from the difference of two diffraction patterns. For a diatomic molecule, the autocorrelation is directly related to the molecular image. For more complex molecules, an image of the structure can be reconstructed using phase retrieval algorithms[60]. In the autocorrelation functions depicted in Fig. 4c,d, the positive regions indicate an increase in population, and the negative regions indicate a decrease. Specifically, Fig. 4c,d indicates that the population of molecules that are lying perpendicular to the laser polarization has decreased, whereas the population parallel to the polarization (vertical in Fig. 4) has increased, that is, more molecules are aligned along the vertical direction. Similarly, Fig. 4e,f shows the measured and simulated diffraction pattern for the oblate distribution at half revival, corresponding to a time delay of 4.1 ps. Figure 4g,h are the Fourier transforms of Fig. 4e,f, which show that the molecular ensemble is aligned in the horizontal plane.

Bond length measurement from diffraction patterns. We have used a fitting method to extract the N_2 bond length from diffraction patterns of aligned and anti-aligned molecules, using the static diffraction pattern as a calibration. The details of the fitting are given in the section Bond Length Fitting from Aligned Patterns in the Methods. The extracted bond length is $1.091 \pm 0.036\,\text{Å}$ for the prolate distribution and $1.096 \pm 0.056\,\text{Å}$ for the oblate distribution, in good agreement with $1.098\,\text{Å}$, the bond length of the ground-state N_2. The small uncertainties indicate that changes in interatomic distances could be measured very accurately with this method. The precision of determining the bond length, 0.036 and $0.056\,\text{Å}$, should not be confused with the $0.76\,\text{Å}$ spatial resolution. The spatial resolution gives the capability to distinguish two bond lengths that are very close to each other, whereas the precision gives how accurate a single bond length can be determined. In real space, the resolution is determined by the width of the peak, whereas the precision is determined by how accurate one can find the centre of the peak. Generally, for well-separated peaks, the centre can be determined to much higher accuracy than its width.

Spatial resolution of the molecular images. The spatial resolution of the 2D diffraction patterns shown in Fig. 4 can be determined in two different ways. First we can use equation (6), with $s_{max} = 8.3\,\text{Å}^{-1}$ we get the spatial resolution $\delta = 0.76\,\text{Å}$. We can also determine spatial resolution directly from the autocorrelation images (Fig. 4c,g). For example, in Fig. 4c, the spatial resolution can be determined by converting the image into polar coordinates $o(r, \theta')$, then using a Gaussian function to fit to along the r dimension. The FWHM of the Gaussian fit was $0.76\,\text{Å}$ using this method, consistent with the spatial resolution obtained using equation (5) and $s_{max} = 8.3\,\text{Å}^{-1}$.

Angular distribution based on 2D images of aligned molecules. The angular distribution of the molecules can be extracted from the patterns in Fig. 4c,g. The resulting distributions for prolate and oblate molecular ensembles are shown in Fig. 5a,b, respectively (see Angular Distribution section in the Methods). The prolate distribution peaks at $\alpha = 0$ and $180°$, in the direction

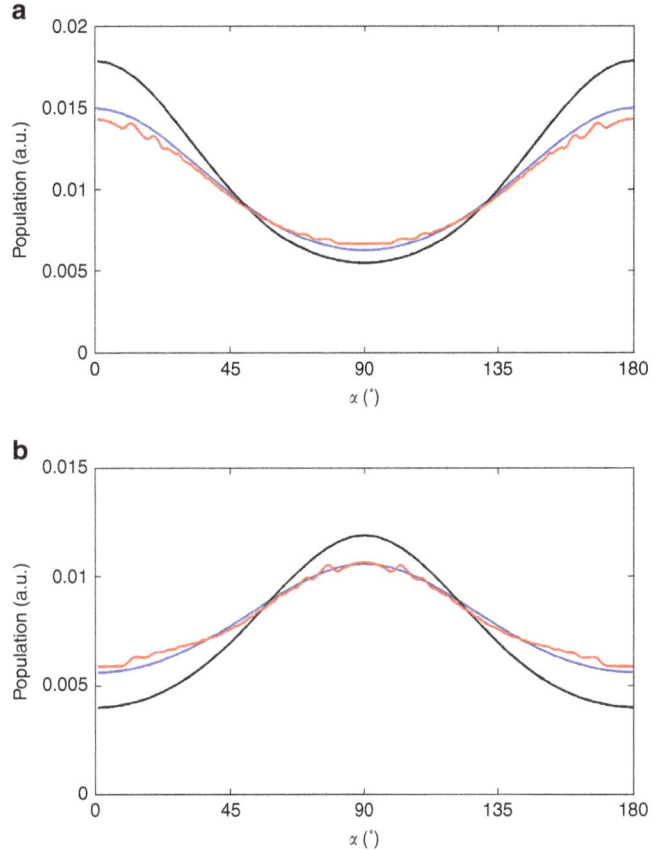

Figure 5 | Angular distributions. (a) Prolate angular distribution at the revival: experimental (red), simulated (black) and simulated convolved with 100 fs RMS temporal resolution (blue). The effect of the temporal resolution is to reduce the measured degree of alignment at the peak, due to time averaging. The measurement agrees very well with the simulation after the convolution. The $<\cos^2\alpha>$ values for red, black and blue curves are 0.41, 0.45 and 0.42, respectively. **(b)** Oblate angular distribution: Experimental (red), simulated (black) and simulated convolved with 100 fs RMS temporal resolution (blue). The $<\cos^2\alpha>$ values for red, black and blue curves are 0.28, 0.25 and 0.28, respectively. The experimental curves are measured from Fig. 4c,g, respectively.

of the laser polarization. The oblate distribution peaks at $\alpha = 90°$, in the direction perpendicular to the laser polarization. The degree of alignment is commonly measured by the quantity

$$\langle\cos^2\alpha\rangle = \frac{\int_0^\pi f(\alpha) \cdot \cos^2\alpha \cdot \sin\alpha \cdot d\alpha}{\int_0^\pi f(\alpha) \cdot \sin\alpha \cdot d\alpha} \quad (6)$$

where $f(\alpha)$ is the angular distribution and α is the angle between the molecular axis and the laser polarization. A value of the alignment parameter $<\cos^2\alpha> = 1$ corresponds to perfect alignment, whereas random orientation gives $<\cos^2\alpha> = 1/3$. Any value between 1/3 and 1 indicates alignment, whereas any value below 1/3 indicates anti-alignment. In Fig. 5, three curves are displayed: $f(\alpha)$ extracted from data (red), $f(\alpha)$ obtained from simulation using the fluence and temperature extracted from the fitting routine (black), and the simulation results convolved with the 100 fs RMS temporal resolution (blue). The temporal resolution is included by convolving a Gaussian pulse with the simulated temporal evolution of the angular distribution, and it has a significant effect because the distribution changes from prolate to oblate in 300 fs. The extracted angular distributions are in good agreement with the simulation for both prolate and oblate distributions, after taking into account the effect of the temporal resolution.

The measured angular distribution can also be used to provide additional confirmation of the temporal resolution. Using the laser parameters and the initial temperature as determined by the fitting from the data in Fig. 3, we ran an optimization to find the temporal convolution that results in a best fit between the measured and simulated angular distributions. From this, we obtain a resolution of 107 fs RMS for the prolate angular distribution and 102 fs RMS for the oblate distribution, in good agreement with the value of 100 fs RMS obtained before.

Discussion

In summary, we have shown that MeV UED can simultaneously reach a temporal resolution of 100 fs RMS and spatial resolution of 0.76 Å, which allows us to characterize the ultrafast evolution of a rotational wavepacket in N_2, measuring the angular distribution and the acquired molecular images with atomic resolution.

This achievement opens the door to a new class of experiments where changes in molecular geometry during a chemical reaction can be followed in space and time. The current results demonstrate not only excellent temporal resolution, but also that sufficient signal-to-noise ratio for sub-Angstrom spatial resolution can be acquired with MeV electrons. Previous picosecond UED experiments with similar spatial resolution have successfully retrieved the structure of transient states[15]. Nitrogen has a low atomic number, Z, and therefore a low scattering cross-section, so we expect that the method will be successful for observing molecular dynamics in a large class of molecules. RF-compressed electron pulses[63,64] and compact UED guns[65–67] have achieved similar temporal resolution in condensed matter experiments, but have not yet been successfully applied to gas-phase experiments. For future developments, a number of upgrades can potentially improve both spatial and temporal resolution significantly. For example, a RF compression cavity could be used to compress the electron beam longitudinally[68], which can potentially lead to a temporal resolution on the order of 10 fs together with orders-of-magnitude increase in charge per pulse. Time jitter and drift could be addressed by time-stamping techniques similar to those developed for X-Ray free electron lasers (XFELs)[69]. Spatial resolution is limited by the signal-to-noise ratio (SNR) in the scattering at larger angles, so RF compression would also improve the spatial resolution by increasing the number of electrons per pulse. Complementary metal-oxide semiconductor (CMOS) active pixel sensor could potentially achieve single electron, single shot detection and thus eliminate detector noise[70], which would also improve the SNR.

Methods

Experimental setup. The electron gun used in this experiment is a replica of the photo-injector used at the LCLS facility at SLAC National Accelerator Laboratory. A 15 fC per pulse charge is generated at the photocathode and 6 fC per pulse is delivered on target. A 50-nm-thick Si_3N_4 membrane is used to separate the sample chamber and preserve the high vacuum in the electron gun. Roughly 60% of the charge is lost after the Si_3N_4 membrane and a 200-µm diameter collimator. The electron beam and pump laser are generated with a repetition rate of 120 Hz.

The N_2 gas is delivered into the chamber through a pulsed valve at a 120-Hz repetition rate. The backing pressure is 0.7 bar, and the nozzle orifice is 100 µm in diameter. During operation, the chamber pressure is $\sim 6 \times 10^{-5}$ torr. The interaction region is roughly 350 µm away from the nozzle exit. The width of the laser beam, the electron beam and the gas jet are all around 200 µm FWHM at the interaction region. The sample-to-detector distance is calibrated with diffraction from a single crystal gold sample. There is a ± 2.5% uncertainty in the distance calibration, due to the quality of the gold sample and quality of the electron beam after the Si_3N_4 membrane. The Si_3N_4 window has a strong diffraction and makes the emittance of the beam much worse. The electrons scattered inelastically from Si_3N_4 also diffract from the gold sample, making the Bragg spots larger. Spatial alignment of the gas jet, the electron beam and the pump laser is obtained by

positioning the focus of the laser approximately underneath the tip of the nozzle, and maximizing for the electron beam deflection by the plasma. Time zero can then be found, to within ~ 200 fs, by adjusting the delay between the pump and the probe beams and observing plasma deflection effects on the unscattered beam[20]. The laser focus is adjusted to 50 µm FWHM spot size for plasma lensing, giving a peak intensity of 8×10^{14} W cm^{-2}. For the alignment experiments, the lens is driven 4 mm out of the focus to reach a spot size of 200 µm in the interaction region. The peak intensity for the experiment is 5×10^{13} W cm^{-2}; only very faint plasma is observed in the interaction region for this intensity.

The detector comprises a phosphor screen, a ring-shaped mirror at a 45° angle to the beam, an f/0.85 lens and an electron multiply charge-coupled device. A 4-mm diameter hole is drilled in the centre of the phosphor screen to allow the unscattered electron beam to pass through. Data in the region $s < 3.5$ Å^{-1} are not captured due to the hole in the phosphor screen. This can be improved by replacing the phosphor screen by a similar one with a smaller hole. We have seen that by replacing the phosphor screen the missing region can be reduced to 1.6 Å^{-1}.

The beam divergence σ is 28 µrad RMS. Using the definition $\lambda/2\pi\sigma$, we get a transverse coherence length of 1.7 nm.

Bond length determination from static pattern. The following steps are used to determine the bond length from the static N_2 diffraction pattern[20]. (i)Starting from an initial guess of the bond length the modified diffraction intensity sM is calculated. (ii) A series of zero points is determined from the simulated sM. (iii) The background is determined by fitting an exponentially decaying function at the zero points. (iv) The experimental sM is obtained by subtracting the background. (v) The experimental sM is compared with the simulated sM and the error is defined as the square of the difference. (vi) Steps i–iv are repeated for different bond lengths, until the minimum error is found.

Data processing of diffraction patterns and anisotropy. The experimental pattern is symmetrized over four quadrants, and the central region $(s < 3.5$ Å$^{-1})$ is extrapolated from existing data by letting the pattern smoothly go to zero towards the centre.

The diffraction pattern of the prolate distribution is integrated over 90 min, and that of the oblate distribution is integrated over 60 min. The anisotropy is calculated by dividing the total counts in a horizontal cone by the total counts in a vertical cone for each diffraction pattern. The region used for this calculation is between $s = 3$ Å^{-1} and $s = 4.5$ Å^{-1}. The horizontal cone has a half angle of 35° and the vertical cone has a half angle of 55°.

Alignment simulation. The simulation of impulsive alignment is calculated using a linear rigid rotor interacting with a non-resonant pulse described by the time-dependent Schrodinger equation[59]. The temporal resolution is implemented by a convolution with a Gaussian function in time.

Diffraction pattern simulation. The simulations of diffraction patterns of a given angular distribution are calculated using an incoherent weighted sum of diffraction patterns from single molecules with different orientations. In the simulations, the atoms are assumed to be stationary at their equilibrium positions without vibrations[60].

Temporal evolution fitting. The temporal evolution fitting shown in Fig. 3a is obtained by a χ^2 fitting of the simulation and data. The simulated anisotropy is obtained by an alignment simulation that gives angular distribution as a function of time, followed by a diffraction simulation according to the angular distribution. The anisotropy is calculated using the same method as for data, and the temporal resolution is obtained by a convolution of the calculated anisotropy with a Gaussian beam in time. The reduced χ^2 error, plotted in the Fig. 3 inset, is defined as $\chi^2_{red} = \frac{1}{w} \sum \frac{(O-E)^2}{\sigma^2}$, where w is the number of degrees of freedom, O and E are the data and the best fit, σ is the s.e., respectively. Here σ is calculated using the standard deviation of 13 patterns before $t = 0$.

Bond length fitting from aligned patterns. The optimal bond length is found by comparing the experimental diffraction pattern with many simulated patterns with different bond lengths using the least-square error. For the experimental patterns, a vertical cone with a half opening angle of 30° is selected, and this part is averaged radially to generate a one-dimensional curve. For the simulated patterns, diffraction patterns with different bond lengths and the same angular distribution are simulated, and the same region is selected to generate the one-dimensional curve. The prolate/ oblate angular distributions are obtained by the fitting in Fig. 3. To get statistics, ten independent data set (6 min data in each set) were used in the fits separately, and the mean and standard deviation of the ten fitting results are taken as the final result and the standard error. In this fitting, we use the static pattern (shown in Fig. 2) as a calibration of the sample-detector distance and electron beam energy.

Angular distribution. The angular distribution in Fig. 5a,b are extracted from Fig. 4c,g, respectively. The procedure includes three steps: (i) extracting the difference angular distribution from the 2D difference pattern, (ii) adding a baseline that accounts for the angular distribution of the reference pattern to the difference in angular distribution and (iii) normalizing the angular distribution. Step i is implemented by first converting Fig. 4c,g to polar coordinates, then integrating over the radial coordinate within the FWHM of the peak. In step ii, the baseline is simulated using equations (2) and (3), corrected by the rescaling parameter that accounts for the spatial overlap between laser and electrons (0.42, obtained from the fitting for Fig. 3). In step iii, the normalization used is $\int_0^\pi f(\alpha) \cdot \sin\alpha \cdot d\alpha = 1$.

References

1. Pedersen, S., Herek, J. L. & Zewail, A. H. The validity of the 'diradical' hypothesis: direct femtoscond studies of the transition-state structures. *Science* **266**, 1359–1364 (1994).
2. Schoenlein, R. W., Peteanu, L. A., Mathies, R. A. & Shank, C. V. The first step in vision: femtosecond isomerization of rhodopsin. *Science* **254**, 412–415 (1991).
3. Mathies, R. A., Brito Cruz, C. H., Pollard, W. T. & Shank, C. V. Direct observation of the femtosecond excited-state cis-trans isomerization in bacteriorhodopsin. *Science* **240**, 777–779 (1988).
4. Blanchet, V., Zgierski, M. Z., Seideman, T. & Stolow, A. Discerning vibronic molecular dynamics using time-resolved photoelectron spectroscopy. *Nature* **401**, 52–54 (1999).
5. Bisgaard, C. Z. et al. Time-resolved molecular frame dynamics of fixed-in-space CS_2 molecules. *Science* **323**, 1464–1468 (2009).
6. Deb, S. & Weber, P. M. The ultrafast pathway of photon-induced electrocyclic ring-opening reactions: the case of 1,3-cyclohexadiene. *Annu. Rev. Phys. Chem.* **62**, 19–39 (2011).
7. Gessner, O. et al. Femtosecond multidimensional imaging of a molecular dissociation. *Science* **311**, 219–222 (2006).
8. Milne, C. J., Penfold, T. J. & Chergui, M. Recent experimental and theoretical developments in time-resolved X-ray spectroscopies. *Coord. Chem. Rev.* **277**, 44–68 (2014).
9. Küpper, J. et al. X-ray diffraction from isolated and strongly aligned gas-phase molecules with a free-electron laser. *Phys. Rev. Lett.* **112**, 83002 (2014).
10. Ihee, H. et al. Ultrafast x-ray diffraction of transient molecular structures in solution. *Science* **309**, 1223–1227 (2005).
11. Siders, C. W. et al. Detection of nonthermal melting by ultrafast X-ray diffraction. *Science* **286**, 1340–1342 (1999).
12. Minitti, M. P. et al. Imaging molecular motion: femtosecond X-ray scattering of an electrocyclic chemical reaction. *Phys. Rev. Lett.* **114**, 255501 (2015).
13. Gao, M. et al. Mapping molecular motions leading to charge delocalization with ultrabright electrons. *Nature* **496**, 343–346 (2013).
14. Siwick, B. J., Dwyer, J. R., Jordan, R. E. & Miller, R. J. D. An atomic-level view of melting using femtosecond electron diffraction. *Science* **302**, 1382–1385 (2003).
15. Ihee, H. et al. Direct imaging of transient molecular structures with ultrafast diffraction. *Science* **291**, 458–462 (2001).
16. Srinivasan, R., Feenstra, J. S., Park, S. T., Xu, S. & Zewail, A. H. Dark structures in molecular radiationless transitions determined by ultrafast diffraction. *Science* **307**, 558–563 (2005).
17. Blaga, C. I. et al. Imaging ultrafast molecular dynamics with laser-induced electron diffraction. *Nature* **483**, 194–197 (2012).
18. Meckel, M. et al. Laser-induced electron tunneling and diffraction. *Science* **320**, 1478–1482 (2008).
19. Krasniqi, F. et al. Imaging molecules from within: ultrafast angström-scale structure determination of molecules via photoelectron holography using free-electron lasers. *Phys. Rev. A At. Mol. Opt. Phys.* **81**, 1–11 (2010).
20. Srinivasan, R., Lobastov, V. A., Ruan, C.-Y. & Zewail, A. H. Ultrafast electron diffraction (UED): a new development for the 4D determination of transient molecular structures. *Helv. Chim. Acta* **86**, 1763–1838 (2003).
21. Yang, J., Beck, J., Uiterwaal, C. J. & Centurion, M. Imaging of alignment and structural changes of carbon disulfide molecules using ultrafast electron diffraction. *Nat. Commun.* **6**, 8172 (2015).
22. Sciaini, G. & Miller, R. J. D. Femtosecond electron diffraction: heralding the era of atomically resolved dynamics. *Rep. Prog. Phys.* **74**, 096101 (2011).
23. Miller, R. J. D. Mapping atomic motions with ultrabright electrons: The Chemists' Gedanken Experiment Enters the Lab Frame. *Annu. Rev. Phys. Chem.* **65**, 583–604 (2014).
24. Levine, B. G. & Martínez, T. J. Isomerization through conical intersections. *Annu. Rev. Phys. Chem.* **58**, 613–634 (2007).
25. Kochman, M. A., Tajti, A., Morrison, C. A. & Miller, R. J. D. Early events in the nonadiabatic relaxation dynamics of 4-(N, N -dimethylamino)benzonitrile. *J. Chem. Theory Comput.* **11**, 1118–1128 (2015).
26. Ischenko, A. A., Ewbank, J. D. & Lothar, S. Structural kinetics by stroboscopic gas electron diffraction Part 1. Time-dependent molecular intensities of dissociative states. *J. Mol. Struct.* **320**, 147–158 (1994).
27. Elsayed-Ali, H. E. & Mourou, G. a. Picosecond reflection high-energy electron diffraction. *Appl. Phys. Lett.* **52**, 103–104 (1988).
28. Ruan, C. Y. et al. Ultrafast diffraction and structural dynamics: the nature of complex molecules far from equilibrium. *Proc. Natl Acad. Sci. USA* **98**, 7117–7122 (2001).
29. Schultz, T. et al. Mechanism and dynamics of azobenzene photoisomerization. *J. Am. Chem. Soc.* **125**, 8098–8099 (2003).
30. Crespo-Hernandez, C., Cohen, B., Hare, P. & Kohler, B. Ultrafast excited-state dynamics in nucleic acids. *Chem. Rev.* **104**, 1977–2019 (2004).
31. McFarland, B. K. et al. Ultrafast X-ray Auger probing of photoexcited molecular dynamics. *Nat. Commun.* **5**, 4235 (2014).
32. Middleton, C. T. et al. DNA excited-state dynamics: from single bases to the double helix. *Annu. Rev. Phys. Chem.* **60**, 217–239 (2009).
33. Schreier, W. J. et al. Thymine dimerization in DNA is an ultrafast photoreaction. *Science* **315**, 625–629 (2007).
34. Siwick, B. J., Dwyer, J. R., Jordan, R. E. & Miller, R. J. D. Ultrafast electron optics: propagation dynamics of femtosecond electron packets. *J. Appl. Phys.* **92**, 1643–1648 (2002).
35. Dantus, M., Kim, S. B., Williamson, J. C. & Zewail, A. H. Ultrafast electron diffraction. 5. Experimental time resolution and applications. *J. Phys. Chem.* **98**, 2782–2796 (1994).
36. Hastings, J. B. et al. Ultrafast time-resolved electron diffraction with megavolt electron beams. *Appl. Phys. Lett.* **89**, 184109 (2006).
37. Li, R. et al. Experimental demonstration of high quality MeV ultrafast electron diffraction. *Rev. Sci. Instrum* **80**, 083303 (2009).
38. Zhu, P. et al. Femtosecond time-resolved MeV electron diffraction. *N. J. Phys.* **17**, 063004 (2015).
39. Wang, X., Qiu, X. & Ben-Zvi, I. Experimental observation of high-brightness microbunching in a photocathode rf electron gun. *Phys. Rev. E* **54**, R3121–R3124 (1996).
40. Wang, X. J., Wu, Z. & Ihee, H. Femto-seconds electron beam diffraction using photocathode RF gun. *Proc. 2003 Part Accel. Conf.* **1**, 420–422 (2003).
41. Wang, X. J., Xiang, D., Kim, T. K. & Ihee, H. Potential of femtosecond electron diffraction using near-relativistic electrons from a photocathode RF electron gun. *J. Korean Phys. Soc.* **48**, 390–396 (2006).
42. Musumeci, P., Moody, J. T. & Scoby, C. M. Relativistic electron diffraction at the UCLA Pegasus photoinjector laboratory. *Ultramicroscopy* **108**, 1450–1453 (2008).
43. Muro'Oka, Y. et al. Transmission-electron diffraction by MeV electron pulses. *Appl. Phys. Lett.* **98**, 2009–2012 (2011).
44. Fu, F. et al. High quality single shot ultrafast MeV electron diffraction from a photocathode radio-frequency gun. *Rev. Sci. Instrum* **85**, 083701 (2014).
45. Manz, S. et al. Mapping atomic motions with ultrabright electrons: towards fundamental limits in space-time resolution. *Faraday Discuss.* **177**, 467–491 (2015).
46. Reiser, M. *Theory and Design of Charged Particle Beams* (Wiley, 1994).
47. Stapelfeldt, H. & Seideman, T. Colloquium: aligning molecules with strong laser pulses. *Rev. Mod. Phys.* **75**, 543–557 (2003).
48. Rosca-Pruna, F. & Vrakking, M. J. Experimental observation of revival structures in picosecond laser-induced alignment of I_2. *Phys. Rev. Lett.* **87**, 153902 (2001).
49. Hensley, C. J., Yang, J. & Centurion, M. Imaging of isolated molecules with ultrafast electron pulses. *Phys. Rev. Lett.* **109**, 133202 (2012).
50. Reckenthaeler, P. et al. Time-resolved electron diffraction from selectively aligned molecules. *Phys. Rev. Lett.* **102**, 213001 (2009).
51. Chen, Y.-H., Varma, S. & Milchberg, H. M. Space- and time-resolved measurement of rotational wave packet revivals of linear gas molecules using single-shot supercontinuum spectral interferometry. *J. Opt. Soc. Am. B* **25**, B122 (2008).
52. Litvinyuk, I. V. et al. Alignment-dependent strong field ionization of molecules. *Phys. Rev. Lett.* **90**, 233003 (2003).
53. Itatani, J. et al. Controlling high harmonic generation with molecular wave packets. *Phys. Rev. Lett.* **94**, 123902 (2005).
54. McFarland, B. K., Farrell, J. P., Bucksbaum, P. H. & Gühr, M. High harmonic generation from multiple orbitals in N_2. *Science* **322**, 1232–1235 (2008).
55. Cryan, J. P. et al. Auger electron angular distribution of double core-hole states in the molecular reference frame. *Phys. Rev. Lett.* **105**, 083004 (2010).
56. Weathersby, S. P. et al. Mega-electron-volt ultrafast electron diffraction at SLAC National Accelerator Laboratory. *Rev. Sci. Instrum.* **86**, 073702 (2015).
57. Salvat, F., Jablonski, A. & Powell, C. J. Elsepa—Dirac partial-wave calculation of elastic scattering of electrons and positrons by atoms, positive ions and molecules. *Comput. Phys. Commun.* **165**, 157–190 (2005).
58. Huber, K. P. & Herzberg, G. in *Molecular Spectra and Molecular Structure IV. Constants of Diatomic Molecules* (Springer, 1979).
59. Ortigoso, J., Rodriguez, M., Gupta, M. & Friedrich, B. Time evolution of pendular states created by the interaction of molecular polarizability with a pulsed nonresonant laser field. *J. Chem. Phys.* **110**, 3870–3875 (1999).
60. Yang, J., Makhija, V., Kumarappan, V. & Centurion, M. Reconstruction of three-dimensional molecular structure from diffraction of laser-aligned molecules. *Struct. Dyn.* **1**, 044101 (2014).

61. Holmegaard, L. *et al.* Control of rotational wave-packet dynamics in asymmetric top molecules. *Phys. Rev. A* **75,** 051403 (2007).

62. Hagena, O. F. Nucleation and growth of clusters in expanding nozzle flows. *Surface Sci. Lett.* **106,** 101–116 (1981).

63. Chatelain, R. P., Morrison, V. R., Godbout, C. & Siwick, B. J. Ultrafast electron diffraction with radio-frequency compressed electron pulses. *Appl. Phys. Lett.* **101,** 2–6 (2012).

64. Van Oudheusden, T. *et al.* Compression of subrelativistic space-charge-dominated electron bunches for single-shot femtosecond electron diffraction. *Phys. Rev. Lett.* **105,** 264801 (2010).

65. Waldecker, L., Bertoni, R. & Ernstorfer, R. Compact femtosecond electron diffractometer with 100 keV electron bunches approaching the single-electron pulse duration limit. *J. Appl. Phys.* **117,** 13109–81901 (2015).

66. Gerbig, C., Senftleben, A., Morgenstern, S., Sarpe, C. & Baumert, T. Spatio-temporal resolution studies on a highly compact ultrafast electron diffractometer. *N. J. Phys.* **17,** 043050 (2015).

67. Sciaini, G. *et al.* Electronic acceleration of atomic motions and disordering in bismuth. *Nature* **458,** 56–59 (2009).

68. Li, R. K., Musumeci, P., Bender, H. A., Wilcox, N. S. & Wu, M. Imaging single electrons to enable the generation of ultrashort beams for single-shot femtosecond relativistic electron diffraction. *J. Appl. Phys.* **110,** 074512 (2011).

69. Beye, M. *et al.* X-ray pulse preserving single-shot optical cross-correlation method for improved experimental temporal resolution. *Appl. Phys. Lett.* **100,** 1–5 (2012).

70. Battaglia, M. *et al.* Characterisation of a CMOS active pixel sensor for use in the TEAM microscope. *Nucl. Instruments Methods Phys. Res. Sect. A Accel. Spectrometers Detect. Assoc. Equip.* **622,** 669–677 (2010).

Acknowledgements

We thank SLAC management for the strong support. The technical support by SLAC Accelerator Directorate, Technology Innovation Directorate, LCLS Laser Science and Technology division and Test Facilities Department is gratefully acknowledged. This work was supported in part by the US Department of Energy (DOE) Contract No. DE-AC02-76SF00515, DOE Office of Basic Energy Sciences Scientific User Facilities Division, the SLAC UED/UEM Initiative Program Development Fund and by the AMOS program within the Chemical Sciences, Geosciences, and Biosciences Division of the Office of Basic Energy Sciences, Office of Science, US Department of Energy. J.Y. and M.C. were partially supported by the US Department of Energy Office of Science, Office of Basic Energy Sciences under Award Number DE-SC0003931. M.S.R. was supported by the National Science Foundation EPSCoR RII Track-2 CA Award No. IIA-1430519.

Author contributions

J.Y., M.G., T.V., M.S.R., R.L., X.S., T.G., F.W., S.W. and X.W. carried out the experiments. N.H., R. C., J.C., I.M., S.V. and A.F. developed the laser system. M.G. and J.Y. constructed the setup for gas phase experiments. C.H., K.J., A.R. and C.Y. helped on experimental setup. J.Y. performed the data analysis and simulations. The experiment was conceived by M.G., M.C. and X.W. The manuscript was prepared by J.Y., M.C., M.S.R. and M.G. with discussion and improvements from all authors. M.C. and X.W. supervised the work.

Additional information

Competing financial interests: The authors declare no competing financial interests.

Size-dependent phase transition in methylammonium lead iodide perovskite microplate crystals

Dehui Li[1], Gongming Wang[1,2], Hung-Chieh Cheng[3], Chih-Yen Chen[1], Hao Wu[3], Yuan Liu[3], Yu Huang[2,3] & Xiangfeng Duan[1,2]

Methylammonium lead iodide perovskite has attracted considerable recent interest for solution processable solar cells and other optoelectronic applications. The orthorhombic-to-tetragonal phase transition in perovskite can significantly alter its optical, electrical properties and impact the corresponding applications. Here, we report a systematic investigation of the size-dependent orthorhombic-to-tetragonal phase transition using a combined temperature-dependent optical, electrical transport and transmission electron microscopy study. Our studies of individual perovskite microplates with variable thicknesses demonstrate that the phase transition temperature decreases with reducing microplate thickness. The sudden decrease of mobility around phase transition temperature and the presence of hysteresis loops in the temperature-dependent mobility confirm that the orthorhombic-to-tetragonal phase transition is a first-order phase transition. Our findings offer significant fundamental insight on the temperature- and size-dependent structural, optical and charge transport properties of perovskite materials, and can greatly impact future exploration of novel electronic and optoelectronic devices from these materials.

[1] Department of Chemistry and Biochemistry, University of California, 607 Charles E. Young Drive East, Los Angeles, California 90095, USA. [2] California Nanosystems Institute, University of California, Los Angeles, California 90095, USA. [3] Department of Materials Science and Engineering, University of California, Los Angeles, California 90095, USA. Correspondence and requests for materials should be addressed to X.D. (email: xduan@chem.ucla.edu).

The hybrid organic–inorganic methylammonium lead iodide perovskite ($CH_3NH_3PbI_3$, denoted as $MAPbI_3$) is emerging as one of most promising solution-processable light absorber for solar cells and thus has attracted intensive recent interest[1-9]. With long carrier diffusion length[10-12] and low non-radiative recombination rate, solution processed perovskite materials have been demonstrated to deliver a certified power-conversion efficiency as high as 20.1% in the past a few years[13]. The excellent optical properties of $MAPbI_3$ perovskite enables it to be applied in a wide range of optoelectronic devices such as photodetectors[14], lasers[15-17] and light-emitting diodes[18]. Despite the tremendous interest in $MAPbI_3$ perovskite, its charge transport properties remain elusive because of the ion motion, which leads to a very large hysteresis and prevents the observation of the intrinsic field-effect mobility at the room temperature[19,20]. In addition, it has been proven that the solar cell efficiency strongly depends on the size of cuboids of perovskites[21]. Therefore, it is expected that the size could significantly influence the optical and charge transport properties of $MAPbI_3$, yet there is no systematic investigation of size-dependent optical and charge transport properties in perovskite materials.

The structural phase transitions can significantly alter the optical and electronic properties of materials[22], both of which are essential to understand the underlying photophysics[23]. The temperature-dependent studies such as photoluminescence (PL) spectroscopy[23], neutron powder diffraction[24], calorimetric and infrared spectroscopy[25] have been utilized to investigate the structural phase transitions in bulk $MAPbI_3$. The $MAPbI_3$ adopts the simple cubic perovskite structure above 330 K, transits to the tetragonal phase at 330 K (refs 26,27), and further evolves into an orthorhombic phase as the temperature is reduced to 160 K (ref. 27). All those phase transitions have been proven to be of first order[25]. Previous studies have shown that the physical size of a material can be an important variable in determining the phase transition points in addition to pressure, temperature and compositions[28-30]. Therefore, it is important to investigate how the size alters the phase transition points in $MAPbI_3$, which in turn affects its optical and electronic properties[31,32]. Nevertheless, the size-dependent phase transition in $MAPbI_3$ remains elusive. Here we report a systematic investigation of size-dependent structural phase transitions in individual $MAPbI_3$ microplate crystals by using temperature-dependent charge transport measurements, PL spectroscopy and transmission electron microscopy and electron diffraction.

Results

Temperature-dependent electrical measurement. To investigate the fundamental charge transport properties of individual perovskite microplates, we have constructed field-effect transistors (FETs) using the perovskite crystals and measured their transistor characteristics from the room temperature to liquid nitrogen temperature in dark. Figure 1a shows a schematic of a typical FET device configuration we used. The individual perovskite microplates served as the semiconducting channel of FET devices bridging two pre-fabricated Cr/Au electrodes as the source-drain electrodes on 300 nm SiO_2/Si substrate (as both the gate dielectrics and gate electrode). The inset of Fig. 1b displays an optical image of a typical FET device, where the thickness of the perovskite microplate is around 400 nm. Figure 1b shows a set of typical output characteristics (source-drain current I_{sd} versus source-drain voltage V_{sd}) of a single perovskite microplate FET device under various gate voltages (V_g) at 77 K. The large positive gate voltage induces a higher source-drain current, which indicates an n-type conduction behaviour of perovskite microplate

(Fig. 1b). The slight nonlinearity of I_{sd}–V_{sd} curves near zero bias suggests that the contact is not fully optimized. The transfer characteristics exhibit dominant n-type behaviour with a slight p-type conductance at negative gate voltage (Fig. 1c). The maximum on/off ratio is nearly six orders of magnitude, which is better than recently reported perovskite thin-film transistors[19].

Strong hysteresis is commonly observed in perovskite thin-film transistors, which prevents fully understanding the charge transport properties and exact determination of carrier mobility in such perovskite materials. The origin of the hysteresis has been attributed to ferroelectricity, ion motion within the perovskite material and trapping/de-trapping of charge carriers at the interfaces[19]. However, no conclusive explanation is available to date. In our microplate devices, considerable hysteresis has been observed for all temperatures from 296 to 77 K, which reduces with decreasing temperature. It has been proven that the ion motion in halide perovskite is a thermally activated process[33] and the ion migration rate exponentially reduces as the temperature decreases. The contribution from ion motion to hysteresis is expected to negligibly small at lower temperatures (for example, 77 K). The presence of hysteresis at 77 K suggests that the ion motion only partly contributes to the hysteresis and other factors such as trap states and surface dipoles may play important roles as well[19]. It is important to note that the hysteresis in our microplate device only increases slightly when the temperature is increased from 77 to 296 K (Supplementary Fig. 1), and the hysteresis at 296 K is considerably smaller than that observed in thin-film perovskite FET devices, where the presence of huge hysteresis prevents the observation of field-effect behaviour above 258 K (ref. 19).

Based on the transfer characteristics, the field-effect carrier mobility can be extracted. The existence of hysteresis in transfer characteristics may lead to systematic errors in mobility

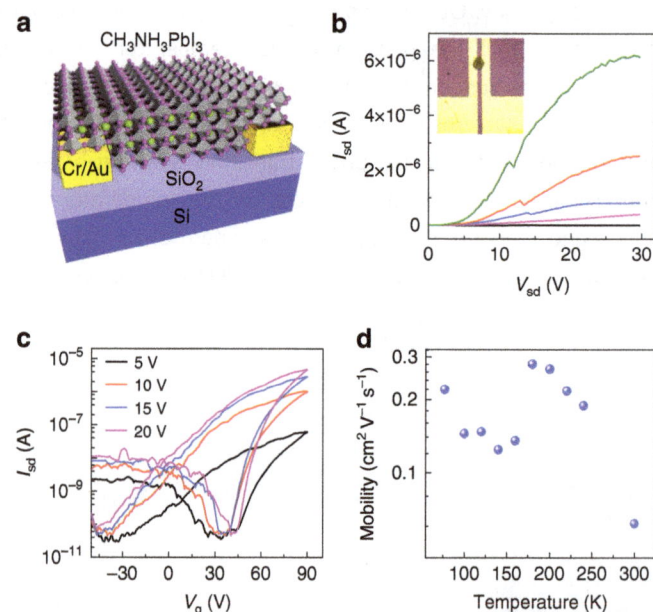

Figure 1 | Field-effect transistors based on individual perovskite microplate. (**a**) Schematic of the bottom-gate, bottom-contact halide perovskite microplate field-effect transistor fabricated on a 300-nm SiO_2/Si substrate with 5 nm Cr/50 nm Au as contact. (**b,c**) The output ($V_g = 0$, 20, 40, 60, 80 V; from bottom to top; **b**) and transfer ($V_{sd} = 5$, 10, 15, 20 V; **c**) characteristics of a field-effect transistor based on a perovskite crystal microplate at 77 K. The inset of **b** shows an optical image of a typical device. The channel length is around 8 μm. (**d**) The temperature-dependent field-effect electron mobility measured with a source-drain voltage of 20 V.

determination, with possible underestimation in positive sweeping direction, overestimation in negative sweeping direction and scan rate dependence (Supplementary Fig. 2). To this end, we have determined carrier mobility based on both the positive and negative sweepings. Nevertheless, both the positive and negative sweepings give the exactly same trend of the mobility versus temperature. For the simplicity of discussion, we focus on the mobility values derived from negative sweeping here. The field-effect electron mobility continuously increases with the decreasing temperature from 300 to 180 K, and then shows a sudden decrease when the temperature is reduced from 180 to 160 K (Fig. 1d). Afterwards with further decreasing temperature, the field-effect electron mobility starts to increase again. Qualitatively similar temperature-dependent characteristics have been observed in all devices except that the transition temperature varies with the thickness of microplates, which we will discuss below in detail. As there is a structural phase transition from the tetragonal phase to the orthorhombic phase at 160 K (ref. 23), we attribute this sudden decrease of field-effect mobility to the structural phase transition. The structural phase transition would induce the change of effective mass and dielectric constant[19,34,35], both of which could contribute to the change of field-effect mobility[19,36].

The theoretical calculation based on semi-classical Boltzman transport theory predicates that the mobility of orthorhombic phase should be larger than that of tetragonal phase[19]. In contrast, we observed a sudden decrease of the mobility when the MAPbI$_3$ transits from the tetragonal phase to the orthorhombic phase. Previous optical studies indicate that there are small inclusions of the tetragonal phase domains within the orthorhombic phase even when the temperature is much lower than the tetragonal phase to orthorhombic phase transition temperature (T_{t-o}), likely due to the strain imposed by the thermal expansion and change of the in-plane lattice constant during the phase transition[23]. Our PL studies also demonstrate the presence of such small inclusions (see below). It is very likely that such small inclusions introduce more boundaries and thus increases the carrier scattering, which also contribute to the sudden decrease of the mobility upon the phase transition. Although the improvement of field-effect mobility with the decreasing temperature can be attributed to the electron–phonon interaction and ion drift under applied electric field within the tetragonal phase[19,37], the origins of the rapid increase of the field-effect mobility within the orthorhombic phase are much more complicated. Previous studies have shown that the minimum phonon energy related to the methylammonium (MA) cation is estimated to be 15 meV (refs 38,39). Therefore, the interaction of carriers with phonons associated with MA libration should be quenched below 170 K. It has been shown that the quench of the carrier–phonon interaction related to the MA libration modes led to a weaker temperature dependence of carrier mobility below 198 K in halide perovskite thin-film transistors[19], indicating the strong interaction between the carriers and MA libration modes. Without the contribution from interaction with MA libration modes < 170 K, the increase rate of the mobility with decreasing temperature should be slowed if only carrier–phonon interaction contributes to the decrease of the mobility after the phase transition. On the contrary, we observed a different picture: a more rapid increase with decreasing temperature (Fig. 1d). Therefore, we suggest that in addition to the carrier–phonon scattering, the decrease of the small inclusions of tetragonal phase domains and the reduction of the grain boundaries might partly contribute to the rapid increase of mobility with the decreasing temperature. The inclusions of tetragonal phase near the transition point within the orthorhombic phase are also confirmed by our temperature-dependent selected area electron diffraction (SAED) studies (see below).

It should be noted that the electron field-effect mobility we extracted here is smaller than those measured by THz spectroscopy in perovskite films (~ 8 cm^2 V^{-1} s^{-1} at room temperature)[40], Hall measurement in perovskite single crystals (~ 66 cm^2 V^{-1} s^{-1})[27] and electrical measurement in the space-charge-limited current regime or time-of-flight measurement in perovskite single crystals (~ 2.5–25 cm^2 V^{-1} s^{-1})[10,11], but compares favourably with the field-effect mobility reported in perovskite thin-film transistors (~ 0.1 cm^2 V^{-1} s^{-1})[19]. Time-resolved THz spectroscopy probes short-time dynamics up to a few nanoseconds and thus measures local carrier transport phenomena, whereas electrical measurements focus long-time (μs) or longer conduction processes occurring over several micrometres length-scale of a device[41,42]. Electrical measurements are therefore more sensitive to grain size, boundaries and interfacial effects because of the carrier transport over the device dimension. Thus, it is not surprising that the mobility measured by THz spectroscopy is larger than that measured by electrical measurements. Although Hall measurements, time-of-flight technique and space-charge-limited current method directly measure intrinsic charge transport, the field-effect mobility is extremely sensitive to the dielectric/semiconductor interfaces as well as the source, drain contact resistance[43]. Such extrinsic factors in field-effect measurement can often lead to an underestimation of the carrier mobility in FETs.

Thickness-dependent phase transition. To investigate how the thickness of the microplates influences the field-effect mobility, we systematically carried out the temperature-dependent transport measurement with different microplate thickness, with the transfer curves (120–200 K) of three representative devices shown in Fig. 2a–c. All three devices exhibit dominant n-type behaviour regardless of the thickness. The field-effect electron mobility extracted from the transfer curves shows a common trend with the temperature for devices with different thickness: as the temperature decreases, the electron mobility first increases, suddenly decreases at the structural phase transition point and increases again with further reducing temperature (Fig. 2d). It is noted that the structural phase transition temperature T_{t-o} strongly depends on the thickness of the perovskite microplates: the thicker the microplates are, the higher the structural phase transition temperature T_{t-o} is. For the 30-nm-thick microplate, the structural phase transition temperature T_{t-o} falls around 130 K, which increases to ~ 150 K for the 90 nm microplate, and to ~ 170 K for the 400-nm-thick microplate. Furthermore, we have carried out temperature-dependent transport measurement using different metal contact including Pt and graphene (Fig. 2e), on a hexagonal boron nitride (hBN) substrate (Supplementary Fig. 3 and Supplementary Note 1) and in a device with very long channel length (40 μm; Supplementary Fig. 4). It is found that the structural phase transition occurs with phase transition temperature T_{t-o} relying only on the thickness of perovskite microplates regardless of the contact materials, substrate and channel length, indicating that the structural phase transition is an intrinsic property of the perovskite microplates.

To precisely locate the tetragonal-to-orthorhombic phase transition point, we scan the temperature range with a higher resolution around the phase transition point for a 200-nm-thick microplate device (Fig. 2f). Similar to the hysteretic behaviour observed in the optical density of perovskite thin films[26] and dielectric and resistance measurement of MAPbI$_3$ crystals[27,35], an apparent hysteresis is observed with a temperature span of 15 K between the cooling cycle and heating cycle. The sharp decrease of the mobility near the phase transition temperature and the presence of the broad hysteresis confirm that the tetragonal phase to the orthorhombic phase transition is a first-order solid–solid phase transition[29,44,45].

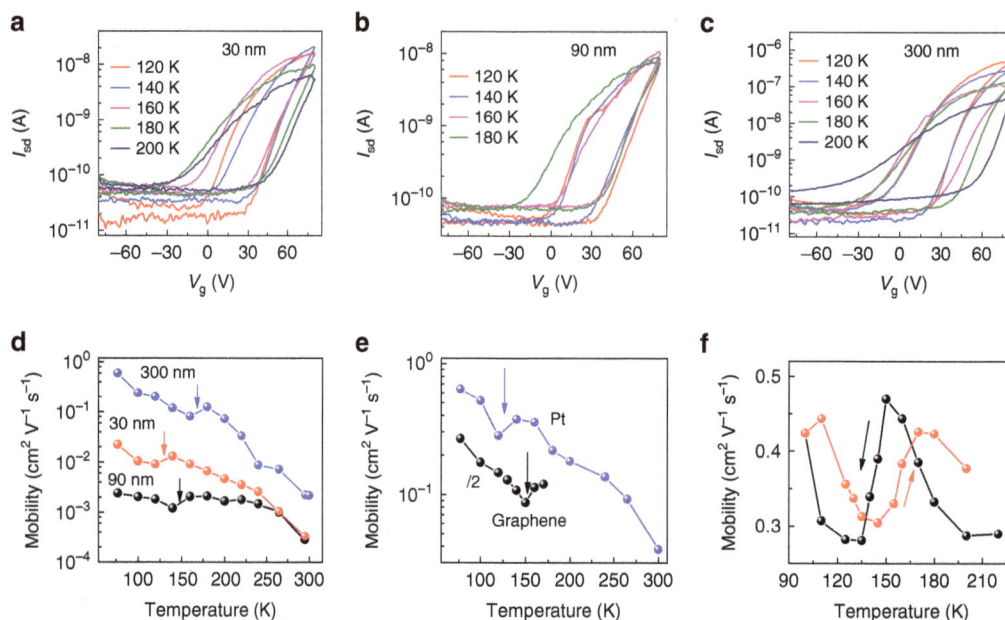

Figure 2 | Size-dependent transition of the orthorhombic phase to tetragonal phase. (**a–c**) The temperature-dependent transfer characteristics of field-effect transistors made of 30 nm (**a**), 90 nm (**b**) and 300 nm (**c**) thick individual perovskite crystal microplates. The applied source-drain voltage is 20 V and the channel length is 8 µm. (**d**) The temperature-dependent field-effect electron mobility for three different thickness devices extracted from **a–c**. The arrows indicate the temperature where the phase transition occurs. The mobility is measured for the heating cycle. (**e**) The temperature-dependent field-effect electron mobility for devices with Pt contact and graphene contact. The thickness of the microplates is around 35 nm and the channel length is 40 µm for Pt contact. For graphene contact, the thickness of the microplates is around 120 nm and the channel length is around 15 µm. The mobility is measured during the heating cycle. (**f**) The temperature-dependent field-effect mobility of a perovskite microplate device (5 nm Cr/50 nm Au as contact) with a thickness of around 200 nm measured for both heating and cooling cycle under a source-drain voltage of 20 V. The channel length of the device is 40 µm.

Temperature-dependent SAED studies. The tetragonal-to-orthorhombic phase transition can be directly confirmed by the temperature-dependent SAED studies. Transmission electron microscopy (TEM) image shows that the converted perovskite microplates we used to acquire the SAED patterns largely retain hexagonal shape similar with the PbI_2 microplate before conversion (Fig. 3a). The SAED pattern at room temperature (Fig. 3b) shows a single set of fourfold symmetric diffraction spots that can be indexed to the tetragonal structure of the perovskite crystals along [001] zone axis. Both first-order and second-order diffraction spots can be clearly distinguished, indicating excellent crystalline quality of the perovskite microplate. Decreasing the temperature to 90 K, the SAED pattern shows a set of fourfold symmetric diffraction spots and a few dispersedly distributed spots indicated by red circles (Fig. 3c). Although the fourfold symmetric diffraction spots can be indexed to the first-order diffraction of the orthorhombic structure along [001] zone axis, the dispersedly distributed spots likely belong to a different set of diffraction patterns. Increasing the temperature to 200 K again, the microplate completely transits from orthorhombic phase back to tetragonal phase. The SAED pattern shows similar features as those obtained at room temperature initially, with the dispersedly distributed spots disappeared. The fourfold symmetric spots can be indexed to the tetragonal structure of the perovskite crystals along [001] zone axis (Fig. 3d), implying that those dispersedly distributed spots at 90 K appear only after the tetragonal-to-orthorhombic phase transition completes. Therefore, those dispersedly distributed spots might be originated from the tetragonal phase along different zone axis because of the inclusions of tetragonal domains within orthorhombic phase or from the orthorhombic domains with different crystalline orientation. Nevertheless, based on lattice spacings analysis it is more likely that the inclusions of tetragonal domains contribute to those dispersedly distributed spots. As mentioned

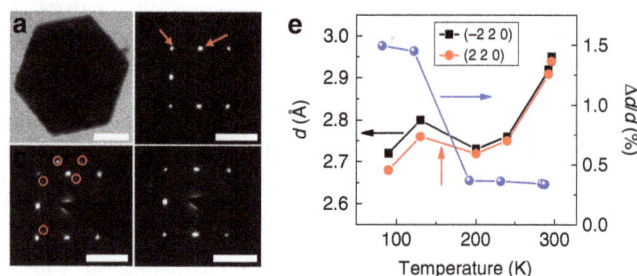

Figure 3 | Temperature-dependent selected area electron diffraction (SAED) patterns. (**a**) Low-resolution TEM image of a perovskite microplate. The scale bar is 1 µm. (**b–d**) The SAED patterns of the microplate at 296 K (**b**), 90 K (**c**) and 200 K (**d**) along [001] zone axis. The scale bar is 5 nm^{-1}. The red circles in **c** indicate the dispersive distributed spots. (**e**) Lattice spacings of (− 2 2 0) plane (black squares) and (2 2 0) plane (red squares) of a perovskite microplate. The percentage of lattice spacing difference between those two planes (defined as $(d(-220) - d(220))/d(220)$) is displayed as well (right axis). The increasing difference between these two lattice spacings with the reducing temperature indicates the transitions from a tetragonal phase to orthorhombic phase.

above, even when the temperature is much lower than the phase transition temperature, there still are inclusions of tetragonal domains within orthorhombic phase, suggesting that single-crystal tetragonal phase may break into smaller grains of tetragonal phase and orthorhombic phase when the phase transition occurs. Although most of those small grains maintain almost same orientation, some of them significantly deviate, leading to the dispersedly distributed spots. The presence of dispersedly distributed spots also indicates the degradation of the crystalline quality after the tetragonal-to-orthorhombic phase transition.

The tetragonal-to-orthorhombic phase transition can also be identified by carefully analysing the temperature-dependent lattice spacing change. We extracted the lattice spacings for ($-2\,2\,0$) plane and ($2\,2\,0$) plane of a perovskite microplate from SAED patterns and found that a sudden change of the lattice spacings between 130 and 200 K, indicating the phase change occurs in this temperature regime (Fig. 3e). As the ($-2\,2\,0$) plane and ($2\,2\,0$) plane are perpendicular to each other, the changing ratio of the lattice spacings between these two planes can be used to identify the crystalline structure as well. Above 200 K, the lattice spacings for those two planes are almost same, indicating that the perovskite microplate has a tetragonal phase. In contrary, there is clear difference between the lattice spacings for those two planes below 200 K, implying the presence of orthorhombic structure (right axis of Fig. 3e). Those results agree with those obtained from transport measurement above.

Temperature- and thickness-dependent PL studies. To further probe the size-dependent tetragonal-to-orthorhombic phase transition, we have also studied temperature-dependent PL. Figure 4a–c displays the PL spectra for four perovskite microplates with various thicknesses at 270, 140 and 77 K. Only one broad emission peak was observed for all four microplates at 270 K and the emission peak shows a blueshift with the reducing thickness (Fig. 4a), which will be discussed in detail below. The PL spectra at 140 K show an extra emission peak at the higher energy for thicker microplates while still exhibits a single peak for the thinner ones (Fig. 4b). Further decreasing the temperature to 77 K, two emission peaks are observed for all microplates but the intensity ratio of the higher energy emission peak ($P2$) to the lower energy emission peak ($P1$) decreases with decreasing microplate thickness (Fig. 4c). The higher energy emission peak can be attributed to the orthorhombic phase, whereas the lower energy emission peak is due to the tetragonal phase domains within the orthorhombic phase[23], consistent with theoretical

calculations that the tetragonal phase has a smaller bandgap than that of the orthorhombic phase[46]. Therefore, the emergence of the two emission peaks signifies the occurrence of phase transition, and our PL studies also suggest that the phase transition temperature T_{t-o} is higher for the thicker microplates (Fig. 4b).

Excitation power-dependent PL spectra. Our excitation power-dependent PL studies also confirms the existence of small tetragonal inclusions below the phase transition temperature T_{t-o}, which is supported by the fact that both the emission peak position and intensity are extremely sensitive to the excitation power. We have collected excitation power-dependent PL spectra for a 20-nm perovskite microplate in orthorhombic phase (77 K), near orthorhombic-to-tetragonal phase transition point (140 K) and tetragonal phase (180 K), respectively, extracted the emission peak energy and plotted against the excitation power for each peak at 77, 140 and 180 K (Fig. 4d–f and Supplementary Figs 5 and 6). At 77 K, the lower energy emission peak P1 originating from tetragonal phase domains shows an obvious blueshift with the increasing of excitation power, whereas the higher energy peak $P2$ from orthorhombic phase shows little change. It is also noted that the $P1$ emission saturates at high excitation power. As the tetragonal phase has a smaller band gap, the photogenerated carriers prefer to occupy the small tetragonal phase inclusions within the orthorhombic phase. As a result, a large number of carriers are trapped and recombine within those small tetragonal inclusions. As the excitation power increases, the quasi-Fermi levels of the photogenerated carriers move into the conduction band and valence band, resulting in a band filling effect. As the size of the tetragonal inclusions is extremely small, the large blueshift of $P1$ emission peak and saturation of $P1$ intensity can be observed. This sort of blueshift of emission peaks has been commonly observed in quantum wells and other confined heterostructures[47,48]. Increasing the temperature to 140 K, similar

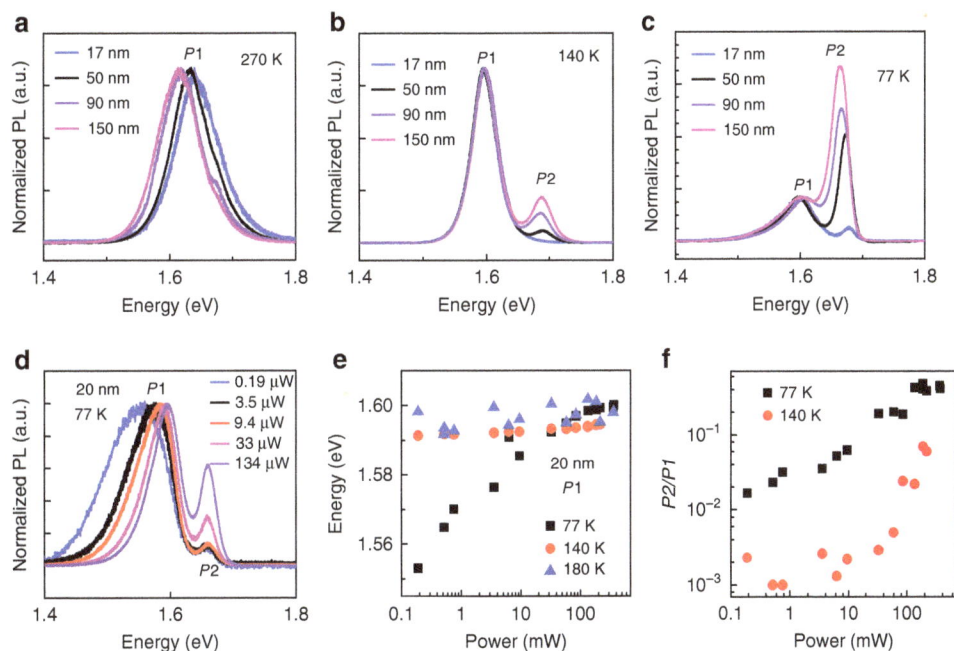

Figure 4 | Thickness and excitation power-dependent photoluminescence studies. (a–c) The photoluminescence spectra for four different thickness halide perovskite microplates at 270 K (**a**), 140 K (**b**) and 77 K (**c**). A 488-nm laser with a power of 3.5 µW was used as the excitation source. All spectra are normalized by the low-energy peak P1 in order to easily compare among each other. (**d**) The excitation power-dependent PL spectra for a 20-nm-thick microplate at 77 K. The spectra have been normalized by the low-energy emission peak P1. (**e**) The excitation power-dependent emission peak position of the tetragonal phase for the 20-nm perovskite microplate. (**f**) The P2/P1 ratios extracted from their corresponding PL spectra under different excitation power at 77 K (black squares) and 140 K (red dots).

trend was observed except that the blueshift of $P1$ becomes negligibly small, which is probably due to the increasing size of tetragonal inclusions. With the increasing size of the tetragonal inclusions, the density of states of the tetragonal inclusions increases accordingly, making it more difficult to observe the band filling effect. At 180 K when the orthorhombic-to-tetragonal phase transition has already completed, no noticeable blueshift was observed. It is expected as the excitation power we used is not big enough such that the band filling effect cannot occur in unconfined systems. The excitation power-dependent $P2/P1$ ratios also clearly demonstrate the band filling effect (Fig. 4f). At 77 K, the $P2/P1$ ratios are very sensitive to the excitation power and show monotonously increases with the increasing excitation power, which indicates the small grains of tetragonal inclusions. At 140 K, the $P2/P1$ ratios are always smaller than that at 77 K and only slightly increase with the excitation power, indicating the increasing size of tetragonal inclusions, which renders the band filling effect hard to be observed. Based on the above discussions, we concluded that the presence of the two emission peaks at low temperatures is due to the small inclusions of tetragonal phase domains within the orthorhombic phase.

The emission peak energy shows a blueshift with the decreasing microplate thickness both for tetragonal phase above 155 K and the orthorhombic phase below 140 K (Figs 4 and 5, and Supplementary Figs 7 and 8), which has been observed in solution-processed perovskite nanocrystals[49,50]. As the thickness of our microplates is much larger than the bulk exciton Bohr radius (2.2 nm) (ref. 51), the quantum confinement effect is unlikely to be the primary factor responsible for this blueshift. Surface effect has been previously proposed to explain such blueshift beyond the quantum confinement regime[52]. In brief, the surface charge-induced depletion electric field near the surfaces or interfaces modifies the confinement potential, leading to a potential well smaller than the actual geometric thickness of the microplates. Our observed blueshift in halide perovskite microplates might be due to the surface effect as well. Nevertheless, the exact underlying mechanism is still unclear and demands further investigation.

Thickness-dependent phase transition in PL spectra. The $P2/P1$ ratios can be used to identify the degree of the phase transition. From $P2/P1$ ratios (Fig. 5a,b and Supplementary Fig. 8), we can conclude that the orthorhombic-to-tetragonal phase transition occurs at a lower transition temperature T_{t-o} in the thinner microplates, which is consistent with the conclusion obtained from the charge transport measurement. The temperature-dependent PL spectra indicates that the portion of tetragonal inclusions within the orthorhombic phase decreases with the decreasing temperature for all four different thickness microplates, which is supported by the increases of the $P2/P1$ intensity ratio with decreasing temperature (Fig. 5a and Supplementary Fig. 8). The transition temperature T_{t-o} decreases with the decreasing thickness: $T_{t-o} < 140$ K for the thickness smaller than 40 nm, and between 140 and 150 K for the thickness around 40–200 nm. Within the respective tetragonal phase (155–290 K) and orthorhombic phase (77–140 K), the emission peak shows a redshift and the full-width at half-maximum (FWHM) narrows as the temperature decreases (Fig. 5c–f). The counter-intuitive redshift of the emission peak with decreasing temperature is strikingly different from the traditional semiconductors, where the emission peak blushifts with the decreasing temperature. This anomalous temperature-dependent band gap remains elusive and demands further investigations. The reducing FWHM with the decreasing temperature can be attributed to the weaker electron–phonon interaction at the lower temperature. For the tetragonal inclusions within the orthorhombic phase, the emission peak energy shows blueshift with the decreasing temperature (Fig. 5c), which is probably due to the increasing quantum confinement effect in the small tetragonal domains. As the temperature decreases, the size of the tetragonal inclusions decreases, leading to a stronger quantum confinement effect and thus a blueshift of emission peak. Furthermore, the FWHM of tetragonal inclusions increases with the decreasing temperature (Fig. 5e), which might be due to the size variation of the tetragonal inclusions. Therefore, the trend of the peak position and FWHM can be used to identify the phase transition points as well (Fig. 5c,e).

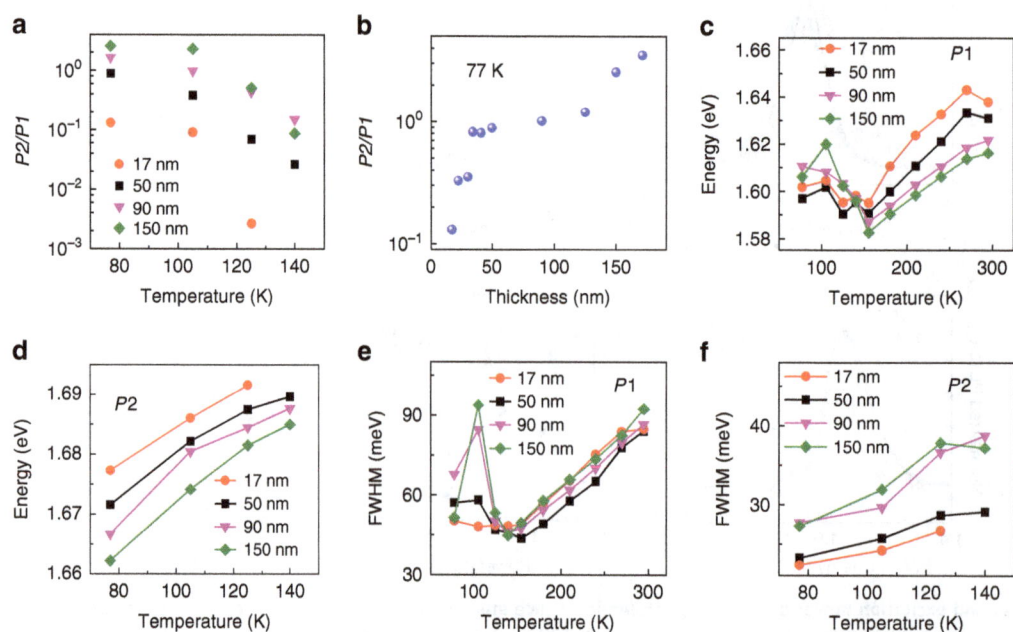

Figure 5 | The temperature- and thickness-dependent photoluminescence studies. (**a**) The temperature-dependent $P2/P1$ ratios for a 17-, 50-, 90- and 150-nm-thick halide perovskite microplates excited by a 488-nm laser with a power of 3.5 μW. (**b**) The $P2/P1$ ratios for halide perovskite microplates with various thicknesses at 77 K. (**c,d**) The temperature-dependent emission energy for tetragonal phase (**c**) and orthorhombic phase (**d**). (**e,f**) The temperature-dependent full-width at half-maximum (FWHM) for tetragonal phase (**e**) and orthorhombic phase (**f**).

Discussion

The size-dependent shift in the phase transition temperature has been extensively investigated and observed in confined systems such as nanocrystal[30,53] and two-dimensional layered materials ($NbSe_2$ (ref. 54) and TaS_2 (ref. 55)). A reduction in the nanomaterial size can lead to the decrease of the structural phase transition temperature. Many competing theories have been proposed to explain this behaviour, which includes the lack of nucleation sites, internal pressure and surface energy difference between polymorphs[53]. The rough surface in our microplates can exclude the possibility of the lack of nucleation sites. The internal strain would lead to the broadening of the emission peak and thus a larger FWHM in the thinner microplates. Nevertheless, the similar or smaller FWHM in the thinner microplates (Fig. 5e,f) implies that the strain effect should not play a dominant role in our case. The thickness-dependent phase change temperature T_{t-o} observed here is more likely due to the surface energy difference between the polymorphs. As the thickness decreases, the surface-to-volume ratio increases, resulting in the lower transition temperature T_{t-o} in the thinner microplates. This explanation is consistent with the blueshift of the PL emission peak with the decreasing of the thickness due to the surface effect (Supplementary Fig. 7).

In summary, we have systematically investigated the size-dependent phase transition in individual MA lead iodide microplate using temperature-dependent PL spectroscopy, charge transport measurement and TEM studies. Our studies demonstrate that the orthorhombic-to-tetragonal phase transition temperature T_{t-o} decreases with the decreasing thickness of the perovskite microplates, and confirm the phase transition is a first-order solid–solid phase transition. In addition to the fundamental importance, the thickness-dependent structural phase transition has important practical implications in the fields ranging from electronics, optoelectronics to materials sciences. Our findings on the thickness- and temperature-dependent optical and electric properties can shed light on the development of electronic and optoelectronic devices not just at room temperature but also at low temperature, which would have important applications in airplanes and satellites[56].

Methods

Sample preparations. A dilute PbI_2 aqueous solution (0.1 g per 100 ml) prepared at 80 °C was cooled to room temperature, which leads to the formation of suspended PbI_2 microplates. For the PL measurement samples, we dipped the substrates (Si substrates with 300 nm SiO_2) with pre-fabricated markers by photolithography into the aqueous solution for a few seconds. After taking the substrate from the solution, we can find various thickness microplates by chance. For the FET samples, the 5 nm Cr/50 nm Au (Pt) electrodes with channel lengths of 8 and 40 μm were defined by photolithography and followed by thermal evaporation and lift-off. Then PbI_2 microplates were grown onto the pre-fabricated electrodes by randomly dispersion. The prepared PbI_2 microplates were converted into $CH_3NH_3PbI_3$ by vapour phase intercalation. The intercalation source (MA iodide powder) was synthesized by a solution method[57]. The MA iodide source was placed at the centre of a quartz tube and the substrate with PbI_2 microplates was placed 5–6 cm away downstream. Before conversion, the tube furnace was vacuumed and refilled with argon for at least three times to completely remove the air in the quartz tube. The conversion was conducted at a pressure of 100 mbar with 100 s.c.c.m. argon flow as carrier gas for several hours. The actual temperature is 140 °C at the MA iodide source region and 120 °C at the PbI_2 micro-plate substrate region measured by a thermocouple probe. Finally, the tube was naturally cooled down to room temperature.

Fabrication of graphene-contact FETs. To fabricate the graphene contact devices, graphene strips (as electrodes) was peeled on clean silicon/silicon oxide (300 nm) substrate, whereas the PbI_2 plates and the top hexagonal boron nitride (hBN) was peeled on polymer stack PMMA/PPC (polypropylene carbonate) spun on a silicon wafer. First, the peeled PbI_2 plate was aligned and transferred onto the graphene strips and the PMMA/PPC was dissolved by using chloroform solution. Then, the PbI_2 plate was converted to $CH_3NH_3PbI_3$ by using the vapour phase intercalation. Afterwards, the top BN was aligned and transferred to protect the perovskite for the following electrode fabrication processes. To make edge contact to graphene, the windows on hBN were first defined by electron-beam lithography exactly upon the graphene stripes and followed by the plasma etching to remove hBN. Afterwards, electron beam lithography was used to pattern the edge contact to graphene and followed by thermal evaporation and lift-off.

Microscopic and optical characterizations. The thickness of the perovskite microplate was determined by tapping-mode atomic force microscopy (Vecco 5,000 system). TEM images and SAED patterns were acquired in an FEI Titan high-resolution transmission microscopy. The PL measurement was conducted under a confocal micro-Raman system (Horiba LABHR) equipped with a 600 g mm^{-1} grating in a backscattering configuration excited by an Ar ion laser (488 nm). For the low-temperature measurement, a liquid nitrogen continuous flow cryostat (Cryo Industry of America) was used to control the temperature from 77 to 300 K.

Electrical measurements. Temperature-dependent FET device measurements were carried out in a probe station ((Lakeshore, TTP4) coupled with a precision source/measurement unit (Agilent B2902A). The scanning rate for the transport measurement is 20 V s^{-1} and the devices were pre-biased at the opposite voltage for 30 s before each measurement.

References

1. Green, M. A., Ho-Baillie, A. & Snaith, H. J. The emergence of perovskite solar cells. *Nat. Photon* **8**, 506–514 (2014).
2. Lee, M. M. *et al.* Efficient hybrid solar cells based on meso-superstructured organometal halide perovskites. *Science* **338**, 643–647 (2012).
3. Liu, M., Johnston, M. B. & Snaith, H. J. Efficient planar heterojunction perovskite solar cells by vapour deposition. *Nature* **501**, 395–398 (2013).
4. Burschka, J. *et al.* Sequential deposition as a route to high-performance perovskite-sensitized solar cells. *Nature* **499**, 316–319 (2013).
5. Zhou, H. *et al.* Interface engineering of highly efficient perovskite solar cells. *Science* **345**, 542–546 (2014).
6. Liu, D. & Kelly, T. L. Perovskite solar cells with a planar heterojunction structure prepared by room-temperature solution processing techniques. *Nat. Photon* **8**, 133–138 (2014).
7. Jeon, N. J. *et al.* Solvent engineering for high-performance inorganic–organic hybrid perovskite solar cells. *Nat. Mater* **13**, 897–903 (2014).
8. Grätzel, M. The light and shade of perovskite solar cells. *Nat. Mater* **13**, 838–842 (2014).
9. Lin, Q. *et al.* Electro-optics of perovskite solar cells. *Nat. Photon* **9**, 106–112 (2015).
10. Dong, Q. *et al.* Electron-hole diffusion lengths > 175 μm in solution-grown $CH_3NH_3PbI_3$ single crystals. *Science* **347**, 967–970 (2015).
11. Shi, D. *et al.* Low trap-state density and long carrier diffusion in organolead trihalide perovskite single crystals. *Science* **347**, 519–522 (2015).
12. Nie, W. *et al.* High-efficiency solution-processed perovskite solar cells with millimeter-scale grains. *Science* **347**, 522–525 (2015).
13. Salim, T. *et al.* Perovskite-based solar cells: impact of morphology and device architecture on device performance. *J. Phys. Chem. A* **3**, 8943–8969 (2015).
14. Dou, L. *et al.* Solution-processed hybrid perovskite photodetectors with high detectivity. *Nat. Commun* **5**, 5404 (2014).
15. Xing, G. *et al.* Low-temperature solution-processed wavelength-tunable perovskites for lasing. *Nat. Mater* **13**, 476–480 (2014).
16. Zhang, Q. *et al.* Room-temperature near-infrared high-Q perovskite whispering-gallery planar nanolasers. *Nano Lett.* **14**, 5995–6001 (2014).
17. Zhu, H. *et al.* Lead halide perovskite nanowire lasers with low lasing thresholds and high quality factors. *Nat. Mater* **14**, 636–642 (2015).
18. Tan, Z.-K. *et al.* Bright light-emitting diodes based on organometal halide perovskite. *Nat. Nanotechnol* **9**, 687–692 (2014).
19. Chin, X. Y. *et al.* Lead iodide perovskite light-emitting field-effect transistor. *Nat. Commun* **6**, 7383 (2015).
20. Mei, Y., Zhang, C., Vardeny, Z. & Jurchescu, O. Electrostatic gating of hybrid halide perovskite field-effect transistors: balanced ambipolar transport at room-temperature. *MRS Commun* **5**, 1–5 (2015).
21. Im, J.-H. *et al.* Growth of $CH_3NH_3PbI_3$ cuboids with controlled size for high-efficiency perovskite solar cells. *Nat. Nanotechnol* **9**, 927–932 (2014).
22. Loi, M. A. & Hummelen, J. C. Hybrid solar cells: perovskites under the Sun. *Nat. Mater* **12**, 1087–1089 (2013).
23. Wehrenfennig, C. *et al.* Charge carrier recombination channels in the low-temperature phase of organic-inorganic lead halide perovskite thin films. *APL Mater* **2**, 081513 (2014).
24. Weller, M. T. *et al.* Complete structure and cation orientation in the perovskite photovoltaic methylammonium lead iodide between 100 and 352K. *Chem. Commun* **51**, 4180–4183 (2015).
25. Onoda-Yamamuro, N., Matsuo, T. & Suga, H. Calorimetric and IR spectroscopic studies of phase transitions in methylammonium trihalogenoplumbates (II)†. *J. Phys. Chem. Solids* **51**, 1383–1395 (1990).

26. Baikie, T. *et al.* Synthesis and crystal chemistry of the hybrid perovskite (CH$_3$ NH$_3$) PbI$_3$ for solid-state sensitised solar cell applications. *J. Phys. Chem. A* **1**, 5628–5641 (2013).

27. Stoumpos, C. C., Malliakas, C. D. & Kanatzidis, M. G. Semiconducting tin and lead iodide perovskites with organic cations: phase transitions, high mobilities, and near-infrared photoluminescent properties. *Inorg. Chem.* **52**, 9019–9038 (2013).

28. Chen, C.-C., Herhold, A., Johnson, C. & Alivisatos, A. Size dependence of structural metastability in semiconductor nanocrystals. *Science* **276**, 398–401 (1997).

29. Tolbert, S. & Alivisatos, A. Size dependence of a first order solid-solid phase transition: the wurtzite to rock salt transformation in CdSe nanocrystals. *Science* **265**, 373–373 (1994).

30. Rivest, J. B. *et al.* Size dependence of a temperature-induced solid–solid phase transition in copper (I) sulfide. *J. Chem. Phys. Lett.* **2**, 2402–2406 (2011).

31. Zhang, Y. Gate-tunable phase transitions in thin flakes of 1T-TaS$_2$. *Nat. Nanotechnol* **10**, 270–276 (2015).

32. Wu, K. *et al.* Temperature-dependent excitonic photoluminescence of hybrid organometal halide perovskite films. *Phys. Chem. Chem. Phys.* **16**, 22476–22481 (2014).

33. Eames, C. *et al.* Ionic transport in hybrid lead iodide perovskite solar cells. *Nat. Commun* **6**, 7497 (2015).

34. Frost, J. M., Butler, K. T. & Walsh, A. Molecular ferroelectric contributions to anomalous hysteresis in hybrid perovskite solar cells. *APL Mater* **2**, 081506 (2014).

35. Onoda-Yamamuro, N., Matsuo, T. & Suga, H. Dielectric study of CH$_3$NH$_3$PbX$_3$ (X = Cl, Br, I). *J. Phys. Chem. Solids.* **53**, 935–939 (1992).

36. Siemons, W. *et al.* Dielectric-constant-enhanced Hall mobility in complex oxides. *Adv. Mater.* **24**, 3965–3969 (2012).

37. Xiao, Z. *et al.* Giant switchable photovoltaic effect in organometal trihalide perovskite devices. *Nat. Mater* **14**, 193–198 (2015).

38. Quarti, C. *et al.* The Raman spectrum of the CH$_3$NH$_3$PbI$_3$ hybrid perovskite: interplay of theory and experiment. *J. Chem. Phys. Lett.* **5**, 279–284 (2013).

39. Brivio, F. *et al.* Lattice dynamics and vibrational spectra of the orthorhombic, tetragonal, and cubic phases of methylammonium lead iodide. *Phys. Rev. B* **92**, 144308 (2015).

40. Wehrenfennig, C. *et al.* High charge carrier mobilities and lifetimes in organolead trihalide perovskites. *Adv. Mater.* **26**, 1584–1589 (2014).

41. Esenturk, O., Melinger, J. S. & Heilweil, E. J. Terahertz mobility measurements on poly-3-hexylthiophene films: device comparison, molecular weight, and film processing effects. *J. Appl. Phys.* **103**, 023102 (2008).

42. Vukmirović, N. *et al.* Insights into the charge carrier terahertz mobility in polyfluorenes from large-scale atomistic simulations and time-resolved terahertz spectroscopy. *J. Chem. Phys. C* **116**, 19665–19672 (2012).

43. Jang, J., Liu, W., Son, J. S. & Talapin, D. V. Temperature-dependent Hall and field-effect mobility in strongly coupled all-inorganic nanocrystal arrays. *Nano Lett.* **14**, 653–662 (2014).

44. Ni, N. *et al.* First-order structural phase transition in CaFe$_2$As$_2$. *Phys. Rev. B* **78**, 014523 (2008).

45. Sethna, J. P. *et al.* Hysteresis and hierarchies: dynamics of disorder-driven first-order phase transformations. *Phys. Rev. Lett.* **70**, 3347–3350 (1993).

46. Even, J., Pedesseau, L. & Katan, C. Analysis of multivalley and multibandgap absorption and enhancement of free carriers related to exciton screening in hybrid perovskites. *J. Chem. Phys. C* **118**, 11566–11572 (2014).

47. Li, D. *et al.* Strain-induced spatially indirect exciton recombination in zinc-blende/wurtzite CdS heterostructures. *Nano Res* **8**, 3035–3044 (2015).

48. Liu, Q. *et al.* Evidence of type-II band alignment at the ordered GaInP to GaAs heterointerface. *J. Appl. Phys.* **77**, 1154–1158 (1995).

49. Di, D. *et al.* Size-dependent photon emission from organometal halide perovskite nanocrystals embedded in an organic matrix. *J. Chem. Phys. Lett.* **6**, 446–450 (2015).

50. D'Innocenzo, V. *et al.* Tuning the light emission properties by band gap engineering in hybrid lead-halide perovskite. *J. Am. Chem. Soc.* **136**, 17730–17733 (2014).

51. Tanaka, K. *et al.* Comparative study on the excitons in lead-halide-based perovskite-type crystals CH$_3$NH$_3$PbBr$_3$ CH$_3$NH$_3$PbI$_3$. *Solid State Commun.* **127**, 619–623 (2003).

52. Li, D., Zhang, J. & Xiong, Q. Surface depletion induced quantum confinement in CdS nanobelts. *ACS Nano* **6**, 5283–5290 (2012).

53. Mayo, M., Suresh, A. & Porter, W. Thermodynamics for nanosystems: grain and particle-size dependent phase diagrams. *Rev. Adv. Mater. Sci.* **5**, 100–109 (2003).

54. Xi, X. *et al.* Strongly enhanced charge-density-wave order in monolayer NbSe$_2$. *Nat. Nanotechnol* **10**, 765–769 (2015).

55. Yu, Y. *et al.* Gate-tunable phase transitions in thin flakes of 1T-TaS$_2$. *Nat. Nanotechnol* **10**, 270–276 (2015).

56. La-o-vorakiat, C. *et al.* Elucidating the role of disorder and free-carrier recombination kinetics in CH$_3$NH$_3$PbI$_3$ perovskite films. *Nat. Commun* **6**, 7903 (2015).

57. Heo, J. H. *et al.* Efficient inorganic-organic hybrid heterojunction solar cells containing perovskite compound and polymeric hole conductors. *Nat. Photonics* **7**, 487–492 (2013).

Acknowledgements

We acknowledge the support from the US Department of Energy, Office of Basic Energy Sciences, Division of Materials Science and Engineering through Award DE-SC0008055.

Author contributions

X.D. and Y.H. designed the experiments. D.L. performed most of the experiments including device fabrication, electric, PL measurement and data analysis. G.W. synthesized the materials. H.-C.C., H.W. and Y.L. contributed to device fabrication. C.-Y.C. conducted the TEM studies. X.D. and D.L. co-wrote the paper. All authors discussed the results and commented on the manuscript.

Additional information

Competing financial interests: The authors declare no competing financial interests.

22

Real-time high dynamic range laser scanning microscopy

C. Vinegoni[1,*], C. Leon Swisher[1,*], P. Fumene Feruglio[1,2,*], R.J. Giedt[1], D.L. Rousso[3], S. Stapleton[1] & R. Weissleder[1]

In conventional confocal/multiphoton fluorescence microscopy, images are typically acquired under ideal settings and after extensive optimization of parameters for a given structure or feature, often resulting in information loss from other image attributes. To overcome the problem of selective data display, we developed a new method that extends the imaging dynamic range in optical microscopy and improves the signal-to-noise ratio. Here we demonstrate how real-time and sequential high dynamic range microscopy facilitates automated three-dimensional neural segmentation. We address reconstruction and segmentation performance on samples with different size, anatomy and complexity. Finally, *in vivo* real-time high dynamic range imaging is also demonstrated, making the technique particularly relevant for longitudinal imaging in the presence of physiological motion and/or for quantification of *in vivo* fast tracer kinetics during functional imaging.

[1] Center for Systems Biology, Massachusetts General Hospital and Harvard Medical School, Richard B. Simches Research Center, 185 Cambridge Street, Boston, Massachusetts 02114, USA. [2] Department of Neurological, Biomedical and Movement Sciences, University of Verona, Strada Le Grazie 8, 37134 Verona, Italy. [3] Center for Brain Science, Department of Molecular and Cell Biology, Harvard University, 52 Oxford Street, Cambridge, Massachusetts 02138, USA. * These authors contributed equally to this work. Correspondence and requests for materials should be addressed to C.V. (email: cvinegoni@mgh.harvard.edu).

The ability to directly visualize cellular and subcellular structures and function has greatly contributed to our knowledge of biological processes[1-3]. Among optical imaging techniques, laser scanning fluorescence microscopy (LSM) is one of the most widely used due to its high sensitivity, resolution, and penetration depth. Two-photon microscopy in particular has enabled major advances in virtually every biological field to which it has been applied to date[4]. Most commonly, LSM techniques are optimized and acquisition parameters are chosen to display a given structure of interest. This approach works well for many applications but is disadvantageous in circumstances where structures of contrasting brightness cannot be displayed simultaneously. This is particularly true for neuronal imaging, where cell bodies are significantly larger than neuronal processes, and where there is heterogeneity in the density of cell populations resulting in high intra-scene dynamic range. Furthermore, images with low signal-to-noise ratio (SNR) will lead to the fragmentation of the neural segments. Conversely, the presence of saturated regions will result in the inability to differentiate cell bodies or processes from neighbouring cells.

Photomultiplier tubes (PMT) are ubiquitous among commercial confocal and two-photon microscopy systems, due to their low cost, high sensitivity and wide coverage of wavelengths. Therefore, a high dynamic range (HDR) imaging method that utilizes PMT technology would provide broad access to microscopists. The PMTs used in LSM have a limited detection dynamic range, typically three orders of magnitude, which determines the range of variance in the detectable fluorescence signal and thus the maximum and minimum intensities that can be simultaneously detected within a field of view[5]. For biological samples, the intra-scene dynamic range (IDR) is determined by the underlying biology and is thus dependent on the distribution and concentration of protein expression or target molecules to be imaged. Because the IDR is typically large compared with the detectable dynamic range of PMTs, images will inevitably have regions with intensities that are either saturated or below the background, leading to information loss and compromised image quality. Moreover, despite the fact that typical microscopy imaging systems provide images with 8 or 12 bits depth, the available IDR acquired from the sensor can be largely reduced by the amount of noise and background resulting in an effective dynamic range with reduced bit depth.

Avalanche photodiode detectors (APD) constitute an alternative option to PMTs, especially when operating at low photon fluxes, where PMTs suffer from a significant amount of dark noise above the shot noise limit[6]. In this regime, pulse counting detection[7] is usually preferred, offering high SNR at low counting rates[6]. However, the dynamic range of single photon counting measurements is relatively low with a limited counting rate on the order of approximately 10^7 counts per second[8], confining its applications in optical microscopy to highly specialized areas where very low number of photons are present. Commercially available single photon counting instruments offer maximum count rates on the order of 10 to 100 Mega-counts per second (ref. 8) but their linearity is still limited to just 1 to 2 MHz (ref. 9). These values are insufficient to produce high-SNR images for pixel dwell times below 10 microseconds or alternatively for pixel acquisition rates higher than 100 kHz (ref. 8). Thus single photon counting is impractical for high-resolution imaging at high SNR, restricting its use to small fields of view and longer dwell times[10]. Another limiting factor is the readout rate (pixel clock rate), which gives the speed at which data can be retrieved from the detection scheme[10]. Only recently has the use of sophisticated photon counting circuity or the implementation of field programmable-gate arrays in combination with statistical

processing substantially improved their dynamic range, extending photon-counting operation to higher-emission rate regimes[11,12]. But these methods are still early in development, far from being commercially available, and have only been applied in a few specialized studies[6,8,11,13].

So far, several approaches have been developed to extend the dynamic range of optical imaging detectors, both hardware and software based. High dynamic range imaging for digital still cameras[14,15], in particular, has reached the mainstream through the use of smart phones and digital single-lens reflex (DSLR) cameras, and is based on the acquisition of several images with progressively increasing exposure times (exposure bracketing). Although these techniques have found a wide range of applications, they lack the resolution and sensitivities necessary to image at the subcellular level. For fluorescence microscopy, hardware-based approaches have also been developed to extend the dynamic range of optical imaging detectors, including adaptive illumination[16]. The adaptive illumination method uses negative feedback loops in combination with analogue optical modulators to hold the average detected power at a constant level[16,17]. Although adaptive illumination is an elegant approach, it requires additional electronics, realignment of the setup and the presence of electro-optics modulators. Statistical approaches can also be effective at extending the linear range in photon-counting measurements during pulsed excitation[18]. Finally, a new class of recently introduced PMT tubes (H13126, Hamamatsu) appears to offer a wide dynamic range up to eight orders of magnitude.

Here, we present a new technical approach for confocal and two-photon microscopy namely, high dynamic range fluorescence laser scanning microscopy (HDR-LSM). The technique is based on the simultaneous or sequential acquisition of progressively saturated images mathematically fused into a composite HDR image. Moreover, we propose a method for simultaneous or sequential acquisition of HDR data, which requires no additional acquisition time (for the simultaneous acquisition case), and can be easily implemented on any commercially available LSM system both in two-photon and/or confocal mode. We show that HDR-LSM improves image segmentation and quantification by applying the method to neural tracing, and on samples with different sizes, anatomy and complexity. Finally, in vivo real-time imaging is demonstrated, allowing for longitudinal HDR imaging in the presence of physiological motion as well as for quantitative imaging of in vivo fast tracer kinetics.

Results

Imaging setup and acquisition pipeline. The acquisition and processing pipeline for HDR-LSM (Supplementary Fig. 1) consists of acquiring, simultaneously or sequentially, a series of images covering the full dynamic range of the sample (Fig. 1a), reconstructing a composite HDR image (Supplementary Note 1) for quantitative signal data analysis, and then remapping the HDR image (rHDR) for display and image feature enhancement for structural data analysis, using a global nonlinear transformation followed by a histogram equalization if further local contrast is required (Supplementary Note 2).

The imaging setup is based on a custom-modified commercial imaging system (Fig. 1b, Supplementary Figs 2–4, 'Methods' section). Here low dynamic range images (LDR) are acquired simultaneously for the real-time acquisition scheme, or sequentially, under different detection conditions (for example, attenuation of the signal before PMT detection) such that different parts of the images progressively result in saturation (Supplementary Figs 2–4). LDR images are then corrected for the detectors' response (Supplementary Fig. 5), and combined into a composite high dynamic range image (HDR; details available in

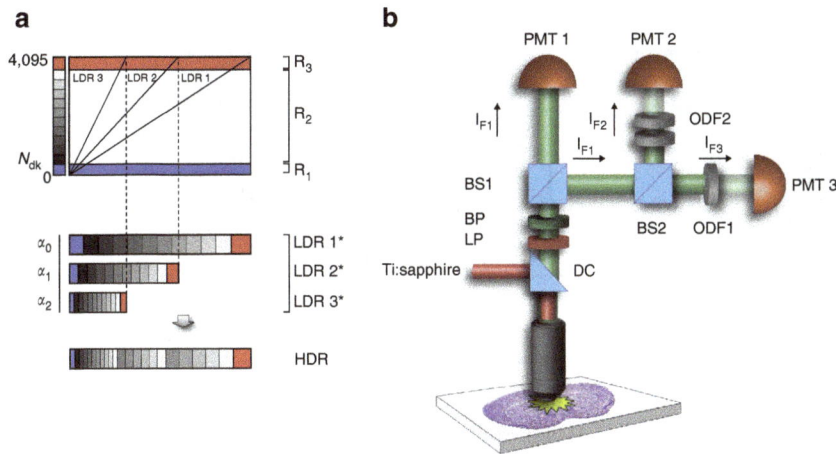

Figure 1 | Imaging setup and image-processing principle for real-time two-photon imaging. (**a**) Principle for HDR imaging. Only a restricted portion of the detector dynamic range can be effectively used for signal quantization (R_2). The dark noise (blue area, R_1) limits the low signal detection, while the high intensity signal near the detector's maximum threshold is saturated (red) and is also disregarded (R_3). By combining multiple images (LDR1, LDR2, LDR3) with different sensitivities (α_0, α_1, α_2) the quantization range can be increased giving rise to a high dynamic range image (HDR). Images are simultaneously acquired from three different detectors (PMT1, PMT2, PMT3). The presence of neutral optical density filters with distinct absorptions gives rise to three images covering different areas of the sample's dynamic range. When no absorption filter is present, the limited quantization range provides high sampling resolution at low signal values while saturated regions values are instead unresolvable. Conversely, the images obtained when increased absorption is present in front of the detector, extend the range of the saturated pixels into the usable quantization interval. Meanwhile, regions within the image with low signal are instead buried into the noise. When the information from all the images is combined together, it provides an extended dynamic range and high SNR image. The dark noise level can be different for other measurements schemes. LDR* represents LDR images weighted by α. (**b**) Schematic representation of the two-photon imaging setup for real-time HDR imaging (Supplementary Fig. 2 and 'Methods' section). BS, beam splitter; BP, bandpass filter; DC, dichroic mirror; LP, longpass filter; ODF, neutral optical density filter; PMT, photomultiplier tubes. Depending on the microscope setup or preferences, BS could be cube beam splitters, plate beam splitters or a combination of both. An arbitrary number of acquisition channels can be used depending on the number of PMTs available and the intra-scene dynamic range. To perform real-time HDR two-photon acquisition, a minimum number of two PMTs is required.

Supplementary Note 1) with a dynamic range greater than the individual LDRs. Other real-time acquisition configurations are possible (Supplementary Note 1), including the use of asymmetric non-polarizing beamsplitters (Supplementary Fig. 4). Here, the whole fluorescence signal contributes to the HDR image reconstruction being the fluorescence distributed by the beam splitters in different ratios to the PMTs.

We first validated the sequential acquisition and reconstruction scheme in a phantom with regions of known fluorescence concentration (Fig. 2a–h, Supplementary Fig. 6, 'Methods' section). In the LDR image (Fig. 2a), the signal from the region with the lowest concentration of fluorophore (γ) was buried below the background noise (β) when the highest concentration of fluorophore (ε) was not saturated. Conversely, when the lowest concentration was above the noise, the highest concentration was saturated (Fig. 2c). However, using HDR-LSM imaging, all three concentrations were within the observation range (Fig. 2i,j). This is accomplished by fusion of the LDR images into a composite HDR image (Supplementary Fig. 7) and then remapping the HDR image (re-mapped HDR, rHDR) for visualization (Fig. 2i)[19–24]. The SNR over the range of the entire image is also greatly improved and imaging is substantially faster compared with the conventional method of averaging (Supplementary Fig. 8).

HDR two-photon imaging. After proof-of-principle validation, we applied our technique for two-photon high-resolution HDR imaging. We utilized a mixture of beads consisting of three different concentrations of fluorophore with fluorescence brightness spanning several orders of magnitude ('Methods' section, Fig. 3a–j). Images (Fig. 3a–c), histograms (Fig. 3d–f, Supplementary Fig. 9) and intensity profiles (Supplementary Fig. 10) show that two-photon HDR-LSM greatly enhances the

dynamic range providing information of the dim beads without losing information from the bright beads (Fig. 3g). We then compared this result to image averaging (Fig. 3h), a common approach to improve SNR. Averaging provides only a modest improvement in SNR, resulting in insufficient SNR and loss of structural information (Fig. 3i,j). HDR, however, significantly improves SNR and maintains structural information for various parameter sets (Supplementary Fig. 11). The technique was also validated on the biological samples using BS-C-1 cells stained for actin (Fig. 4a–f, 'Methods' section). Here a large intra-scene dynamic range is present and both rHDR images (Fig. 4c,f), rHDR and HDR intensity profiles (Fig. 4g,h), reveal structural information over an extended dynamic range with enhancement near saturated pixels. To demonstrate that the information present in the rHDR images does not arise as a result of reconstruction artifacts, a comparison was performed between an HDR reconstruction obtained by acquiring images at a reduced bit depth of the PMT's dynamic range and one acquired utilizing the full dynamic range (Supplementary Fig. 12).

Brain imaging. We then utilized two-photon and confocal HDR-LSM for brain imaging (Fig. 5a–i, 'Methods' section). Remapped HDR images (Fig. 5d,g–i, and Supplementary Figs 13–15) and three-dimensional (3D) rHDR data sets (Supplementary Fig. 16, Supplementary Movie 1–3) were used for visualization and qualitative assessments. Filament Tracer, a module of the commercial software Imaris (Bitplane, St Paul, MN, USA) developed for the detection of neurons, microtubules and filaments in 2D and 3D, was used for segmentation due to its widespread utilization in the scientific literature and the protocols availability (for example high resolution circuit mapping and phenotyping or dendritic spines segmentations)[25,26]. The demonstrated increased accuracy in segmentation

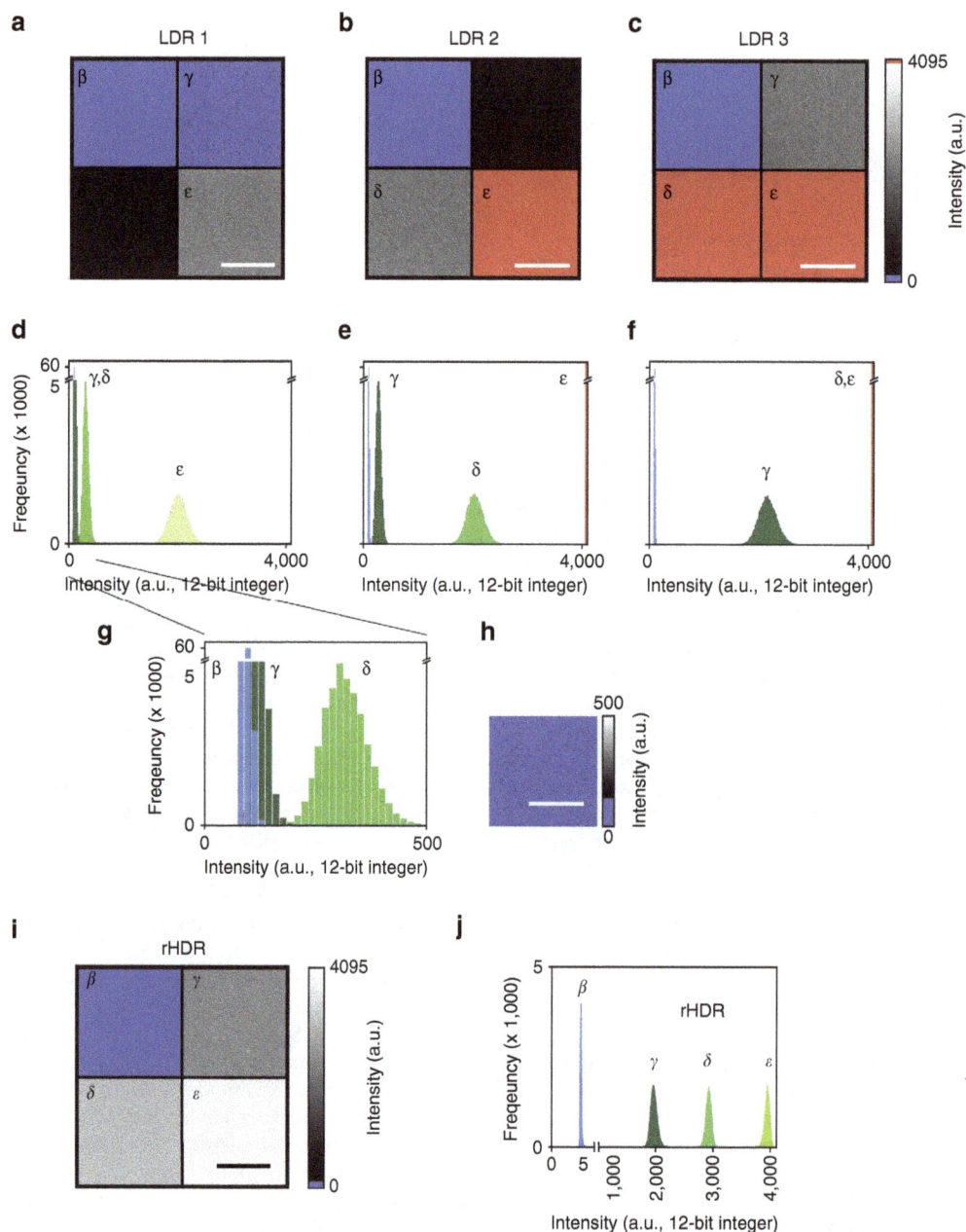

Figure 2 | Imaging phantom. (**a–c**) Three low dynamic range (LDR) fluorescence images with their relative histograms (**d–f**), of an imaging phantom composed of four distinct non-adjacent areas with different fluorophore concentrations (Supplementary Fig. 6 and 'Methods' section), as measured from one detector with its dynamic range accurately centred around the maximum intensity of each specific region (γ, δ, ε). Region γ, dark green. Region δ, green. Region ε, light green. (**g,h**) In the non-saturated image (LDR1), the signal contribution from the region γ is buried within the noise with a low SNR (re-scaled subset image of region γ is shown in **h**. (**i,j**) Remapped HDR image (rHDR, compressed dynamic range for visualization) obtained combining the information from the three different LDR images and displayed with a dynamic range compression mapping algorithm ('Methods' section and Supplementary Note 2), along with its corresponding histogram. The dark noise image (region β) is the same for all phantom's mosaicked images **a–c**. The blue colormap threshold for the dark noise is set at the maximum of the dark noise signal. Image colour bar: blue, dark noise; red, saturation levels. Scale bar, 50 μm.

and quantification (Supplementary Fig. 17) is attributed to the improved SNR and extended dynamic range within the composite and remapped HDR images (Fig. 6). This resulted in a lack of the common artifacts present within LDR images, including noise-induced fragmentation and saturation-induced proximity cell fusion. Moreover, our results show that HDR imaging reveals structures previously unattainable within a single acquisition (Fig. 5e–g, Supplementary Fig. 18). The composite HDR images were used for quantitative measurements of neural

structures (Fig. 7a–e) as they best showcase fluorescence signal over an extended dynamic range.

Improved image segmentation. We also used a trainable Weka (Waikato Environment for Knowledge Analysis) segmentation algorithm (see 'Methods' section), which has been demonstrated in a range of imaging pipelines for many different imaging modalities, including two-photon microscopy. The results of the

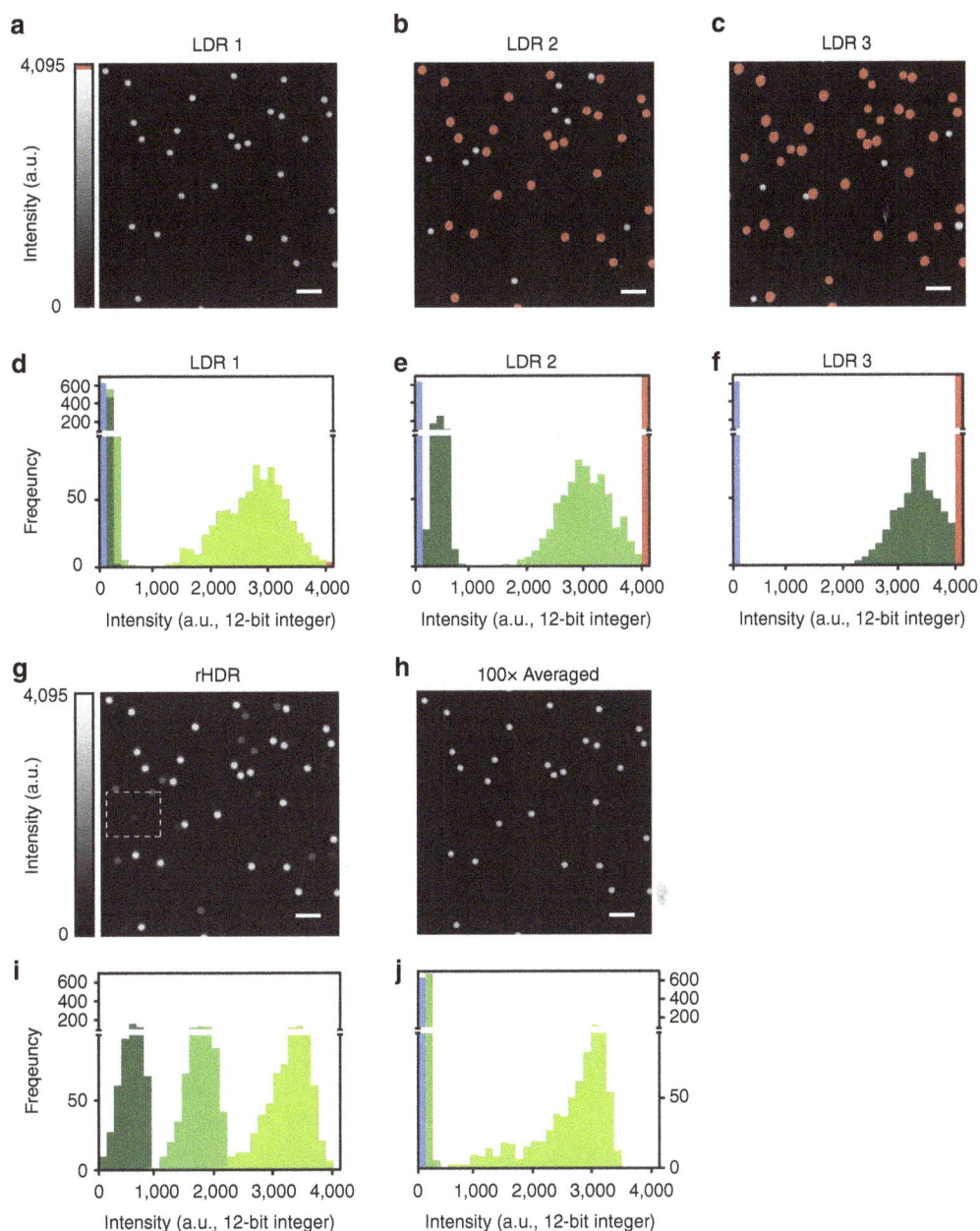

Figure 3 | Microspheres HDR two-photon imaging. (**a–c**) Low dynamic range fluorescence images and (**d–f**) relative histograms of a mixture of three fluorescent microspheres populations with three different discrete values of fluorescence concentrations (see 'Methods' section) centred at different intensity signals within the digitizing range. Bead population 3, light green. Bead population 2, green. Bead population 1, dark green. (**g**) Remapped HDR image (rHDR, compressed dynamic range for visualization) and (**h**) averaged (100 ×) low intensity LDR1 image. (**i,j**) Corresponding histograms. Image colour bar: red, saturation levels. Scale bars, 10 µm.

segmentation approach, including segmentation of cell bodies across different regions of the brain presenting distinct degrees of cell densities, is shown in Fig. 8 and Supplementary Fig. 19. To determine the improvement in performance of the segmentation approach across the different images, a direct comparison was made between automatic and manual (here used as a reference) segmentation approaches applied to both the LDR and HDR images. Higher accuracy was achieved using the automated segmentation algorithm when applied to the HDR images rather than the LDR images (Fig. 8). Specificity, sensitivity and accuracy of cell detection were computed based on the number of false positives (that are incorrectly classified as cell bodies), false negatives (that are undetected cells) and the total number of cell bodies (Supplementary Fig. 19).

Confocal HDR imaging for different sizes and anatomy and complexity. In addition to cellular imaging and segmentation of brain samples, we addressed the imaging and quantification of reconstruction performance of the HDR imaging platform using samples with different sizes, anatomy and complexity. First, we focused on subcellular HDR imaging of mitochondrial structures presenting a high degree of morphological complexity. Recent evidence has illustrated that mitochondria are dynamic networks, which rapidly and continuously remodel themselves[27]. Owing to their morphological complexity, attempts to study mitochondrial networks and their morphology *in vitro* have led to emerging image processing techniques to segment mitochondria labelled with fluorescent dyes or genetic reporters. Unfortunately, highly heterogeneous fluorescent expression found in many

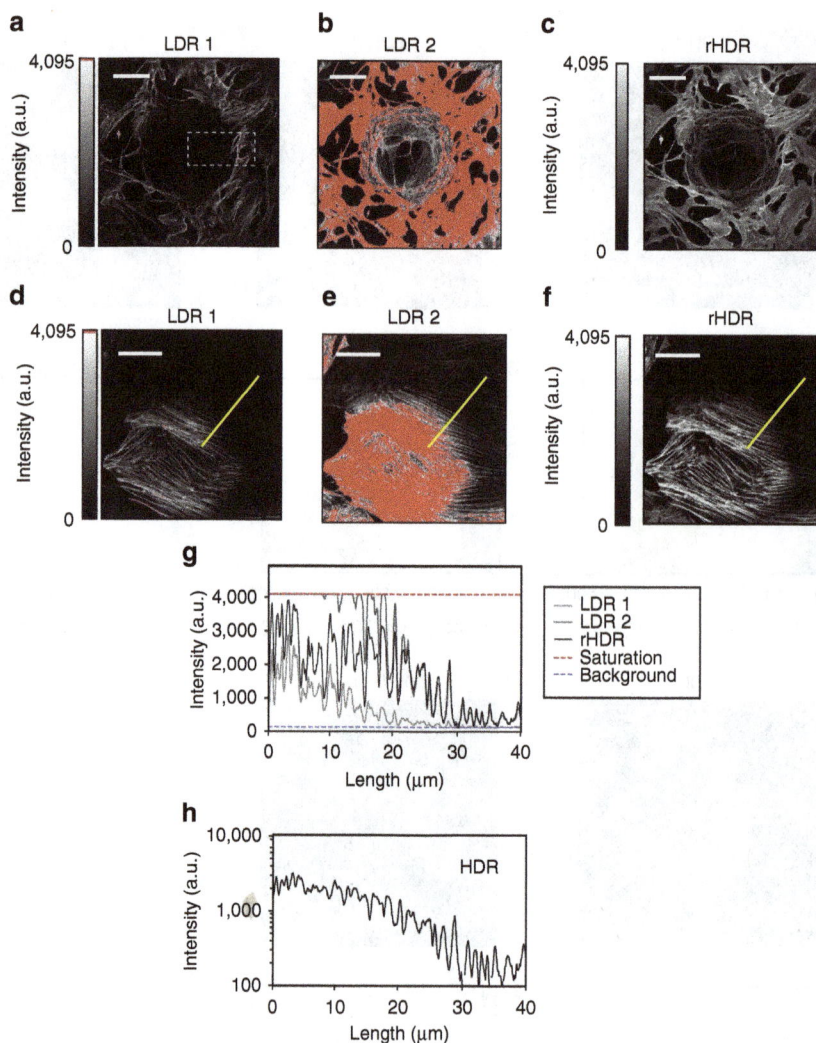

Figure 4 | Cellular HDR two-photon and confocal imaging. (a–f) Images of BS-C-1 cells stained for actin (see 'Methods' section). Low (**a,d**) and high (**b,e**) LDR images and rHDR (**c,f**) images. Emphasis of the extended range at reduced scale is shown within the dashed box in Supplementary Fig. 27. (**g,h**) Actin fluorescence intensity along the profile indicated in **d** for for both LDRs, rHDR (**g**) and HDR (**h**) images. Scale bar (**a**), 100 μm. Scale bar (**d–f**), 20 μm.

reporters affects the overall image quality. Standard LDR images (Supplementary Fig. 20) often contain cells with saturated signal or signal below or near the detection limit (that is, low SNR), making it impossible to accurately segment mitochondrial features. By combining previously validated algorithms[28] (see 'Methods' section), we performed segmentation on both LDR and remapped HDR images and demonstrated the ability to identify and accurately segment a larger percentage of mitochondria in rHDR images compared with LDR images (Supplementary Fig. 20).

After imaging at the subcellular level, we also tested our imaging platform and reconstruction algorithm at the macroscopic level by imaging the vasculature network in cleared organs, including the brain and the heart. The cerebral vascular structure is of fundamental importance in several brain-specific pathologies, such as glioblastoma where vessels are tortuous and disorganized and present large diameters and thicker basement membranes[29]. In the heart, the vascular network also plays a critical role in the delivery of oxygen and nutrients to the cardiomyocytes. A better understanding of the coronary network dysfunctions caused by coronary artery disease, or vascular remodelling of the endocardium following cardiac infarction is required to study disease progression. Therefore, the ability to

perform high fidelity imaging and quantifications of the vascular network in these organs is in great need.

Following DiI staining (see 'Methods' section), we imaged the cleared brain (Supplementary Fig. 21) and heart (Fig. 9) using both LDR and HDR imaging. We then quantified features of the vasculature network, including the number of vascular branches in the heart (Fig. 9, Supplementary Fig. 22). Automated segmentation of rHDR images allowed for the identification of vascular features that agreed with values obtained using ground truth manual segmentation. Conversely, LDR image segmentation resulted in a high degree of vasculature fragmentation (low branch length) due to the low SNR present within the image.

In vivo real-time two-photon HDR imaging. To highlight realtime acquisition capabilities of our HDR imaging platform, we then performed real-time two-photon HDR microscopy for longitudinal imaging in the presence of physiological motion and for quantification of *in vivo* fast tracer kinetics during functional imaging. Imaging was performed in the subcutaneous tissue of mice implanted with a dorsal window chamber (see 'Methods' section). During intravital microscopy imaging, both cardiac and respiratory cycles compromise the ultimate spatial and temporal imaging resolution. If the images are acquired sequentially,

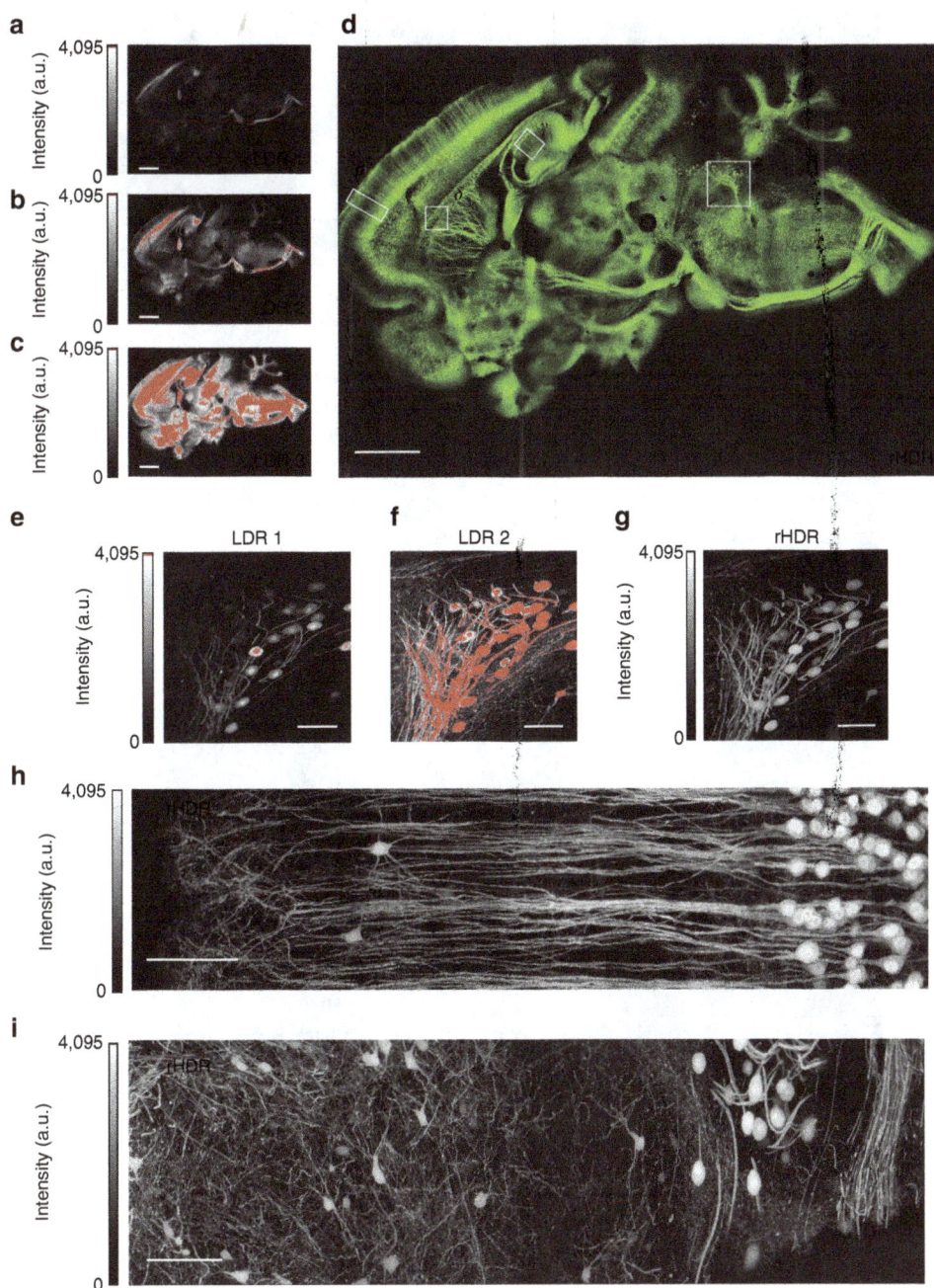

Figure 5 | Brain HDR two-photon and confocal imaging. (**a–c**) Low dynamic range (LDR) images of a whole brain section with (**d**) the corresponding computed remapped HDR image (rHDR, compressed dynamic range for visualization). Red, saturation. Magnified LDRs (**e,f**) and rHDR (**g**) maximum intensity projection (MIP) images of the boxed region α in **d**. (**h,i**) Magnified rHDR MIP images of the boxed region β in **d** and the boxed region of Supplementary Fig. 15a. (**a–d**) Scale bar, 250 μm. (**e–g**) Scale bar, 100 μm. (**h**) Scale bar, 200 μm.(**i**) Scale bar, 150 μm.

physiologically induced motion-artifacts degrade the quality of the HDR reconstructed images. Different areas of sequentially captured images may be misaligned in consecutive frames, giving rise to severe ghosting artifacts in the final reconstructions. In the real-time acquisition modality these artifacts do not occur as the pixels used in the HDR reconstruction are acquired simultaneously via multiple PMTs (Supplementary Fig. 23).

Another important application of real-time HDR microscopy is the possibility to obtain *in vivo* accurate quantitative assessments of the time intensity variations that represent the kinetics of a probe across multiple tissue compartments. This is particularly relevant for studying the intravascular extravasation and extravascular pharmacokinetics of fluorescently labelled drugs.

Single-cell analysis of drug pharmacokinetics requires the ability to quantify drug concentration kinetics in the vascular, interstitial and cellular compartments[30]. However, conventional LDR microscopy imaging does not have sufficient dynamic range to handle the substantial spatio-temporal variations in drug signal intensity, making it challenging to quantify drug pharmacokinetics at the single-cell level[30]. As a proof of concept, we characterized the vascular kinetics following tail vein injection of a bolus of different molecular weight FITC-Dextrans, in a dorsal window chamber mouse model.

A bolus of 2 MDa FITC-Dextran was injected intravenously followed by a bolus of a 4 kDa FITC-dextran (see 'Methods' section). The temporal resolution of the two-photon real-time

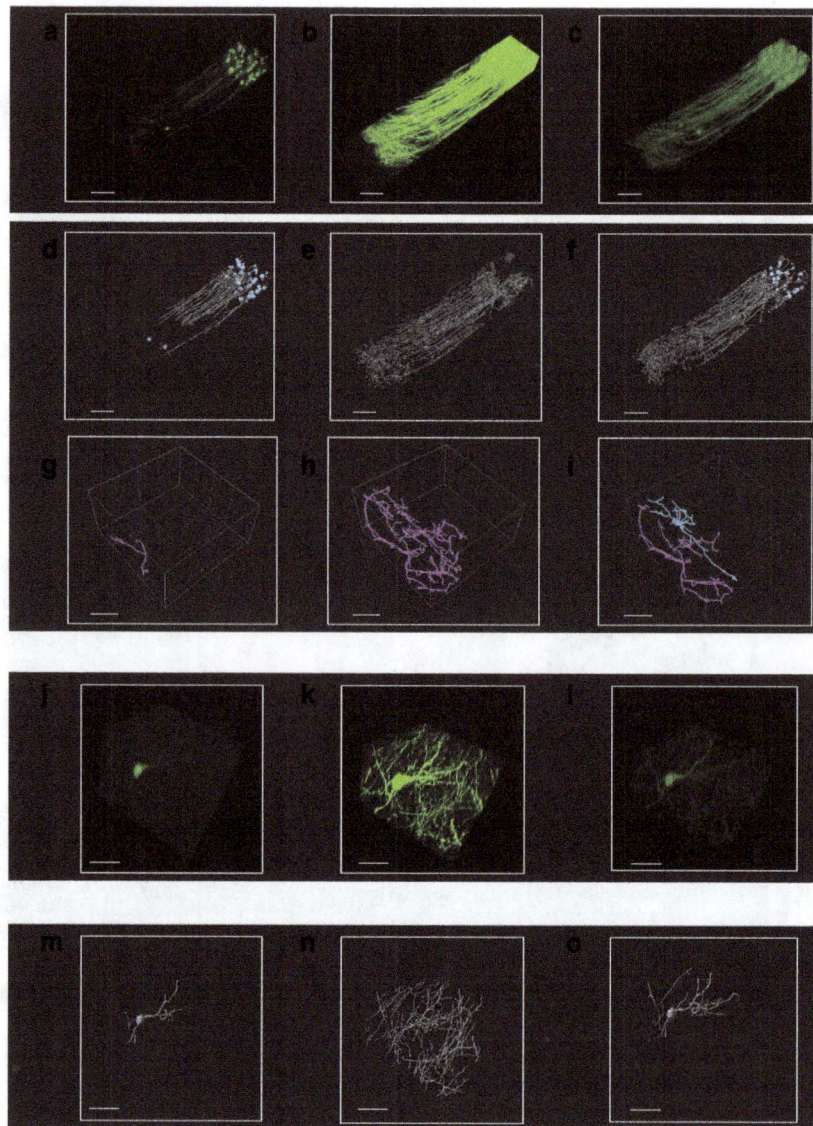

Figure 6 | Brain HDR two-photon imaging facilitates neural segmentation. Volumetric LDRs (**a,b**) and rHDR (**c**) reconstructions of the boxed region β of Fig. 5d. (**d–f**) 3D segmentations of the different cell populations present in **a–c**. LDRs (**g,h**) and HDR (**i**) automatic 3D segmentation of two adjacent cells (blue, purple) within the boxed regions η of Supplementary Fig. 15. Volumetric LDRs (**j,k**) and rHDR (**l**) reconstructions of the boxed region θ of Supplementary Fig. 15. LDRs (**m,n**) and HDR (**o**) automatic 3D segmentation of the cell in **j–l**. White, dendrites and processes. Blue, cell bodies. (**a–f**) Scale bars, 100 μm. (**g–o**) Scale bars, 50 μm.

acquisition was sufficient to capture the vascular kinetics of the 2 MDa probe (Fig. 10) and the extravasation of the 4 kDa probe into the interstitial tissue (Supplementary Fig. 24). Regions of interest were selected in the vascular (Fig. 10) and extravascular (Supplementary Fig. 24) compartments and time–intensity curves were calculated as the mean of the signal within the region of interest as a function of time.

Signal degradation due to photobleaching during HDR imaging, under typical acquisition conditions, was not observed for all the probes used in this study (Supplementary Fig. 25). This was also valid for the cells expressing a green fluorescent protein (GFP) genetic reporter of mitochondria, cells stained with AlexaFluor-488 Phalloidin, DiI stained vasculature in both fixed and cleared tissue, and brain tissue sections stained with AlexaFluor-488 conjugated secondary antibodies. AlexaFluor dyes are frequently used in cleared samples for whole-organ imaging, due to their low photobleaching and for their stability in clearing solution over periods of several months and over multiple imaging sessions[31]. Lipophilic tracers such as DiI also

exhibit low photobleaching and high fluorescence intensity making them suitable for laser scanning microscopy in general as well as for imaging in cleared organs[32].

Discussion

The recent introduction of innovative high-throughput and high-resolution imaging modalities along with the concurrent development of novel clearing techniques[33–37] enables sectioning-free imaging of intact brain tissue[38] facilitating mapping of neural connectivity (connectome) of the whole brain at the microscopic level[33]. However, data analysis constitutes the major bottleneck of the analysis pipeline and requires the use of sophisticated unsupervised image-processing techniques for automatic 3D digital reconstruction and tracing of the individual neuron processes. Unfortunately, the presence of a wide range of signal intensities is a common challenge in neuronal imaging, particularly for large specimens such as the entire brain. Some neural processes are extremely fine and difficult to visualize with fluorescent proteins, requiring scans at

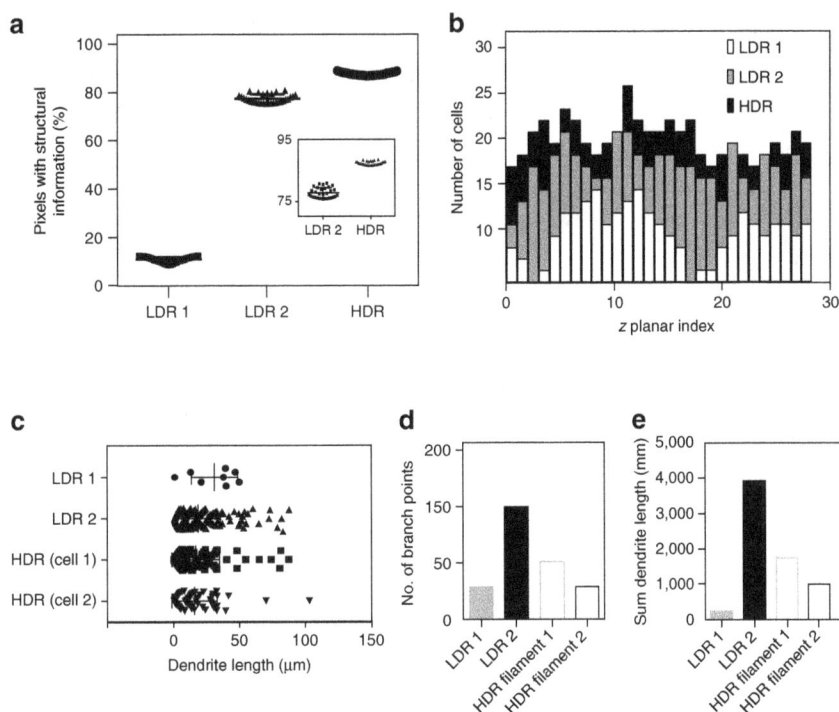

Figure 7 | Structural information in HDR two-photon imaging brain reconstructions. (**a**) Number of pixels containing structural information in the LDRs and HDR images of the boxed regions δ and γ, respectively of Fig. 5d. LDR1, black square; LDR2, black triangle; HDR, black circle. (**b**) Number of counted cells as a function of depth within the LDRs and the respective HDR 3D volumes. Dendrite length (**c**), number of branch points (**d**) and sum of the dendrites length (**e**) calculated for the two cells of Fig. 6g–i, for both LDRs volumes, and the two HDR segmented cells.

high laser power or high gain to resolve their structure. In contrast, larger dendritic varicosities (sites of synaptic contact) often are robustly labelled with fluorescent proteins, requiring lower laser power or gain to preserve structural details (Supplementary Fig. 26). At the microscope level, we are therefore typically forced to strike a balance in the scanning parameters, and it is common to observe loss of data within and among cells (Supplementary Fig. 27), resulting in reduced segmentation accuracy.

Here we have shown that two-photon and confocal HDR-LSM enable visualization and quantification of dim and bright structures within the same field of view, via significantly improved SNR and extended dynamic range. In addition to providing more aesthetically pleasing images through HDR remapping, our HDR fusion algorithm provides more accurate measurements of fluorophore concentration. Moreover, the simultaneous acquisition of multiple LDR images enables real-time HDR-LSM, where non-stationary objects could be imaged in real-time making the technique also suitable for *in vivo* imaging. Specifically, we have illustrated that real-time two-photon HDR imaging provides the ability to remove artifacts caused by physiological motion, to capture data with sufficient temporal resolution to image the tracer kinetics in real-time, and has sufficient dynamic range to capture the substantial signal variations observed between the vascular and extravascular compartments. This demonstrates that real-time two-photon HDR-LSM is particularly useful for *in vivo* systematic analysis of fluorescent drug pharmacokinetics across tissue compartments, and between heterogeneous cell populations in real time[30,39,40].

Moreover, the real-time HDR acquisition allows for accelerated acquisition of large data sets with large dynamic ranges (Supplementary Fig. 8) preventing lengthy imaging sessions. This is particularly relevant for cleared tissue where hours or days are

typically spent for whole-organ imaging with thousands of optical sections collected per single position, in composite stitched images that can cover areas across 1–2 cm in area (approximately one million images per sample).

Compared with other HDR approaches, our technique is simple to implement in any commercially available two-photon imaging system and/or confocal microscope (Supplementary Figs 2–4), at virtually zero cost, and thus can be widely adopted. In addition, our HDR imaging approach may be easily extended to other microscope configurations, including light-sheet, wide-field and spinning-disk microscopy.

We envision a number of other applications where both real-time and sequential two-photon and confocal HDR-LSM would be beneficial such as cell-to-cell communication, detection of fine processes such as filopodia or tunnelling nanotubes[41], imaging of intracellular organelles, network analysis and branching of dendritic and glial cells among others.

Methods

Cell culture and staining. BS-C-1 cells were obtained from ATCC and cultured in Eagles Minimum Essential Media supplemented with 10% FBS and 1% penicillin/streptomycin in a tissue culture incubator. For imaging, cells were seeded on Poly-L-Lysine (Sigma-Aldrich) coated 12-well slides (Ibidi) and cultured overnight.

The cells were fixed in 4% paraformaldehyde (Electron Microscopy Sciences) for 15 min and washed for 3 × 5 min in TBS. Following fixation, the cells were permeabilized using a solution of 0.1% Triton-X in TBS and blocked for 30 min in Odyssey blocking buffer (LI-COR Biosciences). The cells were then incubated with AlexaFluor-488 Phalloidin (Life Sciences), diluted 1:20 in phosphate-buffered saline (PBS) for 15 min and washed for 3 × 5 min in TBS. Finally, the slides were affixed with coverslips before imaging.

Mitochondria expressing cells. The OVCA-429 cells were transduced with a GFP genetic reporter of mitochondria using a CellLight Fluorescent Protein Labeling kit (ThermoFisher). The cells were transduced 2 days before imaging according to the manufacturer's instructions. On the day of imaging, the cells were fixed in a 4% solution of paraformaldehyde in PBS for 10 min, and sealed with a coverslip.

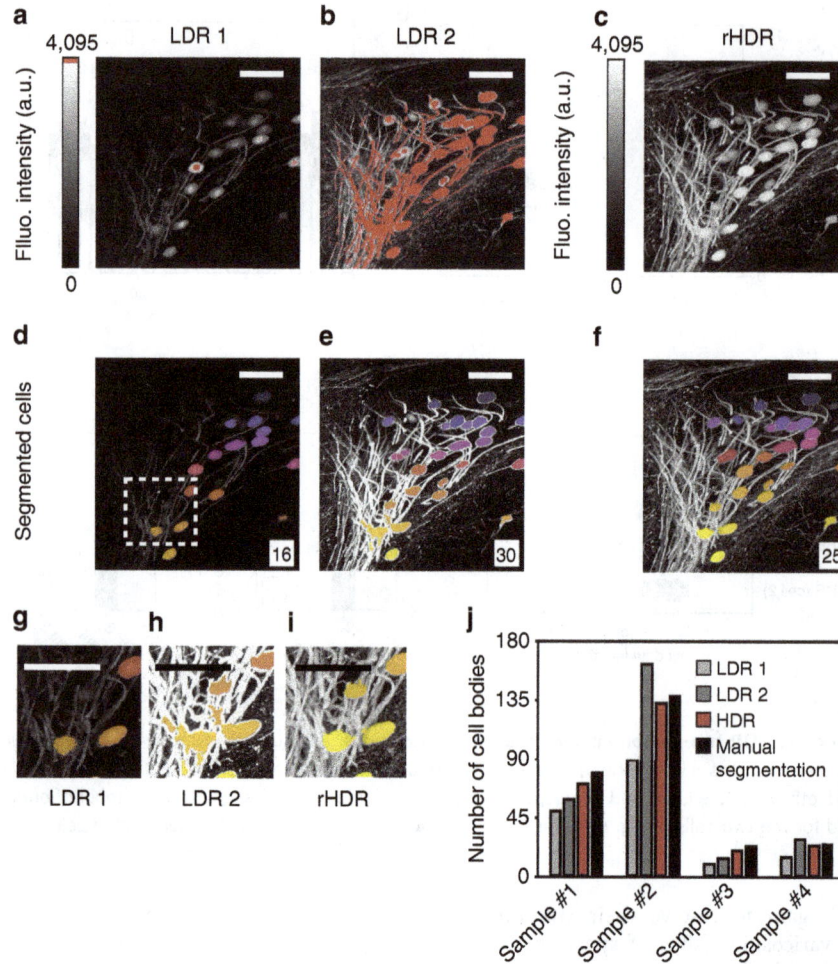

Figure 8 | Cell body segmentation. HDR imaging allows for accurate quantification of cell bodies in sparsely populated fixed tissue specimens. LDRs (**a,b**) and rHDR (**c**) images of the neural cells shown in Fig. 5d (region α). (**d-f**) The cell bodies were segmented for each LDR and HDR image using a trainable Weka algorithm (see 'Methods' section). In each image is indicated the number of body cells identified by the automatic segmentation algorithm. (**g-i**) Magnified image of the box area shown in **d**. Colours are used to help to visualize and distinguish among the different cell bodies present within the field of view. (**j**) Comparison of segmentation performance (defined as total number of cell bodies detected) between LDRs and HDR images over four different images areas. Manual segmentation and counting is used to establish the ground truth. The barplots demonstrate the improved performance of the automatic segmentation algorithm when applied to the HDR images, compared with LDR image, versus manual segmentation and counting. Image colour bar: red saturation levels. Scale bars, 100 μm. Fluo., fluorescence.

Tissue section preparation. Adult Thy1-YFP-H (YFPH)1 mice were anaesthetized with pentobarbital and transcardially perfused with PBS followed by 4% paraformaldehyde (PFA) in PBS. Whole brains were dissected and post-fixed overnight in 4% PFA at 4 °C, washed in PBS, then immersed in 30% sucrose in PBS at 4 °C overnight for sectioning using a cryostat (100–500 μm). Sections were incubated with chicken anti-GFP primary antibodies (Abcam) in blocking buffer (1% horse serum, 0.1% Triton X-100, 0.05% azide in PBS) for 4 days, followed by a wash with PBS and incubation with donkey anti-chicken Alexafluor-488 conjugated secondary antibodies (Jackson Immunoresearch) in blocking buffer. Antibody incubation periods were performed at room temperature[42].

Vessel staining. The brain and heart vasculature were stained using a fluorescent lipophilic dye, DiIC18(3), which accumulates at high concentration in endothelial cell membrane. The staining procedure is very rapid and efficient and the resulting bright fluorescence signal is characterized by very low photobleaching[43,44] making it particularly suitable for laser scanning microscopy. A protocol similar to the one indicated in ref. 32 has been followed. Briefly, after being euthanized the mouse heart was made accessible through thoracotomy and the mouse was perfused by inserting a needle into the left ventricle and with the right atrium cut open. The heart was injected with a solution consisting of 2 ml of PBS, followed by 5 ml of DiI solution and 5 ml of 4% PFA at a rate of 1 ml min^{-1}. The samples were then excised, cut and imaged. To reach higher penetration imaging depth, some specimens were also fixed in 4% PFA overnight and then treated with a clearing solution allowing for whole-organ imaging.

Optical clearing. Tissue clearing and imaging were performed using a slightly modified version of the CUBIC (clear, unobstructed brain imaging cocktails and computational analysis) method[33], which is based on the immersion of fixed tissue in a chemical mixture containing aminoalcohols. CUBIC has been proven to enable rapid whole-brain multicolour imaging of fluorescent proteins or immunostained samples.

One-millimeter fixed brain sections were immersed in a solution obtained by mixing 25 wt% urea (Fisher Scientific, U16–3), 25 wt% N,N,N',N'-tetrakis (2-hydroxypropyl) ethylenediamine (Fisher Scientific 50-014-48142), and 15 wt% Triton X-100 (Life Technologies, 85111). Sectioned slices remained immersed for 2 days at 37 °C, while gently shaken. The cleared slices were then mounted on a custom-made sample holder for microscopy imaging. Alternatively, tissue stained with DiI were cleared using Rapiclear 1.49 (ref. 45), a clearing agent compatible with various endogenous fluorescence proteins and lipophilic tracers such as DiI, following overnight immersion in solution.

Two-photon microscope configuration. The two-photon microscopy setup, illustrated in Fig. 1b and in Supplementary Figs 2 and 4 is based on a custom modified Olympus FV1000-MPE (Olympus, USA) laser scanning microscopy system equipped with an upright BX61-WI microscope (Olympus, USA). Excitation light (red beam) from a Ti:sapphire laser is focused onto the imaged sample with a × 25 1.05 NA water immersion objective (XL Plan N, 2 mm working distance) or a × 25 1.00 NA ScaleView immersion objective (XL Plan N, 4 mm working distance). The emitted fluorescent light (green beam) is epi-collected through the same focusing objective and reflected by a dichroic filter, DC, (690 nm)

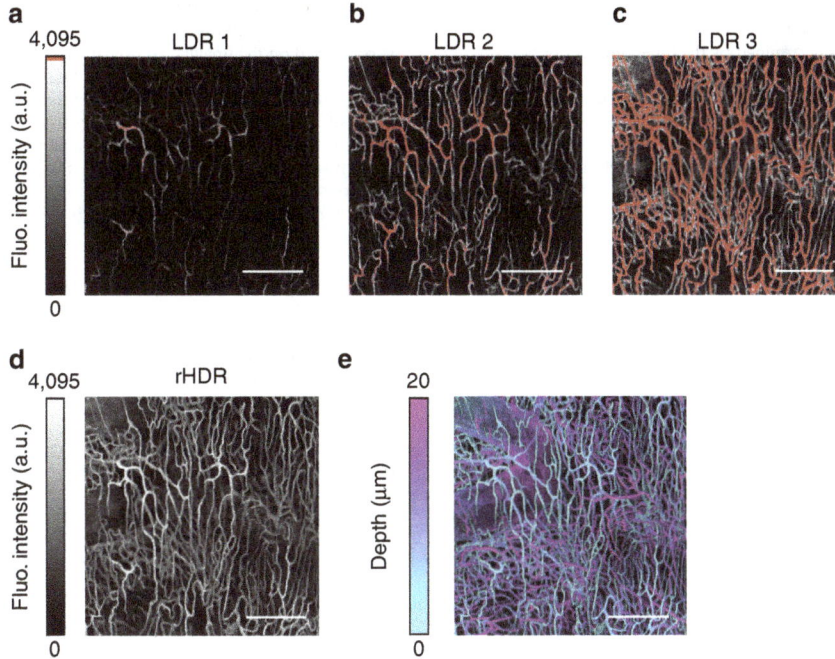

Figure 9 | Volumetric vasculature HDR confocal imaging. LDRs (**a–c**) Images of a cleared DiI stained heart, and (**d**) corresponding rHDR image reconstruction. (**e**) Projection of the three-dimensional rHDR acquisition of the vasculature where colours represent different imaging depths and brightness is related to the fluorescence (Fluo.) signal amplitude. Scale bar, 150 μm.

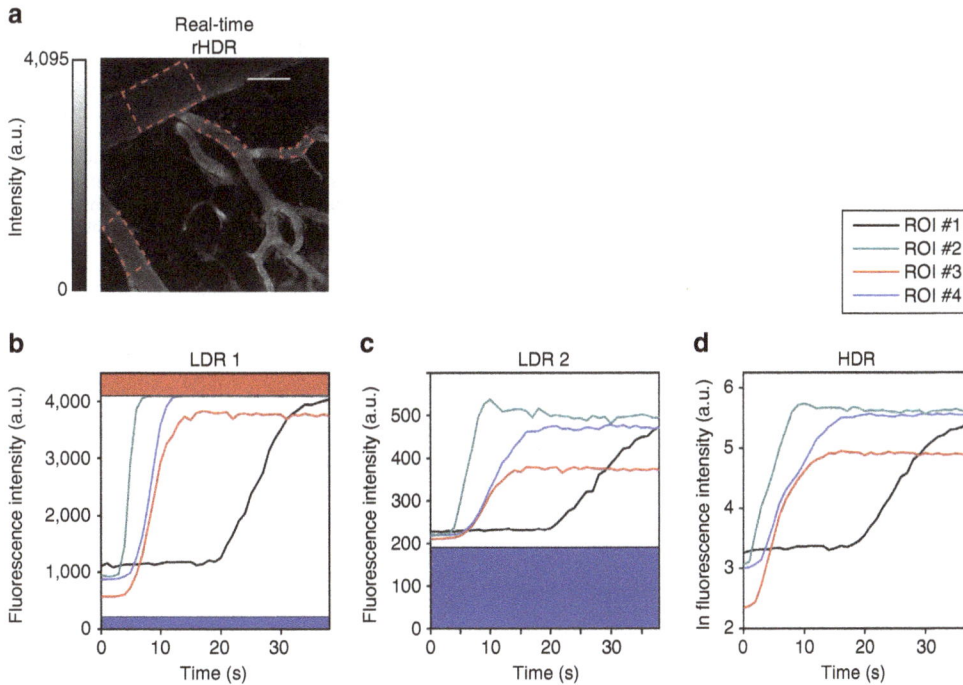

Figure 10 | *In vivo* intravascular dye kinetics. *In vivo* intravascular real-time quantification of the time–intensity variations demonstrating the vascular pharmacokinetics of a fluorescent probe across multiple regions of interests (ROIs). A bolus of 2 MDa FITC-Dextran was injected intravenously through the lateral tail vein (see 'Methods' section) and vascular kinetics were captured by collecting a time sequence of real-time HDR images. (**a**) ROIs were selected in several blood vessels within a dorsal window chamber and time–intensity curves calculated as the mean of the signal within the ROI as a function of time. (**b–d**) Time–intensity curves are plotted for both (**b,c**) LDR and (**d**) HDR time sequences. LDR sequences have a limited ability to capture the full dynamic range of intensities, while HDR sequences present a high SNR and extended dynamic range demonstrating the ability to maintain imaging fidelity for kinetic quantification. Scale bar, 125 μm.

toward a non-descanned detection path. After passing first through a lowpass filter, LP, (685 nm) and a bandpass filter, BP, (490–540 nm), the fluorescent light is split into two beams, IF1, of equal intensity by a 50/50 beam-splitter, BS1. The first component of the beam is directly detected by the first photomultiplier tube, PMT1, with no neutral density filter attenuation (IF1). Meanwhile, the second component of the beam (IF1) is split again by a second 50/50 beam-splitter, BS2, into two new components, IF2 and IF3. Beams IF2 and IF3 are detected by the photomultiplier tubes PMT2 and PMT3, respectively, after passing through two separate neutral density filters ODF1 and ODF2 presenting different optical densities. Three fluorescence signals are then acquired simultaneously and in real time with varying attenuations as determined by the three different optical density filters (typically 0, 0.9 and 1.8 dB). Depending on the range of fluorescence signal present within the samples, two channels can also be acquired instead of three. To avoid bleaching during acquisition, the laser power was always kept well below 10 mW, at typical PMT voltages of 410–650 V. Asymmetric beamsplitters can be used to maximize the number of photons collected at the detectors. A sequential imaging approach can also be implemented for two-photon microscopy, when relying on one PMT only or if accurate dynamic range tuning is desired. Also, two PMTs can be sufficient instead of three, depending on the intra-scene dynamic range, reducing the total number of acquired images.

Confocal microscope configuration. The imaging system used for this work allows for dual confocal and multiphoton microscopy, and can be easily extended for sequential confocal laser scanning microscopy HDR as illustrated in detail in Supplementary Figs 2–4 and 28. Excitation light (blue beam) from a 473-nm diode laser is focused onto the sample using the same lens objectives utilized for two-photon imaging. The emitted fluorescent light (green beam) is epi-collected through the focusing objective and reflected by a dichroic beam splitter DC (SDM560). Light is then bandpass filtered (BA490–540) and directed toward a detection path after being reflected by the galvanometer scanner. Image acquisition at different intensity peak values, necessary for HDR processing, can be obtained by appropriately selecting the excitation laser power or by changing the PMT voltage values, which determine the level of signal amplification. To provide high-quality images, the laser power and PMT voltage should be chosen such that the sample is not bleached and high noise levels are not introduced, which would restrict the dynamic range of the acquired image.

Image acquisition. 3D data sets were all collected in optically cleared tissue sections. Z stacks were collected for both confocal and two-photon microscopy using two motorized stages controlling both planar and axial translations. Typical data set acquisition consisted of two to four z stacks with approximately 100 optical sections (1 μm per section). First, whole-brain HDR images were obtained in confocal mode using a × 10 (UMPLFL, 0.3 NA, 10 mm WD) or alternatively a × 2 air objective (XLFluor, 0.14 NA, 21 mm WD). These preliminary measurements allow for improved identification of interesting structures. For confocal microscopy, the images are collected sequentially by varying the intensity of the excitation light. For real-time two-photon microscopy, optical density filters, or asymmetric beam splitter(s) are added before the detectors (Fig. 1b). The excitation power and voltage were chosen to minimize noise and photobleaching.

Validation phantom experiment. For proof of principle and HDR validation (Fig. 2, Supplementary Fig. 6), fluorescein was prepared with 1:1, 1:10 and 1:100 dilutions in DI water.

Beads sample experiment. For structural validation (Fig. 3), 2.5 μm green-fluorescent microspheres (LinearFlow Flow Cytometry Intensity Calibration Kit, 488 nm excitation/515 nm emission, Life Technologies). The kit's microspheres, originally intended for flow-cytometry calibration, have varying discrete values of fluorescence intensity. Microspheres with 0.3, 3 and 100% relative intensity (bead populations 1, 2 and 3 respectively) were selected and mixed together in equal parts. The beads were then deposited on a glass slide with a cover slip and sealed before imaging. Within each population, the intensity distribution is highly homogenous.

Neural segmentation. High-content analysis in neuroscience makes use of several image-processing tools and resources for digital tracing[46] and all existing software requires a certain amount of user intervention[47]. Within the neuroscience community, Neurolucida, Amira, Eutectic NTS, NeuronJ and Filament Tracer are the most popular software packages and provide similar functionality.

Neural segmentations shown in Fig. 6d–h and Fig. 7c–e were performed with IMARIS filament tracker for automatic detection. Automatic segmentation was chosen over manual segmentation to prevent bias. The parameters were constant when comparing raw LDR images to high dynamic range images. User input parameters were: local contrast threshold = 5; signal threshold = mean noise ± 3σ; starting point diameter = 20 μm, ending point diameter = 2 μm. Disconnected segments were removed. Segmentation for HDR images was done on 32-bit data.

Cellsegm[48] was used for high-throughput 3D cell segmentation of the data presented in Fig. 7b. The Weka automated segmentation algorithm was used to identify regions with cell bodies[49]. The classifier was trained using an equal number of cell bodies from both LDR images. After locating the cell body, the mask was eroded, and a watershed filter was used to separate the neighbouring cell bodies. The cell bodies were counted using a 3D particle counting method previously described by Bolte and Corelieres[50]. The algorithm was used to segment data presented in Fig. 8, and Supplementary Fig. 19.

Mitochondrial segmentation. Mitochondria were segmented by combining previously used algorithms[28]. Briefly, a 'rolling ball' background subtraction was first conducted on the images. To prepare the images for segmentation, a modified version of the previously published algorithm by Giedt et al.[28] was then applied. This consisted of applying a convolution filter with [5 × 5] matrix with a centre element of 24 and the remaining ones equal to − 1, followed by Fast Fourier Transform-based bandpass filtering. A maximum entropy-based thresholding algorithm was applied and size-based filtering was used to eliminate small objects. All analyses were conducted using Matlab. The algorithm was used to segment data presented in Supplementary Fig. 20.

Vasculature segmentation. Vasculature segmentation was performed using a machine learning method based on a trainable Weka segmentation algorithm that has been demonstrated on a wide range of imaging applications and for many different imaging modalities[49], including two-photon microscopy. The algorithm was used to segment data presented in Supplementary Fig. 22.

Animal model. The animal experiments were performed in accordance with the Institutional Animal Care and Use Committee at Massachusetts General Hospital. The surgical procedures were conducted under sterile conditions and facilitated with the use of a zoom stereomicroscope. The mice were anaesthetized by isoflurane vaporization (Harvard Apparatus) at a flow rate of 2 l min^{-1} isoflurane: 2 l min^{-1} oxygen while their body temperature was kept constant at 37 °C during both surgical procedures and imaging experiments. The dorsal skinfold window chambers were implanted 2 days before imaging following a well-established protocol. Because the experiments involved studies of vascular perfusion, it is crucial to insert a metal spacer within the dorsal window chamber to prevent excessive compression of both tissue and vessels.

***In vivo* imaging of vascular perfusion and extravasation.** The mice were anaesthetized as described above with an isoflurane rate of approximately 1 l min^{-1}. A custom stabilizer plate was used to secure the dorsal skinfold window chamber and reducing motion artifacts and axial drifts during imaging. The animals were kept warm with a heating plate to keep their temperature constant at 37 °C. A catheter was inserted into the tail vein and was used for bolus administration of fluorescent dextrans. Two fluorescent probes (fluorescein isothiocyanate-dextran, FITC-Dextran) with different molecular weights (2 MDa and 4 kDa) were used as a fluorescent probe to study vascular perfusion and extravasation. A bolus of 2 MDa FITC-Dextran was injected intravenously (2.5 μM) through the lateral tail vein, followed by a bolus of a 4 kDa FITC-dextran (1.2 mM). Because the perfusion kinetic is very fast, it is important to capture the fluorescence signal at the moment of injection leaving no time to adjust the focal plane. Therefore, a third dye (TRITC-Dextran 4 kDa) at a different emission channel was first used to find a representative imaging area and the correct imaging plane. Two-photon imaging of FITC (excitation 790 nm, emission 490–540 nm) was performed using a × 25 water immersion objective. TRITC was imaged using the HDR confocal microscopy method previously described. A × 2 objective was used to initially identify the best imaging area within the window chamber.

HDR acquisition with real-time reconstructions was performed in free running mode at a 2 Hz frame rate to capture perfusion and extravasation dynamics of the FITC-Dextran. To maximize fluorescence collection efficiency, 90/10 beam splitters were used instead of attenuating filters (Supplementary Fig. 4).

References

1. Pittet, M. J. & Weissleder, R. Intravital imaging. *Cell* **147**, 983–991 (2011).
2. Aguirre, A. D., Vinegoni, C., Sebas, M. M. & Weissleder, R. Intravital imaging of cardiac function at the single-cell level. *Proc. Natl Acad. Sci. USA* **111**, 11257–11262 (2014).
3. Yuste, R. Fluorescence microscopy today. *Nat. Methods* **2**, 902–904 (2005).
4. Svoboda, K. & Yasuda, R. Principles of two-photon excitation microscopy and its applications to neuroscience. *Neuron* **50**, 823–839 (2006).
5. Pawley, J. B. inHandbook of Biological Confocal Microscopy3rd edn (ed Pawley, J. B.Ch. 4Springer-Verlag, 2006).
6. Benninger, R. K. P., Ashby, W. J., Ring, E. A. & Piston, D. W. A single-photon counting detector for increased sensitivity in two-photon laser scanning microscopy. *Opt. Lett.* **33**, 2895–2897 (2008).
7. Art, J. inHandbook of biological confocal microscopy3rd edn (ed Pawley, J. B.Ch. 12Springer-Verlag, 2006).
8. Moon, S. & Kim, D. Y. Analog single-photon counter for high-speed scanning microscopy. *Opt. Express* **16**, 13990–14003 (2008).
9. Benninger, R. K. P. & Piston, D. W. Fluorescence microscopy benefits from advances in single-photon detectors. *Laser Focus World* **45**, 59–63 (2009).
10. Wu, X., Toro, L., Stefani, E. & Wu, Y. Ultrafast photon counting applied to resonant scanning STED microscopy. *J. Microsc.* **257**, 31–38 (2014).
11. Hoover, E. E. & Squier, J. A. Advances in multiphoton microscopy technology. *Nat. Photonics* **7**, 93–101 (2013).

12. Driscoll, J. D. *et al.* Photon counting, censor corrections, and lifetime imaging for improved detection in two-photon microscopy. *J Neurophysiol.* **105**, 3106–3113 (2011).
13. Buehler, C., Kim, K. H., Greuter, U., Schlumpf, N. & So, P. T. C. Single-photon counting multicolor multiphoton fluorescence microscope. *J. Fluoresc.* **15**, 41–51 (2005).
14. Debevec, P. & Malik, J. in *Proceedings of the 24th ACM Annual Conference on Computer Graphics and Interactive Techniques*, 369–378 (Los Angeles, CA, USA, 1997).
15. Robertson, M. A. *et al.* Estimation-theoretical approach to dynamic range enhancement using multiple exposures. *J. Electron. Imaging* **12**, 219–228 (2003).
16. Chu, K. K., Lim, D. & Mertz, J. Enhanced weak-signal sensitivity in two-photon microscopy by adaptive illumination. *Opt. Lett.* **32**, 2846–2848 (2007).
17. Chu, K. K., Lim, D. & Mertz, J. Practical implementation of log-scale active illumination microscopy. *Opt. Express* **1**, 236–245 (2010).
18. Kissick, D. J., Muir, R. D. & Simpson, G. J. Statistical treatment of photon/electron counting: extending the linear dynamic range from the dark count rate to saturation. *Anal. Chem.* **82**, 10129–10134 (2010).
19. Zuiderveld, K. in *Graphics Gems* Vol. 4 (ed Heckbert, P.Ch. 8Morgan Kaufmann, 1994).
20. Min, B. S., Lim, D. K., Kim, S. J. & Lee, J. H. A novel method of determining parameters of CLAHE based on image entropy. *Int. J. Softw. Eng. Appl.* **7**, 113–120 (2013).
21. Shan, Q., Jia, J. Y. & Brown, M. S. Globally optimized linear windowed tone mapping. *IEEE Trans. Vis. Comput. Graph.* **16**, 663–675 (2010).
22. Reinhard, E., Stark, M., Shirley, P. & Ferwerda, J. Photographic tone reproduction for digital images. *ACM Trans. Graph.* **21**, 267–276 (2002).
23. Duan, J., Bressan, M., Dance, C. & Qiu, G. Tone-mapping high dynamic range images by novel histogram adjustment. *Pattern Recogn.* **43**, 1847–1862 (2010).
24. Mertens, T., Kautz, J. & Van Reeth, F. Exposure fusion: a simple and practical alternative to high dynamic range photography. *Comput. Graph. Forum* **28**, 161–171 (2009).
25. Treweek, J. B. *et al.* Whole-body tissue stabilization and selective extractions via tissue-hydrogel hybrids for high-resolution intact circuit mapping and phenotyping. *Nat. Protoc.* **10**, 1860–1896 (2015).
26. Dumitriu, D., Rodriguez, A. & Morrison, J. H. High-throughput, detailed, cell-specific neuroanatomy of dendritic spines using microinjection and confocal microscopy. *Nat. Protoc.* **6**, 1391–1411 (2011).
27. Palmer, C. S., Osellame, L. D., Stojanovski, D. & Ryan, M. T. The regulation of mitochondrial morphology: intricate mechanisms and dynamic machinery. *Cell. Signal.* **23**, 1534–1545 (2011).
28. Giedt, R. J., Pfeiffer, D. R., Matzavinos, A., Kao, C. Y. & Alevriadou, B. R. Mitochondrial dynamics and motility inside living vascular endothelial cells: role of bioenergetics. *Ann. Biomed. Eng.* **40**, 1903–1916 (2012).
29. Jain, R. K. *et al.* Angiogenesis in brain tumours. *Nat. Rev. Neurosci.* **8**, 610–622 (2007).
30. Thurber, G. M. *et al.* Single-cell and subcellular pharmacokinetic imaging allows insight into drug action *in vivo*. *Nat. Commun.* **4**, 1504 (2013).
31. Renier, N. *et al.* iDISCO: a simple, rapid method to immunolabel large tissue samples for volume imaging. *Cell* **159**, 896–910 (2014).
32. Li, Y. W. *et al.* Direct labeling and visualization of blood vessels with lipophilic carbocyanine dye DiI. *Nat. Protoc.* **3**, 1703–1708 (2008).
33. Susaki, E. A. *et al.* Whole-brain imaging with single-cell resolution using chemical cocktails and computational analysis. *Cell* **157**, 726–739 (2014).
34. Chung, K. *et al.* Structural and molecular interrogation of intact biological systems. *Nature* **497**, 332–337 (2013).
35. Hama, H. *et al.* Scale: a chemical approach for fluorescence imaging and reconstruction of transparent mouse brain. *Nat. Neurosci.* **14**, 1481–1488 (2011).
36. Ke, M. T., Fujimoto, S. & Ima, T. SeeDB: a simple and morphology preserving optical clearing agent for neuronal circuit reconstruction. *Nat. Neurosci.* **16**, 1154–1161 (2013).
37. Richardson, D. S. & Lichtman, J. W. Clarifying tissue clearing. *Cell* **162**, 246–257 (2015).
38. Kim, S. Y., Chung, K. & Deisseroth, K. Light microscopy mapping of connections in the intact brain. *Trends Cogn. Sci.* **17**, 596–599 (2013).
39. Laughney, A. M. *et al.* Single-cell pharmacokinetic imaging reveals a therapeutic strategy to overcome drug resistance to the microtubule inhibitor eribulin. *Sci. Transl. Med* **6**, 261ra152 (2014).
40. Dubach, J. M. *et al. In vivo* imaging of specific drug-target binding at subcellular resolution. *Nat. Commun.* **28**, 3946 (2014).
41. Gousset, K. *et al.* Prions hijack tunnelling nanotubes for intercellular spread. *Nat. Cell Biol.* **11**, 328–336 (2009).
42. Feng, G. *et al.* Imaging neuronal subsets in transgenic mice expressing multiple spectral variants of GFP. *Neuron* **28**, 41–51 (2000).
43. Honig, M. G. & Hume, R. I. Fluorescent carbocyanine dyes allow living neurons of identified origin to be studied in long-term cultures. *J. Cell. Biol.* **103**, 171–187 (1986).
44. Honig, M. G. & Hume, R. I. DiI and DiO: versatile fluorescent dyes for neuronal labelling and pathway tracing. *Trends Neurosci.* **12**, 333–335 (1989).
45. Seiradake, E. *et al.* FLRT structure: balancing repulsion and cell adhesion in cortical and vascular development. *Neuron* **84**, 370–385 (2014).
46. Dragunow, M. High-content analysis in neuroscience. *Nat. Rev. Neurosci.* **9**, 779–788 (2008).
47. Parekh, R. & Ascoli, G. A. Neuronal morphology goes digital: a research hub for cellular and system neuroscience. *Neuron* **77**, 1017–1038 (2013).
48. Hodneland, E., Kögel, T., Frei, D. M., Gerdes, H. & Lundervold, A. CellSegm—a MATLAB toolbox for high-throughput 3D cell segmentation. *Source Code Biol. Med.* **8**, 16 (2013).
49. Arganda-Carreras, I., Kaynig, V., Schindelin, J., Cardona, A. & Seung, H. S. in *Advances in Brain-Scale, Automated Anatomical Techniques: Neuronal Reconstruction, Tract Tracing, and Atlasing* (ed. Seung, S.) 73–81 (Society for Neuroscience, 2014).
50. Bolte, S. & Cordelières, F. P. A guided tour into subcellular colocalization analysis in light microscopy. *J. Microsc.* **224**, 213–232 (2006).

Acknowledgements

We thank Professor J.R. Sanes, Drs John Dubach, Kevin King, Nicolas De Silva, Michael Cuccarese, Aaron Aguirre and Jonathan Carlson for helpful discussions on applications of High Dynamic Range to fluorescence microscopy as well as Yoshiko Guiles for help with sample preparation and Dr Hunter Elliott for training with IMARIS. This project was funded in part by Federal funds from the National Heart, Lung, and Blood Institute, National Institutes of Health, Department of Health and Human Services (under Contract No. HHSN26820100004xC), NRSA postdoctoral fellowship F32-CA192531, the Natural Sciences and Engineering Research Council of Canada (NSERC) postdoctoral fellowship, CDMRP DOD BCRP postdoctoral fellowship BC134081 and from the Institute of Biomedical Engineering (under R01EB006432). The research leading to these results has also received funding from the EC Seventh Framework Programme under the Grant Agreement nr. 622182.

Author contributions

C.V. and C.L.S. conceived and designed the study, performed the experiments, acquired and elaborated the data and wrote the manuscript. C.V. built the setup. C.L.S. and P.F.F. wrote reconstruction and dynamic range compression algorithms. P.F.F. contributed to the writing of the manuscript and data elaboration. C.L.S., P.F.F. and R.J.G. wrote and implemented segmentation algorithms D.L.R., P.F.F. and R.J.G prepared the samples and contributed to the writing of the manuscript. S.S. contributed to the *in vivo* imaging and writing of the manuscript. R.W. contributed to the experimental planning, data analysis, funding and writing of the manuscript. All the authors reviewed the manuscript drafts, provided input on the content and approved the final version.

Additional information

23

Mode engineering for realistic quantum-enhanced interferometry

Michał Jachura[1], Radosław Chrapkiewicz[1], Rafał Demkowicz-Dobrzański[1], Wojciech Wasilewski[1] & Konrad Banaszek[1]

Quantum metrology overcomes standard precision limits by exploiting collective quantum superpositions of physical systems used for sensing, with the prominent example of non-classical multiphoton states improving interferometric techniques. Practical quantum-enhanced interferometry is, however, vulnerable to imperfections such as partial distinguishability of interfering photons. Here we introduce a method where appropriate design of the modal structure of input photons can alleviate deleterious effects caused by another, experimentally inaccessible degree of freedom. This result is accompanied by a laboratory demonstration that a suitable choice of spatial modes combined with position-resolved coincidence detection restores entanglement-enhanced precision in the full operating range of a realistic two-photon Mach–Zehnder interferometer, specifically around a point which otherwise does not even attain the shot-noise limit due to the presence of residual distinguishing information in the spectral degree of freedom. Our method highlights the potential of engineering multimode physical systems in metrologic applications.

[1] Faculty of Physics, University of Warsaw, Pasteura 5, 02-093 Warsaw, Poland. Correspondence and requests for materials should be addressed to R.C. (email: radekch@fuw.edu.pl).

Quantum phenomena can facilitate and boost the performance of imaging techniques[1-4], sensitive measurements in delicate materials[5,6], as well as detection schemes probing subtle physical effects such as gravitational waves[7]. These strategies rely on preparing collective superposition states of multiple probes (for example, photons, atoms) to achieve precision enhancement beyond standard limits[8-17]. In the optical domain, a common strategy for collective state preparation is to realize multiphoton interference in linear circuits, for example, free space or integrated interferometers[18-21], fed with non-classical states of light. Attainable precision can be, however, markedly vulnerable to residual distinguishing information between interfering photons. Standard methods to improve indistinguishability based on filtering are often inadequate, in particular, introducing attenuation that may easily diminish the overall benefit of collective state preparation.

The purpose of this paper is to analyse the interplay between degrees of freedom with different experimental accessibility in two-photon interferometry, which is a canonical example of a quantum-enhanced measurement[22]. We demonstrate that detrimental effects caused by distinguishing information present in one degree of freedom that is beyond experimental control or lacks technical means to improve indistinguishability, can be alleviated by mode engineering in another degree of freedom, even though these two remain completely uncorrelated. This feature is investigated in the case of local phase estimation, whose precision becomes strongly dependent on the operating point if the two photons feeding the interferometer exhibit residual distinguishability. It is shown that a carefully designed preparation and detection scheme for a degree of freedom other than the one causing distinguishability allows one to restore quantum-enhanced precision in the entire operating range of the interferometer. We attribute this effect to non-trivial combination of one- and two-photon interference that turns out to augment phase sensitivity beyond the shot-noise limit. We also present an example indicating that similar enhancement occurs also at higher photon numbers. Prospectively, the results reported here may provide another class of strategies to mitigate effects of imperfections and environmental noise in quantum-enhanced metrology[23-29]. The theoretical analysis is complemented with an experiment investigating sensitivity of a balanced Mach–Zehnder interferometer fed with photon pairs. We determine the precision of local phase estimation around the operating point when the photons coalesce pairwise at the interferometer output ports. In this regime, the residual spectral distinguishability within pairs has a markedly deleterious effect on the attainable precision. Building on recent advances in spatially resolved single-photon detection[1,30-39], we demonstrate that by controlling the input spatial structure of interfering photons and extracting complete spatial information at the detection stage it is nevertheless possible to recover the sub-shot-noise precision. This confirms in proof-of-principle settings the feasibility of mode engineering techniques for quantum-enhanced interferometry.

Results

Realistic two-photon interferometer. A generic two-photon Mach–Zehnder interferometer constructed with a pair of balanced 50/50 beam splitters and fed with photon pairs is shown schematically in Fig. 1a. The phase shift θ between the interferometer arms modulates probabilities of detection events at the output ports that can be grouped into two types: either the photons exit through different paths, producing a coincidence event between the detectors monitoring the ports, or both are found in the same output port leading to a double event. If the two photons are indistinguishable at the input, the first beam splitter generates a coherent superposition of both the photons in one or another arm of the interferometer, which is the simplest case of a N00N state providing sensitivity that approaches the Heisenberg limit[10-12].

In the multimode description of the setup one introduces two sets of annihilation operators for upper path modes $\hat{a}_\mu$ and lower path modes $\hat{b}_\mu$ that are matched pairwise. Individual modes in each arm are mutually orthogonal, that is, $[\hat{a}_\mu, \hat{a}_\nu^\dagger] = [\hat{b}_\mu, \hat{b}_\nu^\dagger] = \delta_{\mu\nu}$. The unitary map $\hat{U}(\theta)$ implemented by the interferometer between the input and the output ports transforms pairs of field operators labelled with the same index μ as

$$\hat{U}^\dagger(\theta)\begin{pmatrix} \hat{a}_\mu \\ \hat{b}_\mu \end{pmatrix}\hat{U}(\theta) = \begin{pmatrix} \cos\frac{\theta}{2} & \sin\frac{\theta}{2} \\ \sin\frac{\theta}{2} & -\cos\frac{\theta}{2} \end{pmatrix}\begin{pmatrix} \hat{a}_\mu \\ \hat{b}_\mu \end{pmatrix}. \quad (1)$$

Partial distinguishability of the interfering photons can be modelled by assuming that at the input the photon in the upper path occupies a certain mode $\hat{a}_1$, while the lower path photon is prepared in a combination of a matching mode $\hat{b}_1$ and another orthogonal mode $\hat{b}_2$ with relative weights $\mathcal{V}$ and $1 - \mathcal{V}$, where $\mathcal{V}$ is the visibility parameter specifying the fraction of indistinguishable pairs. To keep the notation concise, we will write the complete two-photon state as pure

$$|\psi\rangle = \hat{a}_1^\dagger\left(\sqrt{\mathcal{V}}\hat{b}_1^\dagger + \sqrt{1-\mathcal{V}}\hat{b}_2^\dagger\right)|\text{vac}\rangle, \quad (2)$$

and trace the final formulas over the index $\mu = 1, 2$ for the initial modes $\hat{b}_\mu$. This is equivalent to taking from the start a reduced density matrix $\mathcal{V}\hat{a}_1^\dagger\hat{b}_1^\dagger|\text{vac}\rangle \langle\text{vac}|\hat{a}_1\hat{b}_1 + (1 - \mathcal{V})\hat{a}_1^\dagger\hat{b}_2^\dagger|\text{vac}\rangle \langle\text{vac}|\hat{a}_1\hat{b}_2$. Here $|\text{vac}\rangle$ is the vacuum state of the entire multimode electromagnetic field satisfying $\hat{a}_\mu|\text{vac}\rangle = \hat{b}_\mu|\text{vac}\rangle = 0$ for any μ. Note that the above model includes the case of partly overlapping wavepackets constructed from a continuum of modes, wherein $\hat{b}_1$ and $\hat{b}_2$ can be identified through the standard algebraic technique of Gram–Schmidt orthogonalization.

The general transformation from equation (1) taken with $\mu = 1, 2$ implies the following expression for the probability of a coincidence event

$$p_c(\theta) = 1 - \frac{1}{2}(1 + \mathcal{V})\sin^2\theta, \quad (3)$$

while the probability of a double event is $p_d(\theta) = 1 - p_c(\theta)$. These formulas combine expressions for fully indistinguishable and completely distinguishable pairs with respective probabilities $\mathcal{V}$ and $1 - \mathcal{V}$. The resulting fringes are depicted in Fig. 1b for $\mathcal{V} = 1$ and 0.93.

As a consequence of the Cramér-Rao bound[40], the minimum uncertainty of any unbiased phase estimate obtained from a measurement using N photon pairs around an operating point θ is given by

$$\Delta^{\text{pair}} = \frac{1}{\sqrt{NF^{\text{pair}}(\theta)}}, \quad (4)$$

where for standard photon counting at output ports of the interferometer the Fisher information $F^{\text{pair}}(\theta)$ is given by a sum of two terms contributed by coincidence and double events[17],

$$F^{\text{pair}}(\theta) = \frac{1}{p_c(\theta)}\left(\frac{dp_c}{d\theta}\right)^2 + \frac{1}{p_d(\theta)}\left(\frac{dp_d}{d\theta}\right)^2. \quad (5)$$

As a reference, we will take the uncertainty of ideal, shot-noise-limited phase measurement $\Delta^{\text{shot}} = 1/\sqrt{2N}$, when $2N$ photons are sent individually to the interferometer. Our figure of merit will be the ratio $\varepsilon = \Delta^{\text{pair}}/\Delta^{\text{shot}}$ of these two uncertainties, with $\varepsilon < 1$ implying that sub-shot-noise precision has been achieved.

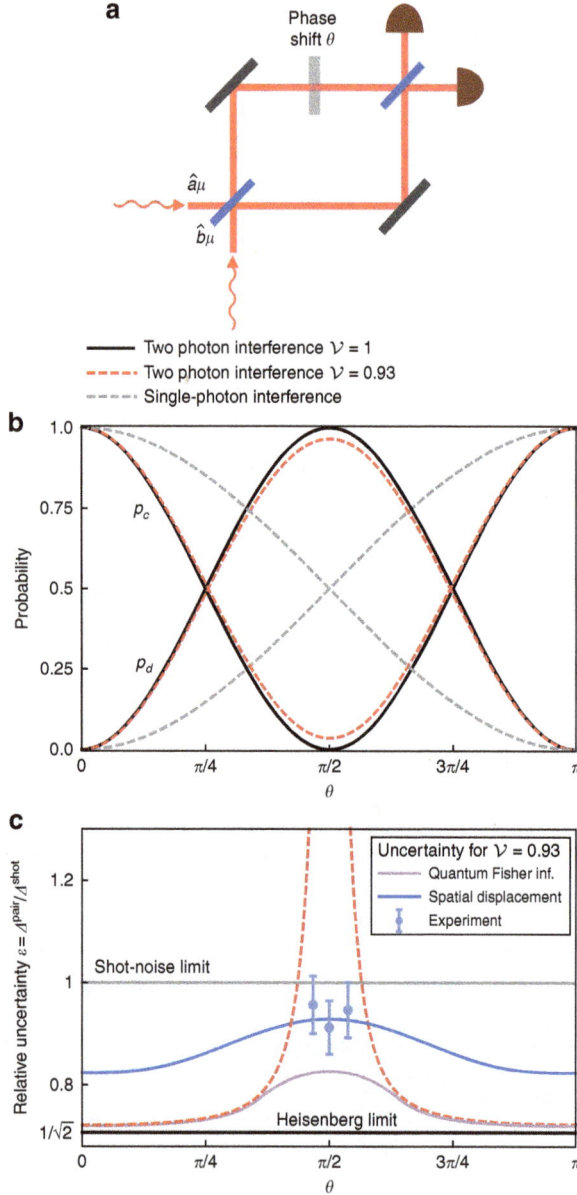

When the interferometer is fed with pairs of perfectly indistinguishable photons characterized by $\mathcal{V}=1$, we have $\varepsilon=1/\sqrt{2}$ independently of the operating point of the interferometer. It can be verified that around $\theta=0$ (equivalent to $\theta=\pi$) and $\theta=\pi/2$ the main contribution to Fisher information defined in equation (5) comes, respectively, from double or coincidence events that occur with vanishing probabilities when approaching these phase values. This is because in the ideal scenario even a small number of rare events provides a sound basis to infer the phase shift. Such a regime corresponds in standard interferometry to dark-fringe operation used, for example, in gravitational wave detectors[7,41].

As seen in Fig. 1c, the precision of phase estimation is affected markedly by the non-ideal indistinguishability of photon pairs. In particular, statistical noise generated by non-vanishing background of coincidence events effectively suppresses information about the phase shift that could be retrieved around $\theta=\pi/2$. We will refer to this operating point as the coincidence dark fringe. An analogous effect would be observed also at $\theta=0$ if any mechanism generating spurious double events was incorporated into calculations.

Restoring quantum enhancement. The analysis presented above assumed that we have no access to the degree of freedom introducing partial distinguishability. For concreteness, we will consider this degree of freedom to be the spectral one, which means that all measurements performed on the photons are integrated in the frequency domain. Suppose now that we can fully control and measure another, uncorrelated degree of freedom of the photon pairs sent to the interferometer. For the clarity of the argument, it will be convenient to use in this role the transverse spatial characteristics of the photons. Let us consider a scenario when in addition to spectral distinguishability characterized by $\mathcal{V}$ we reduce the spatial overlap of the photons by preparing them in nonorthogonal spatial modes. As a result, even in the regime of perfect spectral indistinguishability only a fraction $\mathcal{D}$ of photon pairs would effectively overlap in space. To account for this scenario we will take the mode index to have two components $\mu=i\chi$, where $i=1,2$ refers to the spectral degree of freedom, while $\chi=R,L$ denotes two mutually orthogonal spatial modes. Using this notation, the input state $|\psi_\mathcal{D}\rangle$ is described by an expression analogous to equation (2) with the following substitution of creation operators:

$$
\begin{aligned}
\hat{a}_1^\dagger &\rightarrow \hat{a}_{1R}^\dagger \\
\hat{b}_i^\dagger &\rightarrow \sqrt{\mathcal{D}}\,\hat{b}_{iR}^\dagger + \sqrt{1-\mathcal{D}}\,\hat{b}_{iL}^\dagger, \quad i=1,2
\end{aligned}
\tag{6}
$$

and it explicitly reads

$$
\begin{aligned}
|\psi_\mathcal{D}\rangle = \hat{a}_{1R}^\dagger \Big[&\sqrt{\mathcal{V}}\Big(\sqrt{\mathcal{D}}\,\hat{b}_{1R}^\dagger + \sqrt{1-\mathcal{D}}\,\hat{b}_{1L}^\dagger\Big) \\
&+ \sqrt{1-\mathcal{V}}\Big(\sqrt{\mathcal{D}}\,\hat{b}_{2R}^\dagger + \sqrt{1-\mathcal{D}}\,\hat{b}_{2L}^\dagger\Big)\Big]|\text{vac}\rangle
\end{aligned}
\tag{7}
$$

Note that both the spectral components $\hat{b}_1$ and $\hat{b}_2$ have been subjected to the same spatial transformation. This is in accordance with our assumption that photon manipulations cannot depend on the inaccessible spectral degree of freedom.

Although the spectral and the spatial degrees of freedom are treated on equal footing in equation (7), the crucial difference is the ability to measure the latter at the interferometer output. Therefore, the phase shift θ can be read out from the paths taken by the photons as well as their transverse spatial properties. To find the optimal strategy, we will resort to the concept of quantum Fisher information $F_Q(\theta)$ (refs 42,43), which defines through an expression analogous to equation (4), the minimum uncertainty of a phase estimate inferred from the entire available

Figure 1 | Two-photon interferometer. (a) In the multimode description, two paths are described by families of annihilation operators $\hat{a}_\mu$ and $\hat{b}_\mu$ subjected pairwise to a unitary map dependent on the phase shift θ. **(b)** Probabilities of coincidence p_c and double count events p_d at the interferometer output are noticeably affected around $\theta\approx\pi/2$ by imperfect indistinguishability (red dashed lines for the visibility $\mathcal{V}=0.93$) when compared with the ideal case (black solid lines). Grey dashed lines depict standard single-photon interference fringes, when only one input port is illuminated. **(c)** Residual distinguishability has a dramatic effect on the phase estimation uncertainty, shown with correspondingly coded lines, which diverges at $\theta=\pi/2$ for non-unit visibility $\mathcal{V}$. Engineering overlap in an additional degree of freedom of the interfering photons and implementing optimal measurement allows one to restore the sub-shot-noise precision over the entire operating range, shown with the purple solid line when the fraction of overlapping pairs is optimized individually for each operating point. The solid grey line depicts the shot-noise limit of relative uncertainty. The blue solid line depicts an explicit strategy based on introducing a fixed transverse displacement between interferometer inputs and performing spatially resolved detection. These predictions are confirmed by the experimentally determined estimation precision, depicted as solid circles with two s.d.s. error bars.

characteristics of the physical system used for sensing. When the spatial degree of freedom for coincidence events is taken into account, the explicit expression for quantum Fisher information at the coincidence dark fringe $\theta = \pi/2$, where the effects of spectral distinguishability are most severe, reads

$$F_Q(\pi/2) = 2\frac{1-\mathcal{D}^2}{1-\mathcal{D}\mathcal{V}}. \qquad (8)$$

Detailed derivation of this result is presented in the Methods section. For a given $\mathcal{V}$, the maximum value of the above expression is obtained for $\mathcal{D}^{opt} = \left(1 - \sqrt{1-\mathcal{V}^2}\right)/\mathcal{V}$ and reads $F_Q^{opt}(\pi/2) = 4\left(1 - \sqrt{1-\mathcal{V}^2}\right)/\mathcal{V}^2$. Remarkably, this value gives sub-shot-noise precision for any $\mathcal{V} > 0$. In Fig. 2, we compare the precision implied by numerically computed $F_Q(\theta)$ in the range $0 \le \theta \le \pi$ with $\mathcal{D}$ optimized for an individual operating point to a scenario when no mode engineering has been attempted. A cross-section of these plots for $\mathcal{V}=0.93$ has also been shown in

Fig. 1c. It is seen that the singularity in precision around the coincidence dark fringe is removed and that the sub-shot-noise operation is ensured across the entire range of θ. Note that this result is achieved without any post-selection of two-photon detection events and no filtering or any other manipulation in the spectral domain has been applied.

The standard strategy to improve spectral indistinguishability in two-photon experiments is to restrict the bandwidth of detected photons using narrowband interference filters. However, for any finite filtering bandwidth the visibility parameter would remain below one and consequently the singularity in the uncertainty of phase estimation around $\theta = \pi/2$ would not be removed. Furthermore, filtering reduces the number of detection events that can be used for estimation. Starting from a simple model for partial distinguishability based on gaussian spectral profiles, we verified that if the number of input photon pairs is taken as a benchmark to calculate the relative uncertainty, spectral filtering does not improve the measurement precision at all.

Measurement scheme. To elucidate the origin of the above effect, it is instructive to analyse the operation of the interferometer in the R/L basis of transverse spatial modes. Detecting both the photons in RR modes means that they overlapped spatially at the input and therefore underwent imperfect two-photon interference affected by non-unit spectral visibility $\mathcal{V}$ with fringes shown in Fig. 1b. On the other hand, combinations RL and LR at the output imply that the upper path photon was initially in the mode R and the lower path photon in the mode L. Consequently, both the photons propagated through the interferometer as independent particles exhibiting single-photon interference, which for $\theta = \pi/2$ gives the steepest slope of interference fringes as seen in Fig. 1b. If the lower path photon was prepared in a statistical mixture of R and L modes, only single-photon interference would provide information about the phase shift at $\theta = \pi/2$ and the shot-noise limit could not be surpassed. However, because the lower path photon is sent into the interferometer in a superposition of the modes R and L, information from two- and one-photon interference can be combined in a coherent way through a suitable choice of the measurement basis at the interferometer output. Strikingly, although neither one- nor two-photon interference used separately beats the shot-noise limit itself, their coherent combination restores quantum enhancement of the measurement.

As derived in the Methods section, the explicit form of the optimal measurement attaining quantum Fisher information at the coincidence dark fringe requires discrimination between double events and two types of coincidence events corresponding to the following projections in the spatial degree of freedom:

$$|\pm\rangle = \sqrt{\frac{\mathcal{D}}{1+\mathcal{D}}}|RR\rangle + \frac{1}{2}\left(\sqrt{\frac{1-\mathcal{D}}{1+\mathcal{D}}} \pm 1\right)|RL\rangle$$
$$+ \frac{1}{2}\left(\sqrt{\frac{1-\mathcal{D}}{1+\mathcal{D}}} \mp 1\right)|LR\rangle \qquad (9)$$

In Fig. 3, we depict the resulting interference fringes for all three types of events when $\mathcal{D}$ is optimized for $\mathcal{V}=93\%$ at the operating point $\theta = \pi/2$. It is seen that coincidence events resolved in the $\pm$ basis indeed exhibit both one- and two-photon interference providing a rather steep slope at $\theta = \pi/2$, while their overall probability remains relatively low at this operating point. Combination of these features yields sub-shot-noise sensitivity.

In the limit $\mathcal{V} \to 0$ the optimal $\mathcal{D}$ approaches zero, which means that at the input the two photons are nearly fully

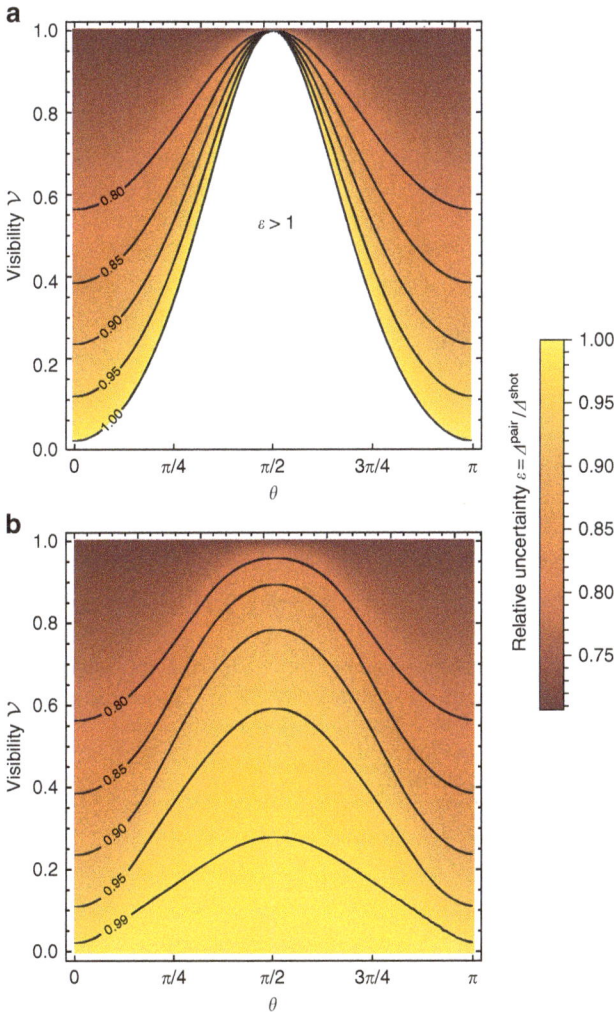

Figure 2 | Enhancement of estimation precision. Relative enhancement $\varepsilon = \Delta^{pair}/\Delta^{shot}$ for (**a**) full spatial overlap $\mathcal{D}=1$ of the two input photons and (**b**) the overlap parameter $\mathcal{D}$ optimized individually for each given spectral visibility $0 \le \mathcal{V} \le 1$ and an operating point $0 \le \theta \le \pi$. The white area in **a** depicts the region $\varepsilon > 1$ where sub-shot-noise sensitivity is lost. It is seen that spatial mode engineering allows one to restore quantum enhancement across the entire parameter range. The uncertainty of the two-photon scheme Δ^{pair} is given by quantum Fisher information derived in equation (26).

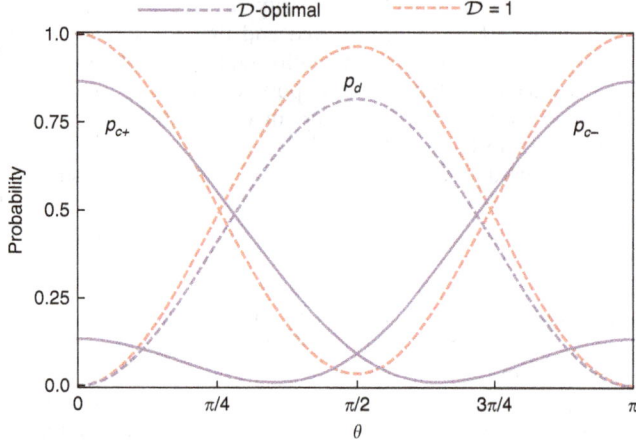

Figure 3 | Optimized interference fringes. The probabilities of double p_d (purple dashed line) and two types of coincidence $p_{c\pm}$ (purple solid lines) events corresponding to projections in the optimal basis $|\pm\rangle$ defined in equation (9) with the overlap parameter $\mathcal{D}$ optimizing quantum Fisher information at $\theta = \pi/2$ for the visibility $\mathcal{V}=0.93$ in the inaccessible degree of freedom. Dashed red lines show standard two-photon interference fringes for the same visibility without engineering an additional degree of freedom.

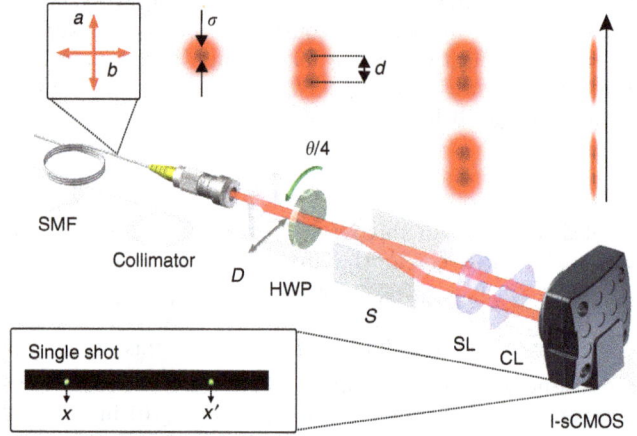

Figure 4 | Experimental setup. Interference between two orthogonally polarized photons a and b delivered by the single-mode fibre (SMF) is realized in the common path configuration with the phase shift θ corresponding to the quadrupled rotation angle of the half-wave plate (HWP). The preceding calcite displacer D introduces transverse displacement between photon modes, which combined with spatially resolved detection restores the sub-shot-noise precision of phase estimation around the coincidence dark fringe. The output ports of the calcite beam separator S are mapped using a spherical lens (SL) and a cylindrical lens (CL) onto the intensified sCMOS camera detecting individual coincidence events with spatial resolution as shown in the inset. The upper right part of the figure depicts the spatial profiles of interferometer modes at the consecutive stages of the setup.

distinguishable in their spatial degree of freedom. In this regime the asymptotic expressions for the optimal measurement basis are $|+\rangle=|RL\rangle$ and $|-\rangle=|LR\rangle$, that is, we need to identify the origin of a photon that has appeared at a given output port. This is equivalent to realizing single-photon interference twice, each time sending a photon into a different input port. The above analysis explains why in the limit of zero visibility $\mathcal{V}=0$ we recover the shot-noise limit as evidenced by Fig. 2b. The described scheme is able to exploit non-classical two-photon interference for any $\mathcal{V}>0$ to achieve sub-shot-noise operation, although unsurprisingly with a diminishing quantum enhancement when $\mathcal{V} \rightarrow 0$.

Spatially resolved detection. The measurement maximizing quantum Fisher information requires implementation of rather exotic projections on two-photon superposition states given in equation (9). This leads to the question about practical realization of the measurement achieving sub-shot-noise precision at a given operating point. Fortunately, as we will now demonstrate, for generic spatial modes sub-shot-noise precision can be restored just by measuring transverse positions of photons emerging from the interferometer. To discuss quantitatively this idea it will be convenient to resort to the paraxial approximation and to introduce two sets of operators $\hat{a}_i(x)$ and $\hat{b}_i(x)$ labelled with a continuous one-dimensional transverse position x. As before, the index $i=1$, 2 labels spectral modes. Suppose now that the photons entering the interferometer along the upper and the lower path are prepared in spatial modes described by respective normalized profiles $u(x)$ and $v(x)$,

$$\hat{a}_{iR} = \int \mathrm{d}x\, u(x)\hat{a}_i(x)$$
$$\sqrt{\mathcal{D}}\hat{b}_{iR}^\dagger + \sqrt{1-\mathcal{D}}\hat{b}_{iL}^\dagger = \int \mathrm{d}x\, v(x)\hat{b}_i(x). \tag{10}$$

The overlap parameter now reads $\mathcal{D} = \left| \int \mathrm{d}x\, u^*(x)v(x) \right|^2$. The probability of detecting photons in two different output ports at positions x and x' is given by

$$p_c(x,x'|\theta) = \sum_{i,j=1,2} \left| \langle \mathrm{vac}|\hat{a}_i(x)\hat{b}_j(x')\hat{U}(\theta)|\psi_\mathcal{D}\rangle \right|^2 \tag{11}$$

where the summation over $i=1$, 2 stems from tracing over the spectral degree of freedom.

As a concrete example we will take two gaussian modes of width σ displaced by d,

$$u(x) = \frac{1}{\sqrt[4]{2\pi\sigma^2}} e^{-(x+d/2)^2/4\sigma^2}, \quad v(x) = u(x-d) \tag{12}$$

that can be readily prepared by simple experimental means. In the phase estimation procedure, we will exploit information contained in the relative position $\xi = x - x'$ of photons detected in coincidence events. The probability distribution for this variable is given explicitly by

$$p_c(\xi|\theta) = \int \mathrm{d}x'\, p_c(x'+\xi, x'|\theta)$$
$$= \frac{1}{2\sigma\sqrt{\pi}} \left(\cos^4 \frac{\theta}{2} e^{-(\xi-d)^2/4\sigma^2} \right.$$
$$\left. + \sin^4 \frac{\theta}{2} e^{-(\xi+d)^2/4\sigma^2} - \frac{1}{2}\mathcal{V}\sin^2\theta e^{-(\xi^2+d^2)/4\sigma^2} \right). \tag{13}$$

When the relative position of the two photons in coincidence events is available, Fisher information defined in equation (5) becomes enhanced by replacing the first term in the sum with the following integral over ξ:

$$\frac{1}{p_c(\theta)} \left(\frac{\mathrm{d}p_c}{\mathrm{d}\theta} \right)^2 \rightarrow F_c(\theta) = \int \mathrm{d}\xi \frac{1}{p_c(\xi|\theta)} \left(\frac{\partial p_c(\xi|\theta)}{\partial \theta} \right)^2. \tag{14}$$

In Fig. 1c, we depict the estimation precision for the ratio $d/\sigma = 1.64$, which was used in the experiment described below. It is seen that the precision is brought to the sub-shot-noise regime over the entire operating range.

Experiment. To verify experimentally sub-shot-noise phase sensitivity of the interferometric scheme described above, we constructed an optical setup shown in Fig. 4. The interferometer is fed with 800 nm photon pairs generated via type-II spontaneous parametric down-conversion process, which are synchronized using a delayed line, spatially filtered through a single-mode fibre and delivered to the setup in two mutually orthogonal linear polarizations corresponding to the two input ports of the Mach–Zehnder interferometer. The photons emerging from the fibre were partly separated in space by inserting a 1.9-mm-long calcite displacer D, which results in the displacement of $d = 200\,\mu m$ between the two ortogonally polarized output paths. The spatial modes can be modelled by gaussian functions defined in equation (12) with $\sigma = 122\,\mu m$. To ensure temporal stability, the interferometer transformation is implemented in the common path configuration as a half-wave plate (HWP), with the rotation angle equal to quadruple the phase shift θ between the interferometer arms, followed by a calcite beam separator S. The equivalence of this setup with the standard Mach–Zehnder interferometer is evidenced by decomposing the interferometer transformation introduced in equation (1) as

$$
\begin{pmatrix} \cos\frac{\theta}{2} & \sin\frac{\theta}{2} \\ \sin\frac{\theta}{2} & -\cos\frac{\theta}{2} \end{pmatrix}
= \begin{pmatrix} \cos\frac{\theta}{4} & -\sin\frac{\theta}{4} \\ \sin\frac{\theta}{4} & \cos\frac{\theta}{4} \end{pmatrix} \begin{pmatrix} 1 & 0 \\ 0 & -1 \end{pmatrix} \begin{pmatrix} \cos\frac{\theta}{4} & \sin\frac{\theta}{4} \\ -\sin\frac{\theta}{4} & \cos\frac{\theta}{4} \end{pmatrix}.
\tag{15}
$$

In the common path configuration the diagonal matrix with entries 1 and -1 describes the HWP in the coordinate system of

its principal axes and the two outer matrices correspond to rotation to the laboratory reference frame by an angle $\theta/4$.

We determined through an independent measurement (see Methods section) the residual spectral distinguishability of the photons to yield $\mathcal{V}=93\%$. Note that this value of visibility has been used in the example presented in Fig. 1b. The transverse distance between the two spatially separated modes corresponding to the output ports of the interferometer is 3.2 mm. The rear surface of the separator S is imaged by means of a spherical lens onto a single-photon-sensitive camera system (see Methods section) capable of spatially resolved detection of coincidence events[38,39]. The profiles of spatial modes propagating in this case through consecutive stages of the setup are shown in the upper right part of Fig. 4.

We recorded spatially resolved two-photon detection events for three values of the phase shift around the dark fringe, $\theta = \pi/2 - 0.1$, $\pi/2$, $\pi/2 + 0.1$, registering $\sim 6 \times 10^3$ events in each case. Experimentally observed spatial distributions $p_c(x, x'|\theta)$ of coincidence events along with their marginals $p_c(\xi|\theta)$ for $\xi = x - x'$ are presented in Fig. 5. It is seen that the joint position distributions are clearly sensitive to the phase shift θ, in particular, the sign of its deviation from $\pi/2$ can be unambiguously inferred from the asymmetry of the distribution with respect to the diagonal.

To quantify information about the phase shift present in spatial distributions, we performed phase estimation from the actual experimental data and determined the estimation precision. In the preliminary step, we verified the applicability of equation (13) as a statistical model for the collected data assuming independently measured mode parameters σ and d. We used the maximum-likelihood method to fit $\mathcal{V}$ and θ, obtaining for the three cases depicted in Fig. 5b respective phase values $\theta = 1.47(2)$, 1.57(2), 1.69(2) and $\mathcal{V} = 0.93$, which is in agreement with the visibility inferred from the independently measured Hong-Ou-Mandel dip.

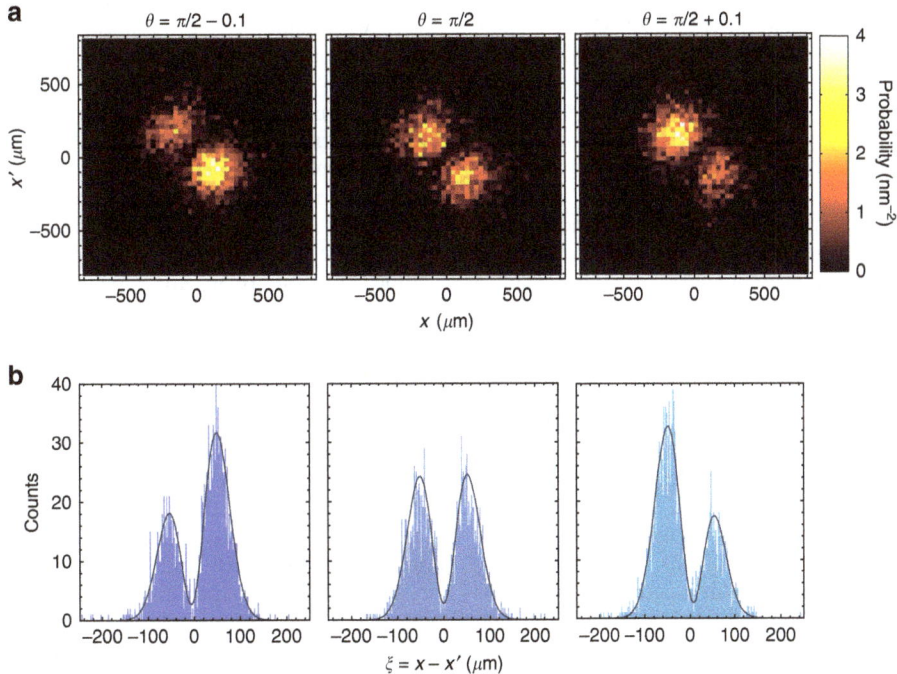

Figure 5 | Spatial distributions of coincidence events. (**a**) Experimentally measured joint distributions $p_c(x, x'|\theta)$ of detecting two photons at positions x and x' in two different output ports of the interferometer for photon pairs characterized by spectral indistinguishability $\mathcal{V}=0.93$ and three phase shifts $\theta = \pi/2 - 0.1$, $\pi/2$, $\pi/2 + 0.1$. (**b**) Useful information about the phase shift can be extracted from the marginal histograms of the relative position between the photons $\xi = x - x'$, shown along with the fitted theoretical model $p_c(\xi|\theta)$ used in the estimation procedure (solid lines).

To determine the actual estimation precision, we divided data obtained for a given HWP setting into ~600 subsets of 10 two-photon detection events and estimated the value of the phase shift separately from each subset. The width of the resulting distribution of individual estimates can be used as a figure for the estimation precision. The choice of the right estimation procedure needs some attention for small sizes of data sets. The maximum-likelihood estimator is known to be asymptotically efficient[40], that is, it saturates the Cramér-Rao bound in the asymptotic limit of infinitely many independent data samples. However, its application to small data sets is not justified owing to potential biasedness. Therefore, we used an estimator that is manifestly unbiased for any data size and yields the precision given by equation (4) at least in the vicinity of a given operation point. Specifically, for an experimentally measured statistical frequency distribution $f(\xi)$ of the relative distance between the two photons, this estimator in the vicinity of $\theta_0 = \pi/2$ is explicitly given by[17]

$$\tilde{\theta}[f] = \theta_0 + \frac{1}{F_c(\theta_0)} \int d\xi \frac{f(\xi)}{p_c(\xi|\theta_0)} \left. \frac{dp_c(\xi|\theta_0)}{d\theta} \right|_{\theta=\theta_0}, \quad (16)$$

where $F_c(\theta)$ has been defined in equation (14). In practice, the integral in the above formula is discretized according to the width of the histogram bins. Applying this estimator to individual data subsets yielded a distribution of phase estimates. The estimation precision was determined as the s.d. of this distribution, shown in Fig. 1c for the three phase shifts along with two-sigma error bars. The obtained values, clearly situated below the shot-noise limit, demonstrate quantum-enhanced operation of the interferometer despite partial spectral distinguishability of the input photons.

The analysis presented above was based on processing only the detected sample of photon pairs. This is a routine approach in proof-of-principle experiments employing 'modest-efficiency detectors'[2,44]. The detected events are generated by a random subset of all photon pairs produced by the source and therefore provide fair statistics to characterize the multiphoton interference underlying the observed quantum enhancement. Current advances in source and detector technology should enable soon unconditional demonstrations of quantum-enhanced metrology also in more complex scenarios such as the one described here.

Discussion

We analysed operation of a two-photon Mach–Zehnder interferometer, which is one of the first[22] and technologically most advanced examples of quantum-enhanced metrology, in a scenario that included two degrees of freedom for interfering photons. It was assumed that one degree of freedom was inaccessible experimentally. Residual distinguishability of interfering photons in this degree of freedom has a markedly deleterious effect on the precision of phase estimation around the operating point where coincidences at the interferometer output are suppressed owing to the Hong-Ou-Mandel effect, producing a coincidence dark fringe. We showed that exploiting another, completely uncorrelated degree of freedom over which one has full experimental control can mitigate this effect, restoring quantum-enhanced precision in the entire operating range. This result is based on a rather subtle interplay between one- and two-photon interference. At the coincidence dark-fringe, imperfect two-photon interference alone does not provide any information about the phase shift, while one-photon interference obviously cannot surpass the shot-noise limit on its own. We demonstrated that combining coherently both types of interference through a suitably designed preparation and measurement scheme in the second degree of freedom yields precision below the shot-noise limit. The feasibility of this approach was confirmed in an experiment using the transverse position of the photons as the controllable degree of freedom that could be measured with high resolution using a single-photon sensitive camera system[38,39]. The presented strategy can be also applied to other scenarios involving two independent degrees of freedom with different experimental accessibility, for example, to mitigate effects of residual spatial distinguishability when only area-integrating detection is available, but photons can be suitably manipulated and measured in the spectral domain.

Beyond two-photon interferometry, one can consider a simple scenario involving a higher photon number when the input ports of a Mach-Zehnder interferometer are fed, respectively, with two photons and a single photon, as shown in Fig. 6a. If all three photons are indistinguishable, counting photons at the interferometer outputs yields fringes depicted in Fig. 6b for which Fisher information $F^{(3)} = 7$ irrespectively of the operating point. This is more than twofold improvement over the shot-noise limit for three photons used independently. When the single photon exhibits residual spectral distinguishability with respect to the two photons (assumed to be identical) feeding the other port, minute changes of the shapes of the interference fringes have a marked effect around points $\theta_0 = 2 \arctan(1/\sqrt{2})$ and $\pi - \theta_0$ on a relative phase estimation uncertainty $\varepsilon = \Delta^{(3)}/\Delta^{shot}$ plotted in Fig. 6c, where $\Delta^{(3)}$ is defined analogously to equation (4). According to Fig. 6b, this can be attributed to a non-vanishing background of events when two photons are detected at one output port and the remaining single photon at the other one. Preparing two photons in the same spatial mode $u(x)$ and the single photon sent individually in a partly overlapping mode $v(x)$ allows one to restore quantum enhancement as seen in Fig. 6c. Interestingly, this effect occurs at different operating points compared with the two-photon case.

A worthwhile candidate to analyse the benefits of mode engineering in a scalable multiphoton scenario may be the celebrated Holland–Burnett scheme[45] employing two Fock states with equal photon numbers. It is easy to verify that sub-shot-noise precision at $\theta = \pi/2$ originates from the suppression of odd photon number events at the interferometer outputs, which again is sensitive to residual distinguishability of input photons. A quantitative analysis of this scenario would require developing an efficient approach to deal with multimode multiphoton states. On the other hand, it is known that in the presence of certain common imperfections, such as photon loss, the ultimate precision follows asymptotically the shot-noise type scaling and the quantum enhancement has the form of a multiplicative factor that can be attained via a repetitive use of finite-size multiparticle superposition states[46]. Therefore, results obtained for a fixed number of photons may also prove useful in the asymptotic limit for realistic scenarios.

The presented results are based on the multimode description of a two-photon interferometer, which goes beyond the simplest models typically used to conceive quantum-enhanced measurement schemes. In practice, the applicability of quantum-enhanced techniques depends crucially on the ability to reduce decoherence effects caused by noise and experimental imperfections. Results presented in this paper suggest that in addition to obvious attempts to suppress decoherence effects in interferometry by improving transmission of optical elements, stabilizing phase reference and so on, exploiting the multimode structure of quantum fields can help to achieve the non-classical regime of operation. If the modal structure of the probes is carefully engineered at the input and suitably detected, such a strategy can offer a noticeable improvement in precision even though

a

Phase
shift θ

b

$\mathcal{V}=1$ $\mathcal{V}=0.93$

p_{30} p_{21} p_{30} p_{21}
p_{12} p_{03} p_{12} p_{03}

c

Spatial displacement
$\mathcal{V}=1$
$\mathcal{V}=0.93$

Shot-noise limit

Heisenberg limit

Methods

Output two-photon state. For two photons prepared initially in the state $|\psi_{\mathcal{D}}\rangle$ defined in equation (7) it will be convenient to write the output state as a sum of two components,

$$\hat{U}(\theta)|\psi_{\mathcal{D}}\rangle = |\psi_c(\theta)\rangle + |\psi_d(\theta)\rangle \tag{17}$$

Here $|\psi_c(\theta)\rangle$ is the conditional state describing coincidence events, when the two photons leave the interferometer through different ports

$$
\begin{aligned}
|\psi_c(\theta)\rangle = &-\Big[\sqrt{\mathcal{D}\mathcal{V}}\hat{a}_{1R}^\dagger\hat{b}_{1R}^\dagger\cos\theta \\
&+ \sqrt{\mathcal{D}(1-\mathcal{V})}\Big(\hat{a}_{1R}^\dagger\hat{b}_{2R}^\dagger\cos^2\frac{\theta}{2} - \hat{a}_{2R}^\dagger\hat{b}_{1R}^\dagger\sin^2\frac{\theta}{2}\Big) \\
&+ \sqrt{(1-\mathcal{D})\mathcal{V}}\Big(\hat{a}_{1R}^\dagger\hat{b}_{1L}^\dagger\cos^2\frac{\theta}{2} - \hat{a}_{1L}^\dagger\hat{b}_{1R}^\dagger\sin^2\frac{\theta}{2}\Big) \\
&+ \sqrt{(1-\mathcal{D})(1-\mathcal{V})}\Big(\hat{a}_{1R}^\dagger\hat{b}_{2L}^\dagger\cos^2\frac{\theta}{2} - \hat{a}_{2L}^\dagger\hat{b}_{1R}^\dagger\sin^2\frac{\theta}{2}\Big)\Big]|vac\rangle,
\end{aligned}\tag{18}
$$

while $|\psi_d(\theta)\rangle$ corresponds to double events, when both the photons emerge at the same output port of the interferometer:

$$
\begin{aligned}
|\psi_d(\theta)\rangle = \tfrac{1}{2}\sin\theta\Big(&\sqrt{\mathcal{D}\mathcal{V}}\Big[\big(\hat{a}_{1R}^\dagger\big)^2 - \big(\hat{b}_{1R}^\dagger\big)^2\Big] \\
&+ \sqrt{\mathcal{D}(1-\mathcal{V})}\big(\hat{a}_{1R}^\dagger\hat{a}_{2R}^\dagger - \hat{b}_{1R}^\dagger\hat{b}_{2R}^\dagger\big) \\
&+ \sqrt{(1-\mathcal{D})\mathcal{V}}\big(\hat{a}_{1R}^\dagger\hat{a}_{1L}^\dagger - \hat{b}_{1R}^\dagger\hat{b}_{1L}^\dagger\big) \\
&+ \sqrt{(1-\mathcal{D})(1-\mathcal{V})}\big(\hat{a}_{1R}^\dagger\hat{a}_{2L}^\dagger - \hat{b}_{1R}^\dagger\hat{b}_{2L}^\dagger\big)\Big)|vac\rangle.
\end{aligned}\tag{19}
$$

The overall probabilities for double and coincidence events are

$$p_d(\theta) = \langle\psi_d(\theta)|\psi_d(\theta)\rangle = \frac{1}{2}(1+\mathcal{D}\mathcal{V})\sin^2\theta \tag{20}$$

and $p_c(\theta) = 1 - p_d(\theta)$. Inserting $\mathcal{D}=1$ gives as a special case the result presented in equation (3).

Let us note that all the terms in $|\psi_d(\theta)\rangle$ exhibit identical dependence on θ. Consequently, resolving double events with respect to the spatial degree of freedom cannot yield more information about the phase shift. Therefore, we will focus our attention on coincidence events described by $|\psi_c(\theta)\rangle$. To analyse information about θ when the spectral degree of freedom cannot be accessed, we will treat it formally as another subsystem Ω, writing $\hat{a}_{i\chi}^\dagger\hat{b}_{i'\chi'}^\dagger|vac\rangle = |\chi\chi'\rangle \otimes |ii'\rangle_\Omega$, where i, $i' = 1, 2$ and χ, $\chi' = R, L$. Tracing the two-photon state over the spectral subsystem yields the reduced density matrix $\hat{\varrho}_c(\theta) = \mathrm{Tr}_\Omega[|\psi_c(\theta)\rangle\langle\psi_c(\theta)|]$, which written in the basis of spatial modes $|RR\rangle, |RL\rangle, |LR\rangle, |LL\rangle$ reads:

$$
\hat{\varrho}_c(\theta) = \begin{pmatrix}
\mathcal{D}\big[1-\frac{1}{2}(1+\mathcal{V})\sin^2\theta\big] & \sqrt{\mathcal{D}(1-\mathcal{D})}(\cos^4\frac{\theta}{2}-\frac{1}{4}\mathcal{V}\sin^2\theta) & \sqrt{\mathcal{D}(1-\mathcal{D})}(\sin^4\frac{\theta}{2}-\frac{1}{4}\mathcal{V}\sin^2\theta) & 0 \\
\sqrt{\mathcal{D}(1-\mathcal{D})}(\cos^4\frac{\theta}{2}-\frac{1}{4}\mathcal{V}\sin^2\theta) & (1-\mathcal{D})\cos^4\frac{\theta}{2} & -\frac{1}{4}(1-\mathcal{D})\mathcal{V}\sin^2\theta & 0 \\
\sqrt{\mathcal{D}(1-\mathcal{D})}(\sin^4\frac{\theta}{2}-\frac{1}{4}\mathcal{V}\sin^2\theta) & -\frac{1}{4}(1-\mathcal{D})\mathcal{V}\sin^2\theta & (1-\mathcal{D})\sin^4\frac{\theta}{2} & 0 \\
0 & 0 & 0 & 0
\end{pmatrix}.
\tag{21}
$$

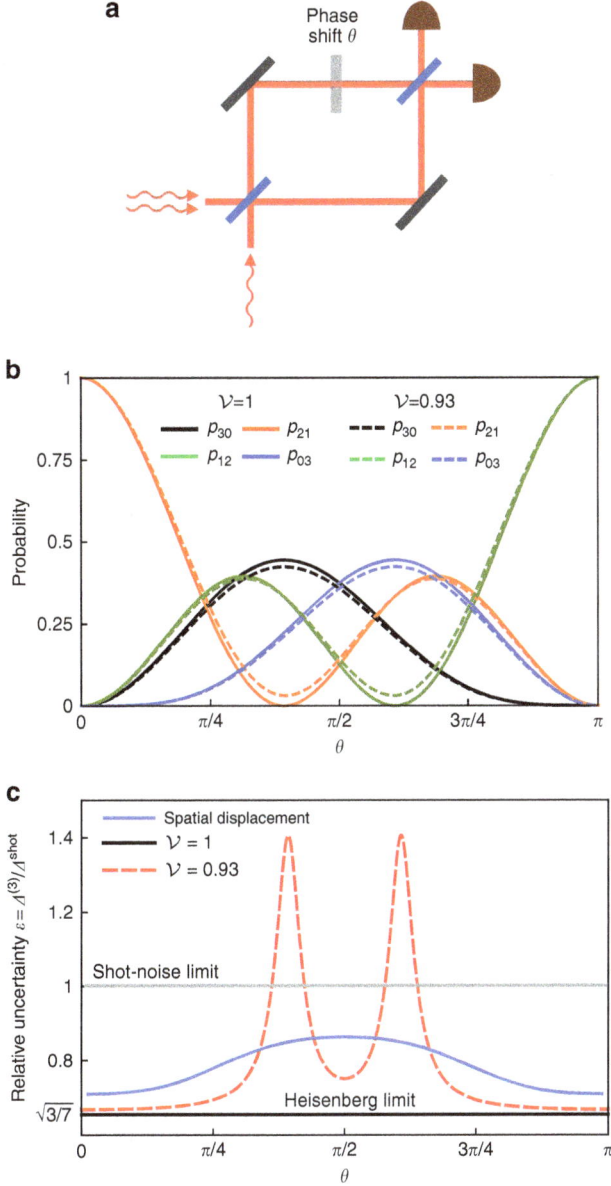

Figure 6 | Three-photon scheme. (**a**) A Mach–Zehnder interferometer fed with a combination of two photons at one input port and a single photon at the other one. (**b**) Probabilities $p_{nn'}(\theta)$ of detecting, respectively, n and n' photons at the interferometer outputs for perfectly indistinguishable photons $\mathcal{V}=1$ and the single photon exhibiting residual distinguishability with respect to the other two mutually identical photons, characterized by $\mathcal{V}=0.93$. (**c**) In the presence of residual distinguishability, the precision of phase estimation, shown with a dashed red line, deteriorates around operating points where dominant contribution to Fisher information comes from the suppression of events $nn' = 21$ or 12. This effect can be mitigated by introducing spatial displacement $d = 1.45\sigma$ between interferometer inputs, resulting in quantum enhancement across the entire operating range (solid blue line). The solid grey line depicts the shot-noise limit of relative uncertainty.

Fisher information. If the spatial degree of freedom of the photons at the interferometer output is projected onto the R/L basis, the complete statistics of measurement results is described by the diagonal elements of the density matrix $\hat{\varrho}_c(\theta)$ for coincidence events and the collective probability $p_d(\theta)$ for all double events. An easy calculation yields the corresponding Fisher information

$$F_{R/L}(\theta) = \frac{2(1+\mathcal{D}\mathcal{V}) - (\mathcal{D}+1)(\mathcal{V}+1)\sin^2\theta}{1-\frac{1}{2}(\mathcal{V}+1)\sin^2\theta}. \tag{22}$$

At the dark fringe we have $F_{R/L}(\pi/2) = 2(1-\mathcal{D})$. This expression, which does not even reach the shot-noise level, can be understood intuitively: information about the phase shift is obtained only from mode-mismatched pairs, when the spatial mode R or L at the output identifies unambiguously the input port of a given photon. The factor $1-\mathcal{D}$ in $F_{R/L}(\pi/2)$ is the overall fraction of these events, while the constant 2 is contributed by single-photon interference exhibited by such pairs.

Clearly, a measurement in the R/L basis neglects information contained in the off-diagonal elements of the density matrix $\hat{\varrho}_c(\theta)$. A measurement strategy that exploits optimally $\hat{\varrho}_c(\theta)$ is described by quantum Fisher information involving the symmetric logarithmic derivative. Its calculation is simplified by switching to a new basis for the spatial degree of freedom,

$$
\begin{aligned}
|\alpha\rangle &= \sqrt{\frac{2\mathcal{D}}{1+\mathcal{D}}}|RR\rangle + \sqrt{\frac{1-\mathcal{D}}{2(1+\mathcal{D})}}(|RL\rangle + |LR\rangle), \\
|\beta\rangle &= \frac{1}{\sqrt{2}}(|RL\rangle - |LR\rangle), \\
|\gamma\rangle &= \sqrt{\frac{1-\mathcal{D}}{1+\mathcal{D}}}|RR\rangle - \sqrt{\frac{\mathcal{D}}{1+\mathcal{D}}}(|RL\rangle + |LR\rangle).
\end{aligned}\tag{23}
$$

decoherence effects due to another degree of freedom that we do not have control over remain at the same level. An interesting question is whether analogous strategies can be generalized to quantum-enhanced interferometry with squeezed states of light[7,41,47], boson sampling with linear multiport devices[18–21] or perhaps benefit other quantum technologies such as optical quantum computing[48,49].

The conditional density matrix $\hat{\varrho}_c(\theta)$ takes the form

$$\langle\alpha|\hat{\varrho}_c(\theta)|\alpha\rangle = \frac{1+\mathcal{D}}{2}\left(1 - \frac{1+\mathcal{V}}{2}\sin^2\theta\right)$$

$$\langle\alpha|\hat{\varrho}_c(\theta)|\beta\rangle = \langle\beta|\hat{\varrho}_c(\theta)|\alpha\rangle = \frac{1}{2}\sqrt{1-\mathcal{D}^2}\cos\theta \qquad (24)$$

$$\langle\beta|\hat{\varrho}_c(\theta)|\beta\rangle = \frac{1-\mathcal{D}}{2}\left(1 - \frac{1-\mathcal{V}}{2}\sin^2\theta\right)$$

while all other elements involving $|\gamma\rangle$ or $\langle\gamma|$ vanish. The symmetric logarithmic derivative $\hat{L}_c(\theta)$, given in general by the implicit formula

$$\frac{d\hat{\varrho}_c}{d\theta} = \frac{1}{2}\left[\hat{L}_c(\theta)\hat{\varrho}_c(\theta) + \hat{\varrho}_c(\theta)\hat{L}_c(\theta)\right] \qquad (25)$$

can now be easily found in the two-dimensional subspace spanned by $|\alpha\rangle$ and $|\beta\rangle$. If double events are not resolved in the spatial degree of freedom, quantum Fisher information can be written as

$$F_Q(\theta) = \text{Tr}\left\{\hat{\varrho}_c(\theta)\left[\hat{L}_c(\theta)\right]^2\right\} + \frac{1}{p_d(\theta)}\left(\frac{dp_d}{d\theta}\right)^2$$
$$= 2\frac{1-\mathcal{D}^2+(1+\mathcal{D}\mathcal{V})^2\cos^2\theta}{1-\mathcal{D}\mathcal{V}+(1+\mathcal{D}\mathcal{V})\cos^2\theta}\sin^2\theta + 2(1+\mathcal{D}\mathcal{V})\cos^2\theta, \qquad (26)$$

where $p_d(\theta)$ is given by equation (20). Specializing the above expression to $\theta = \pi/2$ yields equation (8), whereas its value for $\mathcal{D}$ optimized individually for a given spectral visibility $0 \leq \mathcal{V} \leq 1$ and an operating point $0 \leq \theta \leq \pi$ is presented in Fig. 2b. The symmetric logarithmic derivative has a simple off-diagonal form at $\theta = \pi/2$

$$\hat{L}_c(\pi/2) = -2\frac{\sqrt{1-\mathcal{D}^2}}{1-\mathcal{D}\mathcal{V}}\left(|\alpha\rangle\langle\beta| + |\beta\rangle\langle\alpha|\right). \qquad (27)$$

Quantum Fisher information is saturated by projecting the spatial degree of freedom onto the eigenstates of $\hat{L}_c(\theta)$ (refs 42,43) given explicitly for $\theta = \pi/2$ by $|\pm\rangle = (|\alpha\rangle \pm |\beta\rangle)/\sqrt{2}$. These states are non-trivial superpositions of photon pairs prepared in combinations of R, L spatial modes at two different output ports of the interferometer.

Experimental details. In the experiment we used a photon pair source based on the II-type SPDC process in a 5-mm-long periodically poled KTP crystal (Raicol Crystals) pumped with 8 mW of 400 nm light from a continuous wave diode laser. The produced pairs are transmitted through a 3 nm full-width at half maximum interference filter, carefully synchronized in time using a delay line and spatially filtered using the single-mode fibre. The gaussian-like spatial modes of the photons after the fibre have a flat phase and the half-width $\sigma = 122\,\mu\text{m}$ at $1/e$ height for the intensity distribution measured at the position of the camera system. The residual spectral distinguishability of the photons was determined from the depth of the Hong-Ou-Mandel dip scanned using the delay line and measured with standard avalanche photodiodes for the HWP orientation corresponding to $\theta = \pi/2$.

Our camera system[38,39] begins with an image intensifier (Hamamatsu V7090D) where each detected photon that induces a photoelectron emission produces a macroscopic charge avalanche resulting in a bright flash at the output phosphor screen. The flashes are subsequently imaged with a relay lens onto a fast, low-noise $6.5 \times 6.5\,\mu\text{m}$ pixel size sCMOS sensor (Andor Zyla) and recorded as $\sim$25-px gaussian wide spots, which can be easily discriminated from the low-noise background. The central positions of the spots are retrieved from each captured frame with a subpixel resolution by a real-time software algorithm that provides full information about transverse coordinates of each registered coincidence event as illustrated in the inset of Fig. 4. For the sake of simplicity, we consider only the coordinate x in the horizontal plane of the setup and integrate the signals in the vertical direction. A cylindrical lens with $f = 30\,\text{mm}$ in front of the detector was used to reduce the vertical size of the image, producing effectively a 700×22 px stripe, which significantly decreases frame readout time and allows us to reach 7 kHz collection rate of frames with exposure time 30 ns each. The overall quantum efficiency of the camera system is 23%.

Three-photon scheme. A straightforward but tedious calculation shows that for a three-photon input state of the form $(\hat{a}_{1R}^\dagger)^2\left[\sqrt{\mathcal{V}}\left(\sqrt{\mathcal{D}}\hat{b}_{1R}^\dagger + \sqrt{1-\mathcal{D}}\hat{b}_{1L}^\dagger\right)\right.$ $\left. + \sqrt{1-\mathcal{V}}\left(\sqrt{\mathcal{D}}\hat{b}_{2R}^\dagger + \sqrt{1-\mathcal{D}}\hat{b}_{2L}^\dagger\right)\right]|\text{vac}\rangle$ spatially resolved distributions for events 21 and 12 when the photons are split between the interferometer output

ports into two and one are given by

$$p_{21}(x_1, x_2; x'|\theta) = \mathcal{V}\left|\frac{1}{2}\sin\theta\sin\frac{\theta}{2}(u(x_1)v(x_2)\right.$$
$$+ u(x_2)v(x_1))u(x') - \cos^3\frac{\theta}{2}u(x_1)u(x_2)v(x')\Big|^2$$
$$+ (1-\mathcal{V})\left[\frac{1}{4}\sin^2\theta\sin^2\frac{\theta}{2}(|u(x_1)|^2|v(x_2)|^2 + |v(x_1)|^2|u(x_2)|^2)\right.$$
$$\left.\times|u(x')|^2 + \cos^6\frac{\theta}{2}|u(x_1)|^2|u(x_2)|^2|v(x')|^2\right] \qquad (28)$$

and

$$p_{12}(x; x_1', x_2'|\theta) = \mathcal{V}\left|\sin^3\frac{\theta}{2}v(x)u(x_1')u(x_2')\right.$$
$$- \frac{1}{2}\sin\theta\cos\frac{\theta}{2}u(x)(u(x_1')v(x_2') + u(x_2')v(x_1'))\Big|^2$$
$$+ (1-\mathcal{V})\left[\frac{1}{4}\sin^2\theta\cos^2\frac{\theta}{2}|u(x)|^2(|u(x_1')|^2|v(x_2')|^2\right.$$
$$\left.+ |u(x_2')|^2|v(x_1')|^2) + \sin^6\frac{\theta}{2}|v(x)|^2|u(x_1')|^2|u(x_2')|^2\right] \qquad (29)$$

where we have made use of equation (10). Fisher information taking into account spatially resolved detection of events 21 and 12 is given by

$$F^{(3)}(\theta) = \frac{1}{p_{30}(\theta)}\left(\frac{dp_{30}}{d\theta}\right)^2 + \frac{1}{p_{03}(\theta)}\left(\frac{dp_{03}}{d\theta}\right)^2$$
$$+ \int\frac{dx_1dx_2dx'}{p_{21}(x_1,x_2;x'|\theta)}\left(\frac{\partial}{\partial\theta}p_{21}(x_1,x_2;x'|\theta)\right)^2 \qquad (30)$$
$$+ \int\frac{dxdx_1'dx_2'}{p_{12}(x;x_1',x_2'|\theta)}\left(\frac{\partial}{\partial\theta}p_{12}(x;x_1',x_2'|\theta)\right)^2.$$

Assuming Gaussian spatial modes introduced in equation (12), the above expression was optimized over the displacement d for $\theta_0 = 2\arctan(1/\sqrt{2})$ and $\mathcal{V} = 93\%$. The relative uncertainty of a phase estimate for the obtained value $d = 1.45\sigma$ has been depicted for the entire range $0 \leq \theta \leq \pi$ in Fig. 6c.

References

1. Brida, G., Genovese, M. & Ruo Berchera, I. Experimental realization of sub-shot-noise quantum imaging. *Nat. Photon.* **4**, 227–230 (2010).
2. Ono, T., Okamoto, R. & Takeuchi, S. An entanglement-enhanced microscope. *Nat. Commun.* **4**, 2426 (2013).
3. Schwartz, O. *et al.* Superresolution microscopy with quantum emitters. *Nano Lett.* **13**, 5832–5836 (2013).
4. Tsang, M. Quantum imaging beyond the diffraction limit by optical centroid measurements. *Phys. Rev. Lett.* **102**, 253601 (2009).
5. Wolfgramm, F., Vitelli, C., Beduini, F. A., Godbout, N. & Mitchell, M. W. Entanglement-enhanced probing of a delicate material system. *Nat. Photon.* **7**, 28–32 (2012).
6. Crespi, A. *et al.* Measuring protein concentration with entangled photons. *Appl. Phys. Lett.* **100**, 233704 (2012).
7. LIGO Collaboration. Enhanced sensitivity of the LIGO gravitational wave detector by using squeezed states of light. *Nat. Photon.* **7**, 613–619 (2013).
8. Caves, C. M. Quantum-mechanical noise in an interferometer. *Phys. Rev. D* **23**, 1693–1708 (1981).
9. Yurke, B., McCall, S. L. & Klauder, J. R. SU(2) and SU(1,1) interferometers. *Phys. Rev. A* **33**, 4033–4054 (1986).
10. Bollinger, J. J., Itano, W. M., Wineland, D. J. & Heinzen, D. J. Optimal frequency measurements with maximally correlated states. *Phys. Rev. A* **54**, R4649–R4652 (1996).
11. Boto, A. N. *et al.* Quantum interferometric optical lithography: exploiting entanglement to beat the diffraction limit. *Phys. Rev. Lett.* **85**, 2733–2736 (2000).
12. Giovannetti, V., Lloyd, S. & Maccone, L. Quantum metrology. *Phys. Rev. Lett.* **96**, 10401 (2006).
13. Paris, M. G. A. Quantum estimation for quantum technology. *Int. J. Quantum Inf.* **7**, 125 (2009).
14. Banaszek, K., Demkowicz-Dobrzański, R. & Walmsley, I. A. Quantum states made to measure. *Nat. Photon.* **3**, 673–676 (2009).
15. Maccone, L. & Giovannetti, V. Quantum metrology: Beauty and the noisy beast. *Nat. Phys.* **7**, 376–377 (2011).
16. Tóth, G. & Apellaniz, I. Quantum metrology from a quantum information science perspective. *J. Phys. A* **47**, 424006 (2014).
17. Demkowicz-Dobrzański, R., Jarzyna, M. & Kołodyński, J. Quantum limits in optical interferometry. *Prog. Optics* **60**, 345–435 (2015).
18. Spring, J. B. *et al.* Boson sampling on a photonic chip. *Science* **339**, 798–801 (2013).

19. Spagnolo, N. *et al.* Experimental validation of photonic boson sampling. *Nat. Photon.* **8**, 615–620 (2014).
20. Tichy, M. C., Mayer, K., Buchleitner, A. & Mølmer, K. Stringent and efficient assessment of boson-sampling devices. *Phys. Rev. Lett.* **113**, 020502 (2014).
21. Motes, K. R., Gilchrist, A., Dowling, J. P. & Rohde, P. P. Scalable boson sampling with time-bin encoding using a loop-based architecture. *Phys. Rev. Lett.* **113**, 120501 (2014).
22. Rarity, J. G. *et al.* Two-photon interference in a Mach-Zehnder interferometer. *Phys. Rev. Lett.* **65**, 1348–1351 (1990).
23. Huelga, S. F. *et al.* Improvement of frequency standards with quantum entanglement. *Phys. Rev. Lett.* **79**, 3865–3868 (1997).
24. Dorner, U. *et al.* Optimal quantum phase estimation. *Phys. Rev. Lett.* **102**, 40403 (2009).
25. Knysh, S., Smelyanskiy, V. N. & Durkin, G. A. Scaling laws for precision in quantum interferometry and the bifurcation landscape of the optimal state. *Phys. Rev. A* **83**, 21804 (2011).
26. Jiang, K. *et al.* Strategies for choosing path-entangled number states for optimal robust quantum-optical metrology in the presence of loss. *Phys. Rev. A* **86**, 013826 (2012).
27. Datta, A. *et al.* Quantum metrology with imperfect states and detectors. *Phys. Rev. A* **83**, 63836 (2011).
28. Escher, B. M., de Matos Filho, R. L. & Davidovich, L. General framework for estimating the ultimate precision limit in noisy quantum-enhanced metrology. *Nat. Phys.* **7**, 406–411 (2011).
29. Demkowicz-Dobrzański, R., Guta, M. & Kołodyński, J. The elusive Heisenberg limit in quantum enhanced metrology. *Nat. Commun.* **3**, 1063 (2012).
30. Peeters, W., Renema, J. & van Exter, M. Engineering of two-photon spatial quantum correlations behind a double slit. *Phys. Rev. A* **79**, 043817 (2009).
31. Shin, H., Chan, K. W. C., Chang, H. J. & Boyd, R. W. Quantum spatial superresolution by optical centroid measurements. *Phys. Rev. Lett.* **107**, 083603 (2011).
32. Rozema, L. A. *et al.* Scalable spatial superresolution using entangled photons. *Phys. Rev. Lett.* **112**, 223602 (2014).
33. Abouraddy, A. F., Nasr, M. B., Saleh, B. E. A., Sergienko, A. V. & Teich, M. C. Demonstration of the complementarity of one- and two-photon interference. *Phys. Rev. A* **63**, 63803 (2001).
34. Edgar, M. P. *et al.* Imaging high-dimensional spatial entanglement with a camera. *Nat. Commun.* **3**, 984 (2012).
35. Moreau, P.-A., Devaux, F. & Lantz, E. Einstein-Podolsky-Rosen paradox in twin images. *Phys. Rev. Lett.* **113**, 160401 (2014).
36. Lemos, G. B. *et al.* Quantum imaging with undetected photons. *Nature* **512**, 409–412 (2014).
37. Morris, P. A., Aspden, R. S., Bell, J. E. C., Boyd, R. W. & Padgett, M. J. Imaging with a small number of photons. *Nat. Commun.* **6**, 5913 (2015).
38. Chrapkiewicz, R., Wasilewski, W. & Banaszek, K. High-fidelity spatially resolved multiphoton counting for quantum imaging applications. *Opt. Lett.* **39**, 5090–5093 (2014).
39. Jachura, M. & Chrapkiewicz, R. Shot-by-shot imaging of Hong-Ou-Mandel interference with an intensified sCMOS camera. *Opt. Lett.* **40**, 1540–1543 (2015).
40. Kay, S. M. *Fundamentals of Statistical Signal Processing: Estimation Theory* (Prentice Hall, 1993).
41. Demkowicz-Dobrzański, R., Banaszek, K. & Schnabel, R. Fundamental quantum interferometry bound for the squeezed-light-enhanced gravitational wave detector GEO 600. *Phys. Rev. A* **88**, 41802 (2013).
42. Helstrom, C. W. *Quantum Detection and Estimation Theory* (Academic press, 1976).
43. Braunstein, S. L. & Caves, C. M. Statistical distance and the geometry of quantum states. *Phys. Rev. Lett.* **72**, 3439–3443 (1994).
44. Nagata, T. L., Okamoto, R., O'Brien, J. L., Sasaki, K. & Takeuchi, S. Beating the standard quantum limit with four-entangled photons. *Science* **316**, 726–729 (2007).
45. Holland, M. J. & Burnett, K. Interferometric detection of optical phase shifts at the Heisenberg limit. *Phys. Rev. Lett.* **71**, 1355–1358 (1993).
46. Jarzyna, M. & Demkowicz-Dobrzański, R. True precision limits in quantum metrology. *New J. Phys.* **17**, 013010 (2015).
47. Afek, I., Ambar, O. & Silberberg, Y. High-NOON states by mixing quantum and classical light. *Science* **328**, 879–881 (2010).
48. Knill, E., Laflamme, R. & Milburn, G. J. A scheme for efficient quantum computation with linear optics. *Nature* **409**, 46–52 (2001).
49. Yao, X. C. *et al.* Experimental demonstration of topological error correction. *Nature* **482**, 489–494 (2012).

Acknowledgements

We acknowledge inspiring discussion with M.G. Raymer and I.A. Walmsley. This project was financed by the National Science Centre, No. DEC-2013/09/N/ST2/02229, DEC-2011/03/D/ST2/01941, Polish Ministry of Science and Higher Education Iuventus Plus program for years 2015–2017, No. 0088/IP3/2015/73, European Commission under the FP7 projects SIQS (Grant Agreement No. 600645) and PhoQuS@UW (Grant Agreement No. 316244) co-financed by the Polish Ministry of Science and Higher Education. R.C. was supported by Foundation for Polish Science (FNP).

Author contributions

M.J., R.C. and W.W. designed and performed the experiment. R.D.D. developed theory and analysed the experimental data. K.B. conceived the scheme and interpreted results. All authors contributed to the writing of the manuscript.

Additional information

Competing financial interests: The authors declare no competing financial interests.

24

Plasmonic piezoelectric nanomechanical resonator for spectrally selective infrared sensing

Yu Hui[1], Juan Sebastian Gomez-Diaz[2], Zhenyun Qian[1], Andrea Alù[2] & Matteo Rinaldi[1]

Ultrathin plasmonic metasurfaces have proven their ability to control and manipulate light at unprecedented levels, leading to exciting optical functionalities and applications. Although to date metasurfaces have mainly been investigated from an electromagnetic perspective, their ultrathin nature may also provide novel and useful mechanical properties. Here we propose a thin piezoelectric plasmonic metasurface forming the resonant body of a nanomechanical resonator with simultaneously tailored optical and electromechanical properties. We experimentally demonstrate that it is possible to achieve high thermomechanical coupling between electromagnetic and mechanical resonances in a single ultrathin piezoelectric nanoplate. The combination of nanoplasmonic and piezoelectric resonances allows the proposed device to selectively detect long-wavelength infrared radiation with unprecedented electromechanical performance and thermal capabilities. These attributes lead to the demonstration of a fast, high-resolution, uncooled infrared detector with $\sim 80\%$ absorption for an optimized spectral bandwidth centered around $8.8\,\mu m$.

[1] Department of Electrical & Computer Engineering at Northeastern University, 360 Huntington Avenue, Boston, Massachusetts 02115, USA. [2] Department of Electrical & Computer Engineering at The University of Texas at Austin, 1616 Guadalupe St., UTA 7.215, Austin, Texas 78701, USA. Correspondence and requests for materials should be addressed to M.R. (email: rinaldi@ece.neu.edu) or to A.A. (email: alu@mail.utexas.edu).

Infrared detector technologies were originally developed primarily for military demands, such as night vision, missile tracking, target acquisition and surveillance. In the past few decades, the use of infrared technologies for civilian applications has been steadily growing. Nowadays, infrared detectors can be found in a wide variety of applications, including medical diagnostics, biological and chemical threat detection, electrical power system inspection, infrared spectroscopy, and thermal imaging. Photon detection and thermal sensing are the two main approaches used for the implementation of infrared detectors. Photonic detectors exploit the interaction between photons and electrons in a semiconductor material to produce an electrical output signal upon exposure to infrared radiation[1,2]. They have the advantages of high signal-to-noise ratio (hence high resolution) and fast response time. However, to achieve such a high performance, they typically need cryogenic cooling to prevent thermally generated carriers[3], making them bulky, expensive and power inefficient. On the other hand, thermal detectors rely on the temperature-induced change of material physical properties upon exposure to infrared radiation. They are generally less expensive, more compact and power efficient than semiconductor photon detectors, given their intrinsic capability to operate at room temperature, but they exhibit relatively worse resolution and slower response time. Several uncooled thermal detector technologies have been demonstrated including bolometers[4], thermopiles[5], pyroelectric detectors[6] and more recently, resonant detectors (electromechanical[7] and optical[8]).

Recently, the development of miniaturized, ultra-low-power and low-cost sensor technologies (including uncooled thermal detectors) have attracted great attention for the implementation of highly distributed wireless sensor networks, such as the internet of things, in which physical and virtual object are connected together through the exploitation of sensing and wireless communication functionalities. In this context, micro- and nanoelectromechanical systems (MEMS/NEMS) can have a tremendous impact, since they can simultaneously provide multiple sensing and wireless communication functionalities integrated in a small footprint. The use of MEMS/NEMS has been explored in a large number of applications, spanning from semiconductor-based technology[9] to fundamental science[10]. In particular, MEMS technology has also been employed successfully for miniaturized and ultra-sensitive uncooled infrared detectors such as microbolometers[11] and micromachined thermopile[12]. Among different MEMS/NEMS sensor technologies, the one based on a resonant-sensing mechanism offers significant advantages over other non-resonant approaches. In general, micro–nanoresonant sensors are characterized by a unique combination of high sensitivity to external perturbations, due to the greatly reduced dimensions of sensing element, and ultra-low-noise performance, due to the intrinsically high quality factor (Q) of such resonant systems. Furthermore, resonant sensors use frequency as the output variable, which is one of the physical quantities that can be monitored with the highest accuracy and converted to digital form by simply measuring zero crossings. Among all types of resonant sensors, the one based on MEMS/NEMS resonators can typically deliver the most compact and power efficient sensing solutions thanks to the use of on-chip transduction techniques (such as piezoelectric, electrostatic or thermal) in contrast with the bulky and off-chip optical actuation and readout techniques (requiring the use of power hungry lasers and other bulky optical components and interconnects) typically employed in optical resonators[8,13] or optomechanical resonant systems[14], which are not suitable for many low-power portable applications. The current bottleneck in the development of high-performance MEMS/NEMS resonant infrared detectors is the lack of deeply subwavelength and highly absorbing materials

compatible with standard microfabrication processes and efficient transduction techniques. Indeed, the conventional approach to enable infrared absorptance in MEMS/NEMS resonant structures involves the integration of an infrared absorber (that is, a thin layer of lossy dielectric[8,15,16] or a metal–insulator–metal grating[17]) on top of the vibrating body of the resonant transducer. Although a relatively weak infrared absorptance (<50% and polarization dependent) is typically achieved using this conventional approach, the electromechanical and thermal properties of the resonator (hence detection capability and power efficiency of the infrared sensor) are severely deteriorated due to the electrical and mechanical loading effects of the relatively bulky infrared absorbing material stack attached to the vibrating body of the micro/nanostructure[17].

The fundamental challenge associated with efficient light concentration in planar structures with deeply subwavelength thickness has recently been addressed in the field of nanoplasmonics[18–20]. Metasurfaces with optical properties not found in nature have been synthesized by tailoring plasmonic resonances sustained by arrays of nanostructures with subwavelength dimensions. These effects have been utilized in a wide range of applications, including beam steering[21,22], ultrathin focusing or diverging lenses[23,24], reflectors[25] and absorbers[26,27]. Thanks to these findings, plasmonically enhanced MEMS/NEMS has emerged as a promising research direction towards the development of miniaturized transducers able to convert electromagnetic energy into electric signals by simultaneously exploiting plasmonic, thermal and electromechanical properties[28–31]. In particular, uncooled infrared sensors consisting of a plasmonic absorber attached to a conventional MEMS thermal detector (that is, thermopile or microbolometer) have been demonstrated[32,33], showing enhanced responsivity at specific wavelengths of interest in the infrared range. More recently, the integration of bulk metamaterials and nano-plasmonic infrared/THz absorbers in beam-type nanomechanical structures has also been demonstrated[34]. It has been shown that the plasmonically enhanced light absorption can induce a substantial beam deflection through the intrinsically high thermomechanical coupling of such free-standing nanomechanical structures. Even though these devices exhibit improved detection capabilities compared with some conventional MEMS-based thermal detectors, they require the use of relatively cumbersome and complex off-chip optical readouts to monitor the thermally induced deformation of the nanobeam. In this perspective, the achievement of efficient actuation and sensing of mechanical vibration in a plasmonic nanomechanical structure with intrinsically enhanced light absorption capability (without the need of integrating additional infrared absorbing materials) is highly desirable, being able to combine high sensitivity with power efficient on-chip transduction and readout.

Here we propose an ultrathin (650 nm) piezoelectric plasmonic metasurface forming the vibrating body of a nanomechanical resonator with unprecedented optical and electromechanical performance. By combining plasmonic and piezoelectric electromechanical resonances, we demonstrate efficient transduction of vibration in a nanomechanical structure with a strong and polarization-independent absorption coefficient over an ultrathin thickness, addressing all fundamental challenges associated with the development of performing resonant infrared detectors.

Results

Device design. The proposed plasmonic piezoelectric NEMS resonator, illustrated in Fig. 1, is composed of an aluminum nitride (AlN) piezoelectric nanoplate (500-nm-thick) sandwiched between the two metal layers (Supplementary Fig. 1). The bottom

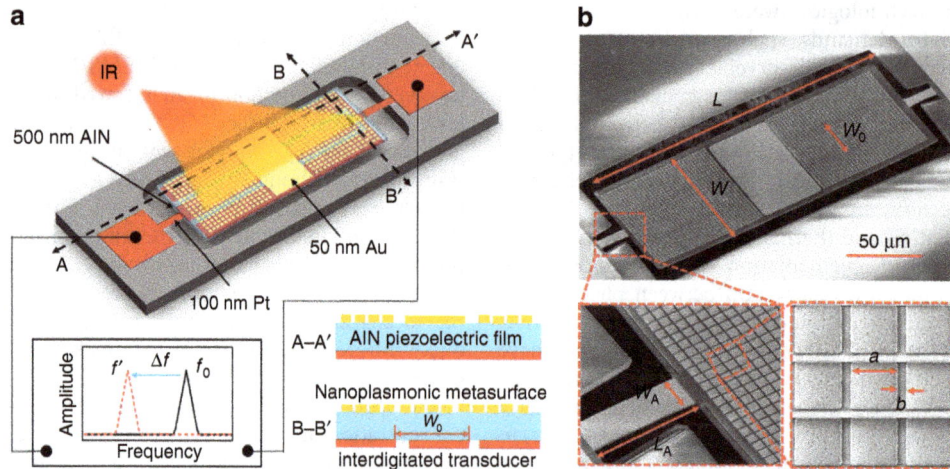

Figure 1 | Overview of the plasmonic piezoelectric nanomechanical resonant infrared detector. (**a**) Mock-up view: an aluminum nitride nanoplate is sandwiched between a bottom metallic interdigitated electrode and a top nanoplasmonic metasurface. The incident IR radiation is selectively absorbed by the plasmonic metasurface and heats up the resonator, shifting its resonance frequency from f_0 to f' due to the temperature dependence of its resonance frequency. (**b**) Scanning electron microscopy images of the fabricated resonator, metallic anchors and nanoplasmonic metasurface. The dimensions of the resonator are as follows: $L = 200\,\mu m$; $W = 75\,\mu m$; $W_0 = 25\,\mu m$ ($19 + 6\,\mu m$); $L_A = 20\,\mu m$; $W_A = 6.5\,\mu m$. The dimensions of the unit cell of the plasmonic metasurface are as follows: $a = 1635\,nm$; $b = 310\,nm$. IR, infrared.

layer (100-nm-thick platinum (Pt)) is patterned to form an interdigitated transducer (IDT) used to actuate and sense a high-order lateral-extensional mode of vibration in the nanoplate[35]; the top electrically floating layer (50-nm-thick gold) is patterned with the goal of confining the electric field induced by the bottom IDT across the piezoelectric nanoplate (Supplementary Fig. 2), while simultaneously enabling absorption of infrared radiation in the ultrathin piezoelectric nanoplate thanks to suitably tailored plasmonic resonances (Supplementary Fig. 3). The nanoplate is released from the silicon (Si) substrate to vibrate freely, and it is mechanically supported by two ultrathin Pt tethers (100-nm-thick, 6.5-μm-wide and 20-μm-long), which also provide electrical contact[36]. Such Pt tethers greatly improve the thermal isolation between the nanoplate and the Si substrate compared with conventional ones composed of an AlN–Pt stack[37]. The mechanical resonance frequency of the plasmonic piezoelectric resonator is defined by the equivalent Young's modulus E_{eq} and density ρ_{eq} of the resonant material stack, and the pitch of the interdigitated electrode W_0 (electrode width plus spacing, see Fig. 1a), given by $f_0 = \frac{1}{2W_0}\sqrt{\frac{E_{eq}}{\rho_{eq}}}$. When an a.c. signal is applied to the bottom IDT of the device, the top electrically floating electrode acts to confine the electric field across the device thickness, and a high-order contour-extensional vibration mode is excited through the equivalent d_{31} piezoelectric coefficient of AlN[35] when the frequency of the a.c. signal coincides with the natural resonance frequency, f_0, of the resonator (Supplementary Note 1). If an infrared beam impinges on the device from the top (Fig. 1a), it is selectively absorbed by the metasurface, leading to a large and fast increase of the device temperature ΔT, due to the excellent thermal isolation and extremely low-thermal mass of the free-standing nanomechanical structure. Such infrared-induced temperature rise results in a shift in the mechanical resonance frequency of the resonator (from f_0 to $f_0 - \Delta f$) due to the intrinsically large temperature coefficient of frequency (TCF) of the device[38]. Therefore, the incident IR power can be readily detected by monitoring the resonance frequency of the device.

As we show in the following, the proposed plasmonic nanomechanical resonant structure provides all the fundamental features necessary for the implementation of uncooled infrared detectors with unprecedented performance (Supplementary

Note 2). First, we need to maximize its absorption within the spectrum of interest: to this end, an array of subwavelength patches (Fig. 1b) is patterned within the top metal electrode of the device. Proper patterning of such plasmonic nanostructures in the top metal layer allows the whole device to behave as a spectrally selective and polarization-independent infrared ultrathin absorber[26], significantly enhancing the electromagnetic field concentration within the AlN dielectric (Supplementary Fig. 3). While a single array of subwavelength patches is fundamentally bound to absorb no more than half of the impinging radiation for symmetry constraints[39], the presence of the piezoelectric nanoplate allows us to go beyond this limit and absorb a large portion of infrared energy at resonance. At the same time, we need to achieve maximum thermal isolation of the resonant body from the heat sink, which is ensured by minimizing the thickness of the tethers used to support the nanoplate. The anchors of piezoelectric MEMS/NEMS resonators are conventionally composed of a thick and thermally conductive piezoelectric layer, directly patterned on the same layer forming the vibrating body of the resonator[35], and a thin metal layer employed to route the electrical signal to the actuation electrode integrated in the body of the resonator. On the contrary, here the relatively thick piezoelectric material is completely removed from the anchors, minimizing their thicknesses (ultimately limited by the need of a thin metal layer for electrical routing), resulting in a resonant thermal detector with markedly enhanced responsivity ($0.68\,Hz\,nW^{-1}$). Third, we need to achieve also a low-thermal time constant ($440\,\mu s$), which is obtained by exploiting the unique properties of high-quality ultrathin AlN films deposited on a Si substrate with a low-temperature sputtering process, enabling low-volume piezoelectric nanoplate resonant structures with a significantly reduced thermal mass. Despite the resonator volume scaling, we are also able to obtain low-noise performance, thanks to the piezoelectric transduction properties of the ultrathin AlN film, which enable efficient on-chip piezoelectric actuation and sensing of a high $Q > 1,000$ and high-frequency bulk vibration mode in a free-standing nanoplate, leading to a resonator with very low noise spectral density ($\sim 1.46\,Hz\,Hz^{-1/2}$ at $100\,Hz$ measurement bandwidth).

The plasmonic metasurface was designed using a transmission line approach, assuming a continuous 100-nm-thick Pt layer beneath the piezoelectric thin film (see Methods). The patch and unit cell dimensions were chosen to provide a Fabry–Perot-like resonance at ~8.8 μm. While a conventional longitudinal resonance would lead to a significant thickness, severely affecting the mechanical and thermal response of the resonator, in our design we tailored the plasmonic metasurface patterned on top of the grounded AlN nanoplate to have a large capacitive surface reactance $X_s = -1/(\omega C_s)$, under a $e^{-i\omega t}$ time convention. Stacked on top of a grounded slab, the dominant resonance is achieved when $X_s = -Z_0 \tan(\beta d)$, where Z_0 and β are the characteristic impedance and propagation constant of the AlN substrate. It confirms that, by tailoring the surface reactance of the plasmonic metasurface to be largely capacitive, it is possible to induce an ultrathin Fabry–Perot resonance in the substrate. It is worth noting that, despite the presence of a small perturbation in the bottom metal layer (interdigitated configuration with two 6-μm-wide gaps rather than a perfectly continuous ground plane), the induced Fabry–Perot-like resonance is mainly determined by the physical dimensions of the gold patches patterned on the top surface of the AlN nanoplate, which are polarization independent[26].

In our device, the plasmonic nanostructures cover 80% of the top metal layer, as a trade-off between large absorption, achieved by coating the entire layer, and high electromechanical transduction efficiency, achieved by removing a portion of the metasurface and replacing it with continuous metal (Supplementary Note 3). We achieved an electromechanical coupling coefficient, $k_t^2 \sim 1\%$ (Supplementary Fig. 4). Figure 2a,b

presents the predicted absorption with and without top subwavelength patches, highlighting how the metasurface can largely increase the absorption at resonance, despite the deeply subwavelength thickness of the device. We also theoretically and experimentally demonstrate that a strong and spectrally selective absorption of long-wavelength infrared (LWIR) radiation, with lithographically determined centre frequency and peak values >85% (over the device area covered by the plasmonic metasurface), can be readily achieved (Fig. 2c,d). Our measurements match very well with the theoretical simulations within the entire band of interest.

Device fabrication and characterization. On the basis of this design, the proposed infrared detector (Fig. 1) was fabricated using a post-complementary metal-oxide-semiconductor (CMOS) compatible microfabrication process involving a combination of photolithography (four masks) and electron-beam lithography (one step). The Fourier transform infrared (FTIR) absorption spectrum of the device was first measured (see Methods) showing that ~80% of the impinging optical power (normal incidence to the device surface) is absorbed at the desired spectral wavelength (Fig. 2b), which validates the performance of the designed piezoelectric metasurface. By removing the plasmonic pattern from the device top electrode would lead to negligible absorption, confirming the uniqueness of the proposed design (Fig. 2a).

We also show that, by altering the lateral size of the nanoplasmonic structure (gold patch), it is possible to accurately

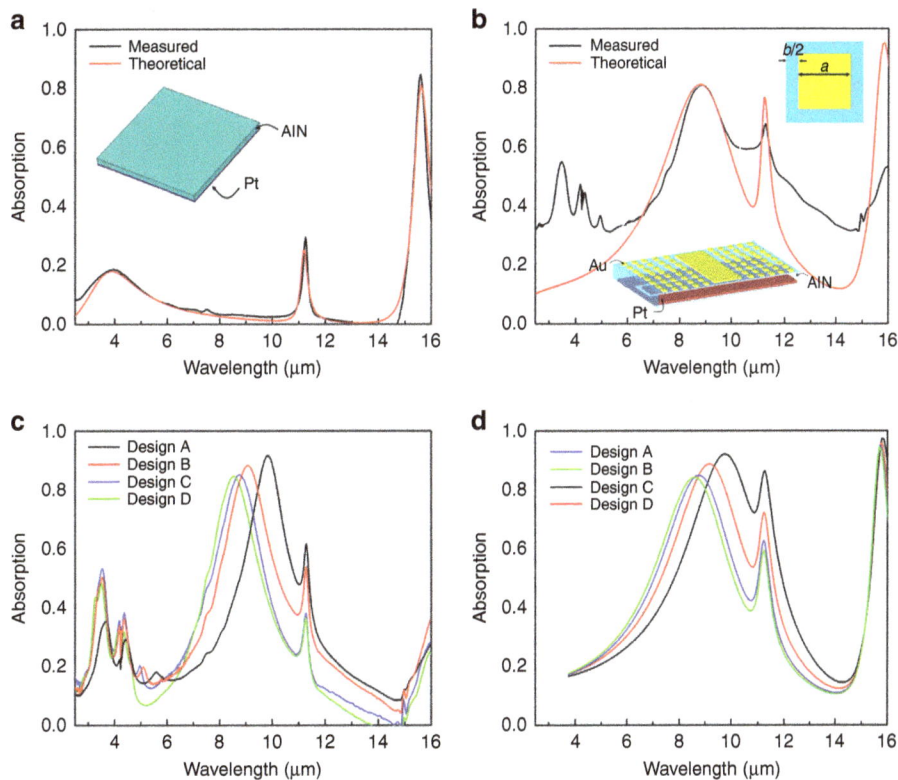

Figure 2 | Absorption properties of the proposed plasmonic piezoelectric nanomechanical resonator. (**a**) Simulated (transmission line theory) and measured (FTIR) absorption spectra of a 500-nm-thick AlN slab grounded by a Pt layer (without plasmonic nanostructures). It shows two intrinsic absorption peaks, associated with AlN at 11.3 μm (888 cm⁻¹) and 15.5 μm (647 cm⁻¹)[40], and one at 4 μm associated with the resonant structure. (**b**) Simulated and measured absorption spectra of the fabricated plasmonic piezoelectric nanomechanical resonator. The dimensions of the Au patches that compose the metasurface are $a = 1,635$ nm, $b = 310$ nm, and the thickness of the Au, AlN and Pt layers are 50, 500 and 100 nm, respectively. (**c,d**) Measured and simulated absorption properties of the piezoelectric plasmonic resonant structure with varied Au patch sizes, demonstrating its functionality of spectrally elective detection of infrared radiation in the LWIR range. The unit cell sizes are as follows: design A: $a = 1,780$ nm, $b = 128$ nm; design B: $a = 1,680$ nm, $b = 253$ nm; design C: $a = 1,640$ nm, $b = 313$ nm; design D: $a = 1,620$ nm, $b = 331$ nm.

tailor the absorption peak over a wide range for spectrally selective infrared detection (Fig. 2c,d). We do note the presence of absorption peaks at around 3–4 µm in Fig. 2b,c, which we attribute to an in-plane resonance arising between the fingers of the interdigitated bottom Pt electrode.

The electromechanical performance of the resonator was characterized by measuring its admittance versus frequency (Fig. 3a). A high $Q = 1,116$ and electromechanical coupling coefficient $k_t^2 = 0.86\%$ were extracted by equivalent model fitting (Fig. 3a; Methods), demonstrating the unique advantages of the proposed design in terms of high electromechanical transduction efficiency and low loss. The thermal properties (thermal resistance, temperature distribution and TCF) of the infrared detector were characterized by both finite element analysis and experimental verification (Supplementary Figs 5–8; Supplementary Table 1; Supplementary Note 4). The response of the fabricated infrared detector in the LWIR band was characterized using a 1,500-K globar (2–16-µm emission) as an infrared source. For the sake of comparison, the incoming infrared radiation was also detected using a conventional AlN MEMS resonator with same frequency sensitivity to absorbed heat but without plasmonic pattern (hence non-enhanced infrared absorptance). Thanks to its properly engineered optical properties, the piezoelectric plasmonic resonator showed fourfold enhanced responsivity (Fig. 3b), despite its absorption band (full width at half maximum of 1.5 µm) being much narrower than the emission band of the source. With a narrowband source at the

frequency of interest, the responsivity would be much larger. The smallest impinging optical power that can be detected was experimentally estimated by measuring the device responsivity, R_s, and noise spectral density (Supplementary Note 5), demonstrating a low noise equivalent power (NEP) ~ 2.1 nW Hz$^{-1/2}$ at the designed spectral wavelength (for which infrared absorptance $\sim 80\%$). The NEP is arguably considered the most important performance metric for an infrared detector (Supplementary Note 5) and the value measured for the realized proof-of-concept detector proposed here is already comparable to the best commercially available uncooled broadband thermal detectors, while providing unique spectral selectivity in the LWIR band. The response time of the detector was also evaluated by measuring the attenuation of the device response when exposed to infrared radiation modulated at increasingly faster rates (Fig. 3c; Methods), showing a low-thermal time constant, $\tau \sim 440$ µs.

Discussion

Differently from more conventional approaches involving the integration of an infrared absorbing material (such as Si_3N_4, SiO_2 or a metal–insulator–metal grating) on top of an optical/mechanical resonant thermal detector[8], our approach uses an individual plasmonic piezoelectric nanostructure acting simultaneously as absorber, resonator and transducer. It is also worth noting that our proposed plasmonic NEMS resonant sensor exploits a high-frequency on-chip piezoelectric

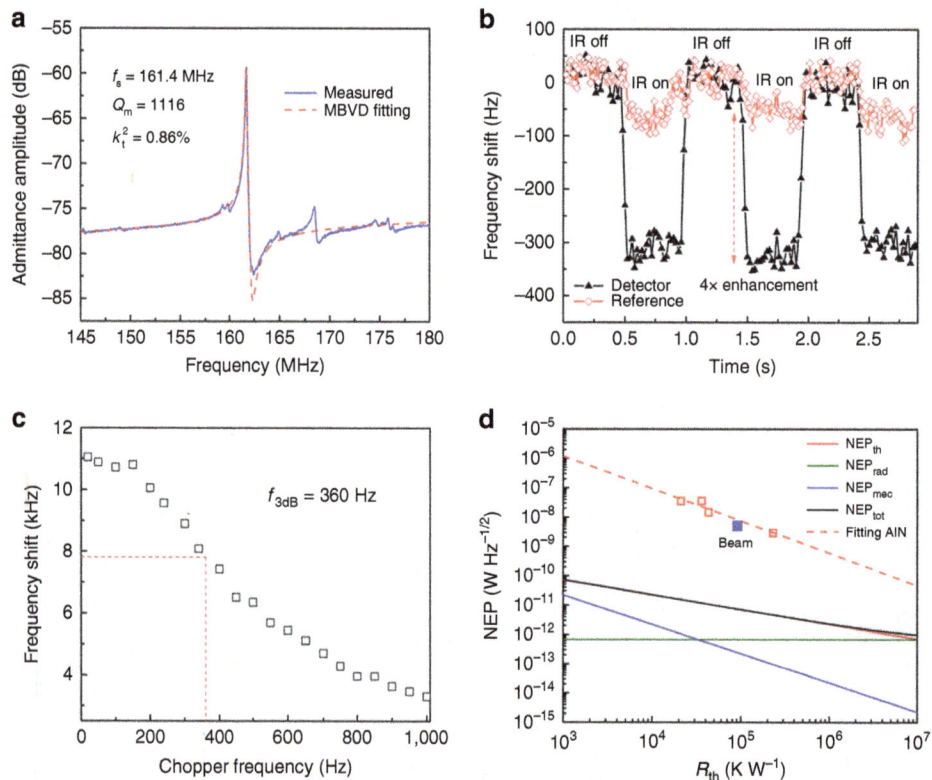

Figure 3 | Device performance. (a) Measured admittance curve versus frequency and MBVD model fitting of the resonator for $Q_{IR} = 0$. The extracted values of the MBVD parameters (see Methods) are as follows: $R_s = 80 \, \Omega$; $R_m = 880 \, \Omega$; $L_m = 1$ mH; $C_m = 1$ fF; $R_0 = 2$ kΩ; $C_0 = 145$ fF. **(b)** Measured response of the plasmonic piezoelectric resonator and a conventional AlN MEMS resonator to a modulated IR radiation emitted by a 1,500-K globar (2–16 µm broadband spectral range). **(c)** Measured frequency response of the detector. The 3dB cutoff frequency, $f_{3\,dB}$, was found to be 360 Hz, resulting in a time constant $\tau = 1/(2\pi f_{3\,dB})$ of 440 µs. **(d)** NEP for different values of thermal resistance (R_{th}). The solid lines indicate the calculated NEP values associated with each of the three fundamental noise contributions (as expressed in Supplementary Equations 5–6), assuming: resonator area $= 200 \times 75$ µm^2, $\varepsilon = 1$, $T_0 = 300$ K, $P_c = 0$ dBm, $|TCF| = 30$ p.p.m. K^{-1}, $Q = 2,000$. The individual data points indicate the measured NEP values of four fabricated AlN resonant plasmonic IR detectors using four different anchor designs (hence four different R_{th} values) and a Si_3N_4 nanobeam (spectrally selective, static measurement using off-chip optical readout)[34]. IR, infrared.

transduction mechanism that has been a key enabler for MEMS technologies. In particular, AlN-based piezoelectric micro-acoustic devices are nowadays the commercial standard used in the radio frequency (RF) front ends of modern smart phones (that is, see Avago Technologies thin-film bulk acoustic resonator (FBAR)). Therefore, the demonstration of a resonant infrared detector based on a bulk-extensional mode plasmonically enhanced piezoelectric resonator, instead of more conventional MEMS, optical resonators or optomechanical structures, is a key advancement over earlier works that really enables the use of these emerging MEMS plasmonic technologies in low-power and miniaturized wireless communication devices. The unique capability of such AlN-based MEMS technology to deliver high-performance and CMOS compatible sensors and RF components makes it the best candidate for the realization of the next-generation miniaturized, low-power, multi-functional and reconfigurable wireless sensing platforms that will be crucial for the development of the internet of things.

Importantly, even though the proposed resonator possesses a unique combination of efficient electromechanical transduction, strong and spectrally selective infrared absorption capability and very low NEP, there is still plenty of room to improve the performance of these piezoelectric plasmonic NEMS devices. First, novel designs could be employed to further improve the absorption of the piezoelectric metasurface to near unity[41]. Second, to reach the thermal fluctuation noise limit (Fig. 3d), all the noise sources contributing to the generation of frequency fluctuations, such as the resonator flicker noise, random walk and drifts[42], need to be carefully investigated and mitigated. Moreover, the volume of the plasmonic piezoelectric resonator can be further reduced (for instance, by scaling the thickness of the piezoelectric nanoplate, see Supplementary Fig. 9) and its design can be optimized, investigating optimal materials and

innovative geometries for the device anchors to increase the thermal resistance up to $\sim 10^7\,\mathrm{K\,W^{-1}}$, which is typical of conventional microbolometers. As a result, we expect this technology to achieve NEP in the order of $\sim 1\,\mathrm{pW\,Hz^{-1/2}}$ (Fig. 3d), thus enabling the implementation of multi-spectral thermal imagers with noise equivalent temperature difference as low as $\sim 1\,\mathrm{mK}$.

In conclusion, we have demonstrated an uncooled NEMS resonant infrared detector with unique spectrally selective infrared detection capability. Sensing and actuation of a high-frequency (162 MHz) bulk acoustic mode of vibration in a free-standing ultrathin piezoelectric plasmonic metasurface have been demonstrated and exploited for the implementation of a NEMS resonator with unique combined optical and electro-mechanical properties. By exploiting the piezoelectric properties of AlN thin films, efficient on-chip transduction of the NEMS plasmonic resonant structure has been achieved, eliminating the need for the cumbersome and complex off-chip optical readouts employed in previous devices. Thanks to properly tailored absorption properties, strong and spectrally selective detection of infrared radiation, for an optimized spectral bandwidth centered around 8.8 μm and high thermomechanical coupling between electromagnetic and mechanical resonances in a single ultrathin piezoelectric nanoplate have been experimentally verified. This work sets a milestone towards the development of a new technology platform based on the combination of nanoplasmonics and piezoelectric nanoelectromechanical systems, which can potentially deliver fast (hundreds of μs), high resolution (NEP as low as $\sim 1\,\mathrm{pW\,Hz^{-1/2}}$) and spectrally selective uncooled infrared detectors suitable for the implementation of high-performance, miniaturized and power efficient infrared/THz spectrometer and multi-spectral imaging systems.

Methods

Fabrication. The plasmonic piezoelectric resonant infrared detector was fabricated using a post-CMOS compatible microfabrication process involving a combination of photolithography (four masks) and electron-beam lithography (one step), as illustrated in Fig. 4. The fabrication started with a high-resistivity (resistivity > 20,000 Ω m) 4-inch silicon wafer: (a) 100-nm-thick Pt was sputter-deposited and patterned by lift-off process to define the bottom IDT; (b) 500-nm-thick high-quality c axis orientated AlN film was sputter-deposited on top of the Pt IDT, and wet etched in H₃PO₄ to open the vias to get access to the bottom electrode and dry etched by inductively coupled plasma in Cl₂-based chemistry to define the shape of the AlN resonator; (c) 100-nm-thick Au was deposited by electron-beam deposition and patterned by lift-off process to define the probing pad; (d) 50-nm-thick Au was deposited by electron-beam deposition and patterned by electron-beam lithography and lift-off process to define the nanoplasmonic metasurface; (e) the Si substrate underneath the resonator was etched by xenon difluoride (XeF₂) to completely release the device.

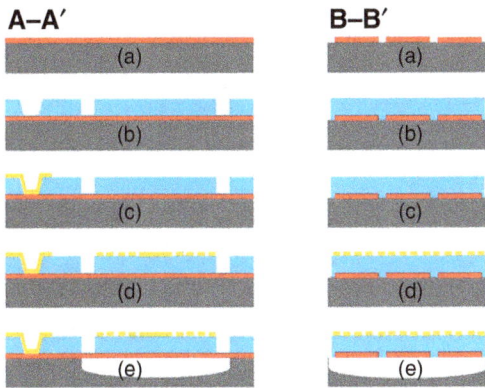

Figure 4 | Microfabrication process for the plasmonic piezoelectric NEMS resonant infrared detector. A–A′ and B–B′ denote longitudinal and transversal axis of the device, as illustrated in Fig. 1.

Modelling. *Nanoplasmonic piezoelectric resonant infrared detector—electrothermal equivalent circuit.* The proposed plasmonic piezoelectric NEMS resonant infrared detector is modelled by a two-port network with both electrical (voltage, V, used to drive the electromechanical resonance) and thermal inputs (infrared power, Q_IR, absorbed in the piezoelectric resonant structure), as shown in Fig. 5a. At the

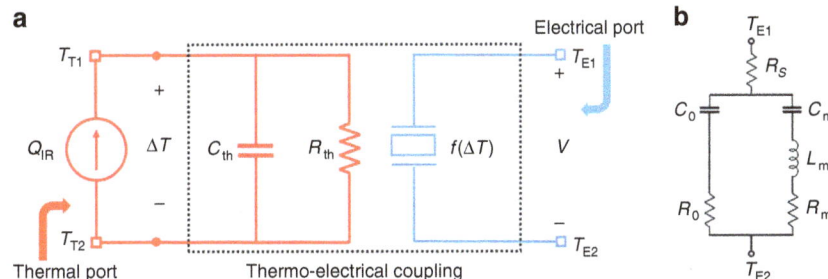

Figure 5 | Electrothermal equivalent circuit of the device. (a) Equivalent thermoelectrical circuit of the nanoplasmonic piezoelectric NEMS resonator. **(b)** Modified Butterworth–Van Dyke (MBVD) equivalent circuit.

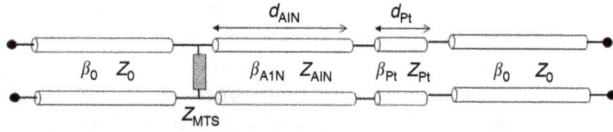

Figure 6 | Electromagnetic circuit model of the piezoelectric resonator.
A normally incident Transverse electromagnetic (TEM) wave impinges on the structure shown in Fig. 1. The outer transmission line sections represent the free space, Z_{MTS}, is the surface impedance of the array of gold patches (metasurface), and the inner transmission line section takes into account the AlN dielectric and the ground platinum layer, respectively. Each section is characterized by its characteristic impedance Z and propagation constant β.

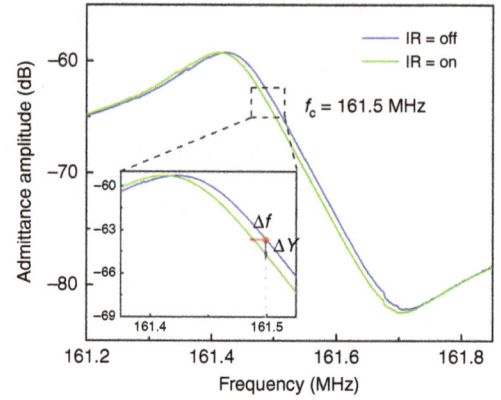

Figure 7 | Measured admittance curves versus frequency of the resonator for IR on and off. The inset shows the zoomed in admittance shift around the resonance frequency. The exciting frequency of 161.5 MHz is also marked at which the frequency response of the device was measured. IR, infrared.

thermal port, the free-standing resonant structure is simply modelled as a thermal mass, with thermal capacitance C_{th}, coupled with the heat sink at a constant temperature T_0 via the thermal conductance G_{th} (thermal resistance, $R_{th} = 1/G_{th}$). The thermal capacitance is intrinsically related to the properties and overall volume of the material stack forming the vibrating structure ($C_{th} = c \cdot v \cdot \rho$; where c is the specific heat capacity, v is the volume and ρ is the material density). The thermal conductance is instead mainly determined by the geometry and material properties of the tethers connecting the free-standing vibrating body of the device to the substrate (Supplementary Note 2). At the electrical port, the resonator is modelled with a modified Butterworth–Van Dyke (MBVD) equivalent circuit model[43], shown in Figure 5b. The MBVD circuit consists of two branches in parallel: an acoustic branch composed of the series combination of the motional resistance, R_m (quantifying dissipative losses), motional capacitance, C_m (inversely proportional to the stiffness) and motional inductance, L_m (proportional to the mass); and an electrical branch composed by the series combination of the capacitance, C_0 (capacitance between the device terminals), and resistance, R_0 (representing the dielectric loss). A series resistance, R_s, is also included in the circuit to represent the electrical loss associated with the metal electrodes and routing (Fig. 5b).

Nanoplasmonic piezoelectric resonant infrared detector—electromagnetic model. This section briefly presents the analysis and design of the proposed plasmonic nanomechanical resonator from an electromagnetic point of view. Specifically, this analysis permits obtaining the frequency-dependent absorption coefficient of the resonator, η, which determines the resonator increase of temperature (ΔT) when infrared radiation is impinging onto the detector (Supplementary Note 2).

Thanks to the structure symmetry and assuming a continuous Pt layer beneath the AlN, the analysis of a single unit cell of the metasurface suffices to investigate the electromagnetic behaviour of the whole device. This analysis is carried out using the equivalent transmission line of Fig. 6. This circuit is totally rigorous[44,45], that is, exactly equivalent to solving Maxwell's equations, assuming that (i) a transverse electromagnetic (TEM) wave is normally impinging on the resonator and (ii) the operation frequency is well below the cutoff frequency of the higher-order modes excited in the metasurface. Note that these two conditions are indeed fulfilled in this case.

The characteristic impedance and propagation constant of the different transmission line sections involved in the model are

$$Z_0 = \sqrt{\frac{\mu_0}{\varepsilon_0}}; \quad Z_{AlN} = \sqrt{\frac{\mu_0}{\varepsilon_0 \varepsilon_{AlN}}}; \quad Z_{Pt} = \sqrt{\frac{\mu_0}{\varepsilon_0 \varepsilon_{Pt}}}$$
$$\beta_0 = \frac{\omega}{c}; \quad \beta_{AlN} = \frac{\omega}{c}\sqrt{\varepsilon_{AlN}}; \quad \beta_{Pt} = \frac{\omega}{c}\sqrt{\varepsilon_{Pt}} \quad (1)$$

where the subscripts '0', 'AlN' and 'Pt' denote free space, aluminum nitrate dielectric and Pt, respectively, 'c' is the speed of light, ω is the radial frequency and ε is permittivity, respectively. In addition, the Pt layer is modelled following a simple Drude model

$$\varepsilon_{Pt} \approx 1 - \frac{\omega_p^2}{\omega^2 + j\omega\tau^{-1}} \quad (2)$$

with plasma frequency $\omega_p = 7.88 \times 10^{15}$ rad s^{-1} and relaxation time $\tau = 7.8$ fs (ref. 46). We characterize the aluminum nitrate substrate using a Lorentzian permittivity model as

$$\varepsilon_{AlN} \approx 4 + \frac{\omega_{p1}^2}{\omega_{t1}^2 - \omega^2 + j\omega\tau_1^{-1}} + \frac{\omega_{p2}^2}{\omega_{t2}^2 - \omega^2 + j\omega\tau_2^{-1}} \quad (3)$$

with $\omega_{p1} = 8.663 \times 10^{13}$ rad s^{-1}, $\omega_{t1} = 1.6894 \times 10^{14}$ rad s^{-1}, $\tau_1 = 0.2367$ ps, $\omega_{p2} = 2.5115 \times 10^{14}$ rad s^{-1}, $\omega_{t2} = 1.2558 \times 10^{14}$ rad s^{-1}, $\tau_2 = 0.3185$ ps. The electromagnetic model of the array of gold patches is very challenging and there are no closed-form expressions available for a general case. However, in case of electrically small unit cells (that is, $a + b \ll \lambda_0$, being λ_0 the operating wavelength) and densely packed planar patches ($b \ll a$), the surface impedance of the array of

gold patches can be approximated by[44]

$$Z_{MTS} = \frac{a+b}{a\sigma_{Au}} - j \frac{\pi}{2\omega\varepsilon_0\varepsilon_{eff}(a+b)\log\left(\csc\left(\frac{\pi b}{2a+b}\right)\right)} \quad (4)$$

where σ_{Au} is the gold conductivity and $\varepsilon_{eff} = (1 + \varepsilon_{AlN})/2$ is the effective permittivity surrounding the metasurface. We have numerically verified that equation 4 provides accurate results in the specific metasurfaces under consideration. However, note that the near field created by the actual thickness of the gold patches (~ 50 nm) slightly decreases the effective distance between the patches. This phenomenon, common in microwave and optics[47], is easily taken into account using an effective separation distance b slightly lower than the physical one.

The proposed circuit is analysed using an analytical transfer-matrix approach, thus providing the absorption coefficient of the resonator versus frequency. The metasurface employed to cover the AlN resonator is designed in two steps: (i) the input impedance and absorption of the AlN slab grounded by the Pt layer are computed using the equivalent circuit of Fig. 6 in the absence of the metasurface (that is, $Z_{MTS} \approx \infty$), and (ii) the dimensions of the unit cell and patches of the metasurfaces are chosen with the help of (equations 2–4) aiming to match the input impedance of the structure to the free-space impedance at the operation frequency, therefore maximizing the absorption coefficient. Finally, the electromagnetic behaviour of the designed structure is confirmed using the full-wave commercial software Computer Simulation Technology (CST) mimcrowave studio.

It is important to point out the good agreement between theory and measurement (Fig. 2a,b), especially taking into account that the fabricated device does not have a continuous Pt layer, but a patterned one (patterned IDT with a partial coverage of $\sim 70\%$, which is needed to excite a 160-MHz contour-extensional mode of vibration in the plasmonic piezoelectric nanoplate). However, the influence of this pattern is limited because (i) most of the unit cells (around 70%) are grounded by a continuous Pt layer and (ii) it basically introduces a small capacitance to the ultrathin Pt layer electromagnetic response. We have further verified with CST full-wave simulations that the influence of this pattern is negligible.

Measurements. The electromechanical performance of the plasmonic piezoelectric resonator was characterized by measuring its equivalent electrical admittance with an Agilent E5071C vector network analyzer after performing an open-short-load calibration on a standard substrate. The scattering parameter, S_{11}, of this one-port electrical network was first measured and then converted into the admittance, Y, according to: $Y = (1/50)(1 + S_{11})/(1 - S_{11})$. The values of the equivalent circuit elements were extracted from the measurement by MBVD model fitting. The mechanical quality factor, Q (inversely proportional to the mechanical damping affecting the resonator), and electromechanical coupling coefficient, k_t^2 (a numerical measure of the conversion efficiency between electrical and mechanical energy in an electromechanical resonator), were extracted according to: $Q = (1/R_m)(L_m/C_m)^{0.5}$, and $k_t^2 = (\pi^2/8)(C_m/C_0)$, respectively[48]. The device admittance amplitude–frequency nonlinearity was measured by monitoring the admittance amplitude versus frequency shift for different levels of RF power. Further details about the measurement and analysis can be found in the Supplementary Note 5.

The reflectance, R, spectra of the fabricated nanoplasmonic metasurface and plasmonic piezoelectric resonator were measured using a Bruker V70 Fourier transform infrared coupled with a Hyperion 1000 microscope. A reflective gold mirror was used as a background for calibration. The absorption spectrum, A, of each sample was then directly obtained as $1 - R$, assuming minimum transmission through the sample, which is a reasonable assumption given the large metal coverage of both the top Au plasmonic structures and bottom Pt electrode.

The response of the fabricated infrared detector in the LWIR band (Fig. 3b) was characterized using a 1,500-K globar (2–16-μm emission) as an infrared source coupled with a Bruker Hyperion 1000 microscope for focusing the infrared beam onto the device. The incident infrared radiation was modulated by an optical chopper at 1 Hz. Measurement bandwidth (intermediate frequency (IF) bandwidth) of the network analyzer was set to 100 Hz to obtain the best noise performance. All experiments were performed in an open environment in ambient temperature and pressure.

The thermal time constant of the detector was evaluated by measuring the attenuation of the device response when exposed to infrared radiation modulated at increasingly faster rates (Fig. 3c). An EOS Photonics $\sim$5-μm continuous-wave quantum cascade laser with a relatively high output power ($\sim$50 mW) was employed as a heat source. The infrared radiation emitted by the quantum cascade laser was modulated using an optical chopper (modulation frequency was varied from 1 Hz to 1 kHz) and focused onto the infrared detector using a zinc selenide (ZnSe) lens (with 70% transmission for infrared radiation in the 0.6–16 μm spectral range). The 3dB cutoff frequency, $f_{3\,dB}$, was found to be 360 Hz, resulting in a time constant $\tau = 1/(2\pi f_{3\,dB})$ of 440 μs (Fig. 3c).

The transient responses of the device were measured by exciting the resonator at a single frequency, $f_c = 161.5$ MHz, for which the slope of admittance amplitude curve versus frequency is maximum (121.3 dB MHz^{-1})[7], and by monitoring the variations over time of the device admittance amplitude (ΔY) using our network analyzer. The admittance amplitude change was then converted to frequency change (Δf) by multiplying the slope of the admittance versus frequency (Fig. 7). The measurement bandwidth (IF bandwidth) of the network analyzer was set to 10 kHz (sampling time of 0.1 ms), which is small enough compared with the thermal time constant of the detector.

References

1. Reine, M. B. HgCdTe photodiodes for IR detector: a review. *Proc. SPIE* **4288**, 266–277 (2001).
2. Kozlowski, Lester J. *et al.* Recent advances in staring hybrid focal plane arrays: comparison of HgCdTe, InGaAs, and GaAs/AlGaAs detector technologies. *Proc. SPIE* **2274**, 93–116 (1994).
3. Rogalski, A. Infrared detectors: status and trends. *Prog. Quant. Electron.* **27**, 59–210 (2003).
4. Chen, C., Yi, X., Zhao, X. & Xiong, B. Characterization of VO$_2$ based uncooled microbolometer linear array. *Sens. Actuators A* **90**, 212–214 (2001).
5. Schaufelbuhl, A. *et al.* Uncooled low-cost thermal imager based on micromachined CMOS integrated sensor array. *J. Microelectromech. Syst.* **10**, 503–510 (2001).
6. Kang, D. H., Kim, K. W., Lee, S. Y., Kim, Y. H. & Keun Gil, S. Influencing factors on the pyroelectric properties of Pb (Zr, Ti) O$_3$ thin film for uncooled infrared detector. *Mater. Chem. Phys.* **90**, 411–416 (2005).
7. Hui, Y. & Rinaldi, M. Fast and high resolution thermal detector based on an aluminum nitride piezoelectric microelectromechanical resonator with an integrated suspended heat absorbing element. *Appl. Phys. Lett.* **102**, 093501 (2013).
8. Watts, M. R., Shaw, M. J. & Nielson, G. N. Optical resonators: Microphotonic thermal imaging. *Nat. Photon.* **1**, 632–634 (2007).
9. Masmanidis, S. C. *et al.* Multifunctional nanomechanical systems via tunably coupled piezoelectric actuation. *Science* **317**, 780–783 (2007).
10. Li, M., Tang, H. X. & Roukes, M. L. Ultra-sensitive NEMS-based cantilevers for sensing, scanned probe and very high-frequency applications. *Nat. Nanotechnol.* **2**, 114–120 (2007).
11. Niklaus, F., Vieider, C. & Jakobsen, H. in Photonics Asia 2007 68360D (International Society for Optics and Photonics, 2007).
12. Graf, A., Arndt, M., Sauer, M. & Gerlach, G. Review of micromachined thermopiles for infrared detection. *Meas. Sci. Technol.* **18**, R59 (2007).
13. Exner, A. T., Pavlichenko, I., Lotsch, B. V., Scarpa, G. & Lugli, P. Low-cost thermo-optic imaging sensors: a detection principle based on tunable one-dimensional photonic crystals. *ACS Appl. Mater. Interfaces* **5**, 1575–1582 (2013).
14. Tallur, S. & Bhave, S. A. A silicon electromechanical photodetector. *Nano Lett.* **13**, 2760–2765 (2013).
15. Hui, Y. & Rinaldi, M. in *The 17th International Conference on Solid-State Sensors, Actuators and Microsystems* (TRANSDUCERS & EUROSENSORS XXVII), 2013 Transducers & Eurosensors XXVII: 968–971 (Barcelona, Spain, 2013).
16. Qian, Z., Hui, Y., Liu, F., Kai, S. & Rinaldi, M. in *18th International Conference on Solid-State Sensors, Actuators and Microsystems* (TRANSDUCERS), 2015 Transducers 1429–1432 (Anchorage, AK, USA, 2015).
17. Gokhale, V. J., Myers, P. D. & Rais-Zadeh, M. in *IEEE Sensors*, 982–985 (Valencia, Spain, 2014).
18. Kildishev, A. V., Boltasseva, A. & Shalaev, V. M. Planar photonics with metasurfaces. *Science* **339**, 1232009 (2013).
19. Yu, N. & Capasso, F. Flat optics with designer metasurfaces. *Nat. Mater.* **13**, 139–150 (2014).
20. Zhao, Y., Belkin, M. & Alù, A. Twisted optical metamaterials for planarized ultrathin broadband circular polarizers. *Nat. Commun.* **3**, 870 (2012).
21. Pfeiffer, C. & Grbic, A. Metamaterial Huygens' surfaces: tailoring wave fronts with reflectionless sheets. *Phys. Rev. Lett.* **110**, 197401 (2013).
22. Esquius-Morote, M., Gomez-Diaz, J. S. & Perruisseau-Carrier, J. Sinusoidally modulated graphene leaky-wave antenna for electronic beamscanning at THz. *IEEE Trans. THz Sci. Technol.* **4**, 116–122 (2014).
23. Pozar, D. Flat lens antenna concept using aperture coupled microstrip patches. *Electron. Lett.* **32**, 2109–2111 (1996).
24. Monticone, F., Estakhri, N. M. & Alù, A. Full control of nanoscale optical transmission with a composite metascreen. *Phys. Rev. Lett.* **110**, 203903 (2013).
25. Yu, N. *et al.* Light propagation with phase discontinuities: generalized laws of reflection and refraction. *Science* **334**, 333–337 (2011).
26. Mosallaei, H. & Sarabandi, K. in *Antennas and Propagation Society International Symposium, 2005 IEEE*, 615–618 (Washington, DC, USA, 2005).
27. Argyropoulos, C., Le, K. Q., Mattiucci, N., D'Aguanno, G. & Alu, A. Broadband absorbers and selective emitters based on plasmonic Brewster metasurfaces. *Phys. Rev. B* **87**, 205112 (2013).
28. Ou, J.-Y., Plum, E., Zhang, J. & Zheludev, N. I. An electromechanically reconfigurable plasmonic metamaterial operating in the near-infrared. *Nat. Nanotechnol.* **8**, 252–255 (2013).
29. Yamaguchi, K., Fujii, M., Okamoto, T. & Haraguchi, M. Electrically driven plasmon chip: active plasmon filter. *Appl. Phys. Express* **7**, 012201 (2014).
30. Dennis, B. *et al.* Compact nanomechanical plasmonic phase modulators. *Nat. Photon.* **9**, 267–273 (2015).
31. Valente, J., Ou, J.-Y., Plum, E., Youngs, I. J. & Zheludev, N. I. A magneto-electro-optical effect in a plasmonic nanowire material. *Nat. Commun.* **6** (2015).
32. Ogawa, S., Okada, K., Fukushima, N. & Kimata, M. Wavelength selective uncooled infrared sensor by plasmonics. *Appl. Phys. Lett.* **100**, 021111 (2012).
33. Talghader, J. J., Gawarikar, A. S. & Shea, R. P. Spectral selectivity in infrared thermal detection. *Light Sci. Appl.* **1**, e24 (2012).
34. Yi, F., Zhu, H., Reed, J. C. & Cubukcu, E. Plasmonically enhanced thermomechanical detection of infrared radiation. *Nano Lett.* **13**, 1638–1643 (2013).
35. Rinaldi, M. & Piazza, G. in *2011 Joint Conference of the IEEE International on Frequency Control and the European Frequency and Time Forum (FCS)*, 1–5 (San Fransisco, CA, USA, 2011).
36. Hui, Y., Qian, Z., Hummel, G. & Rinaldi, M. in *Proceedings of the 2014 Solid-State Sensors, Actuators and Microsystems Workshop (Hilton Head 2014)* 387–391 (Hilton Head Island, SC, USA, 2014) 387–390 (2014).
37. Hui, Y. & Rinaldi, M. in *28th IEEE International Conference on Micro Electro Mechanical Systems (MEMS)* 984–987 (Estoril, Portugal, 2015).
38. Kuypers, J. H., Lin, C.-M., Vigevani, G. & Pisano, A. P. in *2008 IEEE International Frequency Control Symposium*, 240–249 (Honolulu, HI, USA, 2008).
39. Thongrattanasiri, S., Koppens, F. H. & de Abajo, F. J. G Complete optical absorption in periodically patterned graphene. *Phys. Rev. Lett.* **108**, 047401 (2012).
40. Ibáñez, J. *et al.* Far-infrared transmission in GaN, AlN, and AlGaN thin films grown by molecular beam epitaxy. *J. Appl. Phys.* **104**, 033544 (2008).
41. Hendrickson, J., Guo, J., Zhang, B., Buchwald, W. & Soref, R. Wideband perfect light absorber at midwave infrared using multiplexed metal structures. *Optics Lett.* **37**, 371–373 (2012).
42. Rubiola, E. *Phase Noise and Freqeucy Stability in Oscillators* (Cambridge Univ. Press, 2008).
43. Larson, III J. D., Bradley, R., Wartenberg, S. & Ruby, R. C. in *IEEE Ultrasonics Symposium*, 863–868 (San Juan, 2000).
44. Tretyakov, S. *Analytical Modeling in Applied Electromagnetics* (Artech House, 2003).
45. Padooru, Y. R., Yakovlev, A. B., Kaipa, C. S., Medina, F. & Mesa, F. Circuit modeling of multiband high-impedance surface absorbers in the microwave regime. *Phys. Rev. B* **84**, 035108 (2011).
46. Rakic, A. D., Djurisic, A. B., Elazar, J. M. & Majewski, M. L. Optical properties of metallic films for vertical-cavity optoelectronic devices. *Appl. Opt.* **37**, 5271–5283 (1998).
47. Pozar, D. M. *Microwave Engineering* (John Wiley & Sons, 2009).
48. Rinaldi, M., Zuniga, C., Zuo, C. & Piazza, G. in *IEEE Transactions on Ultrasonics, Ferroelectrics and Frequency Control*, Vol. 57, 38–45 (2010).

Acknowledgements

We thank the staff of the George J. Kostas Nanotechnology and Manufacturing Facility, at Northeastern University, and of the Center for Nanoscale Systems, at Harvard University, for their support in device fabrication. This work was supported by the Defense Advanced Research Projects Agency (DARPA) Young Faculty Award N66001-12-1-4221, the National Science Foundation (NSF) Career Award ECCS-1350114, the Awareness and Localization of Explosives-Related Threats (ALERT) Department of

Homeland Security Center of Excellence, the Air Force Office of Scientific Research and the Welch Foundation with grant No. F-1802.

Author contributions

M.R. conceived the concept and initiated the research. Y.H. and M.R. designed the device and the experiments. Y.H. fabricated the devices, performed the experiments and developed the electrothermal model of the device. Y.H. and J.S.G.-D. analysed and processed the data. J.S.G.-D. and A.A. developed the electromagnetic model of the device, performed the analytical, and numerical electromagnetic simulations and fitted the measured infrared absorptance data. Z.Q. contributed to the fabrication of devices and data analysis. M.R. and A.A. coordinated and supervised the research. All authors contributed to the preparation of the manuscript.

Additional information

Competing financial interests: The authors declare no competing financial interests. US Patent Application 14/969,948 has been filed.

25

All-optical design for inherently energy-conserving reversible gates and circuits

Eyal Cohen[1], Shlomi Dolev[1] & Michael Rosenblit[2]

As energy efficiency becomes a paramount issue in this day and age, reversible computing may serve as a critical step towards energy conservation in information technology. The inputs of reversible computing elements define the outputs and vice versa. Some reversible gates such as the Fredkin gate are also universal; that is, they may be used to produce any logic operation. It is possible to find physical representations for the information, so that when processed with reversible logic, the energy of the output is equal to the energy of the input. It is suggested that there may be devices that will do that without applying any additional power. Here, we present a formalism that may be used to produce any reversible logic gate. We implement this method over an optical design of the Fredkin gate, which utilizes only optical elements that inherently conserve energy.

[1]Department of Computer Science, Ben Gurion University of the Negev, 653 Be'er, Sheva 84105, Israel. [2]Ilze Katz Institute for Nanoscale Science and Technology, Ben Gurion University of the Negev, 653 Be'er, Sheva 84105, Israel. Correspondence and requests for materials should be addressed to E.C. (email: koderex@gmail.com).

One of the greatest challenges in computing is to reduce the energy consumption, not only for the sake of reducing the tremendous amount of electric power used by the computer-based industry (for example, Google, Amazon or Facebook) but also to enable computing devices to operate at a higher frequency without melting as a result of the extra heat. A radical change in computer design may possibly lead to the crucial breakthrough needed for achieving more energy-efficient information processing. One promising direction for a break-through is to implement an all-optical[1] universal reversible gate, such as the Fredkin gate[2], and then use the Fredkin gate implementation as a building block in reversible circuits, thus overcoming this significant challenge.

The Fredkin gate is a universal reversible logic gate. Its universality means that any logic operation can be produced using only Fredkin gates. It has been suggested that reversible gates may preserve energy together with the data and reversible devices may be built entirely with energy-conserving elements[3,4]. These elements include a network of directional couplers and controlled nonlinear phased modulators. While the ability to build energy-conserving reversible gates has been discussed, no proof was given for the feasibility of these devices. In addition, no mathematical model has been presented which describes how to design such devices. Therefore, the search for a universal, energy-conserving, reversible logic device is still crucial for reversible computing and its beneficial prospects.

Research in the field of reversible computing is focused on two major aspects which include the implementation of reversible gates on one hand, and their usage for large-scale circuits on the other.

Extensive work has been done trying to utilize the Fredkin gate to perform simple or complex computing primitives. The assumption is that an ideal Fredkin gate is available and used to build efficient circuits. For example, a paradigm named 'Directed Logic'[5] is presented as an energy-efficient computation model used for Boolean gates. Later on, directed logic was used to build automata and circuits[6].

The other major aspect in the research of reversible computing suggests implementations for the Fredkin gate. Publications differ in their technological paradigms. Therefore, we will refer to them as the 'electronic approach', the 'all-optical approach' and the 'hybrid electro-optic approach'. Since electronic and hybrid devices rely on transistors and similar semiconductors paradigms they suffer from inherent energy loss.

Previous publications in the scope of the electronic approach suggested complementary metal-oxide semiconductor (CMOS) implementations[7]. The designs focus on an effort to minimize the energy loss and measure it on every logical state of the gate.

The hybrid electro-optic approach suggests that data and control signals may be partly optical and partly electronic. For example, implementations for 'directed logic' are suggested[8–10], where an electric control may be used for switching the two inputs to the appropriate two outputs. The control is based on manipulation of the resonance of a ring resonator by a silicon p-i-n junction built over the cross-section of the ring[11,12]. This approach requires the conversion of optical data to electronic controls, or electronic signals to optical ones.

The last approach suggests an all-optical approach, where all inputs and controls are optical. For example, it is suggested that light projection may be used to mechanically manipulate the positioning of molecules[13]. This minute manipulation may be very significant if it is done over a ring or a disc resonator. The movement may tune or detune the resonator and, in essence, implement a switch. This design is still not energetically efficient, since the optical energy is transformed into mechanical energy.

Other examples involve utilizing an effect called four-wave mixing (FWM). Under certain conditions in a nonlinear medium, a strong continuous wave signal on one wavelength may switch signals between two other wavelengths. When the continuous wave is absent no manipulation is done, and the signals continue to propagate with negligible modification. This technique is used to implement a Fredkin gate over the wavelength space[14]. A very high intensity is needed for the continuous wave, in order to exhibit the expected wavelength conversion, which also suffers from losses such as harmonics generation. However, it is possible to lower these losses[15] by utilizing coupled-resonator optical waveguides in order to enhance the FWM interaction. The main issue with FWM implementations is that all signals propagate in the same waveguide which requires additional separation, amplification or wavelength conversion in order to cascade the gates.

A different approach is to attempt to utilize quantum optics in order to produce quantum optical gates[16,17]. Although these devices are relatively simple in design, they require single-photon sources. While single-photon sources are becoming available[18], these devices are limited to work only with single-photon sources, making them difficult to integrate with non-quantum optical elements.

It should be noted that from a logic point of view, reversible gates are quantum gates and quantum gates are reversible gates. The two fields share similar concepts and mathematical approaches. However, while quantum gates require quantum implementations, reversible gates are not limited by that requirement, and our formalism and implementation do not refer necessarily to quantum mechanics or quantum implementations.

Here, we present a design for reversible gates built entirely from energy-conserving elements. First, we present our formalism, that utilizes linear and nonlinear unitary transformations in relation to conserving optical elements such as directional couplers. This formalism may be used by designers to design any reversible logic. Next, we use this formalism in order to design a Fredkin gate. We also simulate the different optical elements used such as directional couplers and graphene-induced silicon waveguide showing negligible energy loss. These results are later used to simulate our design of the Fredkin gate. All this effort was made while taking into account further circuit design. For example, making sure that all inputs and outputs are interchangeable without power or wavelength modification. Also, it will permit any cascading design of the circuits.

Results

Optical elements. When considering a computational machine, theoretical values need to be associated with a physical value. In an optical implementation for example, values used will refer to an amplitude of an electric field. The energy hiding in this field is proportional to the square of the magnitude of the electric field. To preserve the overall energy, all values may interact using unitary transformations which preserve the sum over the squares of the magnitudes.

Supplementary Notes 1 and 2 present a few basic available building blocks that manipulate the optical signals in a unitary fashion while Supplementary Fig. 1 demonstrates a representation of the relevant parameters of a general directional coupler. We are going to use these building blocks in order to design our devices. These elements include a linear waveguide which is used to introduce a phase to a signal, and a nonlinear Kerr-induced waveguide, which is used to introduce a nonlinear phase to a signal. Also, we will explain how a directional coupler or a Mach–Zehnder coupler performs a linear unitary transformation on two signals, while an array of such couplers may be used to perform a linear unitary transformation over a set of signals.

Unitary transform formalism. We would like to propose a general theory that allows the design and building of devices that solve a specific reversible logic gate. For example, we will explain the steps needed to design the reversible Fredkin gate. A truth table of the Fredkin gate is given in Table 1. The inputs of the device are considered in a straightforward manner, where each input channel represents a logical input value of the gate. In binary gates, each of these inputs may hold two complex values E_0 and E_1, which are amplitudes that would represent the states 0 and 1, respectively. Similarly, the outputs of the device may also hold the same values as those determined by the gate logic.

Given a state of the gate, we can arrange the inputs for each input channel in a column vector. The different vectors may be arranged side-by-side to build a matrix. This will be the input matrix, which we will denote with A. Similarly, we can arrange the outputs for each output channel in a column vector and arrange these vectors side-by-side to build an output matrix, which is denoted by B.

Our goal is that when the device receives an input that is a column of A, it will produce an output corresponding to an appropriate column of B. The number of columns on both matrices is the number of states the gate supports. The number of rows in A is the number of input channels and the number of rows in B is the number of output channels. Notice that if the gate is reversible, the number of input and output channels is the same, and so are the dimensions of A and B.

If we can find a unitary matrix M that satisfies $B = MA$, we can stop here and build the device using a coupling array that decomposes the matrix. Details on unitary matrix decomposition is given in Supplementary Notes 3 and 5, while Supplementary Fig. 2 demonstrates a coupling array. However, generally, such unitary matrices may not be found, particularly in the case of Fredkin gate, merely because of the fact that the number of columns is greater than the number of rows of A and B. That would mean that the dimensions of M are too small and the degrees of freedom are not enough to solve the equation.

To overcome this problem, we may add rows to both A and B. We denote the matrix added to A with F' and the matrix added to B with F. F is part of the product of the multiplication, while after a small modification it will turn into F', which is then used as part of the multiplier. The transformation from F to F' is done with no loss. Hence, the prime represents a little modification to F.

In other words, we are looking for a unitary matrix M that solves:

$$\begin{pmatrix} B \\ F \end{pmatrix} = M \begin{pmatrix} A \\ F' \end{pmatrix} \qquad (1)$$

A schematic design of the presented setup is given in Fig. 1. The design illustrated presents three inputs, three outputs and four feedback loops with a nonlinear element. Note that all presented elements relate to the matrices A, B, F and F', which hold the electric field over each channel at different states. The row index of the matrices defines the channel and the column

index represents the state. The gate illustrated in Fig. 1a exhibits coupling between seven channels denoted by M. This seven-channel coupling is designed with 21 adjacent couplings illustrated in Fig. 1b. Also, each F channel is manipulated through a nonlinear effect denoted by γ to become F'.

Note that in Fig. 1b, there is no notation for the different channels, it is not described which inputs hold the A channels or the F' channels or which outputs hold the B channels or the F channels. We are free to choose any permutation of these channels as this permutation can be described by a unitary matrix that may be integrated as part of M. A permutation should be chosen such that the decomposition of M will be the easiest to manufacture, taking into account manufacturing properties and limitations.

Two questions are raised by adding the rows of F' and F to A and B. The first regards how many rows need to be added to A or B in the form of F' and F respectively. The answer is that we cannot predict the dimensions of F, and each case should be investigated for its properties. The second regards whether it is possible to have feedback with no nonlinear elements, or is the nonlinearity obligatory.

We can try to use the relation between A, B, F and F' through the unitary matrix M and get:

$$B^\dagger B - A^\dagger A = F'^\dagger F' - F^\dagger F \qquad (2)$$

Note that in this expression M is eliminated. The left-hand side of the equation is determined only by the state representation of an examined gate. Also, note that if the gate is conserving energy for every state, the diagonal of the left-hand side is filled with zeros. Moreover, if some elements of the left-hand side are not zero, this means that a nonlinear interaction must come into play in order to manipulate F to be F', such that this condition is satisfied.

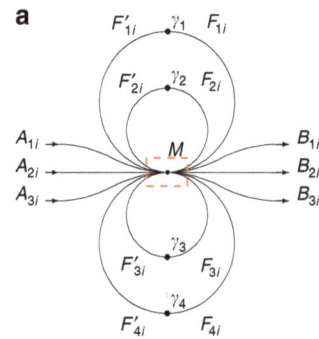

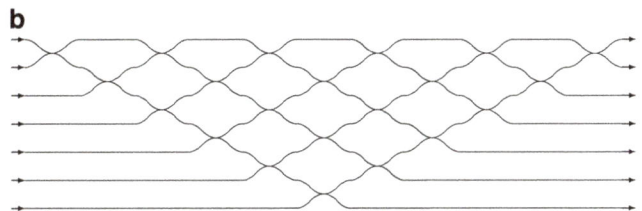

a

b

Figure 1 | A representation of a three channel gate. (**a**) This example shows three inputs denoted by A, and three outputs denoted by B. Four more channels are used for feedback, and denoted by F and F'. The coupling is between seven channels and denoted by M and a red broken box. (**b**) A zoom on the coupling M. This coupling array couples seven channels: the inputs are three A channels and four F' channels. Similarly, the outputs are three B channels and four F channels. This coupling is decomposed to 21 couplings between adjacent channels. When F' is added as an input for the coupling, F is added to the output of the coupling. Each F is then manipulated with an appropriate nonlinear element denoted by γ, which transforms it to F'. Then, F' is fed back to the system. The column index i represents different states.

Table 1 | A truth table of the reversible Fredkin gate.

C	X₁	X₂	C'	Y₁	Y₂
0	0	0	0	0	0
0	0	1	0	0	1
0	1	0	0	1	0
0	1	1	0	1	1
1	0	0	1	0	0
1	0	1	1	1	0
1	1	0	1	0	1
1	1	1	1	1	1

X_1, X_2 and C represent three input bits, while Y_1, Y_2 and C' represent three output bits.

Note, however, that this equation was derived from equation (1), but it holds less information. A solution for equation (1) is therefore more inclusive and we are going to focus on it.

While different solutions for equation (1) may be found, it does not mean that all these solutions may converge to a desired output given an input. In fact, it only means that the desired states are steady states. These states may also be weak steady states, that might diverge after a short time period. Supplementary Note 4 discusses the issue of convergence and its time period. Supplementary Notes 6 and 7 also discusses different representations for logical data using optics, giving examples of the XOR and Fredkin gates.

Numerical solution for Fredkin gate. The terms and conditions formulated were implemented in Matlab, where numerical solutions were found for the different elements of the Fredkin gate. A solution for equation (1) was found, where M is later decomposed to an array of four 3×3 matrices, while the design utilizes three nonlinear elements. This design is given in Fig. 2. Note that feedback was eliminated with this design, which allows the interaction of pulsed light and not necessarily based on continuous wave. The derivation of this design is further discussed in the methods section. The minimal propagation time through this element is proportional to the sum of the gamma values which was minimized to $S_\gamma = 12.5$. Additional information about the derivation of this solution is given in Supplementary Note 8, while Supplementary Fig. 3 shows an intermediate stage in the decomposition of the design.

Optical elements. Finite-difference-time-domain (FDTD) simulations are done over simple elements that include a symmetric waveguide coupler and a waveguide infused with a Kerr nonlinear element. Subsequently, these elements are used as building blocks for photonic integrated circuits.

All the simulations assumed rectangular waveguides of silicon (Si) layered over glass substrate. The waveguide had a width of

400 nm and height of 220 nm. The simulations were done with a wavelength of $\lambda = 1.5\,\mu m$, since it exhibits negligible absorption in Si and low bend loss in single-mode waveguides when utilizing the recent advances in CMOS production[19,20]. Note, however, that Si single-mode waveguides produced utilizing these methods still exhibit a relatively high-propagation loss of $0.3db\,cm^{-1}$. In our discussion, we will remain with typical lengths that are overall smaller than 1 cm. Also, we may consider using different materials, resulting in a lower loss. All simulations refer to the fundamental propagating mode that this waveguide can support. An illustration of the cross-section is given in Fig. 3.

Note that this waveguide profile supports only single-mode propagation. The sources used in the simulation stimulate only this mode, while the monitors that read the corresponding outputs compare their results to this mode. It is assumed that only this mode may propagate and all discussions refer to this mode.

While many simulations, fabrications and tests on directional couplers and Mach–Zehnder interferometers were done and produced significant results, we are still required to produce our own design and results. First, we intend to investigate the losses presented by these elements. Moreover, we choose a specific wavelength and a specific cross-section throughout the design for all waveguides. These waveguides should match the simulations done for the Kerr-induced waveguide. The desired data and parameters may be problematic to obtain based solely on past literature; hence, these elements must be simulated again. Details of these simulations are given in the methods section and Supplementary Note 9. In addition Supplementary Fig. 4 shows the schematic design of the coupler used in our simulations.

A set of directional couplers were simulated to characterize their properties. An example of a propagation process in our design of the directional coupler is given on Supplementary Fig. 5. The results show a loss lower than 0.3% in the energy, or $-0.013db$ for all the range of elements simulated. Supplementary Fig. 6 illustrates the derived coupling values as a function of the bend radius.

This element is later used to build a Mach–Zehnder interferometer by integrating two 50% directional couplers. The results for the Mach–Zehnder interferometer show a loss lower than 0.3% in the energy. An example of a propagation process in our design of the Mach–Zhender coupler is given on Supplementary Fig. 7.

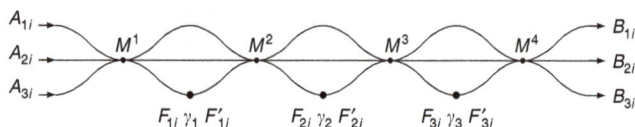

Figure 2 | An implementation of the Fredkin gate where feedbacks are eliminated. This design uses four coupling nodes. All nodes couple three channels. Three channels are used for nonlinear manipulation and denoted by F and F'. Each F is then manipulated with an appropriate nonlinear element denoted by γ, which transforms it to F'. The column index i represents different states.

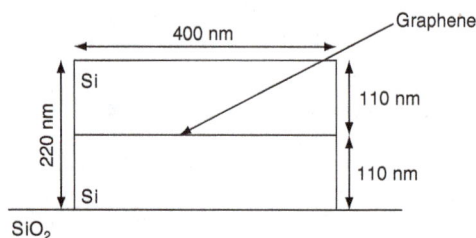

Figure 3 | An illustration of a waveguide cross-section. The waveguide is rectangular silicon over a glass substrate. The width of the waveguide is 400 nm and its height is 220 nm. When needed, a graphene sheet is included into the waveguide centred at the middle with a height (or thickness) of 1 nm. The graphene is included only when the Kerr effect is needed, while it is not included otherwise.

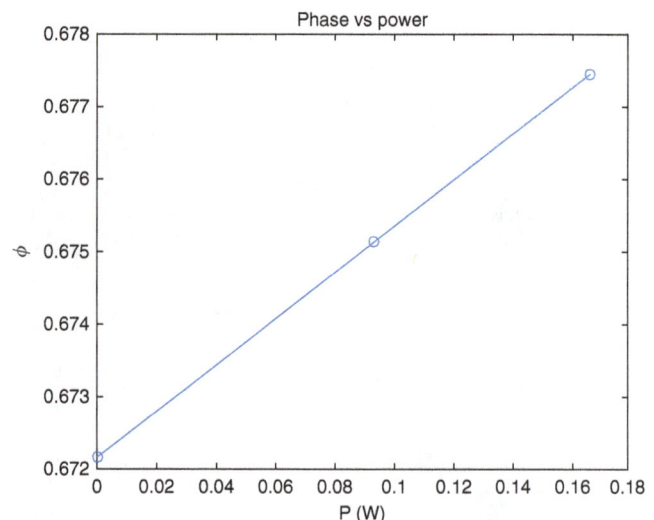

Figure 4 | Phase as a function of the power. A minor phase response is observed when the power propagating through the waveguide is increased. The waveguide includes graphene with a thickness of 1 nm. The length of the Kerr element is 5 μm.

A graphene infused waveguide was also simulated. The length chosen for the Kerr element simulation was $L = 5 \,\mu m$. The results show a negligible loss, lower than 0.07% in the energy, or $-0.003db$. The phase response is given in Fig. 4. It shows a relatively high response giving:

$$\frac{\gamma}{L} = 6.31 \times 10^3 \left[W^{-1} m^{-1} \right] \tag{3}$$

As explained earlier, a sum of the nonlinear values was calculated as $S_y = 12.5$. This result corresponds to a length of $L = 2\,mm$ and a converging time of $T = 2 \times 10^{-11}$ s when using 1 W sources. A convergence energy $E_c = 2 \times 10^{-11}$ J. Note of course that a faster system convergence may be achieved with higher power and designing shorter elements. Also, low-power devices may be produced with longer elements at the expense of the system convergence time.

Integrated circuit. The elements investigated in the FDTD simulation were now used for an integrated circuit simulation. The allows incorporating results acquired with 'Lumerical FDTD Solutions' for a specific basic optical element. These elements may be inter-connected into a simulation of one integrated device. It also provides many built in optical elements such as sources, detectors, Mach–Zehnder coupler, phase and delay lines which may be used together with customized scripted elements, and other integrated compiled elements.

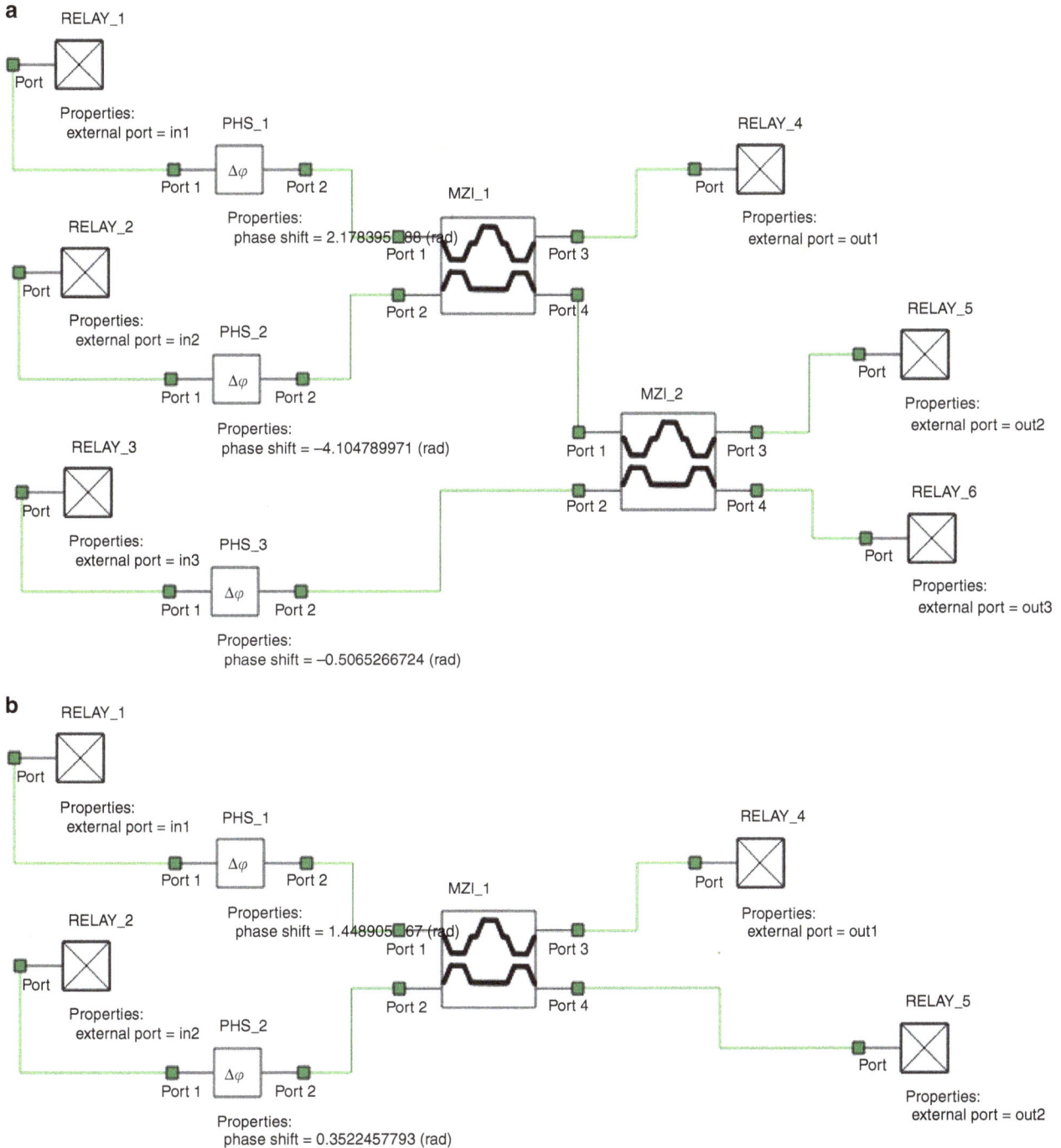

Figure 5 | Compiled elements that perform a unitary transformation. (**a**) An expansion of the compiled element 'mzi-comp' that includes two MZI couplers and three phase elements. (**b**) An expansion of the compiled element 'mzi-ex' that includes one MZI coupler and two phase elements.

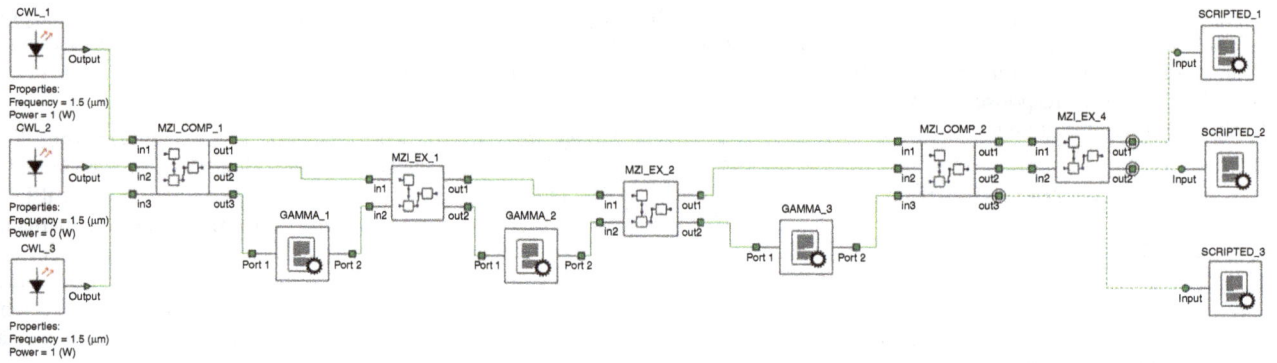

Figure 6 | Schematics of the device suggested. This scheme uses seven Mach–Zehnder couplers, and phase shift elements. This scheme uses compiled elements 'mzi-comp' and 'mzi-ex' in order to simplify the display of the design. Also, the nonlinear Kerr elements were created and are labelled 'gamma'.

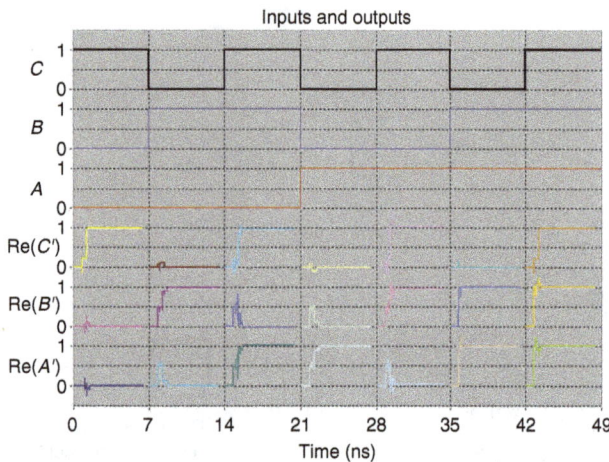

Figure 7 | Timing sequences of the Fredkin gate. Input values and output results are presented across time. The three input values A, B and C are shown together with the real values of the outputs A', B' and C'. As expected from a Fredkin gate, the value of the output C' is very similar to the value of the input C. When C = 0, the values of A' and B' are similar to the values of A and B respectively. When C = 1, the values of A' and B' are similar to the values of B and A, respectively. Note the time delay of the outputs as a result of the propagation delay over the optical elements.

First, in order to simplify the scheme of the design we compile and script a 2×2 unitary transformation by integrating a Mach–Zehnder coupler with phase elements into one compiled element titled 'mzi-ex'. The script selects the proper values for the coupler and phase elements such that the compiled element performs a user defined unitary transformation. Similarly, we compile and script a 3×3 unitary transformation by integrating two Mach–Zehnder couplers with phase elements into one compiled element titled 'mzi-comp'. These elements are displayed in Fig. 5. All couplings were achieved with Mach–Zehnder coupler elements with two 50% couplers and typical phase lengths of 15 μm.

The schematics of the Fredkin gate device are given in Fig. 6. These schematics include the compiled elements from Fig. 5. Three sources were used as input with the power values of zero or 1 W representing the logical 0 and 1. The nonlinear Kerr elements were scripted to produce a phase shift proportional to the power. Output values were collected with scripted elements (titled 'scripted') allowing us to collect phase information together with the intensity.

The integrated circuit was tested with all possible inputs of the Fredkin gate. First, a simulation was done using ideal elements with no loss. The results show almost no deviation from the desired output.

Next, losses were introduced, according to the results of the FDTD simulations, where 0.3% of energy loss was applied over all Mach–Zehnder interferometers. The result of the integrated circuit simulation are displayed on Fig. 7 which displays the timing sequence for the different possible signals of the Fredkin gate where A, B and C are three input signals and A', B' and C' are the three output signals. These results exhibit a small deviation from the desired values. The highest error introduced to the results occurred on the $ABC = 101$ state, where the normalized result of the B' channel was $E_{B'} = 0.961 - 0.044i$ instead of the value 1, which is a 6% deviation over the complex plane.

Discussion

A formalism for the design of reversible circuits was presented. This formalism utilizes only unitary transformations. Linear unitary transformations were represented by a unitary matrix. The definition of unitary transformations was extended to nonlinear transformations by maintaining that the sum over all the squares of the magnitudes is preserved.

The formalism suggested provides a theoretical condition over the reversible gate truth table, to determine if the specific logic may be implemented only by linear elements, or whether it is required to use nonlinear elements.

Optical elements that maintain these unitary properties were presented and tested. A directional coupler or a Mach–Zehnder coupler are used to represent a 2×2 unitary matrix transformation. An array of these couplers may be used for a general unitary matrix transformation.

Also, a Kerr-induced waveguide was simulated. By using graphene as a Kerr material, the effect becomes very strong and shorter elements may be used. This element provides the nonlinear unitary transformation by inducing self phase modulation. Still, since Kerr interactions are usually weak the length needed for these elements is relatively high compared with other elements in the system. This length will determine the propagation time needed in order to achieve the required result.

Simulation over all optical elements show a very low loss, which is almost negligible. This demonstrates the superiority of optical elements in energy efficiency.

We used the formalism presented in order to determine the needed elements for a Fredkin gate. We tried to minimize the length of the nonlinear elements by minimizing the sum over the nonlinear values S_y. We also managed to eliminate feedback loops in the design, allowing the usage of pulsed light and not only continuous waves.

The integrated circuit simulation provided a promising result. The 6% deviation will still allow a cascading usage of the device in larger scale logic devices. Assuming the worst-case scenario where a 6% deviation is accumulated in a geometric progression, it is still possible to cascade the device five times before corrections with thresholds are needed.

This 6% deviation in the amplitude translates to a 7.5% energy loss. If we modify the device to support short pulses, a 1 W source with a pulse time of 50 fs is 10 wavelengths long, and it holds an energy of 50 fJ. The loss in that case is lower than 4 fJ, where we may assume that most of it probably propagates in the wrong channel rather than heating the system. A lower power source may be used to lower the loss even further; however, this will demand a longer propagation in the nonlinear element. Hence a lower power and lower loss may be achieved at the expense of a longer delay.

Also, this deviation may be lowered by modifying the numerical solution. This simulation was intended to assess the deviation of the output from an ideal device given losses. The results showed quasi-random deviation from the desired output. However, the numerical solution considered only ideal elements, and it may be modified to consider losses, such that the outputs will exhibit the same, yet minimal, power loss. These in turn may be used for other Fredkin gates that consider a lower input power as the normalized logical values.

We presented an optical architecture for reversible logic that does not dissipate heat and, while converged to a state, does not waste energy. The new mathematical framework provided here can be implemented for any reversible logic gate. The design was based on three basic energy-conserving elements: linear phase; coupling; and nonlinear manipulation.

The design is not limited to conventional optics, as different physical phenomena may be used. For example, electron optics may be considered. Another physical approach may suggest that the different inputs may be held on the same channel but on different wavelengths. Coupling between different wavelengths is possible through FWM, which was demonstrated efficiently over graphene[21].

The device only utilizes elements that conserve energy, namely waveguides and conserving couplers as linear elements and Kerr effect materials for nonlinear manipulations. The linear coupling may be achieved with different technologies such as a mode-waveguide coupler, or Mach–Zehnder coupling. The Kerr effect was chosen since it is energy-conserving, is widely available with different materials, and is easily formulated. However, the Kerr effect may be replaced by a different nonlinear, energy-conserving effect. Different effects and materials may exhibit significantly stronger interactions, which would allow lower energy loss and higher working frequency.

While converging, the wasted energy is not transformed to heat but rather is removed from the system through the output ports. This energy may be reused. For example, it may be redirected back to induce population inversion which may be used at a laser source.

For ease of manufacturing, a small number of couplings should be used. In our case, the proposed Fredkin gate utilized only seven couplings. A relatively long time is needed to allow propagation through the nonlinear element. This is a result of the extremely weak interaction of the Kerr effect. To achieve better performance, a shorter wavelength should be used. Better performance may be achieved by the discovery and usage of materials with stronger Kerr interactions. Stronger Kerr interactions may be available with future investigation of graphene. It may even be possible to use a totally different effect for the nonlinear interaction.

The example given for the Fredkin gate used a binary representation for each input or output channel. It may be possible to design gates where each channel may hold more than two values. These values may also be complex.

Note that states of the device were defined by a complex value for the electric field and the Kerr effect was dependent on a continuous wave. Future research may try to redesign the device such that it may support short pulses of light. This may be done by carefully choosing the lengths of the elements, or the optical lengths, and allowing signals that interact with each other to arrive and interact with each other at the appropriate point in time and space. By using pulses, however, special attention should be given to the shape of the pulse when dispersion and the Kerr effect may destroy the shape and introduce undesired noise. It may be possible that a soliton solution may be found while considering the presence of these effects.

Although our device introduces long time delay, a device that supports pulses may not necessarily suffer from bandwidth loss. This is because the device may be used asynchronously, while interaction with the pulses does not modify the shape of the pulse.

This device may serve as an intermediate step towards reversible optical quantum computing. The next step may be to design the devices that incorporate energy-efficient elements together with elements that are governed by quantum effects, while supporting one-photon interaction together with two-photon, three-photon or four-photon interactions.

Another approach may use the formalism and design described and integrate it with a semiconductor optical amplifier only to compensate for the minor optical losses. This may lead to a simpler design while maintaining a very low-power consumption.

Methods

Fabrication feasibility. When we deal with the available building blocks, we have to remember that we want to build devices with negligible loss. In other words, the sum of the intensity over all inputs should be nearly the same as the sum of the intensity over all outputs.

Note that negligible loss is possible to maintain providing the advances in fabrication. An 'ultra-high-Q' optical resonator has been presented[22], featuring a design and fabrication process that managed to produce a toroid microcavity resonator on a chip with a Q factor in excess of 100 million. This suggests that loss is negligible and was eliminated by the advanced fabrication methods described.

The formalism presented deals with an ideal theory behind the device. Several parameters and assumptions were chosen and are given in Supplementary Note 10. In the results section, we tested different optical elements, to challenge these assumptions. We tested the feasibility of physically implementing this theory and ultimately demonstrated that these assumptions are almost entirely correct.

Numerical solution. Matlab was used to solve equation 1. A solver was used to find the parameters of the matrix M, under the conditions that it is a unitary matrix. The matrix M was also confined to the convergence conditions given in Supplementary Note 4. The values of F are dependent on the values of M. The values of F' were chosen according to the relation $F'_{jk} = F_{jk} \exp[i\gamma_j |F_{jk}|^2]$, where γ_j is the scaled Kerr coefficient. This solver also provided the values of γ_j while optimizing the solutions such that the sum $\sum_j \gamma_j$ will be minimized.

Another Matlab script was used to decompose the matrix M into arrays of 2×2 unitary transformations. These values were later used in the simulation of the integrated device.

Optical simulations. Simulations were done in two stages. First, the different optical elements were tested using 'Lumerical FDTD solutions'. The simulations included directional couplers, Mach–Zehnder couplers, and Kerr elements. Details about the FDTD simulations of the different optical elements is given in Supplementary Note 9.

Later, a simulation of the integrated circuit using 'Lumerical Interconnect'. 'Lumerical Interconnect' allows incorporating results acquired with 'Lumerical FDTD Solutions' for a specific basic optical element. 'Lumerical Interconnect' also allows the scripting of its elements which was necessary for elements such as general unitary couplers, Kerr elements and the different detectors.

References

1. Caulfield, H. J. & Dolev, S. Why future supercomputing requires optics. *Nat. Photon.* **4**, 261–263 (2010).
2. Fredkin, E. & Toffoli, T. Conservative logic. *Int. J. Theor. Phys.* **21**, 219–253 (1982).
3. Shamir, J., Caulfield, H. J., Micelli, W. & Seymour, R. J. Optical computing and the fredkin gates. *Appl. Opt.* **25**, 1604–1606 (1986).
4. Zavalin, A. I., Shamir, J., Vikram, C. S. & Caulfield, H. J. Achieving stabilization in interferometric logic operations. *Appl. Opt.* **45**, 360–365 (2006).

5. Hardy, J. & Shamir, J. Optics inspired logic architecture. *Opt. Express.* **15,** 150–165 (2007).
6. Anter, A., Dolev, S. & Shamir, J. *Optical Supercomputing.* pp 92–104 (Springer, 2013).
7. Vasudevan, D. P., Lala, P. K., Di, J. & Parkerson, J. P. Reversible-logic design with online testability. *IEEE Trans. Instrum. Meas.* **55,** 406–414 (2006).
8. Tian, Y. *et al.* Proof of concept of directed or/nor and and/nand logic circuit consisting of two parallel microring resonators. *Opt. Lett.* **36,** 1650–1652 (2011).
9. Xu, Q. & Soref, R Reconfigurable optical directed-logic circuits using microresonator-based optical switches. *Opt. Express* **19,** 5244–5259 (2011).
10. Zhang, L. *et al.* Simultaneous implementation of xor and xnor operations using a directed logic circuit based on two microring resonators. *Opt. Express* **19,** 6524–6540 (2011).
11. Xu, Q., Manipatruni, S., Schmidt, B., Shakya, J. & Lipson, M. 12.5 Gbit/s carrier-injection-based silicon micro-ring silicon modulators. *Opt. Express* **15,** 430–436 (2007).
12. Xu, Q., Schmidt, B., Pradhan, S. & Lipson, M. Micrometre-scale silicon electro-optic modulator. *Nature* **435,** 325–327 (2005).
13. Roy, S., Sethi, P., Topolancik, J. & Vollmer, F. All-optical reversible logic gates with optically controlled bacteriorhodopsin protein-coated microresonators. *Adv. Opt. Technol.* **2012,** 727206 (2012).
14. Gui, C., Wang, J., Xiang, C., Liu, Y. & Li, S. *Asia Communications and Photonics Conference* AS3G–AS33 (Optical Society of America, 2012).
15. Morichetti, F. *et al.* Travelling-wave resonant four-wave mixing breaks the limits of cavity-enhanced all-optical wavelength conversion. *Nat. Commun.* **2,** 296 (2011).
16. Tasca, D. S., Gomes, R. M., Toscano, F., Ribeiro, P. H. S. & Walborn, S. P. Continuous-variable quantum computation with spatial degrees of freedom of photons. *Phys. Rev. A.* **83,** 052325 (2011).
17. Milburn, G. J. Quantum optical fredkin gate. *Phys. Rev. Lett.* **62,** 2124 (1989).
18. Lukishova, S. G. *et al.* Organic photonic bandgap microcavities doped with semiconductor nanocrystals for room-temperature on-demand single-photon sources. *J. Mod. Opt.* **56,** 167–174 (2009).
19. Cardenas, J. *et al.* Low loss etchless silicon photonic waveguides. *Opt. Express* **17,** 4752–4757 (2009).
20. Vlasov, Y. & McNab., S. Losses in single-mode silicon-on-insulator strip waveguides and bends. *Opt. Express* **12,** 1622–1631 (2004).
21. Gu, T. *et al.* Regenerative oscillation and four-wave mixing in graphene optoelectronics. *Nat. Photon.* **6,** 554–559 (2012).
22. Armani, D. K., Kippenberg, T. J., Spillane, S. M. & Vahala, K. J. Ultra-high-q toroid microcavity on a chip. *Nature* **421,** 925–928 (2003).

Acknowledgements

We thank Joseph Shamir from the Technion—Israel Institute of Technology for discussions; S.D. is partially supported by the Rita Altura Trust Chair in Computer Sciences, and the Israel Science Foundation (grant 428/11).

Author contributions

E.C. led the design of the formalism, the design of the simulations and the writing of the manuscript, in all of these processes, the other two authors were also consistently contributing ideas, guidance and supervision.

Additional information

Competing financial interests: The authors declare no competing financial interests.

26

Magnetic-free non-reciprocity based on staggered commutation

Negar Reiskarimian[1] & Harish Krishnaswamy[1]

Lorentz reciprocity is a fundamental characteristic of the vast majority of electronic and photonic structures. However, non-reciprocal components such as isolators, circulators and gyrators enable new applications ranging from radio frequencies to optical frequencies, including full-duplex wireless communication and on-chip all-optical information processing. Such components today dominantly rely on the phenomenon of Faraday rotation in magneto-optic materials. However, they are typically bulky, expensive and not suitable for insertion in a conventional integrated circuit. Here we demonstrate magnetic-free linear passive non-reciprocity based on the concept of staggered commutation. Commutation is a form of parametric modulation with very high modulation ratio. We observe that staggered commutation enables time-reversal symmetry breaking within very small dimensions ($\lambda/1{,}250 \times \lambda/1{,}250$ in our device), resulting in a miniature radio-frequency circulator that exhibits reduced implementation complexity, very low loss, strong non-reciprocity, significantly enhanced linearity and real-time reconfigurability, and is integrated in a conventional complementary metal–oxide–semiconductor integrated circuit for the first time.

[1] Department of Electrical Engineering, Columbia University, 1300 South West Mudd, 500 West 120th Street, New York, New York 10027, USA. Correspondence and requests for materials should be addressed to H.K. (email: harish@ee.columbia.edu).

Reciprocity in electronics or, equivalently, the principle of time reversibility in optics is a fundamental property of any linear system or material described by symmetric and time-independent permittivity and permeability tensors[1]. Non-reciprocity, however, enables new applications that span radio frequencies (RF) to optical frequencies. Optical isolators are critical to on-chip all-optical information processing systems for the protection of lasers and amplifiers, and the mitigation of multipath reflections. RF circulators enable full-duplex wireless, an emerging wireless communication paradigm that has been historically considered impractical[2], where the transmitter and the receiver operate simultaneously on the same frequency band, potentially doubling network capacity at the physical layer while offering numerous other benefits at the network layer[3,4].

Non-reciprocal components today are almost exclusively realized through the magneto-optic Faraday effect (Fig. 1a–c)—the application of a magnetic field bias parallel to the direction of propagation rotates the polarization vector of light due to the different propagation velocities of left- and right-circularly polarized waves[5]. Despite significant research efforts in the optical[6,7] and RF domains[8-10], non-reciprocal components based on magneto-optic materials remain incompatible with complementary metal–oxide–semiconductor (CMOS) integrated circuit (IC) fabrication processes due to material incompatibilities and the need for a magnetic field bias, significantly restricting their impact.

As early as 50 years ago, magnetic-free non-reciprocity and circulators had been investigated at microwave frequencies through the use of the inherent non-reciprocal nature of active devices such as direct-current/voltage-biased transistors[11]. More recently, metamaterials with embedded active transistors have been explored at RF and microwave frequencies[12-14]. However, such approaches are fundamentally limited by the noise and

nonlinear distortion generated by the active transistors. Indeed, for non-reciprocal components used at the front end of RF communication systems (such as circulators), and for front-end components more broadly (such as filters and diplexers), passive approaches with superior linearity and noise performance are paramount[15,16]. Techniques for non-reciprocity leveraging nonlinearity[17-21] have also been proposed. These techniques, although potentially useful and extensively explored in optical applications, exhibit non-reciprocity over certain signal power levels only and have limited applicability to scenarios where linearity to the signal is required, such as RF communication. Lorentz reciprocity may also be broken by introducing time dependency into the material or system[1]. Techniques at optical frequencies based on electro-optic phase modulators have been investigated[22,23] but require complex electro-optic modulation networks. Approaches based on spatio-temporal parametric modulation of waveguides have been investigated[24-28]. In such approaches, a travelling-wave modulation of the waveguide's properties produces direction-dependent mode conversion of the desired signal. In the optical domain, the size of the structure is limited by the weak electro-optic or acousto-optic effect[24-26], which results in an extremely low modulation ratio (modulation ratio is defined as the ratio between the maximum and minimum values of the modulated parameter and extremely weak modulation corresponds to a modulation ratio of practically unity). At RF, varactors are able to achieve modulation ratios of around two to four, resulting in structures that are of the order of a wavelength[27]. A second disadvantage of these approaches is that the mode/frequency conversion is undesirable in some applications and necessitates the use of filters[24] or diplexers[27]. In recent times, non-reciprocity through spatio-temporal parametric modulation of resonant rings, resulting in angular momentum biasing[29], has been demonstrated in the acoustic domain[30] and at

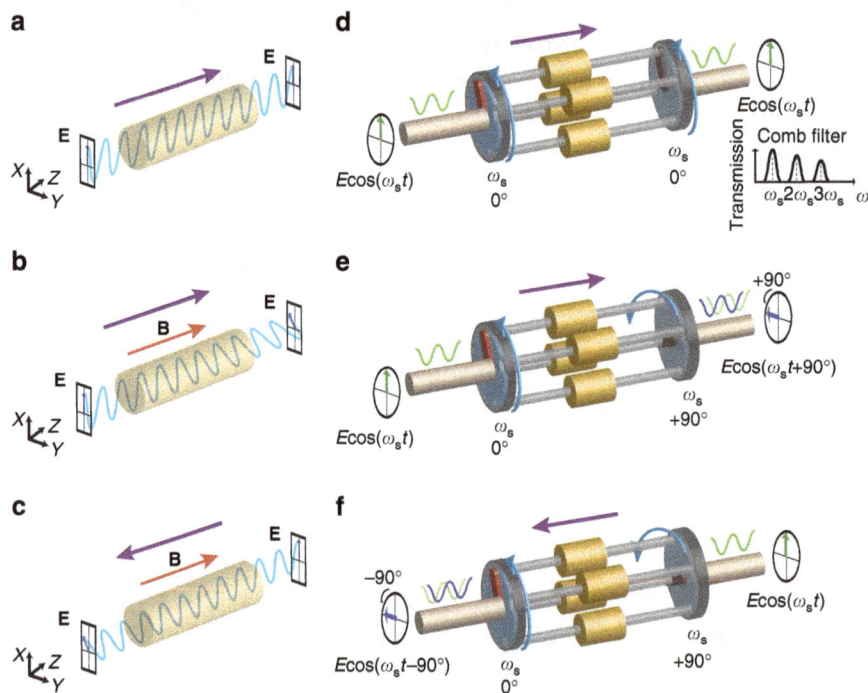

Figure 1 | Comparison between non-reciprocity induced by the magneto-optic Faraday effect and staggered commutation. (a) A wave propagating in a Faraday-active magneto-optic material experiences no Faraday rotation in the absence of magnetic bias. **(b)** In the presence of magnetic bias in the positive z direction, a wave travelling in the positive z direction experiences Faraday rotation due to the difference in the propagation velocities of right-handed and left-handed circularly polarized waves, whereas **(c)** a wave travelling in the negative z direction experiences an opposite rotation. **(d)** A wave propagating through a commutated network with no staggering experiences no phase shift. **(e)** Staggered commutation acts as a bias that breaks time reversibility, producing a phase shift for waves travelling from left to right and **(f)** an opposite phase shift for waves travelling from right to left.

RF[31], resulting in magnetic-free non-reciprocal circulators at or well below the wavelength scale and with low loss[32]. A key challenge with these spatio-temporal parametric modulation approaches in general is that the property that enables modulation (for instance, varactors[27,31] or opto-acoustic interactions[25]) often represents a nonlinearity to the signal itself, in particular when the modulation ratio is high, resulting in nonlinear distortion at higher signal levels. There has also been theoretical work on non-reciprocity and a microwave circulator based on parametric modulation of coupled resonators[33].

Here we introduce magnetic-free, linear and passive phase non-reciprocity based on staggered commutation and a highly-miniaturized RF circulator that embeds the phase-nonreciprocal component within a ring resonator. Commutation may be seen as a form of parametric modulation with very high modulation ratio (practically infinite in our prototype), making the phase-nonreciprocal component effectively a point parametric modulator. This enables its specific placement in the ring relative to the three circulator ports such that the signal across it is suppressed for excitations from one port, yielding significantly enhanced linearity to that port. Furthermore, the need for spatio-temporal modulation in the ring is eliminated, easing modulation complexity. Implementation complexity is further eased as no frequency conversion is seen at the circulator ports. In addition, when compared with prior art employing spatio-temporal parametric modulation of wave-guides[24–27] or electro-optic phase modulation[22,23], these concepts allow miniaturization of the unmodulated ring to deeply subwavelength scales ($\approx \lambda/80$ in our prototype). When compared with prior art employing angular-momentum biasing in parametrically modulated resonant rings[31,32], linearity enhancement is achieved due to the ability to suppress the signal across the point parametric modulator, although it is interesting to consider synergies between the approaches offering the dual benefits of both.

Results

Phase non-reciprocity through staggered commutation.
Commutated networks are a class of linear, periodically time-varying (LPTV) networks where the signal is periodically commutated through a bank of linear, time-invariant (LTI) networks (or media as in Fig. 1d). The first commutated networks relied on mechanical commutation through a rotating brush that periodically contacted a bank of capacitors[34,35] to realize narrow comb filters around harmonics of the commutation frequency. More recently, electronic commutation using passive transistor-based switches has resulted in high quality-factor (Q) comb filters, commonly called N-path filters, which operate at RF, exhibit significantly lower noise and higher linearity when compared with active RF filters, are compatible with conventional CMOS IC

technology and exhibit the potential to replace front-end off-chip surface acoustic wave filters for RF communication applications[15,36–40]. A key requirement is the availability of a high-quality switch with high ON/OFF transmission ratio (or modulation ratio when viewed as parametric modulation) and sufficient switching speed. Modern CMOS transistor switches boast ON/OFF conductance ratios as high as 1,000–100,000 (ref. 41), enough to be practically infinite from the perspective of commutated network operation, and switching speeds that enable such commutated networks to operate well into the RF frequency range and continuously improve with CMOS technology scaling. It should be emphasized that the transistors are used as reciprocal, highly linear, passive switches without direct-current bias. In other words, the switches cannot provide power gain to the desired signal by sourcing power from the modulating clock signals that control the commutation. These clocking signals that control the commutation are easily implemented in CMOS and the associated power consumption is due to the charging and discharging of parasitic capacitance in the clock path. This power consumption also reduces with CMOS technology scaling due to reducing parasitics.

An interesting property of commutated networks that we observe here is that staggering the commutation on either side of the bank of LTI networks results in time-reversal symmetry breaking and phase non-reciprocity (Fig. 1e,f). Waves travelling in the forward and reverse direction see a different ordering of the first and second commutating switches in time and experience opposite shifts in phase, essentially analogous to the magneto-optic Faraday effect. Here, the staggered commutation plays the role of the magnetic Faraday bias.

When the first and second set of switches are staggered by $+90°$, forward and reverse travelling waves at or near the commutation frequency experience phase shifts of $+90°$ and $-90°$, respectively. When a transmission line or waveguide of length $3\lambda/4$ is wrapped around the staggered commutated network, non-reciprocal wave propagation is achieved as waves may propagate in only one direction (Fig. 2a,b). In that direction, the $-270°$ phase delay through the $3\lambda/4$ ring adds with $-90°$ phase shift of the staggered commutated network to satisfy the boundary condition, enabling wave propagation. In the other direction, the $-270°$ phase delay adds with the $+90°$ phase shift of the staggered commutated network to prohibit wave propagation.

Unidirectional wave propagation not only requires phase non-reciprocity but also requires (near-)perfect transmission through the commutated network. In Fig. 3, we examine the requirements on the media across which commutation is being performed. We consider electrically short transmission-line media, in line with our goal of achieving a point parametric modulator. In the depicted simulations, eight-way commutation is considered, the transmission-line characteristic impedance (Z_{medium}), wave velocity (v) and length (l) are varied, and ideal sets of switches are

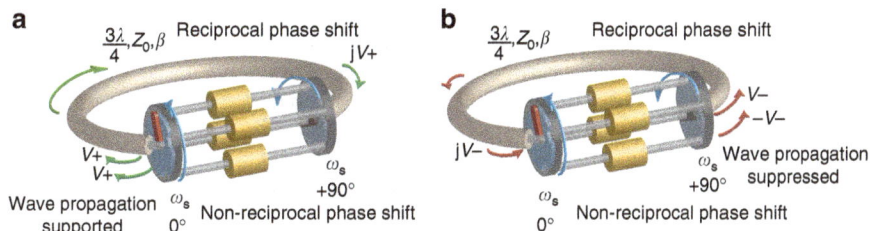

Figure 2 | Embedding the staggered commutated network within a $3\lambda/4$ transmission-line ring results in unidirectional wave propagation. (**a**) When the commutated network with $+90°$ staggering is embedded in a $3\lambda/4$ ring, then in one direction, the $-270°$ phase delay of the ring adds to the $-90°$ phase shift through the commutated network, enabling wave propagation. (**b**) In the other direction, the $-270°$ phase delay adds with the $+90°$ phase shift of the staggered commutated network, prohibiting wave propagation.

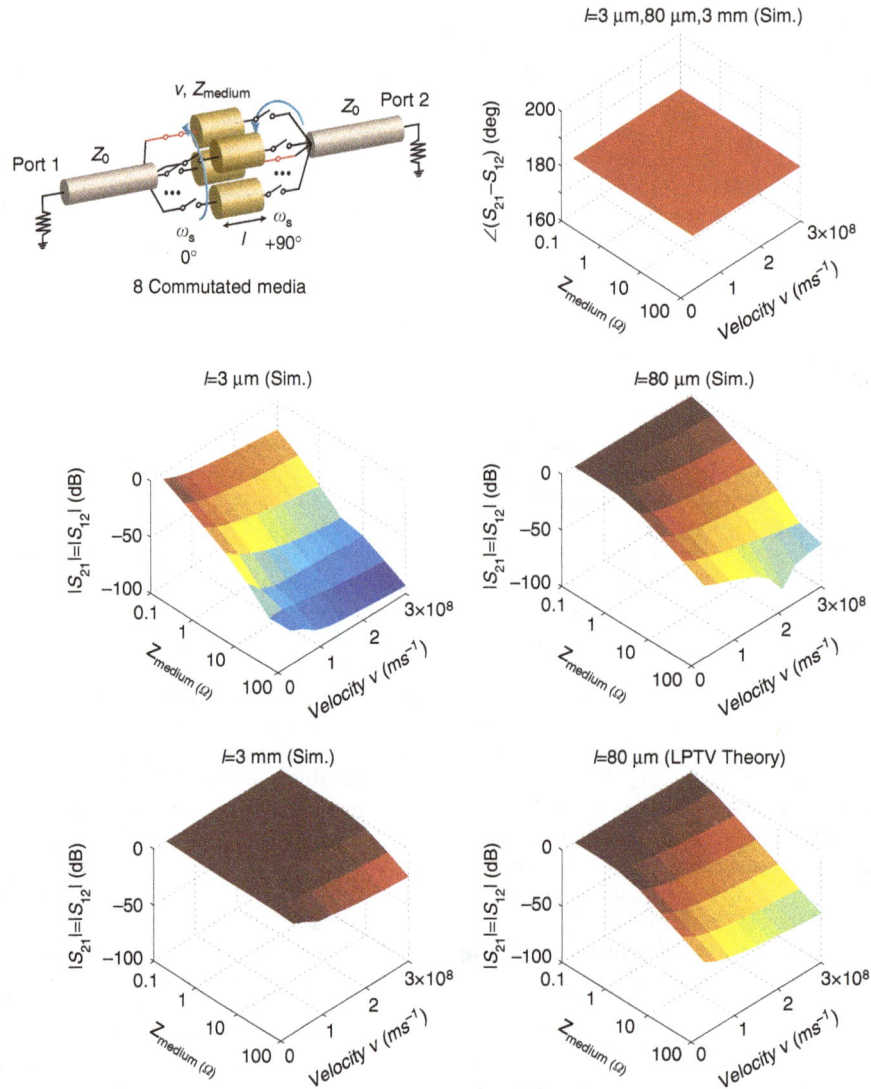

Figure 3 | Signal transmission in a staggered commutated network. Simulated S-parameters of an eight-way staggered commutated network are depicted assuming electrically short transmission line media of varied length *l*, characteristic impedance Z_{medium} and velocity *v*. Ideal switches are commutated at 750 MHz, with each switch active for 12.5% of the time period, and the reference impedance is assumed to be 50 Ω. Theoretical calculations based on the analytical formulation presented in Supplementary Note 1, where the electrically short transmission lines are approximated by their capacitance, are also shown and agree very well with simulations. The phases of S_{21} and S_{12} are always non-reciprocal and differ by 180° for 90° staggering. The magnitudes are always reciprocal; however, for substantial transmission, high *l*, low Z_{medium} and low *v*, or equivalently large capacitance, are required.

used for commutation at 750 MHz frequency, with each switch active for 12.5% of the time period. With +90° staggering, phase non-reciprocity is always observed (phase of the scattering or S-parameters S_{21} and S_{12} at the commutation frequency always have a 180° difference). The magnitudes of S_{21} and S_{12} at the commutation frequency are always reciprocal; however, low Z_{medium}, low velocity *v* and higher lengths *l* are required for significant signal transmission. In other words, the media must have a significant capacitance, similar to the (reciprocal) comb filter implementations described earlier, which use commutation across capacitor banks without staggering.

The need for capacitance may be intuitively understood by the fact with +90° staggering, no direct connection exists between the input and output at any instant assuming at least four paths. Therefore, a certain amount of capacitance is required for the media to store sufficient energy for subsequent transmission. A formal analysis of a commutated network with transmission-line media based on LPTV network theory is challenging. In Supplementary Note 1, with the aid of Supplementary Figs 1

and 2, we have completed an analysis of a staggered commutated network with N capacitors and ideal switches. As electrically short transmission lines are being considered, the approximation with capacitors is expected to be accurate. Figure 3 also depicts the magnitudes of S_{21} and S_{12} at the commutation frequency based on the exact analytical solution in Supplementary Note 1 for the case of $l = 80$ μm and an excellent agreement is seen to the simulations.

The analytical formulation quantifies the capacitance condition for substantial transmission as $C >> \frac{1}{2\pi f_s Z_0}$ where f_s is the commutation frequency and Z_0 is the reference impedance. Under this condition, at f_s for +90° staggering, the S-parameters of the staggered commutated network reduce to

$$S(f_s) \approx \begin{bmatrix} \frac{N^2\left(1-\cos\left(\frac{2\pi}{N}\right)\right)}{2\pi^2}-1 & \frac{N^2\left(1-\cos\left(\frac{2\pi}{N}\right)\right)}{2\pi^2}e^{-j\pi/2} \\ \frac{N^2\left(1-\cos\left(\frac{2\pi}{N}\right)\right)}{2\pi^2}e^{+j\pi/2} & \frac{N^2\left(1-\cos\left(\frac{2\pi}{N}\right)\right)}{2\pi^2}-1 \end{bmatrix} \underset{N\to\infty}{\approx} \begin{bmatrix} 0 & e^{-j\pi/2} \\ e^{+j\pi/2} & 0 \end{bmatrix},$$

$$(1)$$

clearly showing the non-reciprocity in the phase of S_{21} and S_{12}. The (reciprocal) magnitudes of S_{21} and S_{12} approach unity as $N \to \infty$ but even $N = 4$ or $N = 8$ are sufficient to achieve low loss, resulting in -1.8 and -0.45 dB transmission, respectively. The magnitudes of S_{11} and S_{22} approach 0 for $N \to \infty$, implying perfect impedance matching.

A frequency up-/down-conversion and filtering-based explanation. The behaviour may also be explained by viewing each set of commutating switches as an in-phase and quadrature reciprocal modulator that performs a frequency up-conversion and down-conversion, similar to the approach described in prior literature[33]. Indeed, such commutating transistor-based switches are regularly used as frequency converters in RF and microwave integrated electronics[42]. Figure 4 depicts a graphical view of signal propagation through the staggered commutated network in the forward and the reverse directions. In each direction, a sinusoidal input signal, $\cos(\omega_{in}t)$, is assumed at a frequency ω_{in} near the commutation frequency ω_s. Each set of switches is modelled as an in-phase and quadrature reciprocal modulator that multiplies the input signal with cosine and sine versions of the pump signal at ω_s. The second set of pump signals (2 and 4) are assumed to lead the first set (1 and 3) by $+90°$, representing the staggering. The capacitances of the media are assumed to effectively form a

low-pass filter in conjunction with the source and load impedances. The frequency translation of this low-pass filtering to RF by the commutating switches is the basis for the use of unstaggered commutated networks to realize (reciprocal) comb filters[15,36–40]. This low-pass filter effect substantially attenuates the up-converted signal at $\omega_s + \omega_{in}$ and $-\omega_s - \omega_{in}$ after the first modulation. It is seen that this leads to signal transmission with non-reciprocal phase response ($+90°/-90°$) in the forward and reverse directions. It should be noted that although the up-converted frequency components are filtered away, this does not result in any power loss for the desired signal as the commutated media are purely capacitive. Consequently, the staggered commutated network is lossless for $N \to \infty$, as discussed earlier (for finite N, there is loss due to harmonic conversion, albeit small even for $N = 4$ or $N = 8$ as discussed earlier). The reader may also verify that in the absence of low-pass filtering, all frequency components at the outputs cancel and the transmission of the structure in both directions is identically zero, which agrees with the theory and simulations in Fig. 3 that show very weak tranmission for high Z_{medium}, low l and high v (that is, low capacitance in the commutated media), and our expectation given a lack of a direct connection between the input and the output.

A comparison may be made with multi-arm optical isolators based on tandem electro-optic phase modulation[22], where back-to-back phase modulation with quarter-wavelength optical

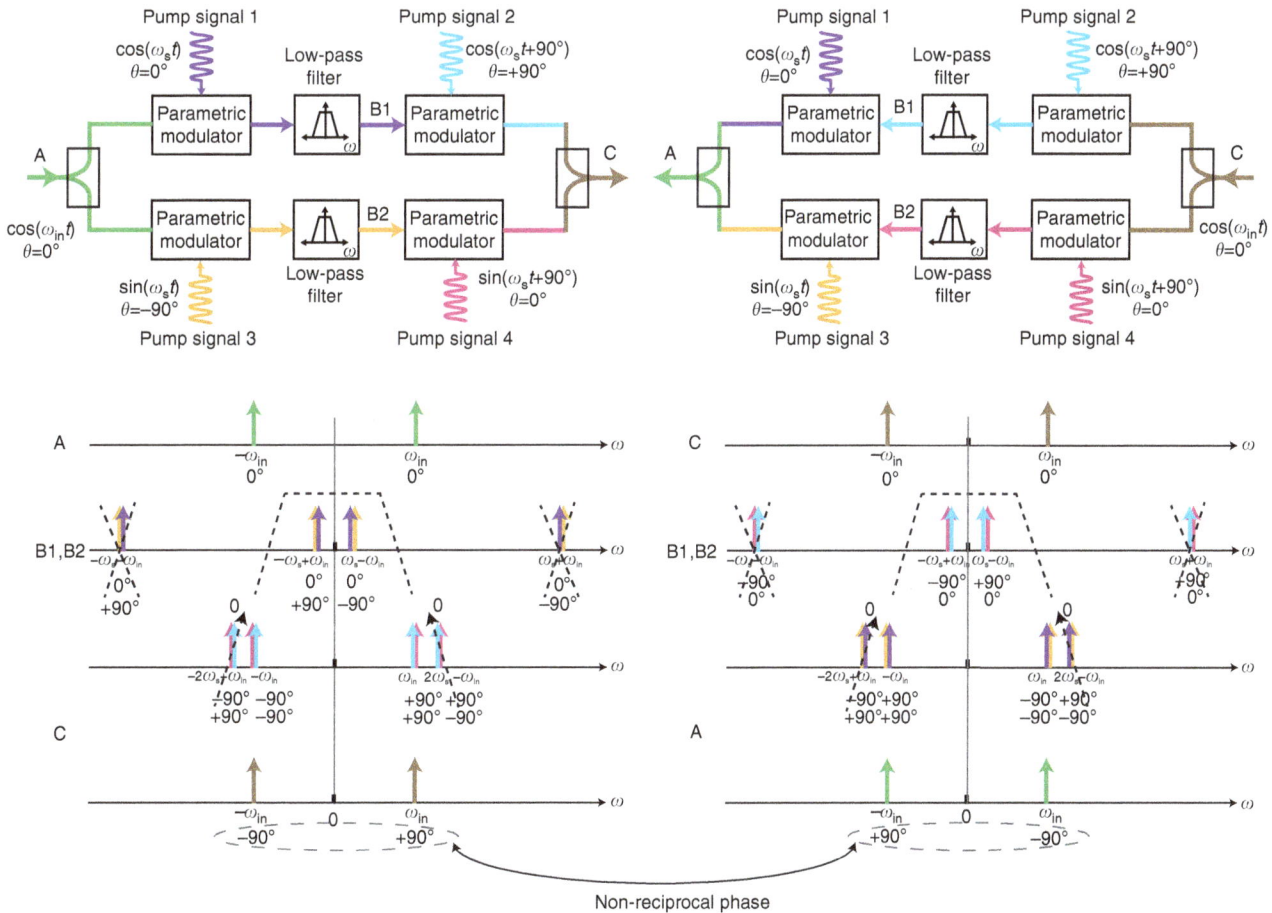

Figure 4 | A frequency up-/down-conversion and filtering-based explanation of phase non-reciprocity in staggered commutated networks. A commutator with at least four paths can be viewed as a reciprocal in-phase and quadrature frequency up-downconverter. The capacitance of the commutated media acts as a low-pass filter that attenuates the up-converted components after the first commutation. As a result, phase non-reciprocity is seen for signals travelling in the forward and reverse directions through a staggered commutated network. It may be verified that in the absence of low-pass filtering, all frequency components at the outputs cancel, leading to zero transmission in either direction, which agrees with the theory and simulations in Fig. 3 that show very weak transmission for high Z_{medium}, low l and high v (that is, low capacitance in the commutated media).

delays in between results in non-reciprocity. In our approach, commutation is essentially amplitude modulation with very high modulation ratio. Staggered commutation results in phase non-reciprocity and the filtering effect that is unique to commutation across media with appreciable capacitance eliminates the need for quarter-wavelength delays, allowing the phase non-reciprocity to be achieved within dimensions as small as $\lambda/1{,}250 \times \lambda/1{,}250$ in our device.

A highly-miniaturized RF circulator with enhanced linearity.
A three-port circulator matched at all ports to a reference impedance of Z_0 may be realized by contacting the $3\lambda/4$ transmission line (also of characteristic impedance Z_0) at three points as long as the three ports are spaced $\lambda/4$ apart along the ring circumference, as shown in Fig. 5a. To analyse this structure, we consider the case of a voltage excitation V_{in} at port 1. Using conventional microwave circuit analysis techniques along with the S-parameters of the staggered commutated network in equation (1) for the case of $N \to \infty$, the various node voltages can be determined to be

$$V_1 = \frac{1}{2}V_{in}, V_2 = \frac{-j}{2}V_{in}, V_3 = 0, V_x = \frac{\sin(\beta l)}{2}V_{in}, V_y = \frac{j\sin(\beta l)}{2}V_{in}, \quad (2)$$

where V_1, V_2 and V_3 are the three port voltages, V_x and V_y are the voltages on the left and right sides of the staggered commutated network, β is the propagation constant of the $3\lambda/4$ ring and l is the circumferential distance between port 3 and the commutated network. Based on these node voltages and by repeating the calculations with excitations at the other two ports, the S-parameter matrix of the structure can be derived to be:

$$S_{circ}(f_s) = \begin{bmatrix} 0 & 0 & -1 \\ -j & 0 & 0 \\ 0 & -j & 0 \end{bmatrix}. \quad (3)$$

This matches the S-parameters of an ideal 3-port circulator. It is interesting to note that the S-parameter matrix does not depend on l, meaning that the position of the staggered commutated network relative to ports 1 and 3 does not have an impact on S_{circ}. However, the voltages seen at the two ends of the staggered commutated network (V_x and V_y) when port 1 is excited are functions of l in equation (2). Interestingly, these voltages become 0 when l is 0. This is a direct consequence of the fact that when l is 0, V_y coincides with port 3 and naturally remains quiet for excitations at port 1 due to the isolation of the circulator (Fig. 5b). The S-parameters of the staggered commutated network force V_x and V_y to have the same magnitude, making V_x also a quiet node (Fig. 5b). As a result, voltage swings across the point parametric modulator are suppressed, yielding very high linearity to excitations at port 1.

A more comprehensive analysis of the circulator may be performed using the S-parameters of the staggered commutated network in equation (1) for the case of finite N. Interestingly, for $l = 0$, we once again arrive at the ideal S-parameter matrix of equation (3) and $V_x = V_y = 0$ for excitations at port 1. This implies that when port 3 coincides with V_y, the presence of a finite number of commutated paths will not limit circulator performance. In the presence of finite switch ON resistance, however, there will be finite voltage swing across the commutating switches for excitations at port 1, leading to finite linearity. However, this linearity to port 1 excitations will be vastly superior to the linearity to excitations at the other ports that do not enjoy this suppression mechanism.

We constructed a prototype at RF using conventional 65 nm CMOS IC technology to implement the electronic commutation across a bank of $N=8$ capacitors, and three lumped C-L-C networks to miniaturize the $3\lambda/4$ transmission line ring (Fig. 6). Miniaturization is eased by the fact that the $3\lambda/4$ ring is unmodulated, resulting in an overall structure that has a maximum dimension of $\approx \lambda/80$ at the operating frequency of 750 MHz. The capacitors of the C-L-C networks are also incorporated in the IC, leaving the three inductors as the only off-chip components. The commutated capacitors are chosen to be 26 pF each, roughly six times $\frac{1}{2\pi f_s Z_0}$ ($Z_0 = 50\,\Omega$), and are each realized on chip as a pair of 80 µm × 80 µm metal–insulator–metal capacitors. The total size of the staggered commutated network is ~ 320 µm × 320 µm ($\lambda/1{,}250 \times \lambda/1{,}250$). The control signals for the staggered commutation are generated using on-chip clock generation circuitry that is described in additional detail in the Methods section.

Experimental results. S-parameter measurements of the three-port circulator were performed using a measurement setup described in the Methods section. In the absence of commutation, with a pair of transistor switches on either side of the capacitor bank permanently closed, the circuit is perfectly reciprocal (Fig. 7a,d,g). In this configuration the high reciprocal isolations from port 3 to ports 1 and 2 are seen because port 3 is shunted to ground by one of the 26-pF capacitors. The quarter-wave transmission lines from port 1 to the staggered commutated network and from port 2 to port 3 transform this near-short-circuit impedance to open circuits at ports 1 and 2, effectively disconnecting port 3 from the rest of the circuit and resulting in a reciprocal structure that exhibits low-loss transmission between port 1 and port 2. Similarly, another reciprocal configuration without commutation is depicted, where all switches are permanently open. Here, simple circuit analysis reveals that port 1 is effectively disconnected from the rest of the circuit, which now exhibits low-loss reciprocal transmission between ports 2 and 3.

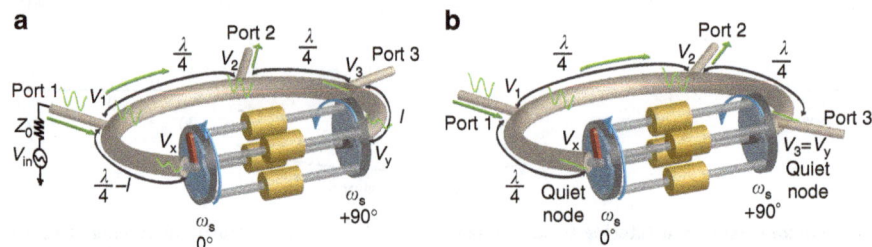

Figure 5 | Circulator architecture. (a) A 3-port circulator may be realized by adding ports at three points along the ring as long as the circumferential spacing between the ports is $\lambda/4$. The S-parameters of such a structure match that of an ideal three-port circulator and are independent of l, the circumferential distance between port 3 and the staggered commutated network. **(b)** However, setting l to 0 suppresses the voltages across the commutated network for excitations at port 1, significantly enhancing the linearity of the structure to port 1 excitations.

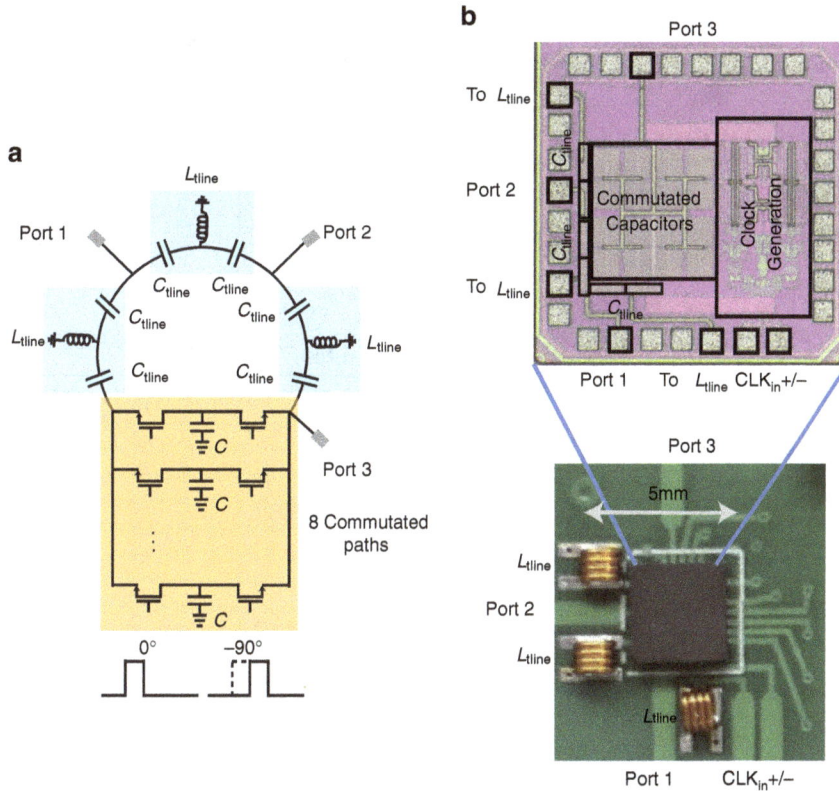

Figure 6 | RF CMOS IC implementation of the circulator. (a) A simplified circuit diagram of the circulator is shown. Electronic commutation across a bank of $N = 8$ capacitors is performed using reciprocal, passive transistor-based switches without direct-current bias. The staggered commutated network enables miniaturization of the unmodulated $3\lambda/4$ ring using three C-L-C sections. **(b)** The microphotograph of the fabricated IC is shown along with a close-up photograph of the fabricated printed circuit board with the IC housed in a quad-flat no-leads (QFN) package and interfaced with the off-chip inductors. The largest dimension of the prototype is 5 mm or $\lambda/80$ at the operating frequency of 750 MHz.

Under commutation at a frequency of 750 MHz with staggering for clockwise circulation, strong non-reciprocity is measured (Fig. 7b,e,h) with low-loss transmission in the direction of circulation (S_{21}, S_{32} and S_{13} are -1.7, -1.7 and -3.3 dB, respectively, at 750 MHz) and strong isolation in the reverse direction (S_{12}, S_{23} and S_{31} are -9.6, -10.4 and -17.4 dB, respectively, at 750 MHz). This represents, on an average, an order of magnitude of non-reciprocity between any two ports and is limited by the imperfect impedance matching at the third port (as is the case with all circulators) caused due to parasitics at the chip-package-board interfaces. If the impedance at the third port is slightly tuned, non-reciprocity of 40–50 dB or four to five orders of magnitude is seen in S_{12}, S_{23} and S_{31} at 750 MHz with negligible impact on transmission in the circulation direction. It should be noted that independent tuning is required at the three ports to achieve very high non-reciprocity in all paths, given the inherent asymmetry in the circulator structure. Figure 7c,f,i depict the simulated S-parameters under commutation, demonstrating an excellent match to the measurements.

Although the analysis in this study has restricted itself to response at the commutation frequency, it is noteworthy that a filtering profile is observed in S_{32} and S_{13} across frequency due to the comb-filter functionality inherent to commutation across a capacitor bank. The LPTV analysis in Supplementary Note 1 may be extended to determine the response of the staggered commutated network and the circulator across frequency, and confirms this observed (and simulated) filtering profile.

Another unique feature of the fabricated prototype is its real-time reconfigurability. By changing the staggering between $+90°$ and $-90°$, the direction of circulation can be altered, as is seen in Fig. 8a. The frequency of operation of the circulator can

also be tuned by changing the frequency of commutation within the limits dictated by the bandwidth of the $3\lambda/4$ transmission line ring. In Fig. 8b, four to five orders of magnitude (40–50 dB) of non-reciprocity in S_{31} is maintained across 700–800 MHz.

In Fig. 8c, we present experimental evidence of the enhanced linearity to excitations at port 1. Two-tone intermodulation distortion tests were performed on the prototype for transmission from port 1 to port 2, and from port 2 to port 3. The input-referred third-order intercept point (IIP3) for transmission from port 1 to port 2 is $+27.5$ dBm (≈ 560 mW), a remarkable number for a 65 nm CMOS IC implementation and nearly two orders of magnitude higher than that from port 2 to port 3 ($+8.7$ dBm or 7.4 mW) due to the suppression of the signal across the point parametric modulator. Aside from intermodulation distortion caused due to third-order non-linearity, other spurious effects to consider include image signals due to quadrature mismatch and modulation feedthrough. The level of the measured spurious image signals produced due to quadrature mismatch for port 1 to port 2 transmission is 51 dB below the main signal. This level of image rejection is high enough so that it is not a serious issue when compared with the third-order intermodulation distortion produced for reasonable port 1 power levels. The measured modulation feedthrough at port 2 is at -57 dBm. Although these levels are already quite low, they can be cancelled further using integrated calibration circuits that are commonly implemented in CMOS RF ICs.

Passive LTI systems have a noise figure (NF) that is equal to their loss. Passive LPTV systems can exhibit noise folding. The measured NF for port 2 to port 3 transmission is 4 dB, higher than the 1.7 dB loss. Simulations indicate that 2 dB degradation arises from modulation path phase noise due to a poor

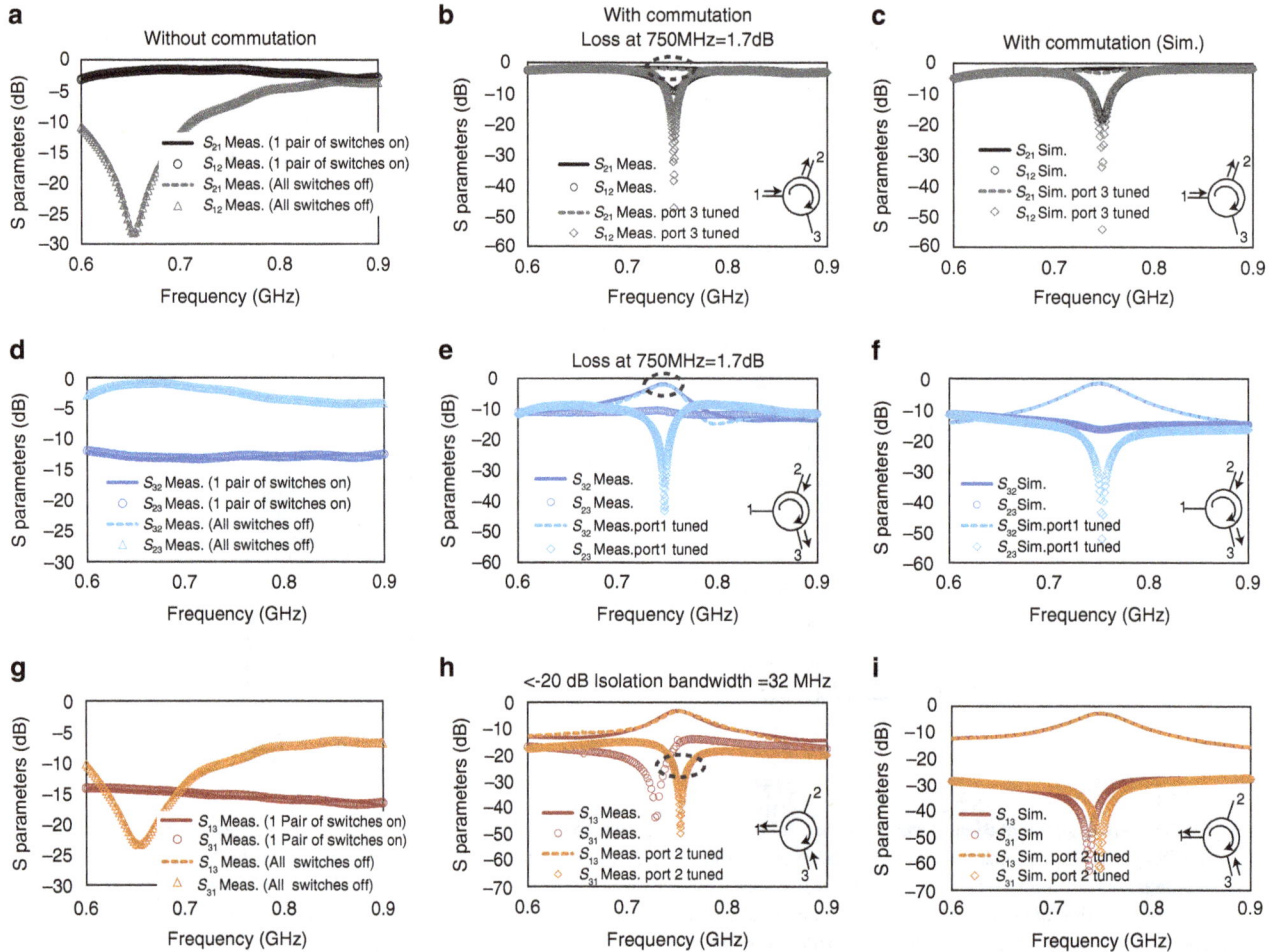

Figure 7 | Circulator S-parameter measurements. Measured circulator S-parameters (**a**), (**d**) and (**g**) without commutation and (**b**), (**e**) and (**h**) with commutation are shown, as well as (**c**), (**f**) and (**i**) simulated S-parameters that show an excellent match to the measurements. Under staggered commutation for clockwise circulation, low loss transmission in the circulation direction (S_{21}, S_{32} and S_{13} are -1.7, -1.7 and -3.3 dB, respectively) and an order-of-magnitude isolation in the reverse direction (S_{12}, S_{23} and S_{31} are -9.6, -10.4 and -17.4 dB, respectively) are measured at the commutation frequency of 750 MHz. When the third port is slightly tuned, non-reciprocity of 40–50 dB is measured in S_{12}, S_{23} and S_{31} at 750 MHz with negligible impact on transmission in the circulation direction. The -20 dB isolation bandwidth in S_{31} after tuning is 32 MHz or 4.3%.

implementation of the modulation path phase shifter. A better implementation of the modulation path phase shifter restores the NF to 2 dB in simulation, close to the 1.7 dB loss level and indicative of only minor NF degradation due to harmonic noise folding.

Discussion

The ability to integrate magnetic-free passive linear non-reciprocal components in CMOS has the potential to revolutionize RF communications. The CMOS circulator described in this study exhibits extremely low insertion loss (<2 dB) and strong non-reciprocity, is compact and features reconfiguration capabilities. Furthermore, the enhanced linearity to port 1 is particularly useful for full-duplex or simultaneous transmit-and-receive communication and radar applications[3,4], where a high-power transmitter (port 1) and a highly sensitive receiver (port 3) operate simultaneously at the same frequency, and must be interfaced with a shared antenna (port 2). In such a configuration, S_{21} (transmitter to antenna loss), S_{32} (antenna to receiver loss) and S_{31} (transmitter to receiver isolation) are the most important performance metrics, where our device performs particularly well. Full-duplex communication is drawing significant interest for emerging 5G communication networks[43]

due to its potential to double network capacity compared with half-duplex communication at the physical layer while offering numerous other benefits at the network layer. There has been active research on fully integrated CMOS transceiver ICs supporting full duplex[44–48]. The ability to include the circulator with the transceiver on the same CMOS IC would significantly reduce the cost and form factor, and enhance the performance of full-duplex systems. The filtering profile in S_{32} is also very useful in protecting the highly sensitive receiver from interference signals outside the frequency band of operation.

For such applications, the ability to reconfigure the device between reciprocal and non-reciprocal operation is also very useful. We demonstrated reconfigurable modes of operation in the absence of commutation where the structure becomes reciprocal with low-loss transmission between the antenna (port 2) and the transmitter (port 1), and the antenna and the receiver (port 3). In other words, the prototype can be reconfigured between operation as a non-reciprocal circulator for full-duplex communication and a reciprocal low-loss transmit/receive switch for half-duplex communication.

Although the design of this prototype is deliberately asymmetric with respect to the three ports to prioritize linearity to port 1 excitations for communication applications, a symmetric design may also be envisioned where a non-reciprocal

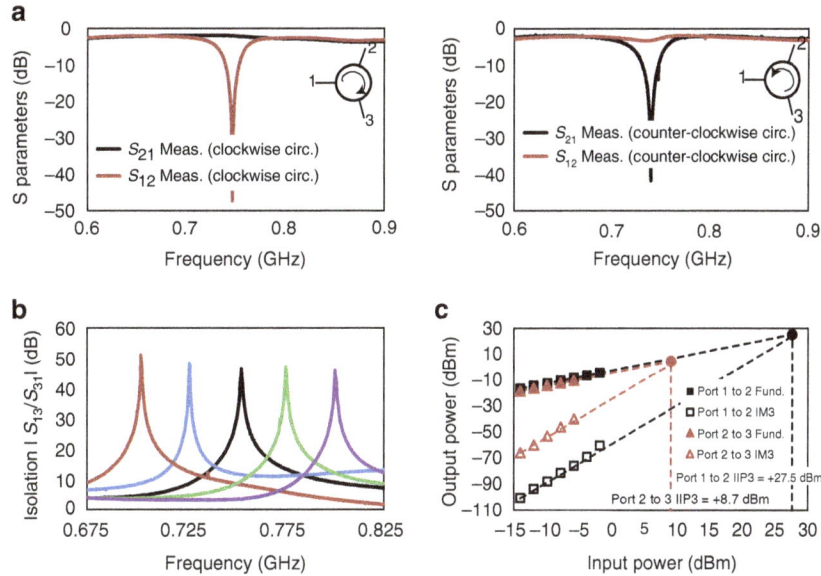

Figure 8 | Experimental evidence of reconfigurability and enhanced linearity to port 1 excitations. (a) The direction of circulation can be altered by changing the staggering between $+90°$ and $-90°$. (b) The frequency of operation of the circulator can be tuned by changing the commutation frequency within the limits dictated by the bandwidth of the $3\lambda/4$ transmission-line ring. Here we present the non-reciprocity in S_{31} across different commutation frequencies ranging from 700 to 800 MHz. In each case, tuning of the port 2 impedance is exploited to achieve 40–50 dB isolation at the commutation frequency. (c) Measured two-tone linearity for transmission from port 1 to port 2 and from port 2 to port 3 are shown when configured for clockwise circulation. Nonlinear systems exhibit intermodulation distortion products when excited with two sinusoidal signals. The input-referred third-order intercept point (IIP3) represents the (extrapolated) input power of each of the two tones at which the third-order intermodulation products (IM3) at the output are as powerful as the fundamental signals. The IIP3 for transmission from port 1 to port 2 is $+27.5$ dBm (≈ 560 mW), nearly two orders of magnitude higher than that from port 2 to port 3 ($+8.7$ dBm or 7.4 mW), owing to the suppression of the signal across the point parametric modulator for port 1 excitations.

phase component is incorporated in all three arms, that is, between all three ports. Such a symmetric circulator structure would enable extension of the concept to non-reciprocal metamaterials that support topologically protected wave propagation modes[49].

The bandwidth of the circulator's isolation is limited by the matching between the frequency responses of the $3\lambda/4$ ring and the staggered commutated network. Currently, the -20 dB isolation bandwidth in S_{31} after tuning is 32 MHz or 4.3% of the operating frequency, useful for commercial cellular LTE and WiFi applications. Dispersion engineering techniques to match their responses over a wider frequency range, thus enhancing the isolation amount versus bandwidth trade-off, represent an interesting topic for future research.

Commutated networks are known to have response at harmonics of the commutation frequency. Response at harmonics can lead to susceptibility to interference near harmonic frequencies and also leads to noise folding. Although we have seen that the effect of noise folding on the NF is very small in this circulator structure, a formal analysis of the harmonic response and noise of the overall circulator structure, as well as techniques to suppress the same, also represent an interesting topic for future investigation.

It is also interesting to consider the application of these concepts to other domains where a high-quality switch is available, such as optical waves. Compact optical switches with one to two orders of magnitude ON/OFF transmission ratio[50,51] open the door to optical non-reciprocity and isolation through commutation-based parametric modulation. The nanosecond-scale switching speed implies GHz-range commutation frequencies, much smaller than the optical carrier frequency, which can be accomodated by commutating across high-Q optical filters that eliminate one of the modulation sidebands, similar to the low-pass filter effect used in our prototype.

Methods

IC implementation details. A block diagram of the IC and component values are provided in Supplementary Fig. 3 and Supplementary Table 1, respectively. The circulator consists of three lumped C-L-C sections that miniaturize the $3\lambda/4$ transmission line and a staggered electronically commutated network of eight capacitors. All components, except for the three inductors, are integrated on the CMOS IC. The switches of the commutated network are implemented using 65 nm CMOS transistors and are driven with two sets of eight non-overlapping clock signals with 12.5% duty cycle. These clock signals are generated from two differential (0/180°) input clocks that run at four times the desired commutation frequency. A divide-by-2 frequency-divider circuit generates four quadrature clocks with 0°/90°/180°/270° phase relationship. These four clock signals drive two parallel paths for the two sets of switches. One of the paths features a programmable phase shifter that allows for arbitrary staggering between the two commutating switch sets. This enables switching between $+90°$ and $-90°$ staggering, which allows dynamic reconfiguration of the circulation direction. The phase shifter also allows for fine tuning of the staggered phase shift, to optimize the transmission loss in the circulation direction and isolation in the reverse direction. This feature has been exploited in the measurements shown in the main text. After phase shifting, another divide-by-2 circuit and a non-overlapping 12.5% duty-cycle clock generation circuit create the clock signals that control the commutating transistor switches.

Experimental setups. Diagrams of the experimental setups are provided in Supplementary Figs 4 and 5, respectively. A list of the equipment used in the two setups is provided in Supplementary Table 2. A 180° hybrid is used to generate two differential (0°/180°) signals from a signal generator to drive the clock inputs of the implemented circulator. A voltage bias signal is added to the clock signals through bias tees to drive the clock inputs. A two-port vector network analyser is used to measure the S-parameters of the circulator two ports at a time, whereas the third port is terminated with a variable impedance tuner to tune the port impedance. For the input-referred third-order intercept point linearity test, a two-way power combiner is used to combine two sinusoidal signals generated from two additional signal generators and feed the input port under consideration. A spectrum analyser is used to monitor the fundamental signals and the third-order intermodulation distortion products of the circulator at the output port under consideration, whereas the third port is terminated with a 50-Ω termination.

References

1. Jalas, D. et al. What is—and what is not—an optical isolator. Nat. Photonics 7, 579–582 (2013).
2. Goldsmith, A. Wireless Communications (Cambridge Univ. Press, 2005).

3. Bharadia, D., McMilin, E. & Katti, S. Full duplex radios. *ACM SIGCOMM Computer Communication Review* vol. 43, 375–386 (ACM, 2013).

4. Sabharwal, A. *et al.* In-band full-duplex wireless: Challenges and opportunities. *IEEE J. Select. Areas Commun.* **32**, 1637–1652 (2014).

5. Pozar, D. M. *Microwave Engineering* (Addison-Wesley, 1990).

6. Bi, L. *et al.* On-chip optical isolation in monolithically integrated non-reciprocal optical resonators. *Nat. Photonics* **5**, 758–762 (2011).

7. Shoji, Y., Mizumoto, T., Yokoi, H., Hsieh, I. -W. & Osgood, R. M. Magneto-optical isolator with silicon waveguides fabricated by direct bonding. *Appl. Phys. Lett.* **92**, 071117 (2008).

8. Adam, J., Davis, L., Dionne, G. F., Schloemann, E. & Stitzer, S. Ferrite devices and materials. *IEEE Trans. Microw. Theory Techn.* **50**, 721–737 (2002).

9. Adam, J. *et al.* Monolithic integration of an X-band circulator with GaAs MMICs. *IEEE MTT-S International Microwave Symposium Digest* **1**, 97–98 (1995).

10. Oliver, S., Zavracky, P., McGruer, N. & Schmidt, R. A monolithic single-crystal yttrium iron garnet/silicon X-band circulator. *IEEE Microw. Guided Wave Lett.* **7**, 239–241 (1997).

11. Tanaka, S., Shimomura, N. & Ohtake, K. Active circulators—the realization of circulators using transistors. *Proc. IEEE* **53**, 260–267 (1965).

12. Wang, Z. *et al.* Gyrotropic response in the absence of a bias field. *Proc. Natl Acad. Sci. USA* **109**, 13194–13197 (2012).

13. Kodera, T., Sounas, D. L. & Caloz, C. Artificial Faraday rotation using a ring metamaterial structure without static magnetic field. *Appl. Phys. Lett.* **99**, 031114 (2011).

14. Kodera, T., Sounas, D. & Caloz, C. Magnetless nonreciprocal metamaterial (MNM) technology: application to microwave components. *IEEE Trans. Microw. Theory Techn.* **61**, 1030–1042 (2013).

15. Mirzaei, A. & Darabi, H. Reconfigurable RF front-ends for cellular receivers. *2010 IEEE Comp. Semiconduct. Integr. Circuit Symp. (CSICS)*, 1–4 (IEEE, 2010).

16. Carchon, G. & Nanwelaers, B. Power and noise limitations of active circulators. *IEEE Trans. Microw. Theory Techn.* **48**, 316–319 (2000).

17. Peng, B. *et al.* Parity-time-symmetric whispering-gallery microcavities. *Nat. Phys.* **10**, 394–398 (2014).

18. Fan, L. *et al.* An all-silicon passive optical diode. *Science* **335**, 447–450 (2012).

19. Soljačić, M., Luo, C., Joannopoulos, J. D. & Fan, S. Nonlinear photonic crystal microdevices for optical isolation. *Opt. Lett.* **28**, 637–639 (2003).

20. Gallo, K., Assanto, G., Parameswaran, K. R. & Fejer, M. M. All-optical diode in a periodically poled lithium niobate waveguide. *Appl. Phys. Lett.* **79**, 314–316 (2001).

21. Mahmoud, A. M., Davoyan, A. R. & Engheta, N. All-passive nonreciprocal metastructure. *Nat. Commun.* **6**, 8359 (2015).

22. Doerr, C. R., Chen, L. & Vermeulen, D. Silicon photonics broadband modulation-based isolator. *Opt. Express* **22**, 4493–4498 (2014).

23. Galland, C., Ding, R., Harris, N. C., Baehr-Jones, T. & Hochberg, M. Broadband on-chip optical non-reciprocity using phase modulators. *Opt. Express* **21**, 14500–14511 (2013).

24. Lira, H., Yu, Z., Fan, S. & Lipson, M. Electrically driven nonreciprocity induced by interband photonic transition on a silicon chip. *Phys. Rev. Lett.* **109**, 033901 (2012).

25. Kang, M., Butsch, A. & Russell, P. S. J. Reconfigurable light-driven opto-acoustic isolators in photonic crystal fibre. *Nat. Photonics* **5**, 549–553 (2011).

26. Yu, Z. & Fan, S. Complete optical isolation created by indirect interband photonic transitions. *Nat. Photonics* **3**, 91–94 (2008).

27. Qin, S., Xu, Q. & Wang, Y. Nonreciprocal components with distributedly modulated capacitors. *IEEE Trans. Microw. Theory Techn.* **62**, 2260–2272 (2014).

28. Zanjani, M. B., Davoyan, A. R., Mahmoud, A. M., Engheta, N. & Lukes, J. R. One-way phonon isolation in acoustic waveguides. *Appl. Phys. Lett.* **104**, 081905 (2014).

29. Sounas, D. L., Caloz, C. & Alù, A. Giant non-reciprocity at the subwavelength scale using angular momentum-biased metamaterials. *Nat. Commun.* **4**, 2407 (2013).

30. Fleury, R., Sounas, D. L., Sieck, C. F., Haberman, M. R. & Alu, A. Sound isolation and giant linear nonreciprocity in a compact acoustic circulator. *Science* **343**, 516–519 (2014).

31. Estep, N. A., Sounas, D. L., Soric, J. & Alu, A. Magnetic-free non-reciprocity and isolation based on parametrically modulated coupled-resonator loops. *Nat. Phys.* **10**, 923–927 (2014).

32. Estep, N., Sounas, D. & Alu, A. On-chip non-reciprocal components based on angular momentum biasing. *2015 IEEE MTT-S Int. Microw. Symp. (IMS)*, 1–4 (IEEE, 2015).

33. Kamal, A., Clarke, J. & Devoret, M. H. Noiseless non-reciprocity in a parametric active device. *Nat. Phys.* **7**, 311–315 (2011).

34. Busignies, H. & Dishal, M. Some relations between speed of indication, bandwidth, and signal-to-random-noise ratio in radio navigation and direction finding. *Proc. IRE* **37**, 478–488 (1949).

35. LePage, W. R., Cahn, C. R. & Brown, J. S. Analysis of a comb filter using synchronously commutated capacitors. *Trans. Am. Inst. Electric. Eng. I Commun. Electron.* **72**, 63–68 (1953).

36. Ghaffari, A., Klumperink, E., Soer, M. & Nauta, B. Tunable high-Q N-path band-pass filters: modeling and verification. *IEEE J. Solid State Circuits* **46**, 998–1010 (2011).

37. Andrews, C. & Molnar, A. Implications of passive mixer transparency for impedance matching and noise figure in passive mixer-first receivers. *IEEE Trans. Circuits Syst. I Regul. Papers* **57**, 3092–3103 (2010).

38. Reiskarimian, N. & Krishnaswamy, H. Design of All-Passive Higher-Order CMOS N-Path Filters. *2015 IEEE RFIC Symp.*, 83–86 (IEEE, 2015).

39. Thomas, C. & Larson, L. Broadband synthetic transmission-line N-path filter design. *IEEE Trans. Microw. Theory Techn.* **63**, 3525–3536 (2015).

40. Gharpurey, R. Linearity enhancement techniques in radio receiver front-ends. *IEEE Trans. Circuits Syst. I Regul. Papers* **59**, 1667–1679 (2012).

41. Tyagi, S. *et al.* An advanced low power, high performance, strained channel 65nm technology. *2005 IEEE Int. Electron Devices Meet. Tech. Digest* 245–247 (2005).

42. Forbes, T., Ho, W.-G. & Gharpurey, R. Design and analysis of harmonic rejection mixers with programmable LO frequency. *IEEE J. Solid State Circuits* **48**, 2363–2374 (2013).

43. Hong, S. *et al.* Applications of self-interference cancellation in 5G and beyond. *IEEE Commun. Mag.* **52**, 114–121 (2014).

44. Zhou, J., Chuang, T.-H., Dinc, T. & Krishnaswamy, H. Receiver with > 20MHz bandwidth self-interference cancellation suitable for FDD, co-existence and full-duplex applications. *2015 IEEE Int. Solid State Circuits Conf. (ISSCC)* 342–343 (IEEE, 2015).

45. van den Broek, D. -J., Klumperink, E. & Nauta, B. A self-interference-cancelling receiver for in-band full-duplex wireless with low distortion under cancellation of strong TX leakage. *2015 IEEE Int. Solid State Circuits Conf. (ISSCC)* 344–345 (IEEE, 2015).

46. Zhou, J., Chuang, T.-H., Dinc, T. & Krishnaswamy, H. Integrated wideband self-interference cancellation in the RF domain for FDD and full-duplex wireless. *IEEE J. Solid State Circuits* **50**, 3015–3031 (2015).

47. Yang, D., Yuksel, H. & Molnar, A. A wideband highly integrated and widely tunable transceiver for in-band full-duplex communication. *IEEE J. Solid State Circuits* **50**, 1189–1202 (2015).

48. Dinc, T., Chakrabarti, A. & Krishnaswamy, H. A 60 GHz same-channel full-duplex CMOS transceiver and link based on reconfigurable polarization-based antenna cancellation. *2015 IEEE RFIC Symp.* 31–34 (IEEE, 2015).

49. Khanikaev, A. B., Fleury, R., Mousavi, S. H. & Alù, A. Topologically robust sound propagation in an angular-momentum-biased graphene-like resonator lattice. *Nat. Commun.* **6**, 8260 (2015).

50. Lira, H. L. R., Manipatruni, S. & Lipson, M. Broadband hitless silicon electro-optic switch for on-chip optical networks. *Opt. Express* **17**, 22271–22280 (2009).

51. Vlasov, Y., Green, W. M. J. & Xia, F. High-throughput silicon nanophotonic wavelength-insensitive switch for on-chip optical networks. *Nat. Photonics* **2**, 242–246 (2008).

Acknowledgements

This work was supported by the DARPA ACT programme under Grant FA8650-14-1-7414. We acknowledge useful discussions with Dr Troy Olsson, Dr Ben Epstein.

Author contributions

N.R. and H.K. developed the concept. N.R. conducted the theoretical analysis and numerical simulations, designed the device and performed the experiments. H.K. directed and supervised the project. N.R. and H.K. wrote the paper.

Additional information

27

Exciton localization in solution-processed organolead trihalide perovskites

Haiping He[1], Qianqian Yu[1], Hui Li[1], Jing Li[1], Junjie Si[1], Yizheng Jin[2], Nana Wang[3], Jianpu Wang[3], Jingwen He[4], Xinke Wang[4], Yan Zhang[4] & Zhizhen Ye[1]

Organolead trihalide perovskites have attracted great attention due to the stunning advances in both photovoltaic and light-emitting devices. However, the photophysical properties, especially the recombination dynamics of photogenerated carriers, of this class of materials are controversial. Here we report that under an excitation level close to the working regime of solar cells, the recombination of photogenerated carriers in solution-processed methylammonium-lead-halide films is dominated by excitons weakly localized in band tail states. This scenario is evidenced by experiments of spectral-dependent luminescence decay, excitation density-dependent luminescence and frequency-dependent terahertz photo-conductivity. The exciton localization effect is found to be general for several solution-processed hybrid perovskite films prepared by different methods. Our results provide insights into the charge transport and recombination mechanism in perovskite films and help to unravel their potential for high-performance optoelectronic devices.

[1] State Key Laboratory of Silicon Materials, School of Materials Science and Engineering, Zhejiang University, Hangzhou 310027, China. [2] Center for Chemistry of High-Performance and Novel Materials and State Key Laboratory of Silicon Materials, Department of Chemistry, Zhejiang University, Hangzhou 310027, China. [3] Key Laboratory of Flexible Electronics (KLOFE) & Institute of Advanced Materials (IAM), Jiangsu National Synergetic Innovation Center for Advanced Materials (SICAM), Nanjing Tech University (NanjingTech), 30 South Puzhu Road, Nanjing 211816, China. [4] Department of Physics, Capital Normal University, Beijing Key Lab for Metamaterials and Devices, and Key Laboratory of Terahertz Optoelectronics, Ministry of Education, Beijing 100048, China. Correspondence and requests for materials should be addressed to H.H. (email: hphe@zju.edu.cn) or to Y.J. (email: yizhengjin@zju.edu.cn) or to Z.Y. (email: yezz@zju.edu.cn).

Recently, organolead trihalide perovskites have been utilized in low-temperature solution-processed photovoltaics[1-6] and light-emitting devices[7-10]. Certified power conversion efficiency approaching 20.1% has been realized[6]. The impressive photovoltaic performance is believed to originate from the long-distance and balanced diffusion of charge carriers[11-12]. Remarkably, the solution-processed perovskite films also exhibit superior luminescence properties. Optically pumped lasing with low thresholds and tunable wavelengths[7], and bright light-emitting diodes[9-10] have been demonstrated.

Despite these remarkable advances, knowledge of the photophysical properties of the perovskites is still lacking. One of the key questions concerns the recombination dynamics of photogenerated charges: whether exciton or free carrier (FC) is the dominant recombination channel in organolead trihalide perovskites. The answer will help to interpret the seemingly counterintuitive facts that organolead trihalide perovskites can act both as extraordinary photovoltaic materials and superior gain mediums for lasing[13]. In general, photovoltaic materials require efficient separation of photocarriers, and lasing materials require high recombination rates. The reported exciton binding energy of the perovskites[14,15] is comparable to the thermal energy at room temperature (RT), which arouses arguments that in such a case excited states will tend to dissociate into FCs rather than recombine radiatively.

Several groups have used photoluminescence (PL) to study the competition between exciton and FC in organolead trihalide perovskites. In a few steady-state PL studies, the RT PL was attributed to exciton recombination,[16-18] but the conclusions lacked solid evidence. Recently, several groups[8,14,19,20] attributed the RT PL to FC recombination. For example, D'Innocenzo et al.[14] argued that excitons generated by low-density excitation are almost fully ionized at RT when the exciton binding energy is moderately larger than the RT thermal energy. The band filling effect[8] and quadratic dependence of the PL intensity on the excitation intensity[19], the two characteristic features of FC recombination, were observed at relatively high excitation levels. We note that the observation of FC recombination at

relatively high excitation levels is not surprising because the reduced exciton binding energy originated from the screening effect of FCs, a phenomenon that has been well established in many semiconductors[21,22].

Here we show that under an excitation level close to the working regime of solar cells, the radiative recombination of photogenerated carriers in solution-processed $CH_3NH_3PbX_3$ perovskites is dominated by excitons localized in band tail states. The excitonic nature of the emission is evidenced by the excellent power-law dependence of the PL intensity on the excitation intensity expected for bound excitons, and is supported by the PL lineshape analysis. The localization effect is supported by the spectral dependence of the PL lifetime and frequency-dependent THz photoconductivity results. We also show that the exciton localization effect is general in several solution-process perovskite films.

Results

Evidence for exciton localization in $CH_3NH_3PbBr_3$ films. We use solution-processed $CH_3NH_3PbBr_3$ films for the PL studies. The films show good crystalline and optical quality (Supplementary Fig. 1). To avoid degradation induced by air exposure, all samples were prepared in a nitrogen-filled glove box, coated with a polymethyl methacrylate (PMMA) layer and were measured in vacuum (10^{-1} Pa). The $CH_3NH_3PbBr_3$ film shows emission at 2.35 eV, in agreement with the reported values[23,24].

Near-band-edge emission in semiconductors may have several origins, including exciton recombination, FC recombination (also known as band-to-band transition), free-to-bound recombination and donor–acceptor pair recombination. To determine which process is dominant in our samples, we measured the PL spectra under various excitation densities close to or lower than the photovoltaic working regime ($\sim 5 \times 10^{14}$ cm^{-3}) (refs 14,25). The PL lineshapes in Fig. 1a are almost identical in the whole range of excitation intensity. The PL intensity shows excellent power-law dependence on the excitation power, with a power-law exponent of 1.179. In direct bandgap semiconductors and

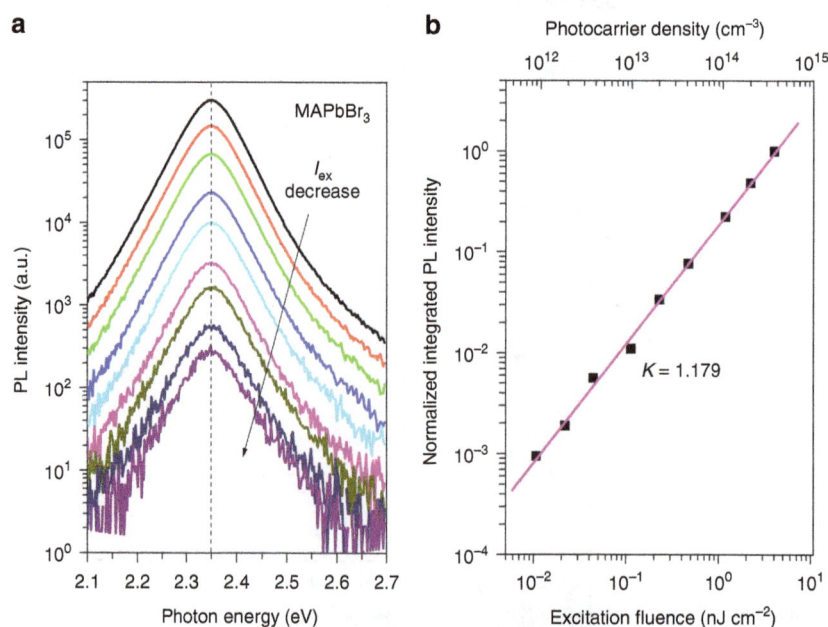

Figure 1 | Excitation density-dependent PL of solution-processed $CH_3NH_3PbBr_3$ films. (**a**) Steady-state PL spectra recorded with excitation density from 0.01 to 4 nJ cm^{-2}. All spectra are measured in vacuum at RT. In all spectra, the peak energy (indicated by the dashed line), lineshape and linewidth are identical within the experimental error. (**b**) Logarithm plot of the integrated PL intensity versus excitation density. The data show a power-law dependence with k = 1.179.

under non-resonant excitation conditions, the integrated PL intensity (I_{PL}) is a power-law function of the excitation density[26],

$$I_{PL} \sim I_{ex}^k \quad (1)$$

with $k=2$ for FC recombination, $1<k<2$ for recombination of excitons (including free excitons and bound excitons) and $k<1$ for free-to-bound recombination and donor–acceptor pair. The model was further refined by Shibata et al[27], who provided an analytical formula to confirm that $1<k<2$, even for free excitons. The physics behind the process is the photo-neutralization of the donors/acceptors, which are present in all semiconductors and result in competitive recombination channels. Our k value agrees well with those reported for excitons in semiconductors[26,27].

Figure 2a shows the RT PL decay curves monitored at different excitation energies. On the low-energy side, the PL lifetime is almost constant for all emission energies. On the high-energy side, the PL lifetime decreases with increasing emission energy. The PL decay curves can be well fitted by the thermalized stretching exponential line shape[28,29]

$$I(t) = I_1 \exp\left(-\frac{t}{\tau_1}\right) + I_2 \exp\left[-\left(\frac{t}{\tau_2}\right)^{\beta}\right], \quad (2)$$

where τ_i is the decay time and I_i is the weight factor of each decay channel. A typical fitting result is plotted in Fig. 2b (for all fitting results, see Supplementary Fig. 2). The fitting curves do not match normal mono-exponential or simple stretched-exponential decay (Supplementary Fig. 2c). Stretched-exponential decay is regarded as evidence of the exciton localization, in which the parameter β is related to the dimensionality of the localizing centres. The former exponential term in equation (2) represents the relaxation of free or extended states towards localized states, whereas the latter stretched-exponential term accounts for the communication between the localized states. We found that all the decay curves can be well fitted with a constant β of 0.43 ± 0.03. Both lifetimes τ_2 and τ_1 show clear spectral dependence (Fig. 2c; Supplementary Fig. 2d). It markedly decreases on the high-energy side, while remaining constant on the low-energy side.

The spectral dependence of τ_2 can be described by a well-established model[30] for excitons localized in the tail states

$$\tau(E) = \frac{\tau_{LE}}{1 + \exp(\frac{E-E_{me}}{E_0})}, \quad (3)$$

where τ_{LE} is the lifetime of localized excitons, E_{me} can be regarded as the mobility edge and E_0 is a characteristic energy of the

density of band tail states, which can be a measure of the localization energy. The best fit of the data gives $\tau_{LE} = 61$ ns, $E_{me} = 2.419$ eV and $E_0 = 41$ meV. The localization energy, 41 meV, is higher than the RT thermal energy, which is consistent with localized excitons being observed even at RT.

We provide evidence to show that the PL in our perovskite films is not due to FC recombination but to a localization effect. The first evidence comes from the PL spectra under high excitation. As shown in Supplementary Fig. 3, increasing the excitation density leads to the occurrence and increase of a higher energy tail at ~ 2.41 eV. The result suggests that the main peak at 2.32 eV does not originate from FC recombination because FC emission has the highest energy among all radiative recombination. With increasing excitation density, the 2.32 eV peak is still dominant over the 2.41 eV FC emission and is almost unshifted. The unshifted luminescence agrees with the reported[31] feature of localized excitons, which means that the many-particle effect of these localized excitons is weak. The second evidence comes from the frequency-dependent THz measurements (the experimental details can be found in Supplementary methods). Supplementary Figure 4 plots the photoconductivity induced by 400-nm pump pulses. The real part of the induced photoconductivity $\Delta\sigma_1(\omega)$ decreases with decreasing frequency, and the imaginary part $\Delta\sigma_2(\omega)$ has a negative value at low frequency. The results do not support the Drude model for FCs[32], which predicts increasing $\Delta\sigma_1(\omega)$ with decreasing frequency and positive $\Delta\sigma_2(\omega)$. The real part increasing with frequency and the negative imaginary part are typical signatures of carrier localization[33,34]. The low conductivity at low energy is consistent with the insulating nature of the charge-neutral excitons[32]. Such features build up immediately after excitation ($\Delta t = 3$ ps) and are maintained within the entire timescale ($\Delta t = 300$ ps).

The above results indicate that the PL decay is dominated by recombination of localized excitons rather than FCs at RT. We also conducted excitation density-dependent PL and spectral-dependent lifetime measurements on $CH_3NH_3PbBr_3$ films at 237 K. These experiments were designed to check whether the assignment of localized excitons is valid for the perovskite films at low temperatures, which have higher quantum efficiency than that of the films at RT. The temperature of 237 K was chosen based on the cubic-to-tetragonal phase transition[35] of $CH_3NH_3PbBr_3$. As shown in Fig. 3, the relative internal quantum efficiency (IQE) at 237 K is estimated to be $\sim 80\%$ based on the temperature-dependent PL intensity (Supplementary Note 1), a method that has been widely used in inorganic compound

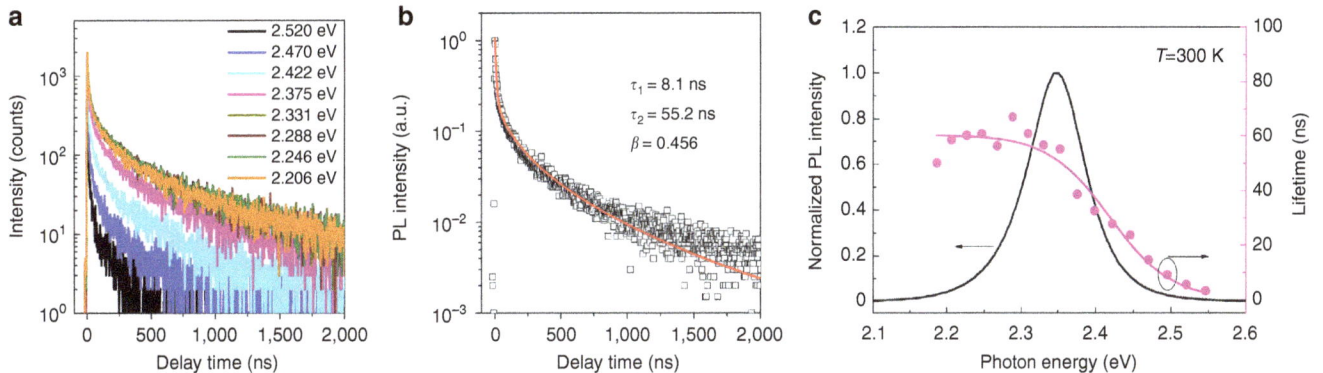

Figure 2 | Spectral-dependent PL decay of solution-processed $CH_3NH_3PbBr_3$ films. (a) PL decay curves monitored at various emission energies. The lifetime decreases markedly on the high-energy side of the emission. **(b)** Typical fitting of a decay curve by the thermalized stretching exponential model described by equation (2). **(c)** The lifetime of localized excitons τ_2 (circles) as a function of emission energy. The data are fitted with equation (3) (magenta line), with the lifetime of localized exciton $\tau_{LE} = 60.5$ ns, mobility edge of 2.419 eV and the localization energy $E_0 = 40.9$ meV.

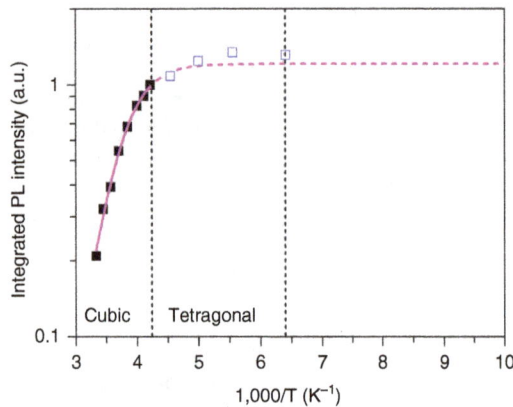

Figure 3 | PL thermal quenching behaviour of solution-processed CH$_3$NH$_3$PbBr$_3$ film. Integrated PL intensity as a function of the reciprocal temperature under low excitation (photocarrier density $\sim 3.7 \times 10^{14}$ cm^{-3}). There is a cubic (RT) to tetragonal phase transition at 236 K, and tetragonal-to-tetragonal phase transition at 155 K. The magenta solid curve represents the fitting result for the cubic data (Supplementary Note 1). The magenta dashed curve is the extrapolation result. The tetragonal data (open square) are plotted for comparison.

semiconductors[36]. The results (Supplementary Fig. 5) show that the k value and PL decay features at 237 K are similar to those at RT, indicating that localized excitons still dominate the PL under the condition of high quantum yield.

Recently, there have been arguments that at RT excitons cannot exist in perovskites because the dielectric constant of perovskite materials is very large[37–40], which results in dielectric screening and consequent dissociation of excitons. However, our evidence for exciton localization suggests that the role of the dielectric screening effect might be more complicated than expected. In fact, excitons do exist in materials with large dielectric constants. In ferroelectric oxides, such as SrTiO$_3$ and FeBiO$_3$, it is accepted that self-trapped excitons[41] and charge transfer vibronic excitons may exist[42]. The mechanism for these excitons can be quite different from the normal one. For example, charge transfer vibronic excitons is correlated electronic and hole polarons induced by the Jahn–Teller type lattice distortion, which are localized and can be quite stable at RT[42]. In view of these facts, we suggest that excitons can also exist in trihalide perovskites, despite the possible large dielectric constant under illumination. As additional experimental evidence, the absorption of exciton resonance is observed in the optical absorption spectra of CH$_3$NH$_3$PbBr$_3$ film (Supplementary Fig. 1b). Recent analysis of the absorption spectra suggested that bound exciton states also exist in CH$_3$NH$_3$PbI$_3$ films[15].

The properties of perovskite films may depend on the preparation methods and/or processing conditions. To test whether the exciton localization is a general mechanism, we prepared CH$_3$NH$_3$PbBr$_3$ films by the two-step method and measured the excitation-dependent PL and spectral-dependent lifetime (Supplementary Fig. 6). The samples also exhibit features similar to that shown in Fig. 2 and Fig. 3, indicating the localized exciton nature.

Exciton localization in CH$_3$NH$_3$PbI(Cl)$_3$ films. The emissions from solution-processed CH$_3$NH$_3$PbI$_3$ and CH$_3$NH$_3$PbI$_{3-x}$Cl$_x$ films are also dominated by localized excitons. The power-law dependence of the PL intensity on the excitation density reveals k values of $\sim$1.5 (Fig. 4a), and the PL decay shows spectral dependence (Supplementary Fig. 7) similar to that observed in CH$_3$NH$_3$PbBr$_3$. The k values are larger than the value obtained in

CH$_3$NH$_3$PbBr$_3$. According to the theory developed by Schmidt et al.[26] and Shibata et al.[27], such a difference mainly represents the different material properties such as the probabilities of radiative recombination and competitive nonradiative recombination. For example, the decrease of crystal perfection is expected to increase the k value. Moreover, the contribution of FC recombination may also lead to a change of k value. The PL lineshape analysis (Supplementary Fig. 8; Supplementary Note 2) indicates a small fraction ($\sim$9.5%) of FC recombination in the emission of CH$_3$NH$_3$PbI$_{3-x}$Cl$_x$ film, which is reasonable because the reported exciton binding energy[43] of CH$_3$NH$_3$PbI$_3$ and CH$_3$NH$_3$PbI$_{3-x}$Cl$_x$ is lower than CH$_3$NH$_3$PbBr$_3$, so the thermal dissociation of excitons is easier. The coexistence of exciton and FC recombination in the PL spectra has been observed in other materials[44] with exciton binding energy comparable to the RT thermal energy. Excitation-dependent PL at low temperature reveals smaller k values (Supplementary Fig. 9), which can be interpreted by reduced thermal dissociation of excitons at low temperature.

It is of interest to determine whether the conclusion of localized excitons in perovskite materials is valid for the same material in a photovoltaic device structure. We construct such a structure using CH$_3$NH$_3$PbI$_{3-x}$Cl$_x$, as shown in the inset of Fig. 4a. The PL intensity still shows power-law dependence on the excitation density, with a k value of 1.547, very close to the result of the bare film. Moreover, the PL decay spectra (Supplementary Fig. 10) show dependence on the emission energy and the lifetime can be well described by equation (3) with $E_0 \sim 17$ meV, as seen in Fig. 4b. The results indicate that the PL of CH$_3$NH$_3$PbI$_{3-x}$Cl$_x$ film in a typical photovoltaic structure is also dominated by the localized excitons rather than FC recombination.

Discussion
The physical picture of the recombination of localized excitons is illustrated in Fig. 5. The density of localized states is approximated by an exponential tail with the form of $\sim$exp $(-E/E_0)$. The excitons can be either partly localized (one carrier is localized with another carrier bound to it by Coulomb attraction) or wholly localized[21]. With increasing energy, the localized excitons may transit to the extended exciton states (approaching free excitons) at the transition region known as the mobility edge. Under low excitation, most of the photocarriers occupy the tail states. The picture can also be understood in the space coordinates as shown in Fig. 5b. In this case, the tail states are represented by the local potential minima in the conduction and/or valence bands. The photogenerated carriers transfer to these potential minima to form localized excitons, which have much longer lifetime than free excitons due to the transfer between localized states[28,29]. The long lifetime of the localized excitons accounts for the observed long PL lifetime. This phenomenon is also observed in inorganic semiconductors. For example, localized exciton lifetime as long as 65 ns at RT has been reported[29] in InGaN.

Tail states are very common in semiconductors and can be induced by doping, compositional changes and structural deformation[21,45]. Although solution-processed perovskites are materials with reasonable crystal quality, structure imperfections are inevitable. For example, the large rotational freedom of the polar CH$_3$NH$_3$$^+$ cation can produce structural disorder[37,46]. Unintentional/intentional doping is possible. The weak bonding between lead and halogens may also produce local disorder, especially in the surface and crystal boundary region. Recent studies[47] revealed that the grain boundaries exhibited faster nonradiative decay. Other results[48] suggest that perovskites with larger grains exhibit better photovoltaic performance. Given these

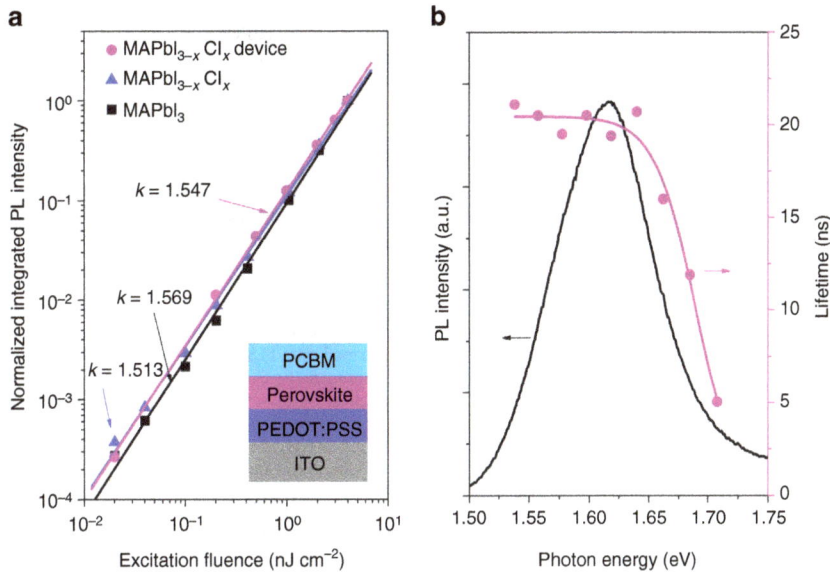

Figure 4 | Steady-state PL spectra and transient PL decay of CH₃NH₃PbI₃ and CH₃NH₃PbI₃₋ₓClₓ. (**a**) Logarithm plot of the integrated PL intensity versus excitation density for $CH_3NH_3PbI_3$ and $CH_3NH_3PbI_{3-x}Cl_x$ films. Insert is a typical photovoltaic structure with a $CH_3NH_3PbI_{3-x}Cl_x$ film sandwiched between two charge-transporting interlayers. The data show good power-law dependence with k values of 1.569, 1.513 and 1.547. (**b**) PL lifetime (solid circles) of ITO/PEDOT:PSS/ $CH_3NH_3PbI_{3-x}Cl_x$/PCBM photovoltaic structure as a function of the emission energy. The data are fitted with equation (3) (magenta line), with the lifetime of localized exciton $\tau_{LE} = 20.5$ ns, mobility edge 1.689 eV and localization energy $E_0 = 17.3$ meV. The PL lifetime is greatly reduced due to the quenching effects of the adjacent layers.

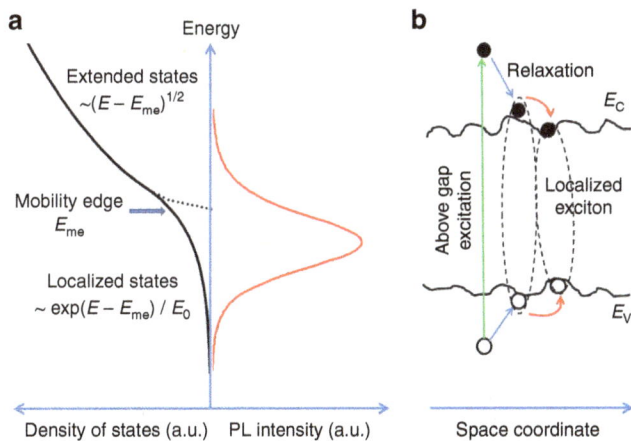

Figure 5 | Physical picture of exciton localization. (**a**) The density of the extended and localized states, and emission in perovskites (schematic). The density of the localized states is approximated by an exponential tail with the form of $\sim \exp(-E/E_0)$. The localized and extended states are divided by the mobility edge. Under low excitation, the excitons mainly occupy the localized states. Under high excitation, the localized states can be filled and the photocarriers also occupy the extended states, leading to emissions of free carriers. (**b**) Schematic drawing of exciton localization in space coordinates. With the presence of structural disorder, the tails of the localized states form local potential fluctuation in the energy bands. These potential minima can localize electrons and holes to form localized excitons. The carriers can transfer between the local potential minima, leading to long PL lifetime.

facts, it is reasonable to suggest the existence of tail states in solution-processed perovskite films.

The energy and intensity of localized state-related emission depend strongly on the nature (for example, energy level) and density of the localized states. In the case of strong localization, such as deep-level centres, the emission energy may be

substantially reduced. The localized states may emit weakly provided their density is low. In the present case, strong near-band-edge emission can be realized, indicating that excitons are weakly localized in the band tail states. The weak localization is evidenced by the relatively small localization energy reflected by the E_0 values in equation (3) and by the small Stokes shift of the luminescence (Supplementary Fig. 1b). The localization effect can even be beneficial to light-emitting devices. Localized exciton may contribute to the optical gain because the localized states can be easily filled, provided their density is not too high[31,49,50]. We performed temperature-dependent PL measurements under moderate excitation to check whether nonradiative channels become dominant when the excitation density increases. Supplementary Fig. 11 shows that the relative IQE of moderate excitation is much higher than that of low excitation, in agreement with the result reported by Deschler *et al*[8]. This result implies that other decay channels such as Auger recombination do not become dominant, which will be of great benefit to low threshold lasing. We conducted PL experiments at high excitation (photocarrier concentration up to $\sim 10^{19}$ cm^{-3}). Amplified spontaneous emission was observed on the low-energy side (at 2.26 eV) of the localized exciton emission with a threshold of ~ 300 µJ cm^{-2} (Supplementary Fig. 3b). It is noteworthy that FC emission (~ 2.41 eV) emerges with increasing excitation density. These results suggest that the amplified spontaneous emission is likely from localized excitons, which indicates that the optical gain can come from the filled localized states, and it is possible to achieve a low threshold by controlling the density of localized states. Moreover, exciton localization may also increase the luminescence efficiency because the oscillator strength of the transitions in the localized states is greatly enhanced[49,51].

We emphasize that the generation of localized excitons at excitation levels close to the working regime of solar cells does not conflict the fact that the hybrid perovskites are superior photovoltaic materials. The localized excitons can diffuse by a thermally activated multiple trapping–escaping process[52,53]. This process leads to an exponential dependence of the diffusion coefficient on the exciton energy (E_{exc}), as well as on the temperature[53]

$$D = D_0 \exp\left(\frac{E_{exc} - E_{me}}{k_B T}\right), \qquad (4)$$

where D_0 is the diffusion coefficient of the extended excitons. Taking the energetic distance $E_{exc} - E_{me} \sim 70$ meV (as extracted from Figs 2c,4b), we estimate the RT diffusion coefficient of localized excitons D as $0.067 D_0$, only approximately one order of magnitude lower than the free exciton. If we invoke the reported[11,12], D value of $0.011 - 0.054 \, \text{cm}^2 \, \text{s}^{-1}$ for the perovskites, a D_0 of $\sim 0.16 - 0.80 \, \text{cm}^2 \, \text{s}^{-1}$ is obtained. This value is comparable to the reported values[53,54] for $CdS_{1-x}Se_x$ ($0.3 \, \text{cm}^2 \, \text{s}^{-1}$) and $In_{1-x}Ga_xN$ ($0.5-1.1 \, \text{cm}^2 \, \text{s}^{-1}$), two alloy semiconductors known as the typical examples for the study of localized excitons. In combination with the very long lifetime, a long carrier diffusion length in perovskites is expected in the scenario of exciton localization.

In summary, we have presented solid evidence that the RT PL in organolead trihalide perovskites is dominated by weakly localized excitons. The evidence includes: (i) the excellent power-law dependence of the PL intensity on the excitation intensity with $1 < k < 2$, (ii) the localization effect indicated by the spectral dependence of the PL lifetime and frequency-dependent THz photoconductivity and (iii) the coexistence of exciton and FC recombination under high excitation. Exciton localization is suggested as the origin of the long PL lifetime in this class of materials. We find that the localization effect is general in solution-process perovskite films due to the presence of crystal imperfections. The localization of excitons strongly influences the transport and recombination properties of perovskite materials. The dominance of the localized exciton in the recombination channels as well as its higher IQE under moderate excitation, strongly suggests that it is possible to utilize these benefits to realize low threshold lasing in perovskites, as has been demonstrated in III–V and II–VI semiconductors and devices. The elaborate tailoring of the localization effect in perovskites is thus highly attractive in designing future high-performance optoelectronic devices.

Methods

Synthesis of perovskite films. All the indium-tin oxide (ITO)-coated glass substrates were cleaned sequentially in deionized water, ethanol, acetone and oxygen plasma before spin-coating. The perovskite $CH_3NH_3PbI_{3-x}Cl_x$ was prepared according to the reported procedure[11]. Methylamine iodide was prepared by reacting 33 wt % methylamine in ethanol (Sigma-Aldrich), with 57 wt % hydroiodic acid in water (Sigma-Aldrich), at RT. Hydroiodic acid was added dropwise while stirring. After drying at 100 °C, the resultant white powder was dried overnight in a vacuum oven and was recrystallized from ethanol before use. To form the $CH_3NH_3PbI_{3-x}Cl_x$ precursor solution, methylammonium iodide and lead (II) chloride (Sigma-Aldrich) were dissolved in anhydrous N,N-dimethylformamide (DMF) in a 3:1 molar ratio of methylamine iodide to $PbCl_2$, with final concentrations 0.88 M lead chloride and 2.64 M methylammonium iodide. The precursor was filtered through a 220-nm polytetrafluoroethylene (PTFE) filter head, then spin-coated at 3,000 r.p.m. for 30 s on ITO-coated glass; finally, it was annealed at 95 °C for ~ 10 min. $CH_3NH_3PbI_3$ was prepared via the sequential deposition route[3]. A PbI_2 (Sigma-Aldrich) solution in DMF (462 mg ml^{-1}) was spin-coated on glass substrate and then kept at 70 °C. After drying, the films were dipped in a solution of CH_3NH_3I in 2-propanol (10 mg ml^{-1}) for ~ 60 s and rinsed with 2-propanol, and then spin-coated to form uniform $CH_3NH_3PbI_3$ thin films. For $CH_3NH_3PbBr_3$ preparation,[24] CH_3NH_3Br was first prepared by mixing methylamine with hydrobromic acid (48% in water; CAUTION: exothermic reaction) in 1:1 molar ratio in a 100-ml flask under continuous stirring at 0 °C for 2 h. CH_3NH_3Br was then crystallized by removing the solvent in an evaporator, washing three times in diethyl ether for 30 min and filtering the precipitate. The material, in the form of white crystals, was then dried in vacuum at 60 °C for 24 h and was then kept in a dark, dry environment until further use. A 20-wt % solution of $CH_3NH_3PbBr_3$ was prepared by mixing $PbBr_2$ and CH_3NH_3Br in a 1:3 molar ratio in DMF. The precursor was spin-coated at 4,000 r.p.m. for 30 s on ITO-coated glass and was annealed at 60 °C for ~ 10 min.

Fabrication of the photovoltaic device structure. The structure of ITO/PED-OT:PSS/perovskite/PCBM was fabricated on patterned ITO-coated glass substrates (sheet resistance: 15 Ω sq^{-1}). The substrates were cleaned sequentially in acetone, ethanol, deionized water and ethanol for 10 min each, followed by oxygen plasma treatment for 15 min. A poly(3,4-ethylenedioxythiophene):poly(styrenesulfonate) (PEDOT:PSS) layer was spin-coated onto the substrates at 4,000 r.p.m. for 60 s and then annealed in air at 140 °C. The sample was transferred into a glove box. Next, the $CH_3NH_3PbI_{3-x}Cl_x$ precursor solution was spin-coated at 3,000 r.p.m. for 45 s, followed by annealing on a hot plate at 100 °C for ~ 60 min. The [6,6]-phenyl-C61-butyric acid methyl ester (PCBM) layers were deposited from a 30 mg ml^{-1} chlorobenzene solution at 2,000 r.p.m. for 45 s. To avoid the degradation induced by air exposure, the devices were packaged with glass.

PL measurements. The steady-state and time-resolved PL measurements were performed on a FLS920 fluorescence spectrophotometer (Edinburgh Instruments). To minimize the effect of air exposure, all the measurements were taken in vacuum with pressure of < 0.01 torr. Pulsed laser diodes (EPL405 and EPL635) with tunable repeating frequency of 20 kHz to 20 MHz were used as the excitation source. For $CH_3NH_3PbI_3$ and $CH_3NH_3PbI_{3-x}Cl_x$ films, EPL635 with a wavelength of 638.8 nm and pulse width of 86.4 ps was used. For $CH_3NH_3PbBr_3$, we used EPL405 with a wavelength of 404.2 nm and pulse width of 58.6 ps. The excitation fluence for both wavelengths was ~ 4 nJ cm^{-2}. For excitation density-dependent measurements at the low level, the lasers operated at 20 MHz and the light fluence was tuned by a neutral attenuator. For moderate excitation experiments, a 355 nm frequency-tripled Nd:YAG laser (FTSS 355-50, CryLaS GmbH) with a pulse width of 1 ns and a repetition rate of 100 Hz was used. For lifetime measurements by time-correlated single-photon counting, the lasers operated at 200 kHz. The temperature-dependent measurements were performed with a closed-cycle helium cryostat.

Calculation of the photocarrier density. The corresponding photocarrier density can be calculated as

ρ_{exc} = light fluence density of a single pulse/(photon energy × optical penetration depth) = 4 nJ cm^{-2}/(3.07 eV × 1.6 × 10^{-19} J × 220 nm) $\sim 3.7 \times 10^{14}$ cm^{-3}

Here the optical penetration depth of $CH_3NH_3PbBr_3$ is taken as ~ 220 nm (ref. 55). In this calculation, we assume a constant excitation because the effective excited volume is remarkably larger than the directly excited one due to the carrier diffusion during the long carrier lifetime[21]. For $CH_3NH_3PbI_3$ and $CH_3NH_3PbI_{3-x}Cl_x$, the photon energy of the excitation laser was 1.94 eV, and the optical penetration depth was 250 nm (ref. 1); hence, the photocarrier density was $\sim 5.2 \times 10^{14}$ cm^{-3}.

References

1. Kojima, A., Teshima, K., Shirai, Y. & Miyasaka, T. Organometal halide perovskites as visible-light sensitizers for photovoltaic cells. *J. Am. Chem. Soc.* **131**, 6050–6051 (2009).
2. Lee, M. M., Teuscher, J., Miyasaka, T., Murakami, T. N. & Snaith, H. J. Efficient hybrid solar cells based on meso-superstructured organometal halide perovskites. *Science* **338**, 643–647 (2012).
3. Burschka, J. Sequential deposition as a route to high-performance perovskite-sensitized solar cells. *Nature* **499**, 316–319 (2013).
4. Jeon, N. J. et al. Compositional engineering of perovskite materials for high-performance solar cells. *Nature* **517**, 476–480 (2015).
5. Bai, S. et al. High-performance planar heterojunction perovskite solar cells: preserving long charge carrier diffusion lengths and interfacial engineering. *Nano Res.* **7**, 1749–1758 (2014).
6. Research Cell Efficiency Records. Available at http://www.nrel.gov/ncpv/ (2015).
7. Xing, G. et al. Low-temperature solution-processed wavelength-tunable perovskites for lasing. *Nat. Mater.* **13**, 476–480 (2014).
8. Deschler, F. et al. High photoluminescence efficiency and optically pumped lasing in solution-processed mixed halide perovskite semiconductors. *J. Phys. Chem. Lett.* **5**, 1421–1426 (2014).
9. Tan, Z. K. et al. Bright light-emitting diodes based on organometal halide perovskite. *Nat. Nanotechnol* **9**, 687–692 (2014).
10. Wang, J. P. et al. Interfacial control toward efficient and low-voltage perovskite light-emitting diodes. *Adv. Mater.* **27**, 2311–2316 (2015).
11. Stranks, S. D. et al. Electron-hole diffusion lengths exceeding 1 micrometer in an organometal trihalide perovskite absorber. *Science* **342**, 341–344 (2013).
12. Xing, G. C. et al. Long-range balanced electron and hole-transport lengths in organic-inorganic $CH_3NH_3PbI_3$. *Science* **342**, 344–347 (2013).
13. Laquai, F. All-round perovskites. *Nat. Mater.* **13**, 429–430 (2014).
14. D'Innocenzo, V. et al. Excitons versus free charges in organo-lead tri-halide perovskites. *Nat. Commun.* **5**, 3586–3591 (2014).
15. Saba, M. et al. Correlated electron-hole plasma in organometal perovskites. *Nat. Commun.* **5**, 5049–5058 (2014).

16. Sun, S. *et al.* The origin of high efficiency in low-temperature solution-processable bilayer organometal halide hybrid solar cells. *Energy Environ. Sci.* **7**, 399–407 (2014).

17. Savenije, T. J. *et al.* Thermally activated exciton dissociation and recombination control the organometal halide perovskite carrier dynamics. *J. Phys. Chem. Lett.* **5**, 2189–2194 (2014).

18. Zhang, W. *et al.* Enhancement of perovskite-based solar cells employing core-shell metal nanoparticles. *Nano Lett.* **13**, 4505–4510 (2013).

19. Yamada, Y., Nakamura, T., Endo, M., Wakamiya, A. & Kanemitsu, Y. Photocarrier recombination dynamics in perovskite $CH_3NH_3PbI_3$ for solar cell applications. *J. Am. Chem. Soc.* **136**, 11610–11613 (2014).

20. Lin, Q. Q., Armin, A., Nagiri, R. C. R., Burn, P. L. & Meredith, P. Electro-optics of perovskite solar cells. *Nat. Photon.* **9**, 106–112 (2015).

21. Klingshirn, C. F. *Semiconductor Optics* (Springer-Verlag, 1997).

22. Cingolani, R. *et al.* Radiative recombination processes in wide-band-gap II–VI quantum wells: the interplay between excitons and free carriers. *J. Opt. Soc. Am. B* **13**, 1268–1277 (1996).

23. Wehrenfennig, C., Eperon, G. E., Johnston, M. B., Snaith, H. J. & Herz, L. M. High charge carrier mobilities and lifetimes in organolead trihalide perovskites. *Adv. Mater.* **26**, 1584–1589 (2014).

24. Edri, E., Kirmayer, S., Cahen, D. & Hodes, G. High open-circuit voltage solar cells based on organic-inorganic lead bromide perovskite. *J. Phys. Chem. Lett.* **4**, 897–902 (2013).

25. Stranks, S. D. *et al.* Recombination kinetics in organic-inorganic perovskites: excitons, free charge, and subgap states. *Phys. Rev. Appl.* **2**, 034007 (2014).

26. Schmidt, T., Lischka, K. & Zulehner, W. Excitation-power dependence of the near-band-edge photoluminescence of semiconductors. *Phys. Rev. B* **45**, 8989–8994 (1992).

27. Shibata, H. *et al.* Excitation-power dependence of free exciton photoluminescence of semiconductors. *Jpn J. Appl. Phys.* **44**, 6113–6114 (2005).

28. Sun, Y. J. *et al.* Nonpolar $In_xGa_{1-x}N$/GaN (1-100) multiple quantum wells grown on γ-$LiAlO_2$ (100) by plasma-assisted molecular-beam epitaxy. *Phys. Rev. B* **67**, 041306 (2003).

29. Onuma, T. *et al.* Localized exciton dynamics in nonpolar (11-20) $In_xGa_{1-x}N$ multiple quantum wells grown on GaN templates prepared by lateral epitaxial overgrowth. *Appl. Phys. Lett.* **86**, 151918 (2005).

30. Gourdon, C. & Lavallard, P. Exciton transfer between localized states in $CdS_{1-x}Se_x$ alloys. *Phys.Stat. Sol. (b)* **153**, 641–652 (1989).

31. Majumder, F. A., Shevel, S., Lyssenko, V. G., Swoboda, H. E. & Klingshirn, C. Luminescence and gain spectroscopy of disordered $CdS_{1-x}Se_x$ under high excitation. *Z. Phys. B* **66**, 409–418 (1987).

32. Kaindl, R. A., Carnahan, M. A., Hagele, D., Lovenich, R. & Chemla, D. S. Ultrafast terahertz probes of transient conducting and insulating phases in an electron-hole gas. *Nature* **423**, 734–738 (2003).

33. La-o-vorakiat, C. *et al.* Elucidating the role of disorder and free-carrier recombination kinetics in $CH_3NH_3PbI_3$ perovskite films. *Nat. Commun.* **6**, 7903–7909 (2015).

34. Nemec, H., Kuzel, P. & Sundström, V. Charge transport in nanostructured materials for solar energy conversion studied by time-resolved terahertz spectroscopy. *J. Photochem. Photobiol. A.* **215**, 123–139 (2010).

35. Mashiyama, H., Kurihara, Y. & Azetsu, T. Disordered cubic perovskite structure of $CH_3NH_3PbX_3$ (X = Cl, Br, I). *J. Korean Phys. Soc.* **32**, S156–S158 (1998).

36. Okamoto, K. *et al.* Surface-plasmon-enhanced light emitters based on InGaN quantum wells. *Nat. Mater.* **3**, 601–605 (2004).

37. Juarez-Perez, E. J. *et al.* Photoinduced giant dielectric constant in lead halide perovskite solar cells. *J. Phys. Chem. Lett.* **5**, 2390–2394 (2014).

38. Frost, J. M. *et al.* Atomistic origins of high-performance in hybrid halide perovskite solar cells. *Nano Lett.* **14**, 2584–2590 (2014).

39. Menendez-Proupin, E., Palacios, P., Wahnon, P. & Conesa, J. C. Self-consistent relativistic band structure of the $CH_3NH_3PbI_3$ perovskite. *Phys. Rev. B* **90**, 045207 (2014).

40. Even, J., Pedesseau, L. & Katan, C. Analysis of multivalley and multibandgap absorption and enhancement of free carriers related to exciton screening in hybrid perovskites. *J. Phys. Chem. C* **118**, 11566–11572 (2014).

41. Kan, D. *et al.* Blue-light emission at room temperature from Ar^+-irradiated $SrTiO_3$. *Nat. Mater.* **4**, 816–819 (2005).

42. Vikhnin, V. S., Eglitis, R. I., Kapphan, S. E., Borstel, G. & Kotomin, E. A. Polaronic-type excitons in ferroelectric oxides: microscopic calculations and experimental manifestation. *Phys. Rev. B.* **65**, 104304 (2002).

43. Tanaka, K. *et al.* Comparative study on the excitons in lead-halide-based perovskite type crystals $CH_3NH_3PbBr_3$ $CH_3NH_3PbI_3$. *Solid State Commun.* **127**, 619–623 (2003).

44. Wang, H., Wong, K. S. & Wong, G. K. L. in *Proceedings of SPIE 3624*, 13–24 (International Society for Optical Engineering, San Jose, CA, USA, 1999).

45. Permogorov, S. & Reznitsky, A. Effect of disorder on the optical-spectra of wide-gap II-VI semiconductor solid-solutions. *J. Lumin.* **52**, 201–223 (1992).

46. Wasylishen, R. E., Knop, O. & Macdonald, J. B. *Solid State Commun.* **56**, 581 (1985).

47. deQuilettes, D. W. *et al.* Impact of microstructure on local carrier lifetime in perovskite solar cells. *Science* **348**, 683–686 (2015).

48. Nie, W. Y. *et al.* High-efficiency solution-processed perovskite solar cells with millimeter-scale grains. *Science* **347**, 522–525 (2015).

49. Satake, A. *et al.* Localized exciton and its stimulated emission in surface mode from single-layer $In_xGa_{1-x}N$. *Phys. Rev. B* **57**, R2041–R2044 (1998).

50. Chen, R. *et al.* Exciton localization and optical properties improvement in nanocrystal-embedded ZnO core-shell nanowires. *Nano Lett.* **13**, 734–739 (2013).

51. O'Donnell, K. P., Martin, R. W. & Middleton, P. G. Origin of luminescence from InGaN diodes. *Phys. Rev. Lett.* **82**, 237–240 (1999).

52. Chichibu, S. F. *et al.* Recombination dynamics of localized excitons in cubic $In_xGa_{1-x}N$/GaN multiple quantum wells grown by radio frequency molecular beam epitaxy on 3C–SiC substrate. *J. Vac. Sci. Technol. B* **21**, 1856–1862 (2003).

53. Schwab, H., Pantke, K. H., Hvam, J. M. & Klingshirn, C. Measurements of exciton diffusion by degenerate four-wave mixing in $CdS_{1-x}Se_x$. *Phys. Rev. B* **46**, 7528–7532 (1992).

54. Okamoto, K. *et al.* in *Proceedings of SPIE (International Society for Optical Engineering)* **4278**, 157–150 (Bellingham, WA, USA, 2001).

55. Kumawat, N. K., Dey, A., Narasimhan, K. L. & Kabra, D. Near infrared to visible electroluminescent diodes based on organometallic halide perovskites: structural and optical investigation. *ACS Photon.* **2**, 349–354 (2015).

Acknowledgements

This work was supported by the Natural Science Foundation of China (nos 51372223, 91333203, 51522209, 11474164, 61405091, 11474249 and 91433204), the Program for Innovative Research Team in University of Ministry of Education of China (no. IRT13037), the National Basic Research Program of China- Fundamental Studies of Perovskite Solar Cells (2015CB932200), the Natural Science Foundation of Jiangsu Province, China (BK20131413, BK20140952), the National 973 Program of China (2015CB654901), the Synergetic Innovation Center for Organic Electronics and Information Displays and the Fundamental Research Funds for the Central Universities (nos. 2014FZA4008 and 2015FZA3005). We thank Mr Yunzhou Deng (Zhejiang University) for his help in the numerical fitting of the PL spectra.

Author contributions

H.H. and Z.Y. supervised the study. H.H., Q.Y., H.L. and J.L. contributed to the PL measurements and analysis. Q.Y., J.S. and N.W. contributed to the synthesis and characterization of the materials. J.H., X.W. and Y.Z. contributed to the THz measurements and analysis. Y.J. and J.W. provided input to the data analysis and discussed the results. H.H. and Y.J. wrote the manuscript. All authors assisted in manuscript preparation.

Additional information

Statistical moments of quantum-walk dynamics reveal topological quantum transitions

Filippo Cardano[1], Maria Maffei[1], Francesco Massa[1,†], Bruno Piccirillo[1], Corrado de Lisio[1,2], Giulio De Filippis[1,2], Vittorio Cataudella[1,2], Enrico Santamato[1] & Lorenzo Marrucci[1,3]

Many phenomena in solid-state physics can be understood in terms of their topological properties. Recently, controlled protocols of quantum walk (QW) are proving to be effective simulators of such phenomena. Here we report the realization of a photonic QW showing both the trivial and the non-trivial topologies associated with chiral symmetry in one-dimensional (1D) periodic systems. We find that the probability distribution moments of the walker position after many steps can be used as direct indicators of the topological quantum transition: while varying a control parameter that defines the system phase, these moments exhibit a slope discontinuity at the transition point. Numerical simulations strongly support the conjecture that these features are general of 1D topological systems. Extending this approach to higher dimensions, different topological classes, and other typologies of quantum phases may offer general instruments for investigating and experimentally detecting quantum transitions in such complex systems.

[1]Dipartimento di Fisica, Università di Napoli Federico II, Complesso Universitario di Monte Sant'Angelo, via Cintia, Napoli 80126, Italy. [2]CNR-SPIN, Complesso Universitario di Monte Sant'Angelo, Via Cintia, Napoli 80126, Italy. [3]CNR-ISASI, Via Campi Flegrei 34, Pozzuoli (NA) 80078, Italy. †Present address: Faculty of Physics, University of Vienna, Boltzmanngasse 5, Vienna 1090, Austria. Correspondence and requests for materials should be addressed to L.M. (email: lorenzo.marrucci@unina.it).

The presence of topological order in matter is responsible for fundamental phenomena, such as the fractional and integer quantum Hall effects[1,2] or the protected surface states observed in topological insulators[3,4]. Non-trivial topological phases are related to symmetries and can be characterized by specific topological invariants, defined in terms of the energy band eigenstates. Besides physical systems that naturally exhibit such features, the same topological scenarios have been reproduced in easy-to-access artificial systems (quantum simulators), which represent convenient platforms for the investigation of the intriguing properties associated with these states of matter[5-9]. In particular, a variety of simulators based on photonic architectures have been realized[10-16].

In the field of quantum simulation of topological phenomena, quantum walks (QWs) are emerging as a versatile tool[17-19]. In its simplest version, a QW is the discrete-time evolution of a particle (the walker) on a one-dimensional (1D) lattice[20]. At each step, the walker can move to either one of the two nearest-neighbour sites of the lattice, as determined by the configuration of an internal two-state quantum system (the coin). Between consecutive steps, the coin state undergoes a rotation that determines the relative probability amplitudes of the subsequent walker move, thus providing the quantum version of the random choice process (the coin toss) characterizing the familiar classical random walk. Among simulators of topological physics, QWs are attracting a wide attention since this simple quantum dynamics can realize all topological phases occurring in one- and two-dimensional systems of non interacting particles[18]; as a first application of this concept, the formation of topologically protected bound states was observed in a photonic QW[10].

In this paper, we report the study of a QW process that exhibits two distinct topological phases, associated with chiral symmetry in a 1D bipartite lattice. Remarkably, identical symmetries and topological features characterize the Su–Schrieffer–Heeger model (SSH), describing for example the mobile electron dynamics in the poly-acetylene chain[21], and the effective theory for spin-less 1D superconductors showing p-wave pairing[22]. By studying the QW protocol introduced here, we found that the probability distribution moments for the particle position, in the limit of an infinitely long temporal evolution, show a different asymptotic behaviour for the two topological phases: varying an external control parameter, whose value determines the phase of the system, they exhibit a slope discontinuity at the quantum transition. These effects are found to occur also in other 1D topological systems, including models belonging to different topological classes; hence, we conjecture here that they are fully general in 1D. We simulated experimentally such phenomena and proposed a theoretical interpretation based on the dispersion relations and the geometric features of the system eigenstates. Remarkably, our analysis takes into account bulk dynamics only, and not the physical effects manifesting at the edges of the system. With a similar approach topological phases have been recently detected in a non-Hermitian QW,[17,11] while all the models we investigate here are Hermitian.

Results

Quantum-walk protocol based on light spin–orbit interaction. In recent years, QWs have been implemented in a variety of physical architectures[23]. In our photonic platform[24], the walker is encoded in the orbital angular momentum (OAM) of light[25]: discrete positions on the lattice are associated with states $|m\rangle$, where m is an integer, describing a photon carrying $m\hbar$ of OAM along its propagation axis. The coin is encoded in the polarization degree of freedom, that is, in the spin angular momentum (SAM) of photons: vectors $|L\rangle$ and $|R\rangle$, representing left and right

circular polarizations (SAM of $\pm\hbar$ per photon), respectively, are the two internal states that determine opposite shift directions in the lattice. The quantum state of the photon after N steps of QW is denoted as $|\psi_N\rangle = \hat{U}_0^N|\psi_0\rangle$, where $|\psi_0\rangle$ is the initial (input) state and $\hat{U}_0$ is the unitary evolution operator of a single QW step. In our protocol, this is realized by cascading two optical elements: a quarter-wave plate (QWP) and a q-plate (QP). A q-plate is essentially a liquid-crystal birefringent cell having the optic axis arranged in a singular pattern[26], with topological charge q (in our case $q = 1/2$). This patterned birefringence gives rise to an engineered optical spin–orbit coupling that induces the polarization-controlled shift of OAM. As specified below (see equation 2), besides the charge q, the action of this device is determined by the value of the optical retardation δ, which can be tuned with an applied electric field[27]; this allows in turn for a fine control of the light spin–orbit interaction mediated by the plate. In our protocol, the QWP is oriented at 90° with respect to the horizontal direction; its action is described by the operator $\hat{W}_{qwp}$, that transforms the polarization states as follows:

$$\hat{W}_{qwp}|L, m\rangle = \frac{1}{\sqrt{2}}(|L, m\rangle - i|R, m\rangle),$$
$$\hat{W}_{qwp}|R, m\rangle = \frac{-i}{\sqrt{2}}(|L, m\rangle + i|R, m\rangle). \tag{1}$$

The q-plate action is described by the operator $\hat{Q}_\delta$, defined as

$$\hat{Q}_\delta|L, m\rangle = \cos(\delta/2)|L, m\rangle + i\sin(\delta/2)|R, m + 2q\rangle,$$
$$Q_\delta|R, m\rangle = \cos(\delta/2)|R, m\rangle + i\sin(\delta/2)|L, m - 2q\rangle. \tag{2}$$

Here we can observe that the spin–orbit interaction introduced by the plate, proportional to $\sin(\delta/2)$, results in an exchange of angular momenta, that is a conversion from SAM to OAM. The combination of $\hat{Q}_\delta$ and $\hat{W}_{qwp}$ realizes the operator for a single QW step, that is

$$\hat{U}_0 = \hat{Q}_\delta \hat{W}_{qwp}. \tag{3}$$

A topological characterization of QWs with chiral symmetry may be defined in terms of the eigenstates of the operator $\hat{U}_0$ which, due to the lattice translation symmetry, are associated with two quasi-energy bands parameterized by a quasi-momentum k, as shown in Fig. 1a. As for momentum k, a periodic quasi-energy variable E replaces here the usual energy, being the time (= step number) a discrete variable. Accordingly, the dispersion relation for our QW is given by

$$E_\delta(k) = \pm \cos^{-1}\left[\frac{\cos(\delta/2) + \sin(\delta/2)\cos(k)}{\sqrt{2}}\right], \tag{4}$$

where the two signs correspond to the two bands. As a consequence, the expression of the associated group velocity is

$$V_\delta(k) = \frac{dE_\delta(k)}{dk} = \mp \frac{\sin(\delta/2)\sin(k)}{\sqrt{2 - [\cos(\delta/2) + \sin(\delta/2)\cos(k)]^2}}. \tag{5}$$

The band eigenstates are given by the product of a walker part, that is plane waves with quasi-momentum k in the walker space, and a coin part, denoted as $|\phi_{s,\delta}(k)\rangle$, where $s \in \{1, 2\}$ is the band index. Conveniently, these can be represented as points on the Poincaré sphere for light polarization, individuated by a three-dimensional unit vector that we refer to as $\mathbf{n}_\delta(k)$; here we

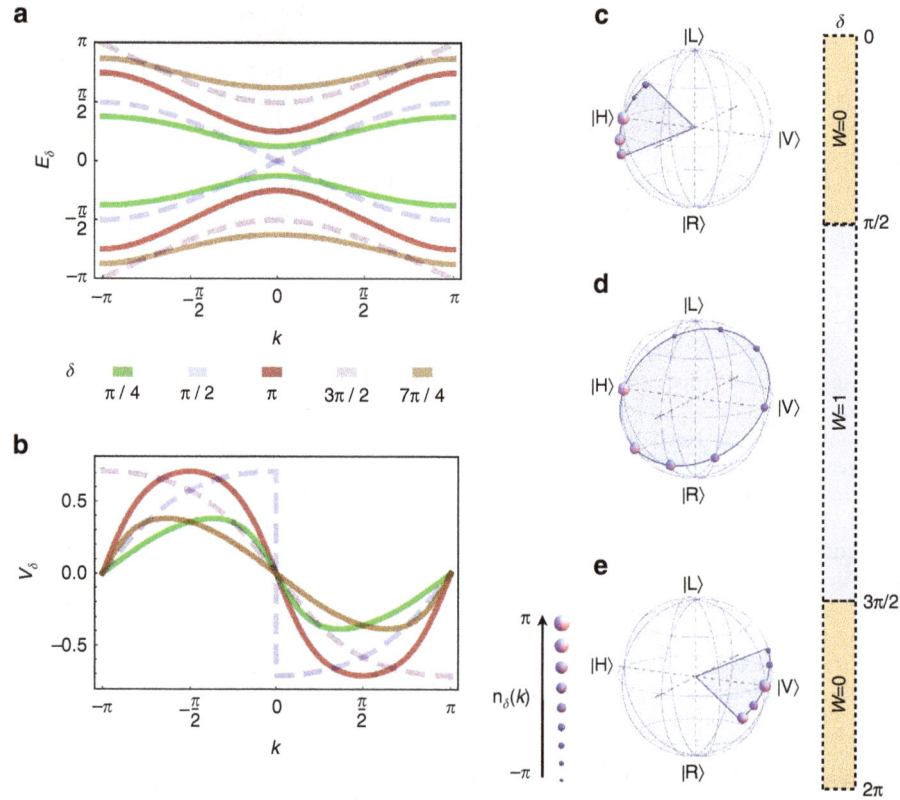

Figure 1 | Topological characterization of the QW process. (a,b) Dispersion relation for the quasi-energy $E_\delta(k)$ and the group velocity $V_\delta(k)$ of the Bloch bands in the QW; for the latter quantity, we plot the lower band only. The dispersion curves depend on the external parameter δ; the temporal coordinate is a discrete variable, hence the quasi-energy is defined in a Brillouin zone $\{-\pi, \pi\}$. Here we report few examples; colours of the dispersion curves are associated with specific values of δ, as shown in the panel legend. Generally, a finite gap separates the energies of the two bands; only at the two points $\delta = \delta_1 = \pi/2$ and $\delta = \delta_2 = 3\pi/2$ (blue and purple dashed lines, respectively) the gap vanishes for $k = 0$ and $k = \pi$, respectively, causing the presence of a discontinuity in the group velocity (see panel **b**). As shown in **c** points δ_1 and δ_2 represent the boundaries between two topological phases. **(c-e)** Topology of the Bloch eigenstates. States $|\phi_s(k)\rangle$, representing the polarization (coin) part of the eigenstates of the system, can be represented as points on the surface of a Poincaré sphere, where they are individuated by $\pm \mathbf{n}_\delta(k)$. Here we represent these states as solid spheres, whose radius is related to the value of k, as shown in the legend. As the quasi-momentum k spans the Brillouin zone $\{-\pi, \pi\}$, chiral symmetry forces these states to lie on a great circle of the sphere. The associated winding number W depends on δ, and determines the existence of a non-trivial and a trivial topological phases; the former ($W = 1$) occurs when $\delta_1 < \delta < \delta_2$, while the latter ($W = 0$) is obtained in the remaining part of the interval $\{0, 2\pi\}$. As an example, in **c-e** we represent the coin eigenstates of a QW with $\delta = \pi/4, \pi, 7\pi/4$, respectively. For $\delta = \pi$ these form a closed loop along the great circle, while in the other cases they go back and forth along a finite arch, whose length depends on the value of δ (when $\delta = 0$ or 2π, the arch reduces to a single point).

report the explicit expressions of its components, that is

$$n_x(k) = \frac{\cos(\delta/2) - \sin(\delta/2)\cos k}{\sqrt{2\{1 - \cos^2[E_\delta(k)]\}}},$$

$$n_y(k) = -\frac{\sin(\delta/2)\sin k}{\sqrt{2\{1 - \cos^2[E_\delta(k)]\}}}, \qquad (6)$$

$$n_z(k) = -n_y(k).$$

Chiral symmetry constrains vectors $\mathbf{n}_\delta(k)$ to lie on a great circle of the sphere. When k varies throughout the Brillouin zone $\{-\pi, \pi\}$, the number of closed loops (winding number) described by $\mathbf{n}_\delta(k)$ is a topological invariant, labelled here as W. As shown in Fig. 1c the value of δ in the range $\{0, 2\pi\}$, that is the strength of the spin–orbit interaction, determines the existence of two phases with a different winding number: a phase with $W = 2q = 1$ occurs when $\delta_1 < \delta < \delta_2$, while $W = 0$ in the remaining regions. As pointed out in refs 19,28, for time-periodic systems with chiral symmetry a complete topological classification would require introducing two topological invariants, so W alone is not enough. Nevertheless, this is not necessary for the purposes of this work, as we focus our attention on the phase transitions only, while we

do not need to identify completely the nature of the topological phases, and it can be shown that all topological phase transitions are associated with a change of W[19,28]. Therefore, to keep the discussion simpler we will use only the winding number W in our discussion. In this sense, we will refer to the phases $W = 0$ and 1 as trivial and non-trivial phases, respectively. In Fig. 1a it can be noted that when $\delta = \delta_1$ and $\delta = \delta_2$ the two bands are touching and the dispersion is locally linear at $k = 0$ and $k = \pi$, respectively, as typically occurs for quantum transitions between topological phases[3]. Interestingly, comparing the expressions of group velocity $V_\delta = dE_\delta/dk$ (equation 5) and vector $\mathbf{n}_\delta(k)$ (equation 6), we can observe that $n_z = -n_y = |V_\delta|$. Thus in the non-trivial phase, where the trajectory of $\mathbf{n}_\delta(k)$ is fixed, the maximum of the group velocity is independent of δ (see the Supplementary Note 1 and the Supplementary Fig.1 for further details). A recently introduced QW protocol, that is the split-step QW, has very similar properties[10,18].

Dynamical moments and topological phases. The features of the energy bands and associated eigenstates have profound consequences on the system dynamics, which shows marked differences in the two topological phases[11,17]. Here we characterize such different behaviour through the analysis of the moments of the probability distribution $P(m)$ associated with the walker position, defined as $M_j = \sum_m m^j P(m)$. In particular, we consider a photon starting its walk in the position $m = 0$ with an arbitrary polarization, that is $|\psi_0\rangle = |0\rangle \otimes |\phi_0\rangle$, where $|\phi_0\rangle$ is a generic state in the coin Hilbert space. Simulations of the QW evolution in the large step-number limit show that the moments M_j assume a constant value (independent of δ) in the non-trivial phase, and undergo abrupt slope variations at the phase transitions, that is at $\delta = \{\delta_1, \delta_2\}$. In the infinite-steps-limit, we proved that these moments have simple asymptotic expressions in terms of the energy band dispersion relations; in particular, those relative to the first and second moments M_1 and M_2 are the following (see Methods for a proof):

$$M_1/N = (s_y - s_z)L(\delta) + O(1/N), \quad (7)$$

$$M_2/N^2 = L(\delta) + O(1/N^2), \quad (8)$$

where $s_i = \langle \phi_0 | \hat{\sigma}_i | \phi_0 \rangle$, with $i \in \{x, y, z\}$, are the reduced Stokes parameters, calculated as the expectation values of the Pauli operators $\hat{\sigma}_i$ for the coin initial state $|\phi_0\rangle$. Interestingly, we observe that M_2 is independent of the initial coin state. The quantity $L(\delta)$ appearing in equations 7 and 8 is equal to the square of the group velocity V_δ, and hence to the square of n_y, averaged over the Brillouin zone:

$$L(\delta) = \int_{-\pi}^{\pi} \frac{dk}{2\pi} [V_\delta(k)]^2 = \int_{-\pi}^{\pi} \frac{dk}{2\pi} [n_y(k, \delta)]^2. \quad (9)$$

For our QW process, the integral in equation 9 admits a closed form, that is

$$L(\delta) = \begin{cases} 2\sin^2(\delta/4) & \text{for} \quad 0 \leq \delta \leq \pi/2 \\ 1 - \frac{1}{\sqrt{2}} & \text{for} \quad \pi/2 \leq \delta \leq 3\pi/2 \\ 2\cos^2(\delta/4) & \text{for} \quad 3\pi/2 \leq \delta \leq 2\pi \end{cases} \quad (10)$$

In Fig. 2a–c, this expression is plotted as a function of δ; as a consequence of the discontinuity present in the group velocity, $L(\delta)$ is a piecewise function, and manifests abrupt slope variations at $\delta_1 = \pi/2$ and $\delta_2 = 3\pi/2$. As shown in Fig. 3 and thoroughly

discussed in the Supplementary Note 2 and in the Supplementary Figs 2–6, by applying a similar approach to other 1D topological systems we were able to observe very similar features. For our QW, in the non-trivial phase $L(\delta)$ is constant; this result can be qualitatively understood by looking at how the dispersion law of the group velocity is modified by a change of δ (Supplementary Fig. 1). On the other hand, in the trivial phase $L(\delta)$ increases or decreases in the two regions $\{0, \pi/2\}$ and $\{3\pi/2, 2\pi\}$, respectively. Relying on a numerical analysis, we find that such features are robust to the presence of disorder arising from possible experimental imperfections in our setup (details are provided in Methods and in Supplementary Fig. 7). In a QW with a finite number of steps, statistical moments M_1 and M_2 have a continuous behaviour, converging to that given by equations 7 and 8 asymptotically as $N \to \infty$. For M_2, this convergence is rapid and visible for values of N that are small enough to be achieved in an experimental simulation, and it is independent of the coin initial state, whereas for M_1 such process is much slower (see Fig. 2). Thus, as an indicator for identifying the quantum transitions, we chose to use $\sqrt{M_2}/N = \frac{\sqrt{\langle \hat{m}^2 \rangle}}{N}$, corresponding to the width of the OAM statistical distribution, normalized to the number of steps. In particular, we implemented a six-step QW and measured the OAM distribution width when tuning the spin–orbit interaction in each q-plate, that is δ, so as to realize the topological quantum transition.

Experimental results. The layout of the apparatus is shown in Fig. 4. A standard heralded single-photon source is realized at the input of the QW system, as discussed in Methods. Before undergoing the QW evolution, a single photon is prepared in localized initial state ($m = 0$), with its polarization (the coin part) described by $|\phi_0\rangle = \alpha |L\rangle + \beta |R\rangle$. The value of the two complex coefficients α and β (with $|\alpha|^2 + |\beta|^2 = 1$) can be tuned by suitably orienting two waveplates that are positioned at input of the setup (see the caption of Fig. 4 for more details). After the QW dynamics, the photon state is analysed in both polarization and OAM so as to determine the output probabilities and the

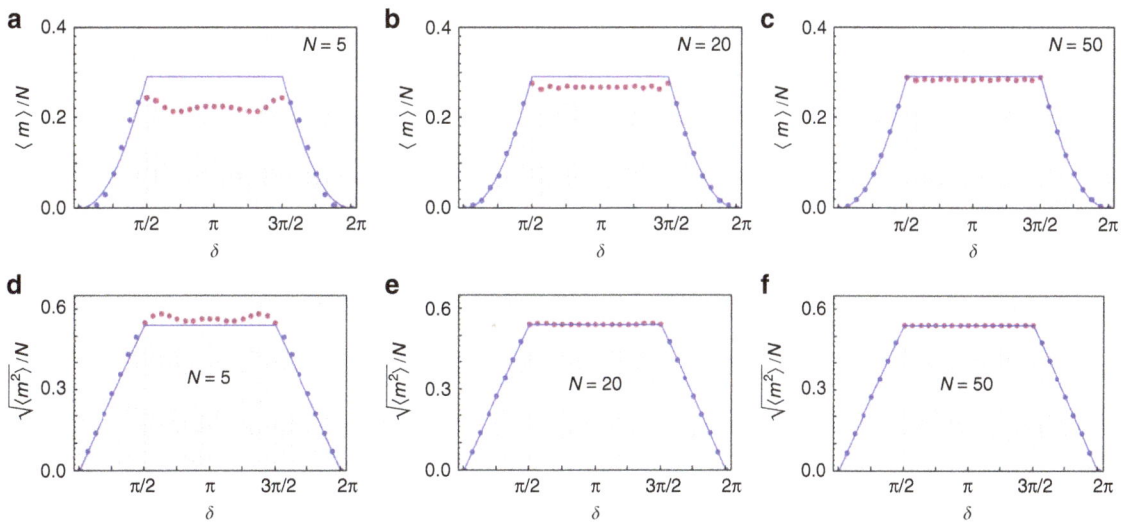

Figure 2 | Analysis of statistical moments in a QW. The system, initially localized at $x = 0$ with a generic coin state $|\phi_0\rangle$, undergoes a QW described by the step operator (3). For every plot, purple points are obtained from a numerical simulation, when varying δ with steps of $\pi/16$ in the range $\{0, 2\pi\}$; continuous blue lines represent the quantity $L(\delta)$ (**a–c**), or $\sqrt{L(\delta)}$ (**d–f**) (see equation 10). For the simulation, we prepared the coin in the state $\{\alpha, \beta\} = \{1, 0\}$, corresponding to $\{s_1, s_2, s_3\} = \{0, 0, 1\}$. (**a–c**) First-order moment M_1, divided by the number of steps of the walk, as a function of the parameter δ, for a walk of 5, 20 and 50 steps, respectively (as specified in each panel). As N increases, simulated data converge to the values predicted by equation 7. (**d–f**) Square root of the second-order moment, divided by the number of steps N. The figures are organized as in **a–c**. In this case, we can observe that simulated data converge much faster to the asymptotic values reported in equation 8, with a discontinuity emerging even for a walk of few steps.

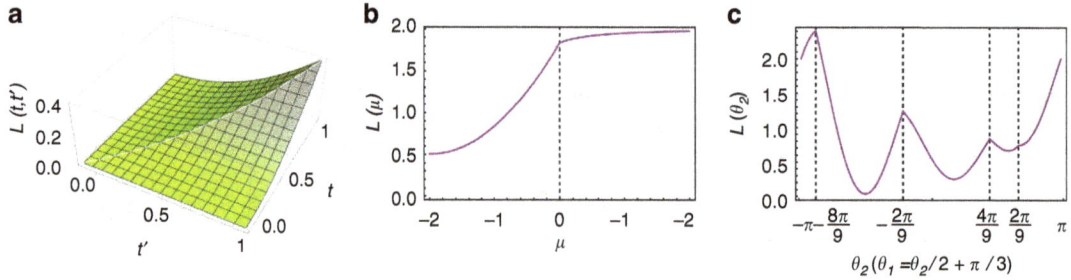

Figure 3 | Dynamical moments reveal topological phase transitions in other 1D topological systems. (a,b) We plot the expected asymptotic values of the second-order moment for a single particle whose evolution is described by the SSH model and the p-wave theory for superconductors, respectively, which have the same symmetries as our QW. In the SSH model (**a**), the value of the topological invariant is determined by the parameters (t, t'). In agreement with our conjecture, non-analyticities of the second-order moment can be observed at the phase change, that is for $t = t'$. In the p-wave theory, topological features are defined in terms of the chemical potential μ. Again, an abrupt slope variation can be appreciated at the transition point $\mu = 0$. (**c**) Here we consider a QW model having particle–hole symmetry only, in order to show that our results are independent of the specific symmetries characterizing the system. In this model, topological phases are determined in terms of two angles θ_1 and θ_2. These two angles are varied along the trajectory given by the equation $\theta_1 = \theta_2/3 + \pi/3$; similarly to **a,b**, the second-order moment after a long temporal evolution shows non-analyticities at the phase changes, identified by vertical dashed lines in the plot. In all simulations, we considered an initial state localized at a specific lattice site. Properties of these models, along with the analysis of other theories, are discussed in detail in the Supplementary Note 2.

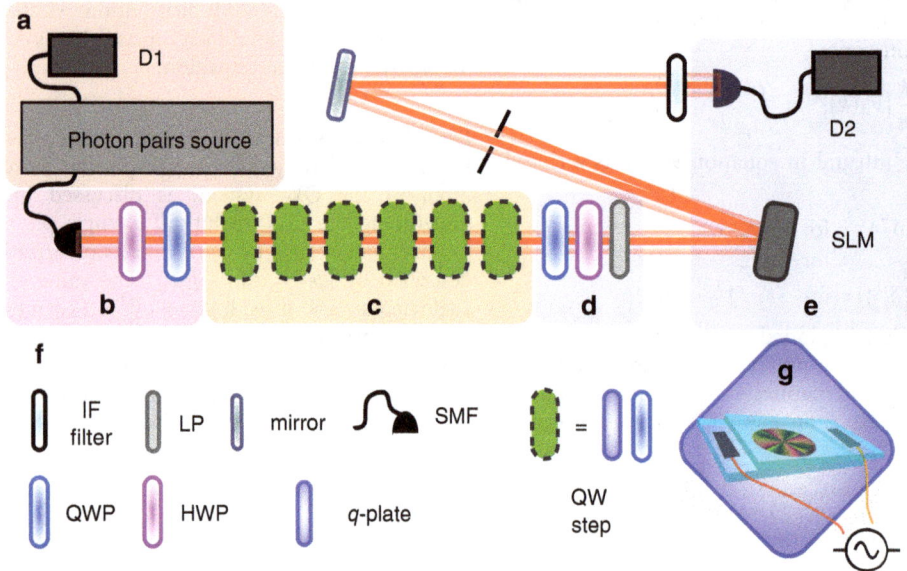

Figure 4 | Layout of the experimental setup. (a) As described in Methods section, pairs of indistinguishable and temporally correlated photons are generated at the input of the setup. For each pair, they are split by exploiting their orthogonal polarization: one photon is directly sent to an APD D1 and acts as a trigger for the detection of its correlated partner. The latter indeed, after coupling into a single-mode optical fibre (SMF), is sent to the QW setup and finally detected at D2, in coincidence with detection signals from D1. (**b**) The photons undergoing the QW dynamics exits a SMF in the OAM state $m = 0$, and then its polarization is prepared in the state $|\phi_0\rangle = \alpha|L\rangle + \beta|R\rangle$; the values of the two complex coefficients α and β (with $|\alpha|^2 + |\beta|^2 = 1$) are set by using a half-wave plate (HWP) and a QWP (apart from an unimportant global phase). (**c**) After the initial state preparation, the photon goes through the six-step QW, with the single step consisting of a QWP oriented at 90° and a q-plate. For each q-plate, the value of the optical retardation δ is controlled by the amplitude of an alternating electric field, induced by an external generator[27]. (**d**) At the exit of the QW, a polarization projection is realized using a second HWP-QWP set followed by an LP. (**e**) The OAM state is then analysed by diffraction on a SLM, followed by coupling into a SMF that is directly connected to the detector D2, consisting in a APD. Before photons are detected, IF centred at 800 nm and with a bandwidth of 3.6 nm are used for spectral cleaning. The latter was required since the photons wavelength strongly affects the action of the devices implementing the QW (q-plate and QWP). (**f**) List of all optical elements used in our setup. (**g**) Illustrative picture of a q-plate, whose optical retardation is controlled by means of an external oscillating electric field. IF, interference filters; LP, linear polarizer; SLM, spatial light modulator.

associated moments. Letting δ vary in the range $\{0, \pi\}$ with steps of $\pi/16$, we determined the corresponding probability distribution of the walker after a six-step QW; a typical example is reported in Fig. 5a. In Fig. 5b,c we plot the measured values for M_1/N and $\sqrt{M_2}/N$, respectively, as a function of δ. Experimental data match well the quantum predictions and, as expected, they are close to the asymptotic limit reported in equations 7 and 8 only in the

case of $\sqrt{M_2}/N$. For the latter, the emergence of an abrupt slope variation at $\delta = \pi/2$ can be appreciated; the observed non-analyticity is a signature of the underlying quantum transition. In addition, M_2 is observed to be locked to a constant value in the non-trivial phase. On one hand, this result reflects the features of the group velocity and the strong link between the latter and the geometry and topology of the system eigenstates.

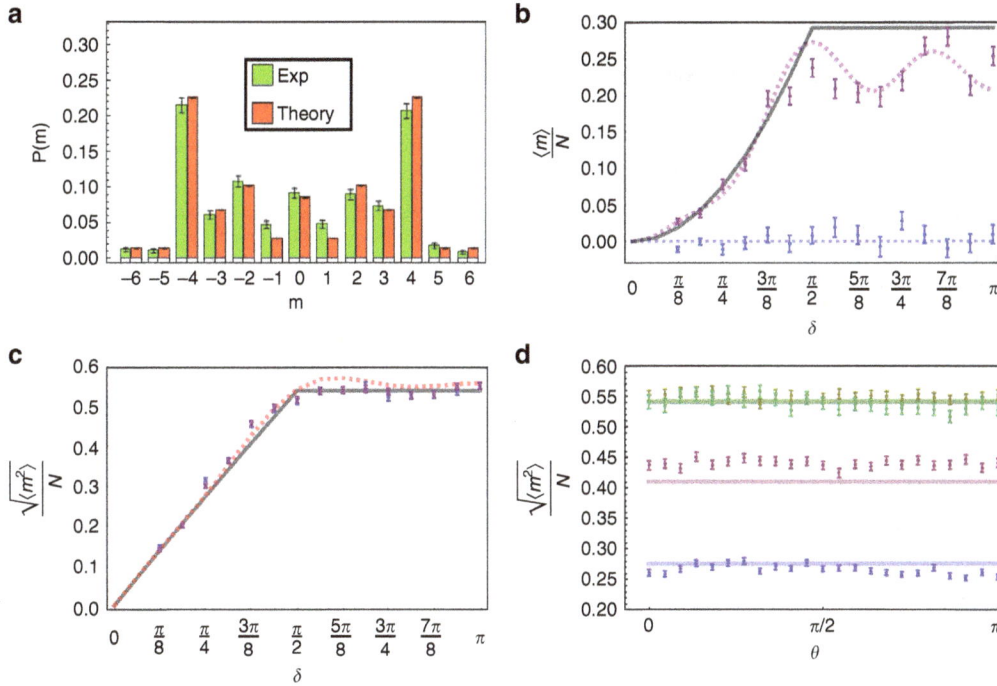

Figure 5 | Theoretical predictions and experimental results. (**a**) Example of a probability distribution for the walker: measured (green, left) and expected (red, right) probability distributions after a six-step QW of a photon initially prepared in the state $m = 0$ and $\{\alpha, \beta\} = 1/\sqrt{2}\{1, 1\}$, for $\delta = 2.95$. (**b,c**) Measured values of statistical moments M_1/N and $\sqrt{M_2}/N$, respectively, when varying δ in the range $\{\pi/8, \pi\}$ with steps of $\pi/16$, for initial states $\{\alpha, \beta\} = 1/\sqrt{2}\{1, 1\}$ (blue points), and $\{\alpha, \beta\} = \{0, 1\}$ (purple points). Dotted lines represent the corresponding results of numerical simulations. The continuous lines are the asymptotic limits given by equation 8. (**d**) Measured values of $\sqrt{M_2}/N$ when the initial polarization corresponds to the state $\cos(\theta/2)|L\rangle + \sin(\theta/2)|R\rangle$, varying the polar angle $\theta \in [0, \pi]$ with steps of $\pi/22$; data are collected in correspondence of $\delta = \pi/4$ (blue points), $\delta = 3\pi/8$ (purple points), $\delta = 3\pi/4$ (yellow points) and $\delta = \pi$ (green points). For each of these configurations, continuous lines give the corresponding asymptotic limit obtained from equation 8. When $\delta = 3\pi/8$, experimental data are slightly higher then the associated continuous line; nevertheless, the discrepancy is compatible with the results of numerical simulations for a finite number of steps (see **c**). These data confirm the prediction that M_2 is asymptotically independent of the input coin/polarization state. In all plots (**a–d**), the error bars represent statistical errors at 1 s.d., calculated assuming Poissonian fluctuations on single counts.

On the other hand, this is not a general feature of 1D topological systems; in the Supplementary Note 2 we show that for other models, such as the SSH, this plateau can be observed in either of the two phases, depending on the specific trajectory followed in the parameters space. Yet, it is possible to show that for our QW (equation 9) and the SSH model (Supplementary Note 3), the value of M_2 is proportional to the integral of n_y^2 over the Brillouin zone and that the latter always shows a constant value, that is a plateau, when $W = 1$. Preparing the initial state of the coin ($m = 0$ for the walker) in two non-orthogonal polarizations we also verified that M_2 (unlike M_1) is independent of the coin initial conditions, as expected from equation 8 (see Fig. 5b,c). This aspect was further investigated by measuring $\sqrt{M_2}/N$ for several initial polarizations, corresponding to specific points along a meridian of the Poincaré sphere, and repeating the experiment for two values of δ in each topological sector. The final data, reported in Fig. 5d, match well the predicted results and clearly show that the value of the second-order moment is independent of the coin initial state; this guarantees the robustness of our results with respect to imperfections in the initial state preparation.

Discussion

The main result of this work is the idea that a signature of quantum transitions between distinct topological phases is present in the behaviour of suitable dynamical observables that are easily accessed experimentally. In the context of 1D topological systems, we propose that the statistical moments of the particle probability distribution in space (in the large step-number limit) provide a convenient choice of such observables. We experimentally validated this proposal by simulating this topological environment within a specific photonic QW architecture. The measured asymptotic moments as a function of a control parameter show slope discontinuities at the phase change, and hence can be used as direct indicators of the phase change occurrence. It is remarkable that these observables reflect bulk properties only, in contrast with the common strategy that relies on investigating topological physics exclusively through edge effects. Numerical simulations show that these features are robust to experimental imperfections and confirm their presence in several other 1D topological systems, thus strongly supporting the conjecture of their general validity in 1D. Moreover, we are currently investigating the possibility to apply similar concepts to 2D systems. Localization induced by spatial disorder is of course expected to affect the final asymptotic behaviour of the dynamics. However, if the disorder is not very strong, it is plausible that an intermediate time regime will still exist, in which the quasi-nonanalytic behaviour of the moments can be detected. In prospect, our approach based on dynamical moments could be applicable to the investigation and to the detection of other classes of quantum transitions, and to topological phases associated with more complex symmetries or with a higher dimensionality, thus helping to shed new light on the physics of topological phenomena and on their simulation in suitable experimental architectures.

Methods

Asymptotic expressions of first and second moments. The large step-number limit for the walker distribution moments (equations 7 and 8) can be derived conveniently by evaluating M_1 and M_2 in momentum representation, where they are defined as follows:

$$M_1 = \int_{-\pi}^{\pi} \frac{dk}{2\pi} \langle \phi_0 | \left(\hat{U}_0^{\dagger} \right)^N (-i) \frac{d}{dk} \hat{U}_0^N | \phi_0 \rangle,$$

$$M_2 = \int_{-\pi}^{\pi} \frac{dk}{2\pi} \langle \phi_0 | \left(\hat{U}_0^{\dagger} \right)^N (-i)^2 \frac{d^2}{dk^2} \hat{U}_0^N | \phi_0 \rangle. \tag{11}$$

Here N is the step-number, $\hat{U}_0$ is the single-step evolution operator (3), and $|\phi_0\rangle$ is the coin initial state. Expanding the evolution operator as

$$\begin{aligned} \hat{U}_0^N &= \text{Exp}\{-iN E_\delta(k)[\mathbf{n}_\delta(k) \cdot \hat{\sigma}]\} \\ &= \cos[N E_\delta(k)]\hat{I}_2 - i\sin[N E_\delta(k)](\mathbf{n}_\delta(k) \cdot \hat{\sigma}) \end{aligned} \tag{12}$$

where $\hat{I}_2$ is the identity matrix in 2D, $\hat{\sigma} = \{\hat{\sigma}_x, \hat{\sigma}_y, \hat{\sigma}_z\}$ and the components of $\mathbf{n}_\delta(k)$ are those reported in equation 6, it is straightforward to obtain the following equations:

$$M_1/N = \int_{-\pi}^{\pi} \frac{dk}{2\pi} V_\delta(k) \langle \phi_0 | (\mathbf{n}_\delta(k) \cdot \hat{\sigma}) | \phi_0 \rangle + O(1/N), \tag{13}$$

$$M_2/N^2 = \int_{-\pi}^{\pi} \frac{dk}{2\pi} [V_\delta(k)]^2 + O(1/N^2), \tag{14}$$

It is important to observe that equations 13 and 14 are valid for any quantum system whose Hilbert space has the same structure as our QW, that is the direct product of an infinite discrete lattice and a two-state quantum degree of freedom. Equation 13, reporting the expression for the first-order moment, can be evaluated considering that $V_\delta = n_z = -n_y$. Applying this substitution, the same equations reads

$$M_1/N = (s_y - s_z) \int_{-\pi}^{\pi} \frac{dk}{2\pi} [V_\delta(k)]^2 + O(1/N). \tag{15}$$

First- and second-order moments reported in equations 14 and 15, respectively, coincide with the equations 7 and 8 of the main text. In the specific case of our QW model, the integral appearing in equations 14 and 15 can be solved analytically. Using the expression for the group velocity reported in equations 5 and 9 yields

$$L(\delta) = \frac{1}{2\pi} \int_0^{2\pi} V_\delta^2 dk = \frac{1}{2\pi} \int_0^{2\pi} \frac{\sin^2(\delta/2)\cos^2(k)}{2 - [\cos(\delta/2) + \sin(\delta/2)\sin(k)]^2} dk. \tag{16}$$

This integral can be calculated by the residue theorem passing to the complex variable $z = e^{ik}$. Then we obtain $L(\delta) = \oint f_\delta(z) dz$, where the integral is along the unit circle in the complex z-plane and $f_\delta(z)$ is given by

$$f_\delta(z) = \frac{i(1+z^2)^2 \sin^2(\delta/2)}{\pi z [(1+z^2)^2 \cos(\delta) - z^4 - 10z^2 - 1 - 4iz(z^2 - 1)\sin(\delta)]}. \tag{17}$$

The poles of $f_\delta(z)$ are located on the imaginary axis at $z_k = \{0, i(\sqrt{2} - 1)\cot(\delta/4),$ $i(\sqrt{2} + 1)\tan(\delta/4), -i(\sqrt{2} + 1)\cot(\delta/4)$ and $-i(\sqrt{2} - 1)\tan(\delta/4)\}$ $(k = 1, \ldots 5)$ and the residues of $f_\delta(z)$ at the poles are given by $2\pi i r_k = \{1, \frac{1}{4}[-\sqrt{2} + 2\cos(\delta/2)],$ $\frac{1}{4}[\sqrt{2} - 2\cos(\delta/2)], \frac{1}{4}[\sqrt{2} + 2\cos(\delta/2)]$ and $\frac{1}{4}[-\sqrt{2} - 2\cos(\delta/2)]\}$, respectively. Apart from the pole at $z = 0$, when δ varies, the locations of the poles move in the complex plane entering and exiting the unit circle, but only the residues of the poles inside the unit circle contribute to $L(\delta)$. Thus, we find

$$L(\delta) = \begin{cases} 2\pi i(r_1 + r_3 + r_5) = 2\sin^2(\delta/4) & \text{for } 0 \leq \delta \leq \pi/2 \\ 2\pi i(r_1 + r_2 + r_5) = 1 - \frac{1}{\sqrt{2}} & \text{for } \pi/2 \leq \delta \leq 3\pi/2. \\ 2\pi i(r_1 + r_2 + r_4) = 2\cos^2(\delta/4) & \text{for } 3\pi/2 \leq \delta \leq 2\pi \end{cases} \tag{18}$$

In Supplementary Note 3 and in Supplementary Fig. 6 we show how with the same approach the analytical expression for the function L can be found for the SSH model.

Heralded single-photon source. In the heralded single-photon source (the optical components used to realize this source are not shown in the setup scheme of Fig. 4, for details see ref. 24), laser pulses (100 fs) at 800 nm generated in a titanium–sapphire source (Ti:Sa) with repetition rate 82 MHz shine on a type-I β-barium borate crystal for second harmonic generation; frequency-doubled pulses at 400 nm, with 110 mW average power and linear-horizontal polarization, pump a type-II β-barium borate crystal, cut for collinear and degenerate spontaneous parametric down-conversion. Signal and idler photons, generated in horizontal and vertical linear polarizations, respectively, are spatially separated by means of a polarizing beam-splitter and then coupled into single-mode fibres; the idler photon is directly sent to an avalanche photodiode (APD D1), while the signal one is sent to the QW system. After the QW evolution, such photon is analysed in polarization and OAM and finally detected by APD D2, in coincidence with D1. In this way, the idler photon is used as a trigger, which heralds the presence of the correlated particle passing through the QW setup.

Experimental imperfections and role of disorder. Experimental realizations of quantum protocols and quantum dynamics exploit specific devices whose behaviour may present deviations with respect to their ideal features. Here we individuate possible imperfections that may arise in our system, discussing in turn how they would affect our QW protocol and, as a consequence, the dynamical moments associated with the walker wavefunction. Interestingly, we find that abrupt variations at the phase change are robust to these kinds of disorder.

A first type of imperfection can result from a bad orientation of the QWPs performing the coin rotation. We considered two different coin input, namely $|H\rangle$ and $|R\rangle$ polarizations already used in the experiment, and simulated a 80 steps QW, where at each step a QWP is oriented at $\pm 2°$ with respect to the correct angle; here we assumed that at $0°$ a QWP implements the operator reported in equation 1. The $\pm$ sign is randomly selected. In Supplementary Fig. 7, we report the results of our numerical analysis. Data clearly show that the second-order moment have no significant differences with respect to the ideal QW; accordingly the phase change can be appreciated still, thus proving the robustness of such indicator. For $|R\rangle$ as initial state, the mean value has a similar behaviour, whereas something different happens for the $|H\rangle$ polarization; in the latter case, ideal QW leads to a vanishing mean value. When QWPs have random bad orientations, this is not true anymore, even though deviations from ideal case are smaller than typical experimental errors affecting our data. A second source of errors can arise from the bad alignment of q-plates. In particular, their centre could be displaced with respect to the propagation axis of the beam. As a first approximation, the q-plate operator reported in equation 2 would be transformed as follows:

$$\begin{aligned} \hat{Q}'_{\delta=\pi}|L, m\rangle &= i(a_m|R, m+1\rangle + b_m|R, m\rangle + c_m|R, m+2\rangle), \\ \hat{Q}'_{\delta=\pi}|R, m\rangle &= i(a_m|L, m-1\rangle + b_m|L, m\rangle + c_m|L, m-2\rangle), \end{aligned} \tag{19}$$

where $\hat{Q}'$ is the operator describing the q-plate action when a displacement of its centre is taken into account; here it has been considered the simple case $\delta = \pi$ (the term proportional to $\cos(\delta/2)$ is the identity operator, which is not affected by such displacement). Moreover, we are explicitly neglecting the coupling to spatial modes of light with different radial distributions, as for instance Laguerre–Gauss modes with radial index $p \geq 1$. We can note that in equation 19 non-vanishing values for c_m are responsible for next-nearest-neighbour couplings in the OAM lattice, corresponding to an 'off-diagonal' disorder[29]. Coefficients $\{a_m, b_m, c_m\}$ can be computed numerically; interestingly, the effects of the q-plate displacement are not equal for all OAM states; in particular, deviations from the ideal case becomes negligible for OAM modes with high $|m|$; for such states, $a_m \simeq 1$ and $b_m \simeq c_m \simeq 0$. Similarly to the previous analysis we implemented a numerical simulation of a 80 steps QW for $|H\rangle$ and $|R\rangle$ coin input; at each step the q-plate is displaced randomly along the x or the y axis (the z axis corresponds to the propagation direction of the light beam). We numerically computed coefficients $\{a_m, b_m, c_m\}$, assuming a displacement equal to 5% with respect to the beam transverse dimensions (that is the radius at the waist of the input Gaussian beam). Data reported in Supplementary Fig. 7 show that the effect of such imperfections is negligible. Importantly, the abrupt variation at the phase change can be still appreciated, thus proving the robustness of such features. Even for this case the mean value for $|H\rangle$ has no significant differences when compared with the ideal case, being it very close to zero for all values of δ (within our experimental uncertainties). In conclusion, we want to compare such imperfections with the presence of static disorder, which is typically responsible for localization phenomena. An imperfect orientation of the waveplates leads to a modification of the coin operator that is equal for all OAM states, thus it cannot produce any localization. Effects associated with the q-plate displacement are not uniform in the OAM space, but still they are quite different with respect to the typical static disorder, that consists in random site-dependent phase shifts, equal at each step. Moreover, in both cases alterations of the step operator are different at every step of the walk, thus they represent a temporal disorder. Importantly, our simulations show the robustness of moments non-analyticities as indicators of quantum phase changes in such disordered scenarios. Although the spatial static disorder is certainly not significant in our experimental implementation, it can be obviously important for the generalization of our results to other systems. Localization induced by this kind of disorder is expected to affect the final asymptotic behaviour of the dynamics. However, if the disorder is not very strong, it is plausible that an intermediate time regime will still exist in which a quasi-nonanalytic behaviour of the moments can be detected. If 'localization time' is sufficiently larger than the measurement time, the disorder will not affect the kinks (slightly rounded in real experiments because of the finite measurement time) that signal the occurrence of the topological transitions. Hence, we believe that our main results on the use of moments for detecting topological transitions can be useful also in the context of disordered systems, as long as the disorder is not too strong.

References

1. Thouless, D. J., Kohmoto, M., Nightingale, M. P. & den Nijs, M. Quantized hall conductance in a two-dimensional periodic potential. *Phys. Rev. Lett.* **49**, 405–408 (1982).

2. Zhang, Y., Tan, Y. W., Stormer, H. L. & Kim, P. Experimental observation of the quantum hall effect and berry's phase in graphene. *Nature* **438**, 201–204 (2005).

3. Qi, X. & Zhang, S. Topological insulators and superconductors. *Rev. Mod. Phys.* **83,** 1057–1110 (2011).

4. Hasan, M. Z. & Kane, C. L. *Colloquium* : topological insulators. *Rev. Mod. Phys.* **82,** 3045–3067 (2010).

5. Gomes, K. K., Mar, W., Ko, W., Guinea, F. & Manoharan, H. C. Designer dirac fermions and topological phases in molecular graphene. *Nature* **483,** 306–310 (2012).

6. Atala, M. *et al.* Direct measurement of the zak phase in topological bloch bands. *Nat. Phys.* **9,** 795–800 (2013).

7. Genske, M. *et al.* Electric quantum walks with individual atoms. *Phys. Rev. Lett.* **110,** 190601 (2013).

8. Hauke, P. *et al.* Non-abelian gauge fields and topological insulators in shaken optical lattices. *Phys. Rev. Lett.* **109,** 145301 (2012).

9. Cayssol, J., Dóra, B., Simon, F. & Moessner, R. Floquet topological insulators. *Phys. Status Solidi RRL* **7,** 101–108 (2013).

10. Kitagawa, T. Observation of topologically protected bound states in photonic quantum walks. *Nat. Commun.* **3,** 882 (2012).

11. Zeuner, J. M. *et al.* Probing topological invariants in the bulk of a non-hermitian optical system. *Phys. Rev. Lett.* **115,** 040402 (2015).

12. Lu, L., Joannopoulos, J. D. & Soljačić, M. Topological photonics. *Nat. Photon.* **8,** 821–829 (2014).

13. Hafezi, M. Measuring topological invariants in photonic systems. *Phys. Rev. Lett.* **112,** 210405 (2014).

14. Ozawa, T. & Carusotto, I. Anomalous and quantum Hall effects in lossy photonic lattices. *Phys. Rev. Lett.* **112,** 133902 (2014).

15. Hu, W. *et al.* Measurement of a topological edge invariant in a microwave network. *Phys. Rev. X* **5,** 011012 (2015).

16. Mittal, S., Ganeshan, S., Fan, J., Vaezi, A. & Hafezi, M. Observation of the Chern-Simons gauge anomaly. *Nat. Photon.* **10,** 180–183 (2016).

17. Rudner, M. & Levitov, L. Topological transition in a non-hermitian quantum walk. *Phys. Rev. Lett.* **102,** 065703 (2009).

18. Kitagawa, T., Rudner, M. S., Berg, E. & Demler, E. Exploring topological phases with quantum walks. *Phys. Rev. A* **82,** 033429 (2010).

19. Asbóth, J. K. Symmetries, topological phases, and bound states in the one-dimensional quantum walk. *Phys. Rev. B* **86,** 195414 (2012).

20. Venegas-Andraca, S. E. Quantum walks: a comprehensive review. *Quantum Inf. Process.* **11,** 1015–1106 (2012).

21. Su, W., Schrieffer, J. & Heeger, A. Solitons in polyacetylene. *Phys. Rev. Lett.* **42,** 1698–1701 (1979).

22. Read, N. & Green, D. Paired states of fermions in two dimensions with breaking of parity and time-reversal symmetries and the fractional quantum Hall effect. *Phys. Rev. B* **61,** 10267 (2000).

23. Wang, J. & Manouchehri, K. *Physical Implementation of Quantum Walks* (Springer, 2013).

24. Cardano, F. *et al.* Quantum walks and wavepacket dynamics on a lattice with twisted photons. *Sci. Adv.* **1,** e1500087 (2015).

25. Yao, A. M. & Padgett, M. J. Orbital angular momentum: origins, behavior and applications. *Adv. Opt. Photon* **3,** 161–204 (2011).

26. Marrucci, L., Manzo, C. & Paparo, D. Optical spin-to-orbital angular momentum conversion in inhomogeneous anisotropic media. *Phys. Rev .Lett.* **97,** 163905 (2006).

27. Piccirillo, B., D'Ambrosio, V., Slussarenko, S., Marrucci, L. & Santamato, E. Photon spin-to-orbital angular momentum conversion via an electrically tunable q-plate. *Appl. Phys. Lett.* **97,** 241104 (2010).

28. Asbóth, J. K. & Obuse, H. Bulk-boundary correspondence for chiral symmetric quantum walks. *Phys. Rev. B* **88,** 121406 (2013).

29. Zhao, Q. & Gong, J. From disordered quantum walk to physics of off-diagonal disorder. *Phys. Rev. B* **92,** 214205 (2015).

Acknowledgements

We thank Maciej Lewenstein, Pietro Massignan and Alessio Celi for reading a preliminary version of this manuscript and for providing useful comments, Takuya Machida for alerting us to some mistakes appearing in a preprint early version of this paper, and Domenico Paparo and Antonio Ramaglia for lending us some equipment. This work was partly supported by the Future Emerging Technologies FET-Open Program, within the 7th Framework Programme of the European Commission, under Grant No. 255914, PHORBITECH.

Author contributions

F.C., M.M., F.M., E.S. and L.M. devised various aspects of the project and designed the experimental methodology. F.C. and M.M., with contributions from C.d.L., carried out the experiment and analysed the data. B.P. prepared the q-plates. F.C., M.M, G.D.F., V.C., E.S. and L.M. developed the theoretical aspects. F.C., M.M. and L.M. wrote the manuscript, with contributions from E.S., G.D.F and V.C. All authors discussed the results and contributed to refining the manuscript.

Additional information

Estimation of a general time-dependent Hamiltonian for a single qubit

L.E. de Clercq[1], R. Oswald[1], C. Flühmann[1], B. Keitch[1,†], D. Kienzler[1], H.-Y. Lo[1], M. Marinelli[1], D. Nadlinger[1], V. Negnevitsky[1] & J.P. Home[1]

The Hamiltonian of a closed quantum system governs its complete time evolution. While Hamiltonians with time-variation in a single basis can be recovered using a variety of methods, for more general Hamiltonians the presence of non-commuting terms complicates the reconstruction. Here using a single trapped ion, we propose and experimentally demonstrate a method for estimating a time-dependent Hamiltonian of a single qubit. We measure the time evolution of the qubit in a fixed basis as a function of a time-independent offset term added to the Hamiltonian. The initially unknown Hamiltonian arises from transporting an ion through a static laser beam. Hamiltonian estimation allows us to estimate the spatial beam intensity profile and the ion velocity as a function of time. The estimation technique is general enough that it can be applied to other quantum systems, aiding the pursuit of high-operational fidelities in quantum control.

[1] Institute for Quantum Electronics, ETH Zürich, Otto-Stern-Weg 1, 8093 Zürich, Switzerland. † Present address: Department of Engineering Science, University of Oxford, Parks Road, Oxford OX1 3PJ, UK. Correspondence and requests for materials should be addressed to J.P.H. (email: jhome@phys.ethz.ch).

Estimation of the underlying dynamics, which drive the evolution of systems is a key problem in many areas of physics and engineering. This knowledge allows control inputs to be designed, which account for imperfections in the physical implementation. For closed quantum systems, the time dependence of a system is driven by the Hamiltonian through Schrödinger's equation. If the Hamiltonian is static in time, a wide range of techniques have been proposed for recovering the Hamiltonian[1-4], which have been applied to a variety of systems including chemical processes[5] and quantum dots[6,7]. These methods often involve estimation of the eigenvectors and eigenvalues of the Hamiltonian via spectroscopy, or through pulse–probe techniques for which a Fourier transform of the time evolution gives information about the spectrum.

These methods are not directly applicable to time-dependent Hamiltonians, which are becoming increasingly important as quantum engineering pursues a combination of high-operational fidelities and speeds, often involving fast variation of control fields, which are particularly susceptible to distortion before reaching the quantum device[8-12]. The time-varying case has thus far been studied in cases where the variation is along a single dimension in the Hilbert space, which for the commonly studied spin is a single spatial direction. In the case that the measured fields dominate the evolution (strong field limit), measurement of the system evolution as a function of time suffices for the reconstruction. For fields which are weaker than other available control fields (weak-field limit) the latter can be used to modulate the effect of the signal Hamiltonian on the quantum system[13-15], providing an excellent signal-to-noise ratio. A further complication arises when a time-varying Hamiltonian contains non-commuting terms (for example, time-variation along two spin axes), because the evolution of the quantum system depends not only on their separate influences, but also on products arising from the non-commutativity. For unspecified time-dependent coefficients, no analytical solution to Schrödinger's equation exists[16,17]. In the weak-field limit, strong control fields can be used to separate out the different components using modulation, however, when the Hamiltonian itself is strong (as is the case in fast quantum control) these techniques cannot be applied.

In this article, we propose and demonstrate a method for reconstructing a general time-dependent single qubit Hamiltonian with non-commuting terms. The technique involves observing the evolution of the spin projection on the z-axis, while applying a static offset to one of the terms of the Hamiltonian. By varying the static offset, we build up data sets, which contain sufficient information to extract the full time-dependent Hamiltonian. Parameterizing the two time-dependent terms using basis splines (B-splines), we introduce an iterative fitting technique, which finds the Hamiltonian that best matches the data. We benchmark the reconstruction method experimentally by transporting a single trapped ion through a static laser beam, a technique suited to scaling up trapped-ion quantum information processing[18,19]. We perform two consistency checks on the Hamiltonian estimation using four separate reconstructions. For the first two, we compare two cases which use the same ion velocity profile, but different laser beam positions. For the second consistency check, we use the same laser beam, but change the velocity profile between the two by using different sets of time-varying control potentials. The method produces consistent experimental parameters in both cases, indicating the success of the reconstruction technique. Our method is applicable to spin Hamiltonians of the general form $\hat{H} = \sum_i f_i(t)\hat{\sigma}_i$, where the $f_i(t)$ are arbitrary time-dependent functions and $\hat{\sigma}_i$ are the Pauli operators.

Results

Hamiltonian estimation method. In our experiments, a Hamiltonian with two non-commuting time-dependent terms arises when we perform quantum logic gates by transporting an ion through a static laser beam[18,19]. In this case, the Hamiltonian describing the interaction between the ion and the laser can be written in an appropriate rotating frame as

$$\hat{H}_I(t) = \frac{\hbar}{2}\left(-\Omega(t)\hat{\sigma}_x + \delta(t)\hat{\sigma}_z\right) \tag{1}$$

which includes a time-varying Rabi frequency $\Omega(t)$, and an effective detuning $\delta(t)$, which is related to the first-order Doppler shift of the laser in the rest frame of the moving ion (see Methods for details).

To reconstruct the Hamiltonian, we make use of two additional capabilities. First, we can switch-off the Hamiltonian at time t_{off} on a timescale much faster than the qubit evolution. Subsequently measuring in the $\hat{\sigma}_z$ basis, we can obtain $\langle\hat{\sigma}_z(t_{\text{off}})\rangle$. On its own, this does not allow us to separate the contributions from $\Omega(t)$ and $\delta(t)$. To do this, we use a second capability, which is the ability to add a controlled offset $\hat{H}_s = \hbar\,\delta_L\hat{\sigma}_z/2$ to the Hamiltonian, resulting in $\hat{H}_I(t) + \hat{H}_s$. The resulting spin measurement is now dependent on both t_{off} and the set value of δ_L. Repeating the experiment for a range of values of δ_L but with otherwise identical settings, we obtain an estimate of the expectation value which we denote as $\langle\hat{\sigma}_z^{\text{meas}}(t_{\text{off}}, \delta_L)\rangle$.

Hamiltonian extraction involves finding the functions $\delta(t)$ and $\Omega(t)$, which generate spin populations $\langle\hat{\sigma}_z^{\text{sim}}(t_{\text{off}}, \delta_L)\rangle$ that most closely match the data. We minimize the reduced χ^2 cost function.

$$J = \frac{1}{\nu}\sum_{t_{\text{off}}}\sum_{\delta_L}\left[\frac{\langle\hat{\sigma}_z^{\text{meas}}(t_{\text{off}}, \delta_L)\rangle - \langle\hat{\sigma}_z^{\text{sim}}(t_{\text{off}}, \delta_L)\rangle}{\sigma^{\text{meas}}(t_{\text{off}}, \delta_L)}\right]^2 \tag{2}$$

where $\nu = N - n - 1$ is the number of degrees of freedom, with N the number of data points, n the number of fitting parameters and $\sigma^{\text{meas}}(t_{\text{off}}, \delta_L)$ the s.e. on the estimated $\langle\hat{\sigma}_z^{\text{meas}}(t_{\text{off}}, \delta_L)\rangle$. This is subject to the initial condition $|\Psi(t=0, \delta_L)\rangle = |0\rangle$, and the following restrictions, which are imposed by quantum mechanics

$$i\hbar\frac{\partial}{\partial t}|\Psi(t, \delta_L)\rangle = \left(\hat{H}_I(t) + \hat{H}_s\right)|\Psi(t, \delta_L)\rangle,$$
$$\langle\hat{\sigma}_z^{\text{sim}}(t, \delta_L)\rangle = \langle\Psi(t, \delta_L)|\hat{\sigma}_z|\Psi(t, \delta_L)\rangle \tag{3}$$

for all δ_L.

One challenge in obtaining an estimate for the Hamiltonian is that we must optimize over continuous functions $\delta(t)$ and $\Omega(t)$. To address this, we represent $\delta(t)$ and $\Omega(t)$ with a linear combination of B-spline polynomials, which allow the construction of smooth functions using only a few parameters[20]. Any smooth function $S(t)$ can be written in terms of B-spline polynomials $B_{i,k}(t)$ and a set of weights α_i as

$$S(t) = \sum_{i=0}^{n}\alpha_i B_{i,k}(t). \tag{4}$$

The polynomial B-spline functions $B_{i,k}(t)$ are of order k with each polynomial centred at a time t_i, which is parameterized by the index i. Further details and an example can be found in Methods. Using the B-spline form for $\delta(t)$ and $\Omega(t)$, the cost function is minimized by adjusting the weights of the B-spline decomposition. Solving this optimization problem in general is hard, because it is non-linear and non-convex due to the nature of Schrödinger's equation and the use of projective measurements. This produces a non-trivial relation between the weights and the spin populations as discussed in Methods. To overcome this challenge, we have implemented a method which we call 'Extending the Horizon Estimation' (EHE) in analogy to a well-established technique called 'Moving Horizon Estimation' (MHE)[21].

The key idea is that because our measurement data arises from a causal evolution, we can also estimate the Hamiltonian in a causal way. Instead of optimizing J over the complete time span at once, we first restrict ourselves to a small, initial time span reaching only up to the start of the qubit dynamics $0 < t_{off} < T_0$, where we denote T_0 as the time horizon for the first step. Optimizing J over this short time span requires fewer optimization parameters and is simpler than attempting to optimize over the full data set. Once we have solved this small sub-problem, we extend the time horizon to T_1 where $T_1 = T_0 + \tau$ and re-run the optimization, extrapolating the results of the initial time span into the extended region to provide good starting conditions for the subsequent optimization. This procedure is iterated until the time span extends over the whole data set ($T_{n_{max}} = \max(t_{off})$). The method allows us to reduce the number of B-spline functions used to represent $\delta(t)$ and $\Omega(t)$, and also reduces the amount of data considered in the early stages of the fit, when the least is known about the parameters. This facilitates the use of non-linear minimization routines, which are based on local linearization of the problem and converge faster near the optimum. More details regarding the optimization routine can be found in Methods.

Conceptually EHE is very similar to MHE. The main difference is that in MHE the time span has a fixed length and thus its origin is shifted forward in time along with the horizon. In EHE, the origin stays fixed at the expense of having to increase the time span under consideration. MHE avoids this by introducing a so-called arrival cost to approximate the previous costs incurred before the start of the time span. This keeps the computational burden fixed over time, which is very important as MHE is usually used to estimate the state of a system in real-time, often on severely constrained embedded platforms. Since neither constraint applies to our problem, we decided to extend the horizon rather than finding an approximate arrival cost. This is advantageous since finding the arrival cost in the general case is still an open problem[22]. Due to the similarity between MHE and EHE, we anticipate future improvements by adapting techniques used in MHE to EHE. This might be used to reduce the data-processing required for reconstruction, which for EHE scales as $T_{n_{max}}^2$.

Experimental implementation. To test the ability of the method to reliably extract a Hamiltonian from data, we apply it to the Hamiltonian for an trapped-ion qubit during transport through a near-resonant laser beam. Our qubit is encoded in the electronic states of a calcium ion, which is defined by $|0\rangle \equiv |^2S_{1/2}, M_J=1/2\rangle$ and $|1\rangle \equiv |^2D_{5/2}, M_J=3/2\rangle$. This transition is well-resolved from all other transitions, and has an optical frequency of $\omega_0/(2\pi) \simeq 411.0420\,\text{THz}$. The laser beam points at $45°$ to the transport axis, and has an approximately Gaussian spatial intensity distribution. The time-dependent velocity $\dot{z}(t)$ of the ion is controlled by adiabatic translation of the potential well in which the ion is trapped. This is implemented by applying time-varying potentials to multiple electrodes of a segmented ion trap, which are generated using a multi-channel arbitrary-waveform generator, each output of which is connected to a pair of electrodes via a passive third-order low-pass Butterworth filter. The result is that the ion experiences a time-varying Rabi frequency $\Omega(t)$ and a laser phase which varies with time as $\Phi(t) = \phi(z(t)) - \omega_L t$, where $\phi(z(t)) = k_z(z(t))z(t)$ with $k_z(z(t))$ the laser wave vector projected onto the transport axis at position $z(t)$ and ω_L the laser frequency. The spatial variation of $k_z(z(t))$ accounts for the curvature of the wavefronts of the Gaussian laser beam. To create a Hamiltonian of the form of equation (1), we

work with the differential of the phase, which gives a detuning $\delta(t) = \delta_L - \dot{\phi} = \delta_L - \left(k_z'(z)z + k_z(z)\right)\dot{z}$ with $\delta_L = \omega_L - \omega_0$ the laser detuning from resonance. For planar wavefronts, $k_z'(z) = 0$, and $\delta(t)$ corresponds to the familiar expression for the first-order Doppler shift (see Methods for details).

The experimental sequence is depicted in Fig. 1. We start in zone B by cooling all motional modes of the ion to $\bar{n} < 3$ using a combination of Doppler and electromagnetically induced transparency cooling[23], and then initialize the internal state by optical pumping into $|0\rangle$. The ion is then transported to zone A, and the laser beam used to implement the Hamiltonian is turned on in zone B. The ion is then transported through this laser beam to zone C. During the passage through the laser beam, we rapidly turn the beam off at time t_{off} and thus stop the qubit dynamics. The ion is then returned to the central zone B to perform state readout, which measures the qubit in the computational basis $\hat{\sigma}_z$ (for more details see Methods). The additional Hamiltonian $\hat{H}_s$ is implemented by offsetting the laser frequency used in the experiment by a detuning δ_L. For each setting of t_{off} and δ_L the experiment is repeated 100 times, allowing us to obtain an estimate for the qubit populations $\langle \hat{\sigma}_z^{meas}(t_{off}, \delta_L) \rangle$.

We first perform a comparison in which the ion velocity is the same but the beam position is changed. Thus we expect to obtain two different profiles for $\Omega(t)$ but the same velocity profile, which is closely related to $\delta(t)$. Experimental data is shown in Fig. 2 alongside the results of fitting performed using our iterative method. The beam positions used for each data set differ by $\sim 64\,\mu\text{m}$ along the transport axis, but the transport waveform used was identical. It can be seen from the residuals that the estimation is able to find a Hamiltonian, which results in a close match to the data.

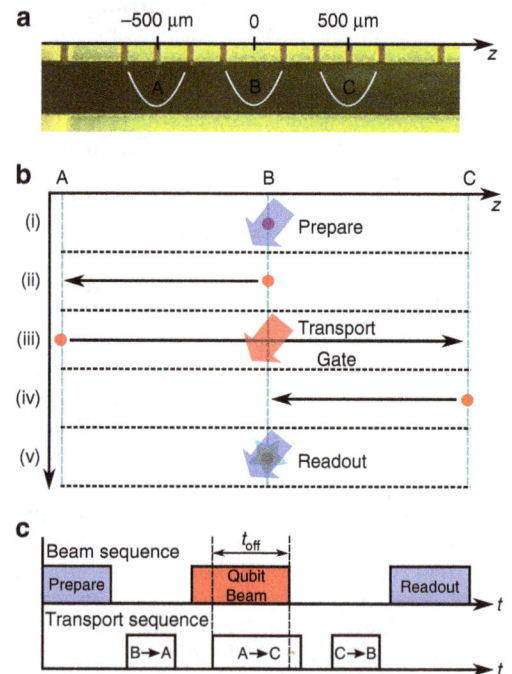

Figure 1 | Experimental sequence and timing. (a) The experiment is carried out in three zones of the trap indicated by A, B and C. **(b)** The experimental sequence involves steps (i) through (v). Preparation and readout are carried out on the static ion in zone B. The qubit evolves while the ion is transported through the laser beam in zone B in a transport operation taking the ion from zone A to zone C. **(c)** Experimental sequence showing the timing of applied laser beams and ion transport, including shutting off the laser beam during transport.

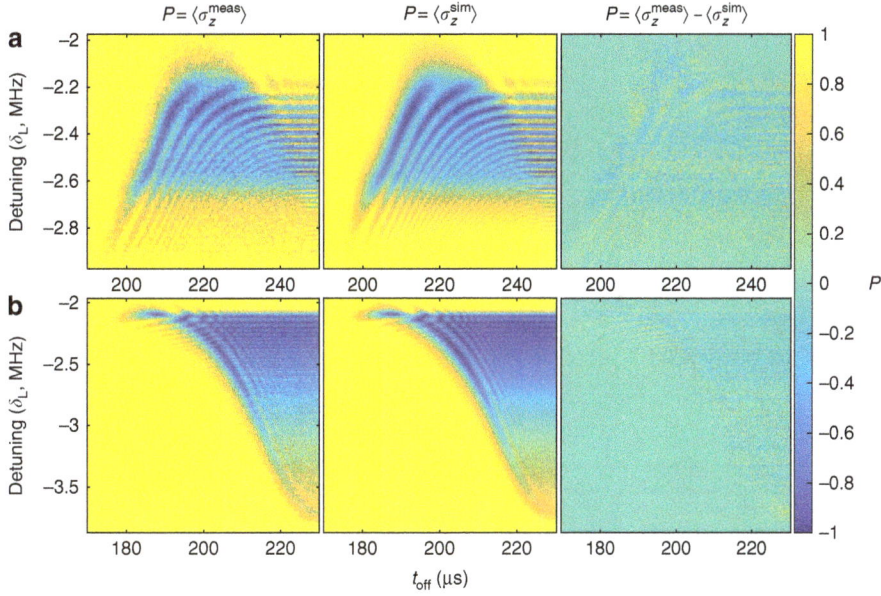

Figure 2 | Measured data, best fit and residuals. Spin population as a function of detuning and switch-off time of the laser beam. **a** is for a laser beam centred in zone B, while for **b** the beam was displaced towards zone C by 64 μm. From left to right are plots of the experimental data, the populations generated from the best fit Hamiltonian, and the residuals. Each data point results from 100 repetitions of the experimental sequence. The data in **a** consist of an array of 100 × 101 experimental settings, while that shown in **b** consists of an array of 201 × 201 settings. This leads to smaller error bars in the reconstructed Hamiltonian for the latter. For the Hamiltonian estimation, the data was weighted according to quantum projection noise.

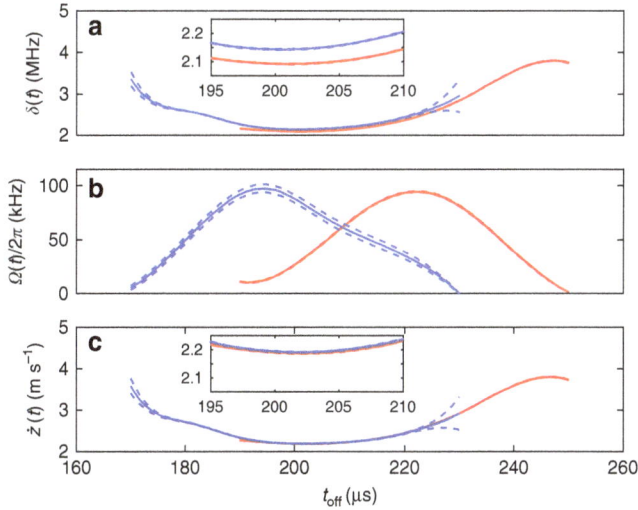

Figure 3 | Estimates of time-dependent co-efficients. (**a**) The effective detuning $\delta(t)$ and (**b**) Rabi frequency $\Omega(t)$ obtained from the two data sets. Blue and red solid lines show data obtained having the beam centred in zone B and with the beam displaced by a few tens of microns. Dashed lines indicate the s.e. on the mean of these estimates, which are obtained using resampling. For **a** the inset shows a close-up of the estimated $\delta(t)$ in the regions where the estimates overlap, showing that these do not give the same value. (**c**) The estimated velocity $\dot{z}(t)$ of the ion obtained after applying wavefront correction. The inset shows that this produces consistent results.

The estimated coefficients of the Hamiltonian extracted from the two data sets are shown in Fig. 3a,b. To estimate the relevant errors of our reconstruction, we have performed non-parametric resampling with replacement, optimizing for the solution using the same set of B-spline functions as was used for the experimental data to provide a new estimate for the Hamiltonian. This is repeated for a large number of samples, resulting in a

distribution for the estimated values of $\delta(t)$ and $\Omega(t)$ from which we extract statistical properties such as the s.e. The error bounds shown in Fig. 3 correspond to the s.e. on the mean obtained from these distributions (see Methods for further details). It can be seen that the values of $\delta(t)$ for the two different beam positions have a similar form but a fixed offset for the region where the reconstructions overlap. We believe that this effect arises from the non-planar wavefronts of the laser beam. Inverting the expression for $\delta(t)$ to obtain the velocity of the ion, we find $\dot{z}(t) = (\delta_L - \delta(t))/\left(k_z'(z)z + k_z(z)\right)$. Using this correction, we find that the two velocity profiles agree if we assume that the ion passes through the centre of the beam at a distance of 2.27 mm before the minimum beam waist, a value which is consistent with experimental uncertainties due to beam propagation and possible mis-positioning of the ion trap with respect to the fixed final focusing lens. The velocity estimates taking account of this effect are shown in Fig. 3c.

Our second comparison involves using two different velocity profiles but with a common beam position. The resolution in both time and detuning were lower in this case than for the data shown in Fig. 2 (see Methods for the data). Figure 4 shows the results of the reconstruction. We observe that the estimated Rabi frequency profiles agree to within the error bars of the reconstruction. One interesting feature of this plot is that the error bars produced from the resampled data sets increase near the peak. We believe that this happens because the sampling time of the data is 0.5 μs, which starts to become comparable to the Rabi frequency (the Nyquist frequency is 1 MHz). To optimize the efficiency of our method, it would be advantageous to run the reconstruction method in parallel with data taking, thus allowing updating of the sampling time and frequency resolution based on the current estimates of parameter values.

Discussion

Our method for directly obtaining a non-commuting time-dependent Hamiltonian uses straightforward measurements of the qubit state in a fixed basis as a function of time and a

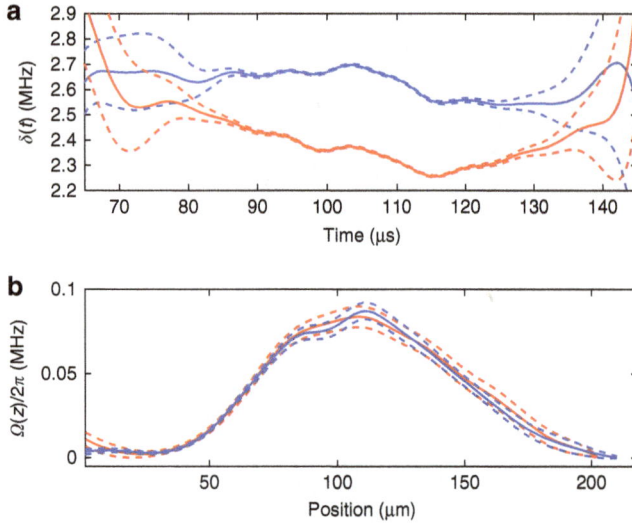

Figure 4 | Time-dependent detuning and spatial Rabi frequency.
(**a**) The estimated $\delta(t)$ obtained from the second pair of data sets (Fig. 8 in Methods). (**b**) The estimated Rabi frequency $\Omega(t)$ for the same two data sets. In each part, the blue and red solid lines show data obtained using different velocity profiles. Dashed lines indicate the s.e. on the mean of these estimates, which are obtained using resampling.

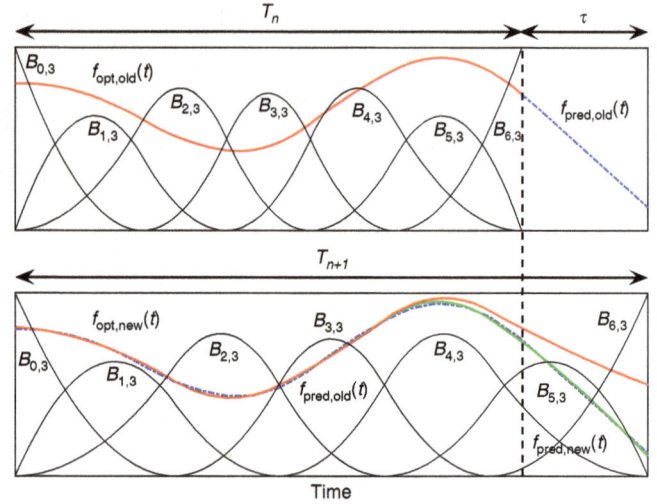

Figure 5 | Extending the horizon estimation. The steps performed when extending the time horizon from T_n to T_{n+1} are illustrated. We first predict in the old basis, then move to the new basis, and finally optimize again. The figure also shows the B-splines $B_{i,k}(t)$.

controlled offset to the Hamiltonian. Unlike schemes based on dynamical modulation or continuous strong driving, it avoids the need for control fields which act more strongly on the qubit than the Hamiltonian to be measured. This is a key advantage in quantum technologies where the Hamiltonian of interest is often already close to the limit of system drive strength. A process-tomography-based approach would require that for every time step multiple input states be introduced, and a measurement made in multiple bases[24–26]. This requires a much greater level of control than the method presented above. An effective modulation of the measurement basis arises in our approach due to the additional detuning δ_L. Nevertheless, it is also worth noting that tomography provides more information than our method: it makes no assumptions about the dynamics aside from that of a completely positive map while we require coherent dynamics. Extensions to our work are required to provide a rigorous estimation of the efficiency of the method in terms of the precision obtained for a given number of measurements, and to see whether a similar approach could be taken for non-unitary dynamics. We have recently used these methods to improve the control over the ion velocity, which is of direct value in optimizing transport operations in scalable trapped-ion quantum information processing[11,12,27], and will be essential for realizing multi-qubit transport gates[18]. We expect them to be applicable across a wide range of physical systems where such control is available, including those considered for quantum computation[4,6,7,28–31].

Methods

Derivation of Hamiltonian. The interaction of a laser beam with frequency ω_L and wave vector $\mathbf{k}(\mathbf{z}(t))$ with a two-level atom with resonant frequency ω_0 and time-dependent ion position $\mathbf{z}(t) = (0, 0, z(t))$ can be described in the Schrödinger picture by the Hamiltonian

$$\hat{H}_S = -\frac{\hbar\omega_0}{2}\hat{\sigma}_z - \hbar\Omega(z(t))\cos(\mathbf{k}(\mathbf{z}(t))\cdot\mathbf{z}(t) - \omega_L t)\hat{\sigma}_x, \quad (5)$$

where the Rabi frequency $\Omega(z(t))$ gives the interaction strength between the laser and the two energy levels. We can define the laser phase at the position of the ion as $\Phi(t) = \phi(t) - \omega_L t$ with $\phi(t) = \mathbf{k}(\mathbf{z}(t))\cdot\mathbf{z}(t) = k_z(z(t))z(t)$ and $k_z(z(t)) = |\mathbf{k}|\cos(\theta(t))$ being the projection of the laser beam onto the z-axis along which the ion is transported. Here $\theta(t)$ is the angle between the wave vector $\mathbf{k}(\mathbf{z}(t))$ and the transport axis evaluated at position $z(t)$. Moving to a rotating frame using the

unitary transformation $U = e^{-\frac{\Phi(t)}{2}}$ and applying the rotating wave approximation with respect to optical frequencies, we obtain

$$\hat{H}_I = \frac{\hbar}{2}\left(-\Omega(t)\hat{\sigma}_x + \left(-\omega_0 - \dot{\Phi}(t)\right)\hat{\sigma}_z\right). \quad (6)$$

Defining a static detuning $\delta_L = \omega_L - \omega_0$, we obtain

$$\hat{H}_I = \frac{\hbar}{2}\left(-\Omega(t)\hat{\sigma}_x + \left(\delta_L - \dot{\phi}(t)\right)\hat{\sigma}_z\right). \quad (7)$$

with

$$\delta(t) = \delta_L - \dot{\phi}(t), \quad (8)$$

which is the expression used in the main text.

B-spline curves and optimization algorithm. The set of polynomial B-spline functions $B_{i,k}(t)$ of order k are recursively defined over the index i over a set of points $\mathbf{K} = \{t_0, t_1, .., t_{n+k}\}$, which is referred to as the knot vector[20].

$$\begin{aligned} B_{i,1}(t) &= \begin{cases} 1 & t_i \leq t \leq t_{i+1} \\ 0 & \text{otherwise} \end{cases} \\ B_{i,k}(t) &= \omega_{i,k}(t)B_{i,k-1}(t) + (1-\omega_{i+1,k}(t))B_{i+1,k-1}(t). \\ \omega_{i,k}(t) &= \begin{cases} \frac{t-t_i}{t_{i+k-1}-t_i} & \text{if } t_i \neq t_{i+k-1} \\ 0 & \text{otherwise} \end{cases} \end{aligned} \quad (9)$$

Figure 5 gives a visualization of the B-splines $B_{i,k}(t)$ and a B-spline curve. The B-spline construction ensures that any linear combination of the B-splines is continuous and has $(k-2)$ continuous derivatives. The knot vector $\mathbf{K}$ determines how the basis functions are positioned within the interval $[t_0, t_{n+k}]$. We notice that for our Hamiltonian the spacing of the B-splines is not critical, which we think is due to the smoothness of the variations in our Hamiltonian parameters $\delta(t)$ and $\Omega(t)$. We therefore used the Matlab function spap2 to automatically choose a suitable knot vector and restricted ourselves to optimizing the coefficients α_i. We collect all coefficients α_i for $\delta(t)$ and $\Omega(t)$ and store them in a single vector $\boldsymbol{\alpha}$.

A detailed algorithmic summary of our implementation of the EHE method is given below.

1. Searching for a starting point: here we reconstruct the Hamiltonian for a first, minimal time horizon such that we can then use this as a starting point to iteratively extend the horizon as described in step 2.

   (a) Choose an initial time horizon such that it contains the region where the first discernible qubit dynamics occur.
   (b) Cut down the number of fitting parameters as much as possible, for example, by using few B-splines of low order. This amounts to choosing empirically a low number of B-splines (and thus the length of $\boldsymbol{\alpha}_0$), which might represent $\delta(t)$ and $\Omega(t)$ over the given region.
   (c) Use a non-linear least-squares fitting routine to minimize J by varying the parameters $\boldsymbol{\alpha}_0$. In the case that the initial fit is poor or no minimum is found, try new initial conditions, change the number of B-spline functions, or manually adjust the function using prior knowledge of the physical system under consideration.

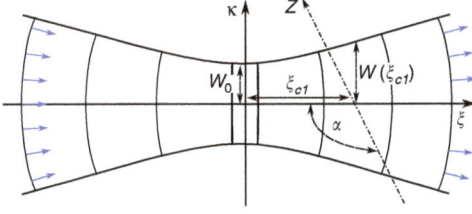

Figure 6 | Beam and ion transport. The beam propagation direction lies along the ξ-axis and the ion is transported along the z-axis lying on the $\kappa\xi$-plane as indicated. Normalized vectors representing $\mathbf{e}_l(\kappa, \xi)$ lying perpendicular to the wavefronts are indicated by the blue arrows.

This procedure is used to provide a starting point for the optimization over the initially chosen window, which is typically performed with a set of higher order B-splines. From this starting point, we iteratively extend the fitting method to the full data set as follows.

2. Extend the horizon: this step is repeated until the whole time horizon is covered. It consists of the following sequence, which is illustrated in Fig. 5.

   (a) Extend the time horizon by τ from T_n to $T_{n+1} = T_n + \tau$.
   (b) Extrapolate $f_{\text{opt,old}}(t)$ within τ, for example, using fnxtr in Matlab.
   (c) Adapt the B-splines to the new time horizon T_{n+1} and represent $f_{\text{pred,old}}(t)$ in the new basis, giving $f_{\text{pred,new}}(t)$. In Matlab one can use spap2 to do this.
   (d) Use $f_{\text{pred,new}}(t)$ as the initial guess for a weighted non-linear least-squares fit over the extended time span up to T_{n+1}.
   (e) Judge the results of the fit based on its reduced χ^2-value χ^2_{red}. If it is below a specified bound, continue with an additional iteration of steps a–d, repeating until the full region of the data is covered. Otherwise, attempt the following fall-back procedures:

     i Reduce τ, the time by which the time horizon is extended, and try again.
     ii Increase the number of B-splines and try again.
     iii Try again using a different starting point.

If all these fail, we have to resort to increasing the bound on χ^2_{red}.

3. Post-processing: the following steps are optional and were performed manually in cases where we wished to improve the fit or examine its behaviour.

   (a) The optimization over the whole time horizon was re-run using different numbers of B-splines for $\delta(t)$ and $\Omega(t)$. This was used to check the sensitivity of the fit.
   (b) The optimization over the whole time horizon was re-run using a starting point based on the previously found optimum plus randomized deviations. This tested the robustness of the final fit.

Wavefront correction. For plane waves, we find that $\dot\phi(t) = \mathbf{k} \cdot \mathbf{v}(t)$, which is the well-known expression for the first-order Doppler shift. For transport through a real Gaussian beam, the wave vector direction changes with position. Taking this into account, the derivative of $\phi(t)$ becomes

$$\dot\phi(t) = \left[k'_z(z(t))z(t) + k_z(z(t)) \right]\dot z(t) \tag{10}$$

where $k'_z = dk_z/dz$ and $\dot z(t)$ is the component of the ion velocity along the z-axis. We extract $\delta(t)$ using our Hamiltonian estimation procedure, thus to obtain the velocity of the ion we use

$$\dot z(t) = \frac{-\delta(t) + \delta_L}{k'_z(z(t))z(t) + k_z(z(t))}. \tag{11}$$

As the ion moves through the beam it experiences the same magnitude of the wave vector $|\mathbf{k}| = 2\pi/\lambda$, but the angle θ between the ion direction and the wave vector changes. Written as a function of this angle, the velocity becomes

$$\dot z(t) = \frac{-\delta(t) + \delta_L}{-|\mathbf{k}|\sin(\theta(z(t)))\theta'(z(t))z(t) + |\mathbf{k}|\cos(\theta(z(t)))} \tag{12}$$

where $\theta'(z(t)) = d\theta(z(t))/dz(t)$. We parameterize our Gaussian beam according to Fig. 6. The phase is given as a function of both the position along the beam axis ξ

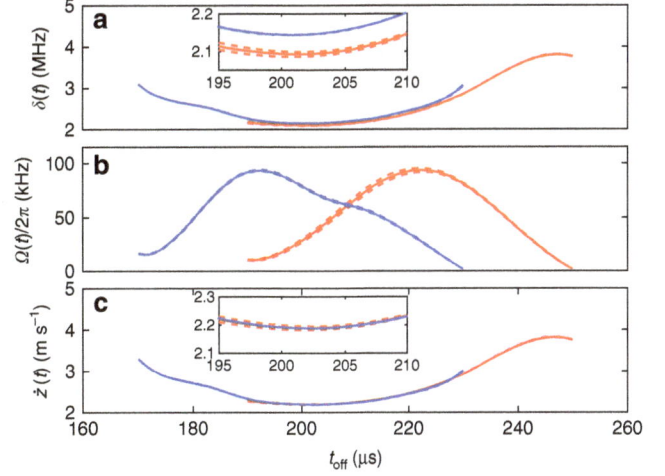

Figure 7 | Parametric bootstrap resampling. Predictions for the effective detuning $\delta(t)$ in **a**, Rabi frequency $\Omega(t)$ in **b** and velocity $\dot z(t)$ in **c**. Blue and red solid lines show data obtained having the beam centred in zone B and with the beam displaced by a few tens of micron. Dashed lines represent the s.e. on the mean of these estimates obtained using parametric bootstrap resampling, assuming quantum projection noise. This can be compared with the error bounds obtained from the non-parametric method, which are shown in Fig. 3 in the main text. The bounds are tighter for the parametric bootstrapping.

and the perpendicular distance from this axis κ by[32]

$$\varphi(\kappa, \xi) = |\mathbf{k}|\xi - \zeta(\xi) + \frac{|\mathbf{k}|\kappa^2}{2R(\xi)}. \tag{13}$$

where the Gaussian beam parameters include the beam waist $W(\xi)$, the radius of curvature $R(\xi)$, the Rayleigh range ξ_R and the Guoy phase shift $\zeta(\xi)$. These are given by the expressions

$$
\begin{aligned}
W(\xi) &= W_0\sqrt{1 + \left(\frac{\xi}{\xi_R}\right)^2} \\
R(\xi) &= \xi\left(1 + \left(\frac{\xi_R}{\xi}\right)^2\right) \\
\zeta(\xi) &= \tan^{-1}\left(\frac{\xi}{\xi_R}\right) \\
\xi_R &= \frac{\pi W_0^2}{\lambda} \\
k &= \frac{2\pi}{\lambda}
\end{aligned}
\tag{14}
$$

where W_0 is the minimum beam waist and λ the laser wavelength. The ion moves along the z-axis as shown in Fig. 6. In the $\kappa\xi$-plane a unit vector $\mathbf{e}_l(\kappa, \xi)$ perpendicular to the wavefronts is given by

$$\mathbf{e}_l(\kappa, \xi) = \frac{\nabla\varphi(\kappa, \xi)}{\|\nabla\varphi(\kappa, \xi)\|} \tag{15}$$

and the unit vector $\mathbf{e}_v$ pointing along the direction of transport is given by

$$\mathbf{e}_v = \begin{bmatrix} \cos(\alpha) \\ \sin(\alpha) \end{bmatrix} \tag{16}$$

The angle $\theta(\xi)$ between the wave and position vector is then given by the dot product

$$\theta(\kappa) = \cos^{-1}(\mathbf{e}_n \cdot \mathbf{e}_v). \tag{17}$$

which can be written in terms of the full set of parameters above as

$$
\begin{aligned}
\theta(\kappa) &= \cos^{-1}(\gamma_1 + \gamma_2) \\
\gamma_1 &= \frac{\cos(\alpha)\left(-2\xi_R(\xi^2 + \xi_R^2) + k\kappa^2(\xi_R^2 - \xi^2) + 2k(\xi^2 + \xi_R^2)^2\right)}{\eta(\kappa)} \\
\gamma_2 &= \frac{\sin(\alpha)2k\kappa\xi(\xi^2 + \xi_R^2)}{\eta(\kappa)} \\
\eta(\kappa, \xi) &= (\xi^2 + \xi_R^2)\left[4\left(\frac{k\kappa\xi}{\xi^2 + \xi_R^2}\right)^2 + \left(-\frac{2\xi_R}{\xi^2 + \xi_R^2} + k\left(2 + \frac{\kappa^2(\xi_R^2 - \xi^2)}{(\xi^2 + \xi_R^2)^2}\right)\right)^2\right] \\
\kappa(t) &= z(t)\sin(\alpha)
\end{aligned}
\tag{18}
$$

where in our experiments $\alpha = 3\pi/4$.

Using equations (12) and (18), we examined the value of ξ_{cl} required for the velocity to match for our two beam positions. We find that they agree for $\xi_{cl} = -2.27$ mm, which is within the experimental uncertainties for our set-up.

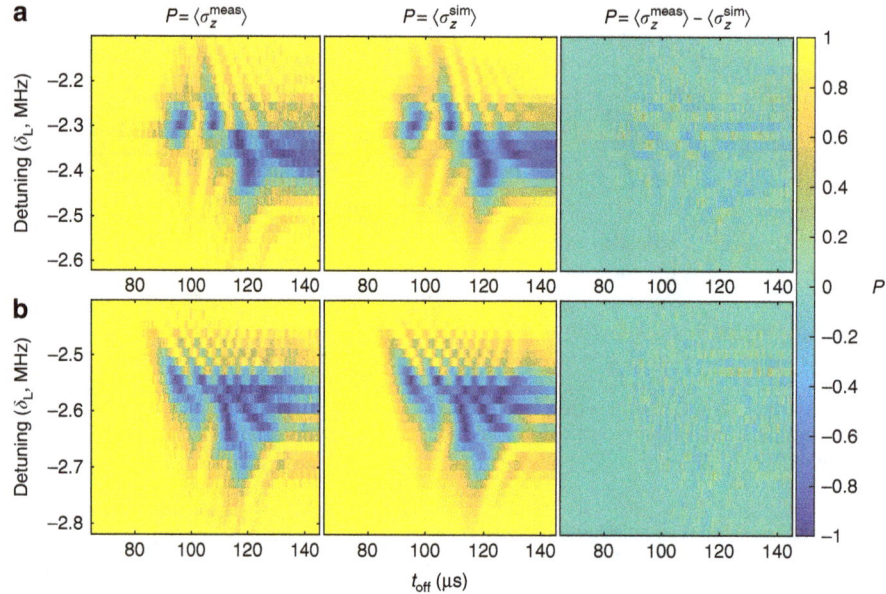

Figure 8 | Measured data, estimation and residuals. Spin population as a function of detuning and switch-off time of the laser beam, for the data sets used to obtain the reconstructed parameters shown in Fig. 4. **a** corresponds to the blue reconstruction while **b** to the red one. Each data point results from 100 repetitions of the experimental sequence. For the Hamiltonian estimation, the data was weighted according to quantum projection noise.

Error estimation. To estimate the errors of the time-dependent functions, we use non-parametric bootstrapping[33]. The process is summarized as follows:

1. Estimate initial solution: estimate the time-dependent functions from the original data using Hamiltonian estimation.
2. Resampling: create N_s sample solutions for all time-dependent functions in the following way:

   (a) Form a sample set by randomly picking with replacement from the photon count data used in qubit detection.
   (b) Re-estimate new time-dependent functions by optimizing over the full time span, using the solution found in (1) as a starting point.
   (c) Record the reduced χ^2-values $\chi^2_{\mathrm{red},r}$ for each sample r along with the B-spline curve coefficients $\boldsymbol{\alpha}_r$.

3. Post-process samples:

   (a) Form a histogram of the χ^2-values $\chi^2_{\mathrm{red},r}$.
   (b) Find and fit a normal-like distribution to the histogram with preference to the spread with lowest lying $\chi^2_{\mathrm{red},r}$ in the case of a multi-modal distribution. From the fit obtain the mean reduced-χ^2-value $\langle\chi^2_{\mathrm{red},r}\rangle$, as well as the s.d. σ_χ.
   (c) Eliminate the outlier samples by removing all $\boldsymbol{\alpha}_r$ with $\chi^2_{\mathrm{red},r}$ values that are 3–5σ_χ from the mean $\langle\chi^2_{\mathrm{red},r}\rangle$.
   (d) Form a matrix Y, where each row vector is a sample set of coefficients $\boldsymbol{\alpha}_r$ that remained after step 3(c).

4. Obtain statistics:

   (a) Find the mean B-spline coefficients $\langle\alpha\rangle$ of equation (4) by taking the mean over the column vectors of Y with each element of the mean given by $\langle\alpha\rangle_i=\langle\alpha_i\rangle$.
   (b) Find the covariance matrix $\Sigma=\mathrm{cov}(\mathbf{Y}^\alpha)$ with $\Sigma_{ij}=\mathrm{E}[(\alpha_i-\langle\alpha_i\rangle)(\alpha_j-\langle\alpha_j\rangle)]$ with E the expectation operator. The s.d. of each of the mean coefficients $\langle\alpha_i\rangle$ is given by $\sigma_{\langle\alpha_i\rangle}=\sqrt{\Sigma_{ii}}$. We record these values in a row vector $\boldsymbol{\sigma}_{\langle\alpha_i\rangle}$.

In evaluating errors using bootstrapping, we use the same set of spline polynomials as were used for the final optimization stage in the data, which makes the reconstruction more reliable in converging to a minimum. We thus expect that the parameter space explored in evaluating the errors is not the same as for the *ab initio* estimation of the Hamiltonian. This is apparent in the regions of the data where the final estimate has large error bounds (for example, in Fig. 4), where an even larger spread might be expected (the dynamics has stopped evolving at this

point). We think that the net effect is to under-estimate the errors in the regions where the Hamiltonian is uncertain, but that the error bars given in the central region (where the Hamiltonian is well-defined) are close to what would be obtained through a full optimization.

We have also applied parametric bootstrapping to obtain the error bounds shown in Fig. 7. The difference to the non-parametric case is that in point (2) the samples are created using the solutions obtained from (1) and adding quantum projection noise. For each sample the Hamiltonian is estimated. The estimates from multiple samples are used to construct error bounds in the same manner as for the non-parametric resampling. We have found that the error bounds obtained from parametric bootstrapping are lower compared with that of the non-parametric case as shown in Fig. 3. We think this is due to the latter exploring deviations around a single minimum in the optimization landscape, while the case resampling arrives at different local minima, which are spread over a wider region.

Single beam profile with two different velocity profiles. To verify that our method can also consistently estimate the Rabi frequency profile, we measure a second pair of data sets in which we take two different velocity profiles using the same beam position. This data is shown in Fig. 8. Also shown are the best-fits obtained from the reconstructed Hamiltonians. The parameter variations obtained from the reconstructed Hamiltonians for these data sets can be found in the main text in Fig. 4. The sampling rate of the data in these data sets was 2 MHz, resulting in a Nyquist frequency of 1 MHz.

References

1. Cole, J. H. *et al.* Identifying an experimental two-state hamiltonian to arbitrary accuracy. *Phys. Rev. A* **71**, 062312 (2005).
2. Mitra, A. & Rabitz, H. Identifying mechanisms in the control of quantum dynamics through hamiltonian encoding. *Phys. Rev. A* **67**, 033407 (2003).
3. de Castro, R. R. & Rabitz, H. Laboratory implementation of quantum-control-mechanism identification through hamiltonian encoding and observable decoding. *Phys. Rev. A* **81**, 063422 (2010).
4. Zhang, J. & Sarovar, M. Quantum hamiltonian identification from measurement time traces. *Phys. Rev. Lett.* **113**, 080401 (2014).
5. de Castro, R. R., Cabrera, R., Bondar, D. I. & Rabitz, H. Time-resolved quantum process tomography using hamiltonian-encoding and observable-decoding. *N. J. Phys.* **15**, 025032 (2013).
6. Devitt, S. J., Cole, J. H. & Hollenberg, L. C. L. Scheme for direct measurement of a general two-qubit hamiltonian. *Phys. Rev. A* **73**, 052317 (2006).
7. Shabani, A., Mohseni, M., Lloyd, S., Kosut, R. L. & Rabitz, H. Estimation of many-body quantum hamiltonians via compressive sensing. *Phys. Rev. A* **84**, 012107 (2011).
8. Zhao, N. *et al.* Sensing single remote nuclear spins. *Nature Nanotech.* **7**, 657–662 (2012).

9. Blais, A., Huang, R.-S., Wallraff, A., Girvin, S. M. & Schoelkopf, R. J. Cavity quantum electrodynamics for superconducting electrical circuits: an architecture for quantum computation. *Phys. Rev. A* **69**, 062320 (2004).

10. Devoret, M. H. & Schoelkopf, R. J. Superconducting circuits for quantum information: an outlook. *Science* **339**, 1169–1174 (2013).

11. Bowler, R. *et al.* Coherent diabatic ion transport and separation in a multizone trap array. *Phys. Rev. Lett.* **109**, 080502 (2012).

12. Walther, A. *et al.* Controlling fast transport of cold trapped ions. *Phys. Rev. Lett.* **109**, 080501 (2012).

13. Kotler, S., Akerman, N., Glickman, Y. & Ozeri, R. Non-linear single-spin spectrum analyzer. *Phys. Rev. Lett.* **110**, 110503 (2013).

14. Magesan, E., Cooper, A., Yum, H. & Cappellaro, P. Reconstructing the profile of time-varying magnetic fields with quantum sensors. *Phys. Rev. A* **88**, 032107 (2013).

15. Cooper, A., Magesan, E., Yum, H. & Cappellaro, P. Time-resolved magnetic sensing with electronic spins in diamond. *Nat. Commun.* **5**, 3141 (2014).

16. Barnes, E. & Das Sarma, S. Analytically solvable driven time-dependent two-level quantum systems. *Phys. Rev. Lett.* **109**, 060401 (2012).

17. Barnes, E. Analytically solvable two-level quantum systems and landau-zener interferometry. *Phys. Rev. A* **88**, 013818 (2013).

18. Leibfried, D., Knill, E., Ospelkaus, C. & Wineland, D. J. Transport quantum logic gates for trapped ions. *Phys. Rev. A* **76**, 032324 (2007).

19. de Clercq, L. *et al.* Parallel Transport-Based Quantum Logic Gates With Trapped Ions. *Phys. Rev. Lett.* **116**, 080502 (2016).

20. de Boor, C. *A Practical Guide to Splines (Applied Mathematical Sciences)*, revised edn (Springer, 2001).

21. Muske, K. R. & Rawlings, J. B. in *Methods of Model Based Process Control* (ed. Berber, R.) (Springer, 1995).

22. Rao, C. V., Rawlings, J. B. & Mayne, D. Q. Constrained state estimation for non-linear discrete-time systems: stability and moving horizon approximations. *IEEE Trans. Automat. Contr.* **48**, 246–258 (2003).

23. Roos, C. F. *et al.* Experimental demonstration of ground state laser cooling with electromagnetically induced transparency. *Phys. Rev. Lett.* **85**, 5547 (2000).

24. Chuang, I. L. & Nielsen, M. A. Prescription for experimental determination of the dynamics of a quantum black box. *J. Mod. Opt.* **44**, 2455–2467 (1997).

25. Poyatos, J. F., Cirac, J. I. & Zoller, P. Complete characterization of a quantum process: the two-bit quantum gate. *Phys. Rev. Lett.* **78**, 390 (1997).

26. Riebe, M. *et al.* Process tomography of ion trap quantum gates. *Phys. Rev. Lett.* **97**, 220407 (2006).

27. Wineland, D. J. *et al.* Experimental issues in coherent quantum-state manipulation of trapped atomic ions. *J. Res. Natl. Inst. Stand. Technol.* **103**, 259–328 (1998).

28. Martinis, J. M. & Geller, M. R. Fast adiabatic gates using only σ_z control. *Phys. Rev. A* **90**, 022307 (2014).

29. Schutjens, R., Abu Dagga, F., Egger, D. J. & Wilhelm, F. K. Single-qubit gates in frequency-crowded transmon systems. *Phys. Rev. A* **88**, 052330 (2013).

30. Barends, R. *et al.* Digital quantum simulation of fermionic models with a superconducting circuit. *Nat. Commun.* **6**, 7654 (2015).

31. Bonato, C. *et al.* Optimized quantum sensing with a single electron spin using real-time adaptive measurements. *Nature Nanotech* **11**, 247–252 (2016).

32. Saleh, B. E. A. & Teich., M. C. *Fundamentals of Photonics* (Wiley, 2007).

33. Murphy., K. P. *Machine Learning: a Probabilistic Perspective* (MIT press, 2012).

Acknowledgements

We thank Lukas Gerster, Martin Sepiol and Karin Fisher for contributions to the experimental apparatus. We thank Florian Leupold for feedback on the manuscript and useful discussions. We acknowledge support from the Swiss National Science Foundation under grant numbers 200021_134776 and 200020_153430, ETH Research Grant under grant no. ETH-18 12–2, and from the National Centre of Competence in Research for Quantum Science and Technology (QSIT).

Author contributions

Experimental data was taken by L.E.d.C., R.O., M.M. and D.N. using apparatus built up by all authors. Data analysis was performed by L.E.d.C. and R.O. The paper was written by J.P.H., L.E.d.C. and R.O., with input from all authors. The work was conceived by J.P.H. and L.E.d.C.

Additional information

Competing financial interests: The authors declare no competing financial interests.

Permissions

All chapters in this book were first published in NC, by Nature Publishing Group; hereby published with permission under the Creative Commons Attribution License or equivalent. Every chapter published in this book has been scrutinized by our experts. Their significance has been extensively debated. The topics covered herein carry significant findings which will fuel the growth of the discipline. They may even be implemented as practical applications or may be referred to as a beginning point for another development.

The contributors of this book come from diverse backgrounds, making this book a truly international effort. This book will bring forth new frontiers with its revolutionizing research information and detailed analysis of the nascent developments around the world.

We would like to thank all the contributing authors for lending their expertise to make the book truly unique. They have played a crucial role in the development of this book. Without their invaluable contributions this book wouldn't have been possible. They have made vital efforts to compile up to date information on the varied aspects of this subject to make this book a valuable addition to the collection of many professionals and students.

This book was conceptualized with the vision of imparting up-to-date information and advanced data in this field. To ensure the same, a matchless editorial board was set up. Every individual on the board went through rigorous rounds of assessment to prove their worth. After which they invested a large part of their time researching and compiling the most relevant data for our readers.

The editorial board has been involved in producing this book since its inception. They have spent rigorous hours researching and exploring the diverse topics which have resulted in the successful publishing of this book. They have passed on their knowledge of decades through this book. To expedite this challenging task, the publisher supported the team at every step. A small team of assistant editors was also appointed to further simplify the editing procedure and attain best results for the readers.

Apart from the editorial board, the designing team has also invested a significant amount of their time in understanding the subject and creating the most relevant covers. They scrutinized every image to scout for the most suitable representation of the subject and create an appropriate cover for the book.

The publishing team has been an ardent support to the editorial, designing and production team. Their endless efforts to recruit the best for this project, has resulted in the accomplishment of this book. They are a veteran in the field of academics and their pool of knowledge is as vast as their experience in printing. Their expertise and guidance has proved useful at every step. Their uncompromising quality standards have made this book an exceptional effort. Their encouragement from time to time has been an inspiration for everyone.

The publisher and the editorial board hope that this book will prove to be a valuable piece of knowledge for researchers, students, practitioners and scholars across the globe.

List of Contributors

C. Schuck, X. Guo, L. Fan, M. Poot and H.X. Tang
Department of Electrical Engineering, Yale University, New Haven, Connecticut 06511, USA

X. Ma
Department of Electrical Engineering, Yale University, New Haven, Connecticut 06511, USA
Institute for Quantum Optics and Quantum Information, Austrian Academy of Science, A-1090 Vienna, Austria

D. Sando
Unité Mixte de Physique, CNRS, Thales, Univ. Paris-Sud, Université Paris-Saclay, 91767 Palaiseau, France
School of Materials Science and Engineering, University of New South Wales, Sydney 2052, Australia

C. Carrétéro, V. Garcia, S. Fusil, M. Bibes and A. Barthélémy
Unité Mixte de Physique, CNRS, Thales, Univ. Paris-Sud, Université Paris-Saclay, 91767 Palaiseau, France

Yurong Yang and L. Bellaiche
Department of Physics and Institute for Nanoscience and Engineering, University of Arkansas, Fayetteville, Arkansas 72701, USA

E. Bousquet and Ph. Ghosez
Theoretical Materials Physics, Université de Liège, B-5, B-4000 Sart-Tilman, Belgium

D. Dolfi
Thales Research and Technology France, 1 Avenue Augustin Fresnel, 91767 Palaiseau, France

Woo Young Kim, Hyeon-Don Kim, Hyun-Sung Park, Kanghee Lee, Hyun Joo Choi, Bumki Min and Jaehyeon Son
Department of Mechanical Engineering, Korea Advanced Institute of Science and Technology(KAIST), Daejeon 305-701, Republic of Korea

Seung Hoon Lee
Department of Mechanical Engineering, Korea Advanced Institute of Science and Technology(KAIST), Daejeon 305-701, Republic of Korea
Department of Applied Physics and Materials Science, California Institute of Technology, California 91125, USA

Teun-Teun Kim
Department of Mechanical Engineering, Korea Advanced Institute of Science and Technology(KAIST), Daejeon 305-701, Republic of Korea
Metamaterial Research Centre, School of Physics and Astronomy, University of Birmingham, Birmingham B15 2TT, UK

Namkyoo Park
Photonic Systems Laboratory, School of EECS, Seoul National University, Seoul 151-744, Republic of Korea

You-Chia Chang
Center for Photonics and Multiscale Nanomaterials, University of Michigan, 2200 Bonisteel Blvd., Ann Arbor, Michigan 48109, USA.
Department of Physics, University of Michigan, 450 Church St, Ann Arbor, Michigan 48109, USA

Che-Hung Liu, Zhaohui Zhong and Theodore B. Norris
Center for Photonics and Multiscale Nanomaterials, University of Michigan, 2200 Bonisteel Blvd., Ann Arbor, Michigan 48109, USA
Department of Physics, University of Michigan, 450 Church St, Ann Arbor, Michigan 48109, USA

Chang-Hua Liu
Department of Electrical Engineering and Computer Science, University of Michigan, 1301 Beal Avenue, Ann Arbor, Michigan 48109, USA

Siyuan Zhang and Seth R. Marder
School of Chemistry and Biochemistry, Georgia Institute of Technology, 901 Atlantic Drive, Atlanta, Geogia 30332, USA

Evgenii E. Narimanov
School of Electrical and Computer Engineering and Birck Nanotechnology Center, Purdue University, 1205 West State Street, West Lafayette, Indiana 47907, USA

Xiao Tao Geng, Seungchul Kim and Dong-Eon Kim
Max Planck Center for Attosecond Science, Max Planck POSTECH/KOREA Res. Initiative, Pohang, Gyeongbuk 376-73, South Korea
Department of Physics, Center for Attosecond Science and Technology (CASTECH), POSTECH, Pohang, Gyeongbuk 376-73, South Korea

Byung Jae Chun and Young-Jin Kim
School of Mechanical and Aerospace Engineering,
Nanyang Technological University (NTU), 50 Nanyang
Avenue, Singapore 639798, Singapore

Ji Hoon Seo and Kwanyong Seo
Department of Energy Engineering, Ulsan National
Institute of Science and Technology (UNIST), Ulsan
689-798, South Korea

Hana Yoon
Energy Storage Department, Korea Institute of Energy
Research (KIER), Daejeon 305-343, South Korea

J. Vieira, E.P. Alves, J.T. Mendonça and L.O. Silva
GoLP/Instituto de Plasmas e Fusão Nuclear, Instituto
Superior Técnico, Universidade de Lisboa, 1049-001
Lisbon, Portugal

R.M.G.M. Trines
Central Laser Facility, STFC Rutherford Appleton
Laboratory, Didcot OX11 0QX, UK

R.A. Fonseca
GoLP/Instituto de Plasmas e Fusão Nuclear, Instituto
Superior Técnico, Universidade de Lisboa, 1049-001
Lisbon, Portugal
DCTI/ISCTE Lisbon University Institute, 1649-026
Lisbon, Portugal

R. Bingham
Central Laser Facility, STFC Rutherford Appleton
Laboratory, Didcot OX11 0QX, UK
Department of Physics, 107 Rottenrow East, Glasgow
G4 0NG, UK

P. Norreys
Central Laser Facility, STFC Rutherford Appleton
Laboratory, Didcot OX11 0QX, UK
Department of Physics, University of Oxford, Oxford
OX1 3PU, UK

Vittorio Peano and Christian Brendel
Institute for Theoretical Physics, University of Erlangen-
Nürnberg, Staudtstr. 7, 91058 Erlangen, Germany

Florian Marquardt
Institute for Theoretical Physics, University of Erlangen-
Nürnberg, Staudtstr. 7, 91058 Erlangen, Germany
Max Planck Institute for the Science of Light, Gu¨nther-
Scharowsky-Strae 1/Bau 24, 91058 Erlangen, Germany

Martin Houde and Aashish A. Clerk
Department of Physics, McGill University, 3600 rue
University, Montreal, Quebec, Canada H3A 2T8

D. Pierangeli, M. Ferraro and E. DelRe
Dipartimento di Fisica, Università di Roma 'La
Sapienza', Rome 00185, Italy

F. Di Mei and G. Di Domenico
Dipartimento di Fisica, Università di Roma 'La
Sapienza', Rome 00185, Italy
Center for Life Nano Science@Sapienza, Istituto
Italiano di Tecnologia, Rome 00161, Italy

C.E.M. de Oliveira and A.J. Agranat
Department of Applied Physics, Hebrew University of
Jerusalem, Jerusalem 91904, Israel

**Satyabrata Kar, Hamad Ahmed, Prokopis
Hadjisolomou, Ciaran L.S. Lewis, Gagik Nersisyan
and Marco Borghesi**
School of Mathematics and Physics, Queen's University
Belfast, Belfast BT7 1NN, UK

**Rajendra Prasad, Mirela Cerchez, Stephanie
Brauckmann, Bastian Aurand, Anna M. Schroer,
Marco Swantusch and Oswald Willi**
Institut für Laser-und Plasmaphysik, Heinrich-Heine-
Universität, Düsseldorf D-40225, Germany

Giada Cantono
Department of Physics E. Fermi, Largo B. Pontecorvo
3, Pisa 56127, Italy

Andrea Macchi
Department of Physics E. Fermi, Largo B. Pontecorvo
3, Pisa 56127, Italy.
Consiglio Nazionale delle Ricerche, Istituto Nazionale
di Ottica, Research Unit Adriano Gozzini, via G.
Moruzzi 1, Pisa 56124, Italy

Alexander P.L. Robinson
Central Laser Facility, Rutherford Appleton Laboratory,
Didcot, Oxfordshire OX11 0QX, UK

Matt Zepf
School of Mathematics and Physics, Queen's University
Belfast, Belfast BT7 1NN, UK.
Helmholtz Institut Jena, 07743 Jena, Germany. 7
Institut für Optik und Quantenelektronik, Universität
Jena, 07743 Jena, Germany

**Austin P. Spencer, Boris Spokoyny, Supratim Ray,
Fahad Sarvari and Elad Harel**
Department of Chemistry, Northwestern University,
2145 Sheridan Road, Evanston, Illinois 60208, USA

Daniela Wolf
Experimental Physics III, University of Bayreuth, Universita''tsstrasse 30, D-95440 Bayreuth, Germany
Max Planck Institute for Solid State Research, Heisenbergstrasse 1, D-70569 Stuttgart, Germany

Thorsten Schumacher and Markus Lippitz
Experimental Physics III, University of Bayreuth, Universita''tsstrasse 30, D-95440 Bayreuth, Germany

Wei Jin Hu, Weili Yu and Tom Wu
Materials Science and Engineering, King Abdullah University of Science and Technology (KAUST), Thuwal 23955-6900, Saudi Arabia

Zhihong Wang
Advanced Nanofabrication Core Lab, King Abdullah University of Science and Technology (KAUST), Thuwal 23955-6900, Saudi Arabia

Andrea Crespi, Roberto Osellame and Roberta Ramponi
Istituto di Fotonica e Nanotecnologie, Consiglio Nazionale delle Ricerche (IFN-CNR), Piazza Leonardo da Vinci, 32, I-20133 Milano, Italy
Dipartimento di Fisica, Politecnico di Milano, Piazza Leonardo da Vinci, 32, I-20133 Milano, Italy

Marco Bentivegna, Fulvio Flamini, Paolo Mataloni, Fabio Sciarrino Nicolò Spagnolo and Niko Viggianiello
Dipartimento di Fisica, Sapienza Università di Roma, Piazzale Aldo Moro 5,
I-00185 Roma, Italy

Luca Innocenti
Dipartimento di Fisica, Sapienza Università di Roma, Piazzale Aldo Moro 5,
I-00185 Roma, Italy
Università di Roma Tor Vergata, Via della ricerca scientifica 1, I-00133 Roma, Italy

Rinaldo Trotta, Javier Martín-Sánchez, Johannes S. Wildmann, Marcus Reindl, Christian Schimpf and Armando Rastelli
Institute of Semiconductor and Solid State Physics, Johannes Kepler University Linz, Altenbergerstr. 69, A-4040 Linz, Austria

Giovanni Piredda, Sandra Stroj and Johannes Edlinger
Forschungszentrum Mikrotechnik, FH Vorarlberg, Hochschulstr. 1, A-6850 Dornbirn, Austria

Eugenio Zallo
Institute for Integrative Nanosciences, IFW Dresden, Helmholtzstr. 20, D-01069
Dresden, Germany
Paul-Drude-Institut für Festkörperelektronik, Hausvogteilplatz 5-7, 10117 Berlin, Germany

Jianbo Yin, HuanWang, Lei Liao, Li Lin, Hailin Peng and Zhongfan Liu
Center for Nanochemistry, Beijing Science and Engineering Center for Nanocarbons, Beijing National Laboratory for Molecular Sciences, College of Chemistry and Molecular Engineering, Peking University, 202 Chengfu Road, Haidian District, Beijing 100871, China

Zhenjun Tan and Xiao Sun
Center for Nanochemistry, Beijing Science and Engineering Center for Nanocarbons, Beijing National Laboratory for Molecular Sciences, College of Chemistry and Molecular Engineering, Peking University, 202 Chengfu Road, Haidian District, Beijing 100871, China
Academy for Advanced Interdisciplinary Studies, Peking University, Beijing 100871, China.

Han Peng and Yulin Chen
Clarendon Laboratory, Department of Physics, University of Oxford, Parks Road, Oxford OX1 3PU, UK

Ai Leen Koh
Stanford Nano Shared Facilities, Stanford University, Stanford, California 94305, USA

Yan Chen, Jiaxiang Zhang, Michael Zopf, Kyubong Jung, Yang Zhang, Robert Keil and Fei Ding
Institute for Integrative Nanosciences, IFW Dresden, Helmholtzstrae 20, 01069 Dresden, Germany

Oliver G. Schmidt
Institute for Integrative Nanosciences, IFW Dresden, Helmholtzstrae 20, 01069 Dresden, Germany
Material Systems for Nanoelectronics, Chemnitz University of Technology, Reichenhainer strasse 70, 09107 Chemnitz, Germany

Dietmar Korn, Matthias Lauermann, Patrick Appel, Luca Alloatti and Robert Palmer
Institute of Photonics and Quantum Electronics (IPQ), Karlsruhe Institute of Technology (KIT), 76131 Karlsruhe, Germany.

Sebastian Koeber, Wolfgang Freude and Christian Koos
Institute of Photonics and Quantum Electronics (IPQ), Karlsruhe Institute of Technology (KIT), 76131 Karlsruhe, Germany
Institute of Microstructure Technology (IMT), Karlsruhe Institute of Technology (KIT), 76344 Eggenstein-Leopoldshafen, Germany

Juerg Leuthold
Institute of Photonics and Quantum Electronics (IPQ), Karlsruhe Institute of Technology (KIT), 76131 Karlsruhe, Germany
Institute of Microstructure Technology (IMT), Karlsruhe Institute of Technology (KIT), 76344 Eggenstein-Leopoldshafen, Germany
Laboratory for Electromagnetic Fields and Microwave Electronics (IFH), Swiss Federal Institute of Technology (ETH), Zürich8092, Switzerland

Pieter Dumon
Department of Information Technology, IMEC, 9000 Gent, Belgium

Marc-Antoine Lemonde and Aashish A. Clerk
Department of Physics, McGill University, 3600 rue University, Montreal, Quebec, Canada H3A 2T8

Nicolas Didier
Department of Physics, McGill University, 3600 rue University, Montreal, Quebec, Canada H3A 2T8
Départment de Physique, Université de Sherbrooke, 2500 Boulevard de l'Université, Sherbrooke, Québec, Canada J1K 2R1

Kent A.G. Fisher, Jean-Philippe W. MacLean and Kevin J. Resch
Institute for Quantum Computing and Department of Physics and Astronomy, University of Waterloo, 200 University Avenue West, Waterloo, Ontario, Canada N2L 3G1

Duncan G. England and Philip J. Bustard
National Research Council of Canada, 100 Sussex Drive, Ottawa, Ontario, Canada K1A 0R6

Benjamin J. Sussman
National Research Council of Canada, 100 Sussex Drive, Ottawa, Ontario, Canada K1A 0R6. 3 Department of Physics, University of Ottawa, 150 Louis Pasteur, Ottawa, Ontario, Canada K1N 6N5

Jie Yang, Matthew S. Robinson and Martin Centurion
Department of Physics and Astronomy, University of Nebraska-Lincoln, 855 N 16th Street, Lincoln, Nebraska 68588, USA

Markus Guehr
PULSE Institute, SLAC National Accelerator Laboratory, Menlo Park, California 94025, USA
Institute of Physics and Astronomy, Potsdam University, Potsdam 14476, Germany

Theodore Vecchione, Renkai Li, Nick Hartmann, Xiaozhe Shen, Ryan Coffee, Jeff Corbett, Alan Fry, Kelly Gaffney, Tais Gorkhover, Carsten Hast,Keith Jobe, Igor Makasyuk, Alexander Reid, Joseph Robinson, Sharon Vetter, Fenglin Wang, Stephen Weathersby, Charles Yoneda and Xijie Wang
SLAC National Accelerator Laboratory, Menlo Park, California 94025, USA

Dehui Li and Chih-Yen Chen
Department of Chemistry and Biochemistry, University of California, 607 Charles E. Young Drive East, Los Angeles, California 90095, USA

Gongming Wang and Xiangfeng Duan
Department of Chemistry and Biochemistry, University of California, 607 Charles E. Young Drive East, Los Angeles, California 90095, USA
California Nanosystems Institute, University of California, Los Angeles, California 90095, USA

Yu Huang
California Nanosystems Institute, University of California, Los Angeles, California 90095, USA
Department of Materials Science and Engineering, University of California, Los Angeles, California 90095, USA

Hung-Chieh Cheng, Hao Wu and Yuan Liu
Department of Materials Science and Engineering, University of California, Los Angeles, California 90095, USA

C. Vinegoni, C. Leon Swisher, S. Stapleton, R. Weissleder and R.J. Giedt
Center for Systems Biology, Massachusetts General Hospital and Harvard Medical School, Richard B. Simches Research Center, 185 Cambridge Street, Boston, Massachusetts 02114, USA

P. Fumene Feruglio
Center for Systems Biology, Massachusetts General Hospital and Harvard Medical School, Richard B. Simches Research Center, 185 Cambridge Street, Boston, Massachusetts 02114, USA
Department of Neurological, Biomedical and Movement Sciences, University of Verona, Strada Le Grazie 8, 37134 Verona, Italy

D.L. Rousso
Center for Brain Science, Department of Molecular and Cell Biology, Harvard University, 52 Oxford Street, Cambridge, Massachusetts 02138, USA

Michał Jachura, Radosław Chrapkiewicz, Rafał Demkowicz-Dobrzański, Wojciech Wasilewski and Konrad Banaszek
Faculty of Physics, University of Warsaw, Pasteura 5, 02-093 Warsaw, Poland

Yu Hui, Zhenyun Qian and Matteo Rinaldi
Department of Electrical & Computer Engineering at Northeastern University, 360 Huntington Avenue, Boston, Massachusetts 02115, USA

Juan Sebastian Gomez-Diaz and Andrea Alù
Department of Electrical & Computer Engineering at The University of Texas at Austin, 1616 Guadalupe St., UTA 7.215, Austin, Texas 78701, USA

Eyal Cohen and Shlomi Dolev
Department of Computer Science, Ben Gurion University of the Negev, 653 Be'er, Sheva 84105, Israel

Michael Rosenblit
Ilze Katz Institute for Nanoscale Science and Technology, Ben Gurion University of the Negev, 653 Be'er, Sheva 84105, Israel

Negar Reiskarimian and Harish Krishnaswamy
Department of Electrical Engineering, Columbia University, 1300 South West Mudd, 500 West 120th Street, New York, New York 10027, USA

Haiping He, Qianqian Yu, Hui Li, Jing Li, Junjie Si and Zhizhen Ye
State Key Laboratory of Silicon Materials, School of Materials Science and Engineering, Zhejiang University, Hangzhou 310027, China

Yizheng Jin
Center for Chemistry of High-Performance and Novel Materials and State Key Laboratory of Silicon Materials, Department of Chemistry, Zhejiang University, Hangzhou 310027, China

Nana Wang and Jianpu Wang
Key Laboratory of Flexible Electronics (KLOFE) & Institute of Advanced Materials (IAM), Jiangsu National Synergetic Innovation Center for Advanced Materials (SICAM), Nanjing Tech University (NanjingTech), 30 South Puzhu Road, Nanjing 211816, China

Jingwen He, Xinke Wang and Yan Zhang
Department of Physics, Capital Normal University, Beijing Key Lab for Metamaterials and Devices, and Key Laboratory of Terahertz Optoelectronics, Ministry of Education, Beijing 100048, China

Filippo Cardano, Maria Maffei, Bruno Piccirillo and Enrico Santamato
Dipartimento di Fisica, Università di Napoli Federico II, Complesso Universitario di Monte Sant'Angelo, via Cintia, Napoli 80126, Italy

Francesco Massa
Dipartimento di Fisica, Università di Napoli Federico II, Complesso Universitario di Monte Sant'Angelo, via Cintia, Napoli 80126, Italy
Faculty of Physics, University of Vienna, Boltzmanngasse 5, Vienna 1090, Austria

Corrado de Lisio, Giulio De Filippis and Vittorio Cataudella
Dipartimento di Fisica, Università di Napoli Federico II, Complesso Universitario di Monte Sant'Angelo, via Cintia, Napoli 80126, Italy
CNR-SPIN, Complesso Universitario di Monte Sant'Angelo, Via Cintia, Napoli 80126, Italy

Lorenzo Marrucci
Dipartimento di Fisica, Università di Napoli Federico II, Complesso Universitario di Monte Sant'Angelo, via Cintia, Napoli 80126, Italy
CNR-ISASI, Via Campi Flegrei 34, Pozzuoli (NA) 80078, Italy

L.E. de Clercq, R. Oswald, C. Flühmann, D. Kienzler, H.-Y. Lo, M. Marinelli, D. Nadlinger, V. Negnevitsky and J.P. Home
Institute for Quantum Electronics, ETH Zürich, Otto-Stern-Weg 1, 8093 Zürich, Switzerland

B. Keitch
Institute for Quantum Electronics, ETH Zürich, Otto-Stern-Weg 1, 8093 Zürich, Switzerland
Department of Engineering Science, University of Oxford, Parks Road, Oxford OX1 3PJ, UK

Index

A

Amplification, 3, 35, 39, 41, 47, 49, 63, 85, 120-123, 125-129, 164, 186

Anisotropic Strain, 13, 94, 96, 106-107, 109

Atomic Vapours, 92, 96

B

Bandwidth Conversion, 130-131, 134

Beam Collimation, 57

C

Chemical Vapour, 16, 19, 21-22, 100, 105, 134

Chiral Inelastic Transport, 42

Composite Ferroelectrics, 50, 55

Compressive Sensing, 64-66, 69, 224

D

Diamond Quantum Memory, 130, 134

Diffractive Imaging, 136

E

Elasto-optic Effect, 8, 12

Electron Pulses, 136-138, 143-144

Electronic Structure, 10, 13, 64, 69, 104

Electroresistance, 55, 75-76, 78-79, 81-83

Energy Selection, 57-58, 62

Energy-conserving Reversible Gates, 185-186

Entangled Photons, 6-7, 34, 91-93, 95-98, 106-107, 109-110, 135, 175

Exciton Localization, 203-209

F

Ferroelectric Tunnel Junctions, 75-76, 82-83

Fourier Transform Chip, 84

Frequency Comb, 28-34, 135

G

Graphene-ferroelectric Metadevices, 15

H

Hilbert Space, 219

Hyperbolic Metamaterials, 21, 26-27

I

Information Technology, 113, 185

Infrared Ellipsometer, 21

Infrared Sensing, 176

Interferometry, 6, 65, 69, 91, 143, 166-167, 172, 175, 225

L

Laguerre-gaussian Lasers, 35

Laser Pulses, 20, 35, 41, 58, 61-63, 65, 68, 71, 90, 115, 119, 131

Laser Scanning Microscopy, 153-154, 160, 162, 164

Laser-driven Ions, 57-58, 62

Linear Field Distribution, 70

Logic-gate Operations, 15-16, 18

Lorentz Reciprocity, 193-194

M

Mach-zehnder Interferometer, 166, 171, 175

Mechanical Amplification, 122, 127

Metal-dielectric Multilayers, 21

Metal-oxide-semiconductor, 179, 193-194

Methylammonium Lead Iodide, 145, 151-152

Methylammonium-lead-halide, 203

Micrometric Lattice Constant, 50, 55

Microplate Crystals, 145-146

Miniature Modular Structure, 57

Mode Engineering, 166-167, 169

Multiphoton Fluorescence Microscopy, 153

N

Nanoplasmonic Spectroscopy, 28-29

Natural Femtosecond Timescale, 136

Neural Segmentation, 153, 160, 164

Nitrogen Molecules, 136

Nonlinear Interactions, 122

Nonlinear Near Field, 70-71

Nonvolatile Memory, 15-16, 19-20

O

Optoelectronic Devices, 22, 92, 145-146, 151, 183, 203, 208

Organolead Trihalide, 152, 203-204, 208-209

P

Photocurrent Generation, 99, 101-104

Photonic Circuits, 1-3, 5-7, 28-29, 110

Photonic Systems, 15, 42-43, 217

Photonic-electronic Integration, 113

Photovoltage, 75, 100, 104-105

Piezoelectric Nanomechanical Resonator, 176, 179

Plasmonic Nanoparticles, 70

Polar Compounds, 8

Q

Quantum Dots, 92-93, 96-98, 106-107, 121

Quantum Interference, 1-3, 6-7, 84, 90-91, 96, 104

Quantum Optomechanics, 122

Quantum States, 65, 84, 123, 131, 175

Quantum-walk Dynamics, 210

R

Radio-frequency Circulator, 193

Reversible Electrochromism, 8

Rotational Wavepacket, 136-137, 139, 142

S

Silicon Chip, 1-2, 6, 107, 202

Silicon Micro-electromechanical System, 106

Silicon-on-insulator (soi), 113-114

Silicon-organic Hybrid Waveguides, 113

Single Photons, 91, 96, 98, 128, 130-131, 134-135

Single Qubit, 218-219

Single-element Detection, 64

Spatial Variation, 220

Staggered Commutation, 193-194, 198, 200

Statistical Moments, 210, 215

Stimulated Raman Scattering, 35, 114

Subnanometric Lattice Constant, 50

Subwavelength Photonic Circuits, 28-29

Super-crystals, 50, 52

Super-resolution Microcopy, 35

Suppression Law, 84-86, 88-91

Surface Plasmon Resonance, 28, 34

T

Time-dependent Hamiltonian, 218-219

Topological Phase Transitions, 42, 44, 214

Topological Quantum Transitions, 210

U

Ultrafast Dynamics, 64-65, 137

Ultrathin Plasmonic Metasurfaces, 176

V

Van Hove Singularity, 99, 104

W

Wavelength-tunable Sources, 92

X

X-ray Diffractometry, 50

www.ingramcontent.com/pod-product-compliance
Lightning Source LLC
Chambersburg PA
CBHW080252230326
41458CB00097B/4277